HISTORY OF THE METEOROLOGICAL OFFICE

Malcolm Walker tells the story of the UK's national meteorological service – now known simply as the Met Office – from its formation in 1854 with a staff of four and a budget of a few thousand pounds, to its present position as a scientific and technological institution of national and international importance with a staff of nearly 2000 and a turnover of nearly 200 million pounds per year. The Met Office has long been at the forefront of research into atmospheric science and technology and is second to none in providing weather services to the general public and a wide range of customers around the world. The history of the Met Office is therefore largely a history of the development of international weather prediction research in general.

Formed as the Meteorological Department of the Board of Trade with a specifically maritime purpose, the Met Office is now an Executive Agency and Trading Fund responsible to the UK government's Department for Business, Innovation and Skills and serves not only the shipping industry but also many other groups of users. It is at the forefront of pure and applied research in meteorology and related sciences and, moreover, cooperates and interacts with the international meteorological community at administrative, operational and research levels. In addition to being a premier forecasting bureau, it is at the forefront of the modelling of climate change in the modern era.

This volume will be of great interest to meteorologists, atmospheric scientists and historians of science, as well as amateur meteorologists and anyone interested generally in weather prediction.

MALCOLM WALKER was an academic at Cardiff University from 1967 to 1998, first as a Lecturer, then, from 1983, as Senior Lecturer and, from 1996, as Deputy Head of the Department of Maritime Studies and International Transport. He was Education Resources Manager of the Royal Meteorological Society from 1998 to 2007. He is a Fellow of the Royal Meteorological Society and a Member of the American Meteorological Society. He co-authored *The Ocean-Atmosphere System* (1977), with A.H. Perry. He chaired the Royal Meteorological Society's History Group from 1989 to 1999 and again from 2007 to the present. He was awarded the Group's Jehuda Neumann Memorial Prize in 2001 and the Royal Meteorological Society's Outstanding Service Award in 2007. Since 1980 he has had a strong scholarly interest in the history of ideas in meteorology and physical oceanography and the people behind these ideas. He has published numerous articles and lectured many times on this subject.

HISTORY OF THE
METEOROLOGICAL OFFICE

MALCOLM WALKER

CAMBRIDGE
UNIVERSITY PRESS

University Printing House, Cambridge CB2 8BS, United Kingdom

One Liberty Plaza, 20th Floor, New York, NY 10006, USA

477 Williamstown Road, Port Melbourne, VIC 3207, Australia

4843/24, 2nd Floor, Ansari Road, Daryaganj, Delhi - 110002, India

79 Anson Road, #06-04/06, Singapore 079906

Cambridge University Press is part of the University of Cambridge.

It furthers the University's mission by disseminating knowledge in the pursuit of education, learning and research at the highest international levels of excellence.

www.cambridge.org
Information on this title: www.cambridge.org/9781108445566

First published 2012
First paperback edition 2017

A catalogue record for this publication is available from the British Library

Library of Congress Cataloging in Publication data
Walker, J. M. (John Malcolm), 1942–
History of the Meteorological Office / J. M. Walker.
p. cm.
ISBN 978-0-521-85985-1 (hardback)
1. Great Britain. Meteorological Office – History.   2. Meteorology – Great Britain – History.   I. Title.
QC989.G69.W35   2012
551.50941–dc23   2011025094

ISBN 978-0-521-85985-1 Hardback
ISBN 978-1-108-44556-6 Paperback

Contents

Illustrations

Foreword

LORD HUNT OF CHESTERTON, FRS

The Met Office is a well-respected and familiar British institution, whose weather forecasts people hear every day. News about its activities and advice are discussed at length in the media and in Parliament. It was founded in 1854 and still goes strong. Queen Victoria had personal weather forecasts from Admiral FitzRoy, the first director, for her three-mile voyage across the Solent to the Isle of Wight. It is not an exaggeration to say that the Office has played an important part in the history of the United Kingdom and of many other countries over this period, in both peace and war. Although it is an institution based on the application of science and technology, its outstanding staff have in fact originated some of the key developments of meteorological science and technology, a tradition that continues today.

This book had its origins in the PhD of the late Dr Jim Burton, a forecaster in the Leeds Weather Centre, who was encouraged by Sir John Houghton, then Director-General of the Met Office. His predecessor, Sir John Mason, helped the project. The necessary financial ingenuity to arrange funding for research into the history of the Office was provided by Martyn Bittlestone, our finance director, when the Office became a Trading Fund in 1996. The Chief Scientist, Professor Julia Slingo, has helped recently with her insights into developments up to 2010. I am very grateful to Malcolm Walker, formerly of Cardiff University, for accepting the invitation not only to write the book but also to listen tactfully to all the inputs from meteorologists and commentators on the text. Stan Cornford's detailed study in 1994 of the Met Office involvement in the D-Day forecast was also a significant contribution.

I was introduced to the familiar names of the shipping forecast areas (Wight, Dover, Thames, etc.) by my grandfather, who liked to sail dinghies in very rough weather, and to the scientific approach to meteorology by my great uncle L F Richardson, who originated numerical weather prediction when he was in the Met Office between 1913 and 1920.

When I was Chief Executive at the Office, I found it instructive to read how FitzRoy was able to push through his scientifically controversial ideas while at the same time dealing with his political masters. His extraordinary personality was vividly portrayed in the play on FitzRoy and Darwin by Juliet Lacey, which the Royal Society funded in 1997. Napier Shaw, who like other directors came straight from a university position, had a longer and happier time; he demonstrated how the Met Office should always make best use of science and technology and adapt its services to changes in society and politics. He emphasized the importance of involving outside experts and 'stakeholders' in these developments and in the international responsibilities of the Met Office.

This book does not hide the many arguments within the Met Office about scientific and technical questions, but, unlike some other technical agencies of government, the Office has been quite open about these disagreements and then decisive about implementing the best practical method that emerged from the discussions, as with numerical weather prediction using computers (where scepticism continued for more than thirty years), satellites, weather radar, automatic weather stations (a surprisingly controversial technology) and now the Internet. However, it has often been commented that the Office has not been successful in stimulating new businesses in the UK to produce and market these inventions. But that was not an objective given to it by government.

Government agencies and even meteorological offices come and go, and sometimes get moved between different parts of government, as for example with the U.S. weather service. This existential aspect of the Met Office continues to be discussed regularly in the media and in Whitehall. After the UK joined the European Economic Community in 1973, the question asked by officials, anxious to save money, was whether the time had come to merge the UK Met Office with those of other countries or even in a European Met Office. Although a very successful European Centre was established in Reading for Medium-Range Weather Forecasts, Ministers have continued to support a strong national Met Office in order for the UK to have strategic independence. But they also supported progressively closer collaboration in various European initiatives.

After starting in the Board of Trade, the Met Office had a period of being run by a committee of governmental and nongovernmental bodies which included the Royal Society. This was too cumbersome for the new applications of meteorology in modern warfare, so, under Winston Churchill's influence, it was transferred after the First World War to the Air Ministry and merged with the meteorological services of the Army and Royal Air Force and in part with that of the Royal Navy in 1919. Endorsed by subsequent inquiries, notably that of Lord Brabazon in 1955, the Met Office remained there satisfactorily for more than ninety years, while also increasing its collaboration with most other government departments.[1] It has been largely funded by the government through various financial arrangements, culminating in 1996 in its

---

[1] The Met Office was transferred to the UK government's Department for Business, Innovation and Skills on 18 July 2011.

becoming a 'Trading Fund', which means that it now negotiates financial and service delivery contracts with about thirty government departments and agencies. Its history suggests that a major reason for its survival is that, even as its structure in government has changed, it has been able to provide up-to-date services needed by government, commerce and the public, through its great success in being one of the world's leading services in the effective and economical use of new science and technology.

Other countries still have two or more weather services, typically for civil aviation and defence, which clearly adds to their costs and does not seem to improve their service. Along with the various changed arrangements, the Met Office's buildings and their locations keep changing. From a single office in 1854, receiving data and delivering warnings by the new technology of telegraph, it kept expanding to more than 6000 staff and many hundreds of offices and observing stations worldwide in the Second World War. It is now down to a little under 2000, with a custom-built headquarters at Exeter and fewer than twenty other installations from the Falkland Islands to Cyprus to the Shetland Islands. This reduced network is sufficient as communications have changed from the electric telegraph in the 1850s to the Internet in the 1990s; also, these developments enabled the range of services and advice, both civil and military, to expand continually.

Met Offices are always under scrutiny, but they generally survive. There is one instance of closure when the director of an early meteorological service in France, the famous scientist J B Lamarck, was dismissed on Napoleon's orders in 1810 – an unfortunate overture to his difficulties in the Russian winter of 1812. In the UK, both Houses of Parliament have regularly inquired about and debated the doings of the Met Office, which are well covered in this book. The long speech by the back-bench Lord Wrottesley in 1853, in which he explained why British shipping needed a Met Office to be able to compete with the United States, was followed by the Office's founding in 1854. The book also covers in some detail the debates that led to the forced cessation of forecasting for the public for thirteen years from 1866, because of its alleged inaccuracy. But storm warnings were accurate enough to reduce the number of shipwrecks, which led to Parliamentary complaints that these forecasts were having a damaging effect on the ship salvage businesses of Cornwall and Devon. We have the same story today from some advocates who argue that forecasts of climate change are damaging business and should be ignored or even suppressed.

However, Parliamentary debate has also prompted technical developments, such as when long-range forecasts were initiated after the very cold winter of 1962–1963 and when Prime Minister Ted Heath in 1973 asserted the need to understand global climatic variations, in particular El Niño, in order to understand variations in the prices of food. Parliament has also monitored the Met Office's work on controversial environmental problems, from London smog, acid rain over Scotland and Scandinavia, the ozone hole and the spread of foot-and-mouth disease to the oil fires of the Gulf War. All recent Prime Ministers have actively promoted research on analysing and predicting climate change by the Hadley Centre, a branch of the Met Office whose leading scientists shared in the award of the Nobel Prize for Peace in 2007 along with other distinguished climate scientists around the world. When Mrs Thatcher was Prime Minister, she devoted most of her speech at the United Nations in 1989 to this subject and made a great impact at the Met Office when she opened the Hadley Centre at Bracknell in 1990.

International collaboration, which has been an essential element of every aspect of the Met Office's work, has been very effective and economical because of the UK's active involvement and financial support of the World Meteorological Organization (WMO), a United Nations agency, which grew out of the International Meteorological Organization set up by directors of Met Offices in 1879. Despite different national policies and political tensions between countries, exchanges of data, weather forecasts and warnings about natural hazards have progressively improved. The passing of 'Resolution 40' brought the agreements up to date at the 1995 WMO Congress, but these arrangements still need extending and also applying to the urgent hydrological problems of warning about floods and droughts, for which the Met Office is also responsible at WMO.

This book reveals how the science of meteorology and its applications has developed over the past 150 years and how it has led to new concepts and techniques. Sometimes developments arose from new technologies and from responses to new user requirements, which have often arisen quite suddenly, such as the recent problem in 2010 of improving the forecasts of how volcanic ash is dispersed in order to advise aircraft where they can fly safely after volcanic eruptions. Some of the innovations of meteorology have had wide ramifications in the worlds of science and technology, notably in the communication of very complex and uncertain information.

It is worth recalling that the Met Office's provision of forecasts and estimations of their accuracy in the 1860s predated the statistical theory and methods that developed later in the century – part of the reason for FitzRoy's problems. L F Richardson had to invent approximate methods for solving the basic equations of fluid flow, starting with uncertain and incomplete measurements in order that forecasts could be made systematically by calculation. Later, Neumann, Lorenz, Charney and others in the United States after the Second World War discovered how to simplify and compute the equations, making use of the new electronic computers, and then proposed a new approach for analysing the chaotic nature of the predicted weather patterns. The computational methods they developed later became the basis of much modern aeronautical design and environmental modelling. Central bankers and economists have finally learnt from meteorology, after about twenty years, how to present the uncertainty of their financial forecasts with multiple curves spreading out into the future.

This history is memorable for its many stories – about great individuals and great team work, some personal tragedies, extraordinary bravery and incredible perseverance – accounts of which, even now, cannot be told in much detail, such as the secret observers of weather data in war zones and the use of the data in breaking codes and forecasting. The best story of all and the greatest achievement of the Met Office came in June 1944, when Group Captain Stagg provided the right twenty-four-hour forecast of moderate winds and low cloud to General Eisenhower for the perilous crossing of the English Channel by the Allied invasion forces. The forecasting office at Dunstable led by C K M Douglas and assisted by the Norwegian Sverre Petterssen collaborated with teams from the United States and Canada in providing the first 'multi-centre' forecast. There is an interesting postscript. Since 2000, with modern technology and scientific understanding, this collaborative approach is now operational, but now the diplomatic role of Stagg has been taken over by connections between sophisticated computer programs – which of course have their problems too.

I am sure readers will enjoy this history, and I hope that when you listen to your next forecast on the radio or TV you will understand a bit better about what the people do who make it possible.

Julian Hunt

Additional Commentary

DR DAVID N AXFORD, CENG, FIET, CMET(RET), FRMETS

It has been an honour to have provided some small help to Malcolm Walker in the production of this important book, which records the history of the Meteorological Office from its gestation as a mainly data-gathering and -analysing institute through its important role forecasting the weather in the Second World War to its present position, where it is enjoined to act as a semi-commercial enterprise within the boundaries set by the Government. The time and effort in researching and summarizing the essence from the multitude of papers drawn together in this book is much to the credit of the author.

Julian Hunt has provided an excellent overview of the book in his Foreword, and I would not wish to repeat his words. Instead I would like to emphasize the contributions to the international status of the UK in the science of meteorology during the second half of the twentieth century by the three Director-Generals (DGs) that I had the pleasure to serve under during my years in the Office from 1958 to 1989.

These were the 'golden years' to my mind, during which forecasting practice progressed from a science based on intuition (and the plotting of sparse data on a map with two ink pens [one blue-black, one red], drawing up the map and making a guess, based on experience, of the way the atmosphere would change over the next twenty-four hours) to a truly physics-based profession with observations from satellites as well as the land and oceans being analysed by super-computers and the analysis being forecast forward using classical physics-based numerical weather prediction programmes. Nowadays the main job of the human forecaster is to interpret the computer output and to add local knowledge where necessary while packaging the output appropriately for the customer.

When I first joined the Office in 1958, Sir Graham Sutton (DG from 1953 to 1965) was in charge. As recorded in this book, he brought together the various branches of the Office which had been scattered in and around London to a new centre in Bracknell, Berkshire, while also focusing on changing the organizational structure of the Office to bring it into line with the new post-war requirements. Following the end of the Second World War, meteorological technology, automation and, from the 1950s, the application of electronic computers made rapid progress, and Sir Graham kept the Office focused on a goal of being at the scientific forefront of meteorology.

Under Sir John Mason (DG from 1965 to 1983) the Meteorological Office became internationally recognized as a Centre of Excellence within which staff were active both in providing weather services to the public, the military and the aviation industry, and in conducting front-edge pure and applied research in meteorological science. All staff were encouraged to achieve their potential through regular review and forward planning by the Directorate. High-achieving scientists of international renown were recruited to ensure that fundamental research was carried out in parallel with the development of new and improved operational services. Sir John Mason himself was a leading light on the international stage, too, ensuring that UK meteorologists were fully involved in the first globally supported scientific experiments such as the Global Atmospheric Research Programme (GARP), GARP Atlantic Tropical Experiment (GATE) and others.

Sir John Houghton (DG from 1983 to 1991) continued the work of his predecessor, in particular in the field of satellite meteorology and, later, in the science related to climate forecasting. There was a growing unease amongst international scientists concerning the likelihood of future climate change. In this respect, his role within the international community of national meteorological services at the World Meteorological Organization in Geneva was seminal. He became the first chairman of the Scientific Working Group of the Intergovernmental Panel on Climate Change (IPCC). I was honoured to be present at a lunch that he hosted in Geneva during which he promoted the setting up of the internationally backed Global Climate Observing System (GCOS) which has been an essential component in the scientific understanding of the Earth's climate, and, at a national UK level, he steered through the establishment of the Hadley Centre, which is now internationally recognized as one of the leading centres of expertise in the understanding of climate and climate change.

Looking to the future, it is to be hoped that the UK Meteorological Office maintains its 'centre of excellence' position amongst the world's meteorological services. The pure scientific research conducted during the second half of the twentieth century in areas such as satellite technology, atmospheric chemistry, climate modelling, cloud physics and many other fields has borne serendipitous fruit so that the Office has been well placed to offer the best advice available to the politicians grappling with the new twenty-first-century environmental problems of air pollution, ozone depletion and climate change. May it continue to hold this position of pre-eminence in the future.

David N Axford
Stanford in the Vale

Acknowledgements

The origins of this book lie in the PhD of the late Jim Burton, whose dissertation presented to The Open University in 1988 focused on the history of the Meteorological Office to the year 1905. The then–Chief Executive of the Office, Professor Julian Hunt (now Lord Hunt of Chesterton), commissioned me in 1996 to extend Jim's work and write a book about the history of the Office from its antecedents to the present day. Generous funding from the Office allowed Cardiff University, my employer at the time, to engage a research assistant for two years. Tim Hunt (no relation of Julian) proved an excellent assistant who was particularly skilled at unearthing in archives important information about the Office's development.

Over the years, several members of staff of the National Meteorological Library and Archive have been generous with their time. I am indebted to them all, particularly Maurice Crewe, Mick Wood, Alan Heasman, Graham Bartlett, Sara Osman, Sarah Pankiewicz, Joan Self and Glyn Hughes. I am especially grateful to Steve Jebson, who has not only helped me find material in the Library on numerous occasions but also digitized most of the pictures in the book.

For permission to reproduce images, I am very grateful to the Meteorological Office, the Royal Meteorological Society, the City of Westminster Archives Centre and the NEODAAS/University of Dundee Satellite Receiving Station.

My grateful thanks go also to Steve Poole, grandson of L H G Dines, great-nephew of J S Dines and great-grandson of W H Dines. He has supplied much information about the Dines family, and he has kindly granted me permission to include Figure 8.3 in the book. Others who have assisted by supplying information or helping me get the story right include a number of former and present members of the Office's staff, notably Stan Cornford, Marjory Roy, Brian Booth, Martin Stubbs and Brian Golding.

It has been a great pleasure to have as advisors for the book two distinguished meteorologists, Julian Hunt and David Axford, both of them former members of the Office's staff. Their guidance and support are hugely appreciated. The encouragement and help of the Office's current Chief Scientist, Julia Slingo, is also much appreciated.

Finally, I wish to record my gratitude to Matt Lloyd of Cambridge University Press, who has been very supportive and patient. He has been my editor from the start and must have wondered at times if I would ever complete the book. Indeed, without the encouragement and increasingly persistent urging of my beloved wife, Diane, the book might still be a work in progress.

Malcolm Walker

Abbreviations

AA	The United Kingdom's Automobile Association
AP1134	The official report on the work of the Meteorological Office during the Second World War
BBC	British Broadcasting Corporation
BIS	Department for Business, Innovation and Skills
British Association	British Association for the Advancement of Science
BRO	British Rainfall Organization
CAA	Civil Aviation Authority
CEGB	Central Electricity Generating Board
CFO	Central Forecasting Office
CMet	Chartered Meteorologist
COMESA	Committee on Meteorological Effects of Stratospheric Aircraft
COST	The EEC's scheme for Cooperation in Science and Technology
Defra	Department for Environment, Food and Rural Affairs
DERA	Defence Evaluation and Research Agency
DETR	Department of the Environment, Transport and the Regions
DSIR	Department of Scientific and Industrial Research
ECMWF	European Centre for Medium Range Weather Forecasts
ECOMET	The Economic Interest Grouping of the National Meteorological Services of the European Economic Area
EEC	European Economic Community
ESSA	Environmental Science Services Administration
EUMETSAT	The European Organization for the Exploitation of Meteorological Satellites
FGGE	First GARP Global Experiment
FRONTIERS	Forecasting Rain Optimized using New Techniques of Interactively Enhanced Radar and Satellite
GARP	Global Atmospheric Research Programme
GATE	GARP Atlantic Tropical Experiment
ICSU	International Council of Scientific Unions

IGY	International Geophysical Year
IIOE	International Indian Ocean Expedition
IMC	International Meteorological Committee
IMD	India Meteorological Department
IMO	International Meteorological Organization
IPCC	Intergovernmental Panel on Climate Change
IPY	International Polar Year
IQSY	International Quiet Sun Year
JASIN	Royal Society Joint Air-Sea Interaction Project
JCMM	Joint Centre for Mesoscale Meteorology
MMU	Mobile Meteorological Unit
MoD	Ministry of Defence
MOLARS	Meteorological Office Library Accessions and Retrieval System
MOSS	Meteorological Office Observing System for Ships
MOWOS	Meteorological Office Weather Observing System
MRC	Meteorological Research Committee
MRF	Meteorological Research Flight
NERC	Natural Environment Research Council
NIO	National Institute of Oceanography
NPL	National Physical Laboratory
NWP	Numerical Weather Prediction
Office	(when used with capital O): Meteorological Office
PMO	Port Meteorological Officer
PWD	Petroleum Warfare Department
RAF	Royal Air Force
RAFVR	RAF Volunteer Reserve
RE	Royal Engineers
RFC	Royal Flying Corps
THUM	Temperature and humidity (in the context of upper-air data obtained by aircraft ascents)
TIROS	Television Infra-Red Observation Satellite
TV	Television
UK	United Kingdom
UM	Unified Model
UN	United Nations
WMO	World Meteorological Organization
WWW	World Weather Watch

1

Seeds Are Sown

One afternoon in February 1854, an announcement was made in the House of Commons. A new government department was to be formed, to collect and digest meteorological observations made on board merchant and Royal Navy ships. Six months later, the Meteorological Office was born.

When the Office took its first tentative steps, it had a staff of four and a budget of a few thousand pounds per year. Since then, Britain's national meteorological service has experienced several major changes in control and organization and is now an Executive Agency and Trading Fund responsible to the United Kingdom (UK) government's Department for Business, Innovation and Skills, with a staff of nearly two thousand and a turnover of nearly 200 hundred million pounds per year. It is a scientific and technological institution of national and international importance, serving not only the shipping industry but also many other groups of users, including the general public. It is also at the forefront of fundamental pure and applied research in meteorology and related sciences and, moreover, cooperates and interacts with the international meteorological community at administrative, operational and research levels. What were the origins of this institution? How did it come to be founded?

The simple answer is that its foundation was an outcome of an international conference held in 1853, but this answer begs a number of questions. Why was the conference held, and why then? Who organized it, and why them? Had there been attempts to form a body resembling a national meteorological institution in the UK before 1854? Did any institutions of this kind exist abroad already? Was the foundation of the Office solely the result of a conference? It is appropriate to review not only the origins of the Office as an institution but also the scientific context. Without an awareness of this context and the preceding discoveries and inventions, neither the Office's foundation nor its work in its formative years can be fully understood.

Meteorology in Ancient Times

Long before the 1850s, there were meteorologists, persons who study the processes and phenomena of the atmosphere. The ancient Greeks gave us the word *meteorology*, from μετέωρος (lofty or raised up), and λογος (discourse). Aristotle (*c.* 384–322 BC) used it in his *Meteorologica*, the earliest known treatise on atmospheric phenomena, and stated that it had been used by his predecessors. Who actually coined the word is not, however, known. Who first tried to understand the ways of the atmosphere is not known either, but there has probably never been a time when people took no notice of the weather.

Our earliest ancestors left no records, so nothing is known of their meteorological knowledge and understanding. The earliest records came with the dawn of civilization, in the form of texts and symbols written on walls or papyrus and inscriptions made on clay tablets. From these we know that the ancient Babylonians attached great significance to clouds, winds, storms and thunder, though many of their observations served more as omens of political and economic events than as signs of weather to come. Most ancients considered meteorology a branch of astronomy, and the Babylonians founded astro-meteorology, a pseudo-science concerned with the alleged influences of celestial phenomena such as comets and planetary conjunctions on weather and climate.[1]

The ancient Greeks were careful observers of nature and devised hypotheses which they tested by means of experiments. Thus their approaches were essentially scientific, though most of their ideas have failed the test of time. An exception is their concept of the hydrological cycle, which was recognized by Anaxagoras of Clazomenai (*c.* 500–*c.* 428 BC) and can hardly be faulted today. Some Greek philosophers turned their attention to weather forecasting. We know from a work by Theophrastos of Eresos (*c.* 372–*c.* 288 BC), for example, *De signis tempestatum* (On weather signs), and from a work by Aratos of Soloi (*c.* 315–*c.* 240 BC), *Diosēmeia* (Weather forecasts), that the Greeks relied on weather lore in the form of proverbs, rhymes and rules based on lunar and planetary influences on the atmosphere, the flowering of plants, the behaviour of animals, the appearance of the sky, and so on.

The Greeks appear to have been the first to make and record meteorological observations regularly. By the fifth century BC, they were making them public by means of parapegmata, which were almanacs fixed to columns. The predominance of wind observations suggests that the information was particularly important to seafarers. To ascertain wind direction, a vane may have been used. This device appears to have existed in ancient times in Japan and China and was widely used in Greece by the first

---

[1] Weather is the state of the atmosphere at any given time and is expressed in terms of temperature, humidity, visibility, wind speed, wind direction, whether or not rain is falling, etc. Climate is the synthesis of the weather at any place and is generally expressed in terms of averages and variability about those averages.

century BC. The Tower of the Winds still stands in Athens today but has lost its vane, a revolving bronze Triton. This structure, built about 40 BC, is properly known as the Horologe of Andronikos Kyrrhestes (after the astronomer who built it) and originally served the triple purpose of sundial, water-clock (clepsydra) and weather-vane.

Several basic elements of a modern meteorological service existed in ancient Greece, albeit in rudimentary or crude form: observations of atmospheric phenomena were made systematically; explanations of these phenomena were sought; forecasts of the weather were attempted; and the effects of the weather on seafarers, farmers and others were matters of concern.

The Dawn of Modern Meteorology

Aristotle's *Meteorologica* did not become available in the West before the late twelfth century, when the first three of its four books were translated into Latin by Gerard of Cremona (1114–1187).[2] The works of Aratos of Soloi were, however, well known to the Romans, as revealed by the *Georgics* of Virgil (70–19 BC), *De rerum natura* by Lucretius (*c.* 95–*c.* 55 BC) and *Naturalis Historia* by Pliny the Elder (AD 23–79). Few works on meteorology appeared in the so-called Dark Ages (the period from the fifth to the tenth centuries AD). Important among them was a compendium of astronomy and meteorology called *De natura rerum*, published by Isidore of Seville in around 620. Important, too, was the first work on meteorology written by an Englishman, the Venerable Bede's own *De natura rerum*. Written in the early eighth century, it drew heavily on Isidore's work and the writings of the classical writers, notably Pliny the Elder.

After the time of Bede, particularly in the period from 1100 to 1300, a considerable number of encyclopædias containing meteorology were published, and many of them were translated into vernaculars, some of them into English. It is evident from these compendia that the science of ancient Greece continued to dominate meteorological thought, especially that expounded in *Meteorologica*, but the tenets of Greek science were eventually questioned.

In around 1270, Roger Bacon (*c.* 1214–1292) wrote a commentary on *Meteorologica* in which he cast doubt on Aristotelian theories. At the time, this was tantamount to blasphemy. Nevertheless, criticisms of Greek science mounted until, in the seventeenth century, the science of the ancients was rejected. Astrological methods of weather forecasting, which had been practised since the days of the Babylonians and had flourished during the Middle Ages, were also questioned increasingly until, again in the seventeenth century, they too were rejected by most scholars. There continued to

---

[2] He translated them from the Arabic of the Moslems. Books I–III of *Meteorologica* are concerned mainly with meteorology but also cover aspects of astronomy, geography, geology and seismology. Book IV deals mainly with chemistry and may be the work of Straton of Lampsacus (died 270 BC), rather than Aristotle. *Meteorologica* has been translated into English by H D P Lee (Loeb Classical Library, 1987, 433 pp.).

be adherents of these methods, however, and in the early years of the Meteorological Office, astro-meteorology became an issue for a while, when its leading practitioners challenged the official approach to weather forecasting (see Chapter 2).[3]

The earliest extant journal of the weather was kept by an English meteorologist and clergyman, William Merle (dates of birth and death unknown), who maintained a systematic written record of the weather from January 1337 to January 1344. A century later, the practice of keeping weather diaries was encouraged by the spread of mass printing, for it brought about an increase in the use of calendars and almanacs, in the margins of which weather notes were often made. Merle's and other early weather diaries have proved useful to modern students of climatic change, but their usefulness is limited by the lack of instrumental measurements.

Other than rainfall, the only weather variable ascertained instrumentally before the late Middle Ages was wind direction. Without measurements, a fully quantitative scientific approach to the study of the atmosphere cannot be achieved, however diligent observers of the weather might be. The sixteenth and seventeenth centuries saw not only the rejection of Aristotelian theories but also, especially in Germany, France, Italy and England, the dawn of modern approaches to science, including the invention of the three most basic instruments in meteorology: the hygrometer, thermometer and barometer.

The credit for inventing an instrument capable of measuring the amount of water vapour in the atmosphere is generally accorded to Nicholas de Cusa (1401–1464), who used a balance and quantity of wool to show that the weight of hygroscopic material increases as the amount of moisture in the air increases and decreases as the air becomes drier. Soon afterwards, around 1485, Leonardo da Vinci (1452–1519) described a balance hygrometer which was similar but relied on cotton rather than wool.

An apparatus which demonstrated that air expands when heated and contracts when cooled was described by Philo of Byzantium in his work *De ingeniis spiritualibus* (On pressure engines), published in the second or third century BC. Hero of Alexandria utilized this property of air in the first century AD in his device for opening temple doors, but neither he nor Philo appears to have realized that it could be applied to measure heat. This step was not taken until the 1590s, when a thermometer which relied on the expansion and contraction of air was constructed by Galileo Galilei (1564–1642), who was familiar with the works of Philo and Hero. Galileo used his thermometers to compare the temperatures of different places and to investigate diurnal and seasonal variations of temperature. He recorded temperatures in degrees,

---

[3] Whether or not the Meteorological Office should ever have become involved in weather forecasting may be a moot point. In the December 1989 issue of *Weather* (Vol. 44, p. 478), Jackie Hoskins quoted from *The Pelican Social History of Britain: 16th-century England* by Joyce Youings, where (on p. 36) it is stated that "an Act of Parliament of 1541 included in its prohibition of all manner of sorcery the forecasting of the weather". Mrs Hoskins wondered if this act had ever been repealed!

but the method he used for graduating the stems of his thermometers is not known. The scales that are commonly used today came later. That of Gabriel Fahrenheit (1686–1736) was published in 1724, and the centigrade scale of Anders Celsius (1701–1744) was published in 1742.

A student of Galileo, Evangelista Torricelli (1608–1647), invented the mercurial barometer. In 1643, assisted by one of his pupils, Vincenzo Viviani (1622–1703), he sealed one end of a glass tube, filled the tube with mercury and inverted it with the open end in a dish of mercury. Finding that the height of the mercury column was less than the length of the tube, he reasoned that the space above the column was occupied by a vacuum and therefore that the weight of the column was balanced by the weight of the atmosphere. He later noticed that the height of the column varied over time and concluded that variations in the pressure exerted by the atmosphere accompanied changes in the weather. In France, Blaise Pascal (1623–1662) repeated Torricelli's experiment with different liquids (one of them red wine!) and verified a prediction of Torricelli: that the pressure of the atmosphere decreases with altitude. On 19 September 1648, he found that the atmospheric pressure near the summit of the Puy de Dôme (1464 m) was about 10% less than it was at the foot of the mountain.[4]

The first network of meteorological stations was set up in 1654. Directed by a Jesuit named Antinori, who was secretary to the patron of the project, Grand Duke Ferdinand II of Tuscany, the network comprised stations at Florence, Vallombrosa, Cutigliano, Bologna, Parma, Milan, Warsaw, Innsbruck, Osnabrück and Paris. Observations were made at specific times of day and written down in special tables called 'formulae'. The network was closely associated with the Accademia del Cimento of Florence, the world's first formal scientific institution, and ceased to function in 1667, when the Academy was disbanded. Founded in 1657 by the Grand Duke and his brother, Prince Leopold, the Academy was devoted to experiment and was much involved in the development of barometers and spirit-in-glass thermometers, many of which were used at stations of the network. Viviani was a member.

Whereas the Accademia del Cimento was a private institution dependent on patronage, the Royal Society of London was established as a corporate body, though also with royal patronage. Founded in 1660 for the pursuit of experimental natural philosophy, it is the longest extant academy. It has influenced the development of meteorology in many ways over the three and a half centuries of its existence and has played an important role in the history of the Meteorological Office, as we will see later.

Of the early members of the Royal Society, two in particular made fundamental contributions to meteorology: Robert Boyle (1627–1691) and his assistant, Robert

---

[4] In honour of Pascal, the unit of pressure in the Système International (SI) is called the pascal (Pa), 1 Pa being a pressure of one newton per square metre. For convenience, the units commonly used in modern meteorology are the hectopascal (hPa) and the millibar (1 mb = 1 hPa = 100 Pa).

Hooke (1635–1703). Together, they developed barometers, thermometers and thermometric scales, and together they investigated the use of the barometer in weather forecasting. In addition, Hooke measured wind strength with a primitive anemometer which relied on the movement of a swinging plate over a scale. Some say that the anemometer was invented by him, though a wind-measuring instrument which employed a swinging plate had been described two centuries earlier by Leon Battista Alberti (1404–1472) in his treatise *On the Pleasures of Mathematics*, published in around 1450.

The habit of making weather observations regularly and systematically was encouraged by the Royal Society, and as early as 1663 Hooke presented to the Society his paper titled 'A method for making a history of the weather', in which he set out precisely what should be included in a weather observation and how, using standard instruments, observations should be made. He stated that he wished "there were divers in several parts of the World, but especially in distant parts of this Kingdom, that would undertake this work, and that such would agree upon a common way somewhat after this manner, that as neer as could be, the same method and words might be made use of". Thus he showed himself aware of not only the need for uniform procedures in the making of weather observations but also the potential value of comparing meteorological observations made simultaneously at different places. His words were apparently heeded, for the practice of making meteorological observations regularly and systematically spread in the ensuing decades, with networks of meteorological stations established in various European countries, notably France and Germany.

The need for uniform procedures in the meteorological observing practices of seafarers was not highlighted until the nineteenth century, when it was an important factor in the foundation of the Meteorological Office.

The Origins and Growth of Marine Meteorology

Seafarers of ancient Greece ventured beyond the Pillars of Hercules (the mountains on either side of the Strait of Gibraltar), and the Romans also travelled far and wide. Indeed, *The Periplus of the Erythraean Sea*, written in the first century AD, provides documentary evidence that Graeco-Roman sailors maintained trade links between the Middle East and India and also understood monsoon winds sufficiently well to sail by direct routes between the Red Sea and India.

By the eighth century, Arabs were trading regularly between the Persian Gulf and China. By the middle of the sixteenth century, they had accumulated a wealth of knowledge of the winds and weather over the Indian Ocean and adjacent lands. This is shown by two works on navigation: *The Book of Useful Instructions and Principles of the Science of the Sea*, written by Omani pilot Ahmad Ibn Majid in the second half of the fifteenth century, and *The Ocean*, written by Turkish admiral Sidi Ali Celebi between 1554 and 1557. Both authors discussed winds, weather, ocean currents and the state of the sea, and both provided specific advice on sailing seasons, defining them in terms of the dates when monsoonal wind reversals and associated changes in the weather normally took place. The Arabs also considered the causes of the phenomena they observed. Indeed, the earliest correct explanation of land and sea breezes can be found in Majid's work.

The Portuguese explorers of the Indian Ocean drew upon the nautical expertise of the Arabs, including their knowledge of winds, weather and sea conditions. No such expertise was available to Columbus when he set out across the Atlantic Ocean in 1492. Nevertheless, he had by then been a seafarer for many years and undoubtedly possessed considerable knowledge of the winds, weather and currents of the North Atlantic. His understanding of atmospheric behaviour was shown in 1502, when, off Haiti, he made use of local weather signs to predict successfully the advance of a hurricane.

By the late seventeenth century, knowledge of marine meteorology had advanced to such an extent that Edmund Halley (1656–1742) was able to produce the first substantial contribution to meteorology since the days of Aristotle. It was published in 1686 in the *Philosophical Transactions of the Royal Society of London* (henceforth abbreviated to *Philosophical Transactions*) and bore the title 'An Historical Account of the Trade Winds and Monsoons, Observable in the Seas between and near the Tropicks; with an Attempt to Assign the Phisical Cause of the Said Winds'. He deduced correctly that thermal contrasts between land and sea are fundamental in the shaping of atmospheric circulation patterns on the scale of trade winds and monsoons, but in attributing the westward course of trade winds to the effect of the sun shifting westward over the ocean, his intuition failed him. It remained for George Hadley (1685–1768) to propose, in a paper published in the *Philosophical Transactions* in 1735, that the westward course is due to the influence of Earth's rotation on air currents flowing towards the equator.

Halley made full use of mariners' observations. So, too, did William Dampier (1652–1715) when compiling his *Discourse of Winds, Breezes, Storms, Tides and Currents*, published in 1699. This work drew on the observations not only of others, but also his own, made on three voyages around the world. In it, Dampier pointed out the resemblance between the patterns of prevailing winds and ocean currents of the globe and suggested that winds drive currents. Thus another concept which had endured since the time of Aristotle was challenged: that the waters of the sea flow from high latitudes, where the evaporation rate is low, to the tropics, where the rate is much greater and sea level therefore lower. Dampier's *Discourse* contained vivid and accurate descriptions of weather phenomena and remained the standard work on marine meteorology for more than a century. He encouraged seafarers to make meteorological observations systematically, and among those who did were the master mariners of the East India Company, though there is no evidence that he was in any way responsible for their doing so.

No one knows who first took meteorological instruments to sea. The marine barometer was, however, invented by Robert Hooke, who in 1667 presented to the Royal Society a paper which contained a discussion of the difficulties caused by the motions of vessels at sea and offered a solution, namely, the introduction of a capillary bore in the barometer tube to dampen oscillations of mercury. Who first took a thermometer to sea is also not known, but as early as 1663 Hooke took part in oceanographic experiments carried out in the Thames Estuary. These included the measurement of temperature at depths of one foot and sixteen fathoms.

The Emergence of Organized Meteorology

During the eighteenth century, the foundations for the meteorological advances of later centuries continued to be laid. Meteorological instruments were invented and improved, the Royal Society continued to take an interest in meteorology, the number of individuals making regular weather observations increased, knowledge of weather systems expanded, meteorology became progressively more organized, and advances in the physics and mathematics which underlie today's numerical models of the atmosphere were made.

In the latter half of the seventeenth century, Gottfried Leibniz (1646–1716) and Sir Isaac Newton (1642–1727) discovered the calculus (independently), and Newton made pioneering contributions to mechanics and other aspects of theoretical physics. Among those who built on their work were Daniel Bernoulli (1700–1782), Leonhard Euler (1707–1783) and Jean d'Alembert (1717–1783). Others who built on it were Joseph Lagrange (1736–1813), who revolutionized analytical mechanics and the theory of equations, and Pierre Laplace (1749–1827), who produced a mathematical expression known as 'Laplace's equation' which has proved invaluable in various fields of physics, particularly hydrodynamics. In turn, these advances paved the way for the fundamental contributions to understanding of fluid flow made in the nineteenth century by Augustin Cauchy (1789–1857) and George Green (1793–1841).

The need for systematization and standardization in meteorological observing practices continued to be stressed throughout the eighteenth century, and attempts to create international networks of weather observers continued to be made. From 1717 to 1727, for example, Johann Kanold (1679–1729) compiled and published observations from Germany and several places abroad (including London) in a quarterly journal, *Breslauer Sammlung*. At the same time, Secretary of the Royal Society James Jurin (1684–1750) attempted to build on the lead given by Hooke in the 1660s by publishing, in 1723, in the Society's *Philosophical Transactions*, 'An Invitation for Making Meteorological Observations', in which he stated that "changes in the weather, especially when great or sudden, have much influence on the health of mankind". He therefore considered it necessary to observe the weather and "discover the causes of these changes". He recommended that, "for the sake of comparison, all observations be made at the same hour of the day" and appealed to "such persons as may be pleased to make the observations" to send copies of them to the Royal Society that "they may be compared with the diary kept in London" and "comparisons and influences" published in the *Philosophical Transactions*.

From 1724 onward, observers in Britain, North America, India and many parts of Europe duly sent their journals to the Royal Society, and the observations they contained were discussed in the *Philosophical Transactions*. Though the supply of such journals dwindled to almost nothing by 1735, the idea of making weather observations at standard times with standard instruments had taken root and so, too, had an essential element of climatology, the need to compare and contrast observations made at different places. However, successful implementation of the recommended practices depended very much on the enthusiasm of individuals such as Hooke, Kanold and Jurin or the support of bodies such as the Accademia del Cimento and the Royal Society. As yet, there were no state meteorological services.

One who responded to Jurin's appeal was an American, Isaac Greenwood (1702–1745), who pressed for an extension of Jurin's idea of a worldwide network of meteorological stations. In 1728, he urged the Royal Society to extract information about winds and weather from the logbooks of ships and encourage mariners to observe the weather systematically. If this information were compiled in tabular form for the different oceans, he argued, there would be benefits for both meteorology and marine navigation. More than a century was to elapse before this proposal became reality.

An idea similar to Jurin's was put forward in 1744 by Roger Pickering (1718–1755) in a paper in the *Philosophical Transactions* titled 'A Scheme of a Diary of the Weather, Together with Draughts and Descriptions of Machines Subservient Thereunto'. Like previous schemes for making, compiling and analysing weather observations, however, Pickering's bore little fruit. It depended so much, like other schemes, on the enthusiasm of individuals or scientific societies, and most of these societies promoted all branches of science, not just meteorology. The Royal Society was no exception, though it did show more interest in meteorology than most societies. Indeed, in 1725, it supplied barometers and thermometers to observers at its own expense.

On 9 December 1773, the Council of the Royal Society approved a scheme drawn up by Henry Cavendish for (as it was put in the minutes) "regulating the manner of making daily meteorological observations by the Clerks of the Society". Thereafter, from 1774 onwards, observations of "barometer, thermometer, rain-gage (*sic*), wind-gage (*sic*), and hygrometer" were made regularly by the Society until the close of 1843, when responsibility for making them was transferred to the Royal Observatory at Greenwich. The observations were published in the *Philosophical Transactions*.

The reliability of the meteorological records was questioned in the 1820s, as we will see later in this chapter. The Académie des Sciences de Paris was also greatly interested in meteorology, and Père Louis Cotte (1740–1815) published his celebrated *Traité de Météorologie* under the auspices of this academy in 1774.

Before 1780, when Elector Palatine Karl Theodor of Bavaria (1724–1799) founded the Societas Meteorologica Palatina (in Mannheim), there was no society devoted solely to meteorology. However, the political turmoil in France and elsewhere in Europe made the society's survival difficult, and it collapsed in 1795. In its short life, however, it had shown what could be achieved with well-organized and well-equipped observational networks. The observers were equipped with calibrated instruments and detailed instructions, together with special forms on which to record observations, and the instruments were supplied free of charge. The Mannheim Society was a model for the national and international meteorological organizations which were formed more than half a century later. Moreover, the data published in its *Ephemerides* proved to be of considerable value in subsequent studies of weather and climate. At its most extensive, in the late 1780s, the Society's network of observers reached from the Urals across Europe to Greenland and eastern North America but never included anyone from Britain.[5]

In England, meanwhile, in the Old Deer Park at Richmond, Surrey, a short distance from the present-day Kew Gardens, an astronomical observatory had been erected. Originally called 'The King's Observatory at Richmond' and later known as 'Kew Observatory', it had been built for King George III and completed in time for the transit of Venus which occurred on 3 June 1769. Meteorological observations were first made there in 1773 and continued to be made until 1980. During the Royal period, which ended in 1841, observations were made at least once a day and included readings of thermometers, hygrometers, a barometer and a rain gauge. From 1842 to 1980, as we see in later chapters, Kew was one of the principal observatories in the world for the study of meteorology and related branches of physics.

Before the middle of the eighteenth century, no one had attempted to use kites or balloons to study the atmosphere aloft. But in 1749, the Professor of Practical Astronomy in the University of Glasgow, Alexander Wilson (1714–1786), attached thermometers to kites, and in 1752, Benjamin Franklin (1706–1790) carried out his famous, but hazardous, experiment with a kite in a thunderstorm. In later life, Franklin was a balloon enthusiast and was, indeed, present on 27 August 1783, when Jacques Charles (1746–1823) made the first ascent over Paris.

Meteorological observations were first made on a balloon flight on 1 December 1783, when Charles ascended to a height of 3467 metres, also over Paris, taking with

---

[5] See 'Meteorology in Mannheim: the Palatine Meteorological Society', by David C Cassidy, published in 1985 in *Sudhoffs Archiv* (Vol. 69, pp. 8–25). See also 'Societas Meteorologica Palatina', by Albert Cappel, published in 1980 in *Annalen der Meteorologie* (Vol. 16, pp. 10–27); and 'The Societas Meteorologica Palatina: an eighteenth-century meteorological society', by J A Kington, published in 1974 in *Weather* (Vol. 29, pp. 416–426).

him a thermometer and a barometer, the latter for estimating altitude. The following year, on 30 November 1784, John Jeffries (1745–1819) and Jean-Pierre Blanchard (1750–1809) made the first balloon ascent specifically for meteorological purposes.[6] During the flight, measurements of temperature and pressure were made, clouds were observed and samples of air were taken. The balloon was launched in central London and taken by westerly winds to a point near Dartford, Kent.

In 1796, Alexander von Humboldt (1769–1859) pointed out the need for meteorological observatories not only in different climatic zones but also at different altitudes. He appears to have been the first to realize that climate varies with both latitude and altitude. He also appears to have been the first to plot isotherms and isobars, which are, respectively, lines connecting points having the same temperature and lines connecting points having the same pressure. The isopleths which were drawn on the first published weather charts were, however, lines connecting places with equal departure of pressure from normal. These so-called synoptic charts were constructed by Heinrich Brandes (1777–1834) and drawn for each day of 1783, utilizing data collected by the Palatine Meteorological Society. They were discussed in *Beiträge zur Witterungskunde*, published in Leipzig in 1820. The credit for the idea of comparing meteorological observations made simultaneously at different places is often awarded to Antoine Lavoisier (1743–1794), but Lavoisier himself said the idea was put forward by Jean de Borda (1733–1799).

Early Attempts to Model Weather Systems

Though invaluable immediately in studies of weather systems, such charts could not yet serve any useful purpose in weather forecasting. Without the electric telegraph, which was invented in the 1830s, there was no way of communicating observations to analysts in time for current charts to be drawn. Among those who drew conclusions from the early synoptic charts was Brandes himself, who deduced from inspection of them that storms are barometrical depressions which advance from west to east and move into Europe from the Atlantic. Nevertheless, he was not the first to realize that the weather systems of middle latitudes are mobile.

In North America one evening in October 1743, clouds over Philadelphia prevented Benjamin Franklin from observing a lunar eclipse, but observers in Boston were more fortunate. For them, viewing conditions were good. The weather became cloudy in Boston an hour after the eclipse ended. Franklin realized that the clouds over Philadelphia were associated with a travelling storm that subsequently reached Boston. Thus he gained a place in history as the person who discovered that weather systems are moving formations. Hitherto, the belief had been that weather developed in situ. However, Columbus may have realized in around 1500 that the hurricanes of the

---

[6] See Crewe, M (1994), 'The first meteorological research flight', *Weather*, Vol. 49, pp. 61–65.

Caribbean are moving formations. If so, then Franklin's contribution may have been the discovery that the depressions of middle latitudes are also moving formations, though this, too, is debatable, because Franklin investigated the depression in question and found that it had once been a hurricane! Whatever the truth of the matter, it is probably safe to say that Franklin was the first to study the movements of weather systems *scientifically*.

An observation as fortuitous as Franklin's was made by William Redfield (1789–1857) on a journey through Connecticut and Massachusetts towards the end of 1821. He noticed that trees blown down by a storm of tropical origin had fallen in a pattern which indicated that the storm had been a vortex rotating counter-clockwise. From an investigation of this storm and a number of hurricanes in the Caribbean, he concluded that the wind directions around storms correspond to the tangents of concentric circles and therefore that the storms are 'progressive whirlwinds' with calm centres. He published his findings in 1831 in the *American Journal of Science and Arts* in a paper titled 'Remarks on the Prevailing Storms of the Atlantic Coast of the North American States'.

His paper was epoch-making. It was the first in a line of scientific contributions which have furthered *understanding* of weather systems, as distinct from *knowledge* of them, and it attracted attention immediately. Indeed, Redfield soon found himself a protagonist in a controversy over his 'Circular Theory'. A prominent American meteorologist, James Pollard Espy (1785–1860), was his chief adversary. According to Redfield, the low barometric pressure at the centre of a storm system was due to "the centrifugal tendency, or action, which pertains to all revolving or rotatory movements". Espy, on the other hand, rejected the concept of rotary motion in a storm system. He stressed the importance of thermal convection and argued that vigorous upward motion at the centre of a system causes air to flow inwards radially from all directions. As is so often the case in controversies, both arguments contained elements of the truth.

Unfortunately, Redfield and Espy did not amalgamate their ideas and, moreover, did not take into account the influence of Earth's rotation, by which a moving body is deflected to the right in the northern hemisphere, left in the southern. This influence had been recognized qualitatively since the days of Hadley and is now called the 'Coriolis Effect', after the French mathematician Gaspard Gustave de Coriolis (1792–1843), who expressed it analytically in 1835.

Another who disagreed with Redfield was Henry Piddington (1797–1858). He too rejected the idea of winds blowing parallel to isobars, saying that his analyses of observations recorded in the logbooks of ships indicated that wind patterns in storms over the Arabian Sea and Bay of Bengal represented a combination of circular and centripetal movements of air. In recognition of this, he coined the term *cyclone*, first using it in print in 1848 in the first edition of *The Sailor's Horn-Book for the Law of Storms*. The term stems from the Greek κύκλωμα, meaning 'coil of a snake', and is a possible allusion to Cyclops, the one-eyed giant of Greek mythology who forged thunderbolts for Zeus.[7] A tropical cyclone possesses an eye.

In the opinion of Heinrich Wilhelm Dove (1803–1879), who had studied meteorology under Brandes, the storm systems of the tropics were essentially symmetrical, and he pointed out, furthermore, that air circulates around their centres clockwise in the southern hemisphere, counter-clockwise in the northern. He did not agree, however, that the weather systems of temperate latitudes were also symmetrical. In his opinion, they appeared to contain two opposing air currents, one warm and moist, the other cold and dry. This was clear from his examinations of the charts of Brandes and was also his personal experience and the experience of seafarers. As he explained in *Ueber das Gesetz der Stürme*, published in 1841, the depressions of middle latitudes result from the continual conflict between two currents of air, "of which one flows from the pole to the equator, the other from the equator to the pole".

A similar view was held by an English master mariner, George Jinman (1829–1884), who argued in his *Winds and Their Courses; or a Practical Exposition of the Laws Which Govern the Movements of Hurricanes and Gales* (first published in 1859) that the weather systems of temperate latitudes were "governed by far different laws from those of the hurricane or typhoon". In the temperate zone of the North Atlantic, he considered, Redfield's circular theory and Piddington's rules were "not only wrong but dangerously misleading". He had concluded that "there never was such a thing as a really circular gale". The model of extratropical weather systems that he put forward contained two distinct currents of air flowing in opposite directions and crossing each other along lines of confluence. His daily life, he said, gave him an opportunity to test "by practical experience *at sea* the laws which closet philosophers were elaborating *on shore*"!

Progress through Organized Science

By the early nineteenth century, meteorological observations were being made in many parts of the world, but scientific progress was slow. Developments in meteorology still depended largely on the enthusiasm of individuals. With the exception of the Palatine Meteorological Society, societies that were devoted solely to meteorology had still not been formed. In general, the literary and philosophical societies that were formed in the capital cities and major provincial towns of Europe, Asia and North America in the eighteenth century and early part of the nineteenth displayed no more interest in meteorology than the learned societies that went before them. An exception was the American Philosophical Society, proposed and established by Benjamin Franklin in 1743.

---

[7] Nowadays, 'cyclone' is used as a generic name for both the hurricanes of the tropics and the depressions of middle and high latitudes.

Another exception was the Manchester Literary and Philosophical Society, which was formed in 1781 and numbered among its most active members the chemist and physicist John Dalton (1766–1844). He joined in 1794 and remained a member for the rest of his life, reading before the Society in those fifty years 116 papers, many of them on meteorological topics. Though best known for his work on atomic theory, he is considered by some to have been first and foremost a meteorologist. His meteorological work inspired many, not only in Britain but also abroad. For example, Espy declared that Dalton's researches on the physics of atmospheric water vapour had made a remarkable impression on him.

Luke Howard (1772–1864) was also a chemist and meteorologist. He made many substantial contributions to meteorology but is best remembered for his essay 'On the Modifications of Clouds, and on the Principles of Their Production, Suspension, and Destruction', published in 1803 by the Askesian Society. In this, he defined and described cloud types and noted that clouds have characteristic forms, for which he proposed "a methodical nomenclature" based on Latin terms that remain familiar today, namely *cirrus* (curl), *cumulus* (heap), *stratus* (layer) and *nimbus* (rainy cloud). His was the first practical cloud classification, and it forms the basis of modern classifications of clouds, including that still used by the Meteorological Office.[8]

Howard was among those who attended a meeting held at the London Coffee House on 15 October 1823. At the meeting, Britain's first meteorological society was formed, and Howard was elected a Council Member. It was called the Meteorological Society of London and flourished for seven months, after which, despite the initial enthusiasm and activity, it languished until 1836, when it was revived, only to fail again in 1843. The aspirations of the newly formed society were eminently scientific, for in early 1824, a committee set up by the Council recommended that "immediate measures be taken to procure correct registers of comparable observations from different parts of Great Britain and its colonies, as well as from other parts of the world, with instruments graduated to the common scales". "To effectuate this purpose with advantage", they considered it "absolutely necessary that the Meteorological Society of London should set the example of the requisite precision by establishing a Meteorological Observatory in the metropolis, or its vicinity".

These recommendations were, in fact, implemented in due course, as we see later, but, alas, not by the society, for it did not remain active long enough to play any significant role. When the society was revived, its objectives were not as clearly defined as in 1824, and in the early 1840s, indeed, the astro-meteorological tendencies

---

[8] The first known cloud classification was that suggested by the French naturalist Jean Baptiste Lamarck (1744–1829), who proposed (in 1802, in Volume 3 of his yearbook *Annuaire Météorologique*) three levels of cloud and five types of cloud ('hazy clouds', 'massed clouds', 'dappled clouds', 'broomlike clouds' and 'grouped clouds'). Though Howard's classification was adopted as the foundation of the modern classification of clouds, Lamarck's tripartite division of the atmosphere is still used today, but in modified form. His yearbooks were published annually from 1800 to 1810 and then discontinued after an unnecessarily public and brutal tirade from Napoleon in which Lamarck was told by the Emperor he should confine his attention to natural history.

of some leading members lowered the society's credibility among the establishment scientists of the day.[9]

Another who served on the Council of the Meteorological Society in 1823–1824 was John Frederic Daniell (1790–1845), inventor of the cell and hygrometer which bear his name and author of *Meteorological Essays and Observations*, published in 1823. In the Preface to this work, he mentioned that his enthusiasm for meteorology had resulted from his friendship with the geophysicist Edward Sabine (1788–1883), who was to become a greatly influential figure in meteorology. Sabine was for many years involved in the running of Kew Observatory, and his roles in the foundation and subsequent development of the Meteorological Office were central, as we will see in Chapters 2 and 3. Moreover, he figured prominently in the formation of meteorological services in Canada and India and the establishment of observatories at Toronto, Hobart, St Helena, Cape of Good Hope, Mauritius and Hong Kong.

Daniell criticized the Royal Society's meteorological work, noting that both Dalton and Howard had "recorded their dissatisfaction" over the state of the society's instruments and the unreliability of the society's meteorological records. His dissatisfaction was but part of a more general concern over the decline of science in Britain which became more insistent through the 1820s and helped bring about both the reform of the Royal Society in the 1830s and 1840s and the formation of the British Association for the Advancement of Science in 1831.[10]

At the British Association's Second Meeting, in 1832, a substantial and highly critical *Report upon the Recent Progress and Present State of Meteorology* was presented by James David Forbes (1809–1868), who later became Professor of Natural Philosophy in the University of Edinburgh. In it, he discussed meteorological instruments and focused on what he considered the unsatisfactory state of meteorology, not just in Britain but all over the world. He did not mince his words.

Meteorological instruments have been for the most part treated like toys, and much time and labour have been lost in making and recording observations utterly useless for any scientific purpose.

He "lamented" that the life of the Palatine Meteorological Society had been so short and commented that the observational system operated by that Society had been "a model of a scheme of combined exertion which the savants of the nineteenth century would do well to imitate". At the Association's Tenth Meeting, in 1840, he presented a *Supplementary Report on Meteorology*. This was as critical as the report he presented in 1832 and almost twice as long. It concluded with suggestions "for the advancement of meteorological science".

---

[9] For the story of this meteorological society and the entirely separate Meteorological Society of London that existed from 1848 to 1850, see 'The Meteorological Societies of London', by J M Walker, published in 1993 in *Weather* (Vol. 48, pp. 364–372).

[10] Hereafter in this book, 'British Association for the Advancement of Science' is abbreviated to 'British Association'.

First, he suggested, well-equipped "public observatories" should be established, "to furnish standards of comparison, to establish the laws of phænomena and to fix *secular*, or normal data". Second, he suggested, "private individuals, fond of science", might set up their own meteorological stations to supplement his envisaged network of public observatories. Finally, he proposed, observations should be made by travellers. At no point did he press for the establishment of a government body responsible for co-ordinating meteorological work, even though moves to set up national meteorological institutes had already been made in a number of countries.

The case for establishing such a body in the UK was first made in 1847, by William Radcliff Birt (1804–1881), who had been employed by Sir John Herschel (1792–1871) from 1839 to 1843 on the reduction, tabulation and graphical representation of barometric data. This work had led Herschel to conclude, as he put it in a letter to Birt dated 28 July 1843, that the atmosphere might be considered "a vehicle for wavelike movements which may embrace in their single swell & fall a whole quadrant of a globe".

To further their investigations of atmospheric waves, Herschel and Birt received generous financial support from the British Association. Despite this, Birt came to believe that even greater support for research on the waves from government sources and scientific bodies was needed, not least to help overcome the chronic lack of data which hampered his investigations. In 1847, in letters to a number of influential figures, among them Sabine, he proposed that meteorological observations, especially measurements of barometric pressure, should be made by officers aboard Royal Navy ships in various parts of the world. Through greater understanding of atmospheric waves, he argued, the weather systems of the tropics and middle latitudes would be better understood. The main beneficiaries would be, in his opinion, the naval, merchant and military services, directors of observatories and members of learned societies. Observations would be standardized, regularly deposited in a central meteorological office and arranged under the superintendence of a keeper.

Herschel agreed. To him, there should be a central government office, with a scientific meteorologist as its director and a number of salaried assistants who carried out calculations, provided, as he put it in a letter to Birt in August 1847, the nation could afford it and provision was made for the regular publication of results. Dr John Lee (1783–1866), a patron of astronomy and meteorology, thought the nation ought to be able to afford the proposed meteorological office and suggested, in a letter to Birt in August 1847, that "if one million [pounds] of the annual expenditure of £20,831,077 now spent on the Army, Navy and Ordnance were spent on the Arts and Sciences the public would readily approve of the appropriation of the funds"!

Birt wrote to the President of the Royal Society in October 1847 to seek support for his ideas and published several papers on atmospheric waves in the *Philosophical Magazine* in the period 1846 to 1850. However, his work was largely ignored by the scientists of the day, even though one so distinguished and influential as Sir John

Herschel supported him. His applications to the Royal Society in 1850 and 1851 for grants to help him continue his investigations of atmospheric waves were turned down, and he became progressively more isolated from the scientific community. Eventually, he himself recognized empirical and conceptual difficulties and took care to make little of atmospheric waves in his last meteorological work, his *Handbook of the Law of Storms; Being a Digest of the Principal Facts of Revolving Storms. For the Use of Commanders in Her Majesty's Navy and the Mercantile Marine* (published in 1853).

Given that Herschel's idea of atmospheric waves and Birt's development of it could have led to the foundation of a meteorological office seven years before one was actually founded, it is a curious quirk of history that the most scathing criticism of the idea came from the first Director of the Office, who stated, in 1863, in his *Weather Book*, that "what are commonly called 'atmospheric waves' are delusive". And it is ironic that seafarers were to have assisted Birt's research on atmospheric waves by making observations and would have been among the main beneficiaries of his work, given that the Office came into being to serve the needs of seafarers and also enlisted seafarers as meteorological observers!

Networks of Observers Develop

Observations made by seafarers are not made at fixed locations. To overcome this difficulty, the idea was put forward that the ocean surface be divided into squares for the correlation of data and identification of oceanic areas. These have long been called 'Marsden squares', after William Marsden, who was Secretary of the British Admiralty from 1804 to 1807. They were first used early in the nineteenth century for showing on a chart the distribution of meteorological data over the oceans and are still used by marine climatologists today.[11] It is not clear, however, who actually conceived the idea of using squares. The credit may belong to Isaac Greenwood, who was mentioned earlier in this chapter. In his paper in the *Philosophical Transactions* in 1728, he proposed a method of tabulating data by using squares of latitude and longitude.[12]

The first systematic attempt to collect and analyse the meteorological observations of seafarers was made in 1831 by Captain Alexander Becher (1796–1865) of the British Hydrographic Office, and he used Marsden squares. However, the work had to cease almost immediately for lack of funds and other resources. So, too, did attempts

---

[11] Marsden squares are each 10° latitude by 10° longitude on a Mercator chart and every square is sub-divided into 100 squares which are each 1° latitude by 1° longitude. The Marsden Squares are numbered systematically, and each sub-square bears a number from 00 to 99. Thus position on the ocean can easily be ascertained to the nearest degree.

[12] See Agnew, D C (2004), 'Robert FitzRoy and the myth of the 'Marsden square': transatlantic rivalries in early marine meteorology', *Notes and Records of the Royal Society of London*, Vol. 58, pp. 21–46.

by others, but a successful system eventually came into being in the United States as a result of an unfortunate accident.

On 17 October 1839, whilst travelling through Ohio, an American naval officer, Matthew Fontaine Maury (1806–1873), was thrown from an overloaded stagecoach when it overturned and was injured so severely that his seagoing career was ended abruptly. In July 1842, on being recalled to duty, he was employed in Washington, first as Superintendent of the U.S. Navy's Depot of Charts and Instruments and subsequently, from October 1844, as Superintendent of the U.S. Naval Observatory (later Naval Observatory and Hydrographical Office).

For much of his time, he extracted from the logbooks of ships information about winds, weather and currents, which he compiled for use by mariners. The fruits of his labour were published in 1847 in the first sheet of his *Wind and Current Charts*, the primary means by which he disseminated to mariners the information he had accumulated. He also prepared special charts and forms on which American naval officers and merchant seamen were asked to record observations (in addition to those recorded in their logbooks). Most were keen to do so, and from the information thus obtained, he was able to revise the charts he had already produced and publish, in 1851, *Explanations and Sailing Directions to Accompany the Wind and Current Charts*. His work led to safer and faster voyages than hitherto. However, he was not satisfied with these contributions to practical navigation. He wanted to establish an international system.

In Britain in the 1840s, meanwhile, James Glaisher (1809–1903) had established a network of reliable meteorological observers, among them Samuel Charles Whitbread (1796–1879) of Cardington, Bedfordshire. Glaisher and Whitbread were two of the ten gentlemen who assembled in the Library of Hartwell House, near Aylesbury, Buckinghamshire, on 3 April 1850, "to form a society the objects of which should be the advancement and extension of meteorological science by determining the laws of climate and of meteorological phenomena in general". The gentlemen called this society 'The British Meteorological Society' and appointed Whitbread its President. Thus he became the first to preside over the body which is known today as the Royal Meteorological Society. He and his gardeners made meteorological observations for many years, and in January 1873 he presented a set of observations to the Meteorological Office. Titled *Fluctuations of Barometer Cardington Observatory, January 1st 1846 to December 31st 1870*, they are held today in the Office's Archive at Exeter, Devon.[13]

Glaisher was Superintendent of the Magnetic and Meteorological Department of the Royal Observatory at Greenwich, responsible for administration of the Department

[13] The National Meteorological Library and Archive at Exeter and the Office's archives in Edinburgh and Belfast are the principal repositories of the UK's meteorological treasures, among them weather diaries, logbooks of ships and early photographs. Committee minutes and other records of the Royal Meteorological Society are in the National Meteorological Archive at Exeter.

and for organizing observations and investigations. He was appointed in 1840 and held the post until 1874, when he resigned. He was Britain's first full-time government-appointed meteorologist.

In 1848, he was commissioned by the proprietors of the *Daily News* to organize the collection of weather reports by electric telegraph for publication in their newspaper. This action was prompted by the rainy and inclement weather which had prevailed throughout that summer and caused concern as the harvest approached. Publication of weather observations, it was considered, would help dispel or allay the fears of the general public and provide practical information for agriculturalists, especially during the period when harvesting was taking place. Accordingly, from 31 August to 30 October 1848, a table showing the "State of the wind and weather" was published daily in the newspaper.

Britain's first telegraphic Daily Weather Report contained observations made at 9 a.m. the previous morning by a network of twenty-nine railway stationmasters (twenty-seven in England, two in Scotland) and was made possible by cooperation between the railway companies and the rapidly expanding Electric Telegraph Company. Such was public and official approval of the weather reports that publication was resumed on 14 June 1849. That day, reports from only thirteen stations were published, but the network rapidly expanded until reports were published from about fifty stations, some in Ireland. From the observations, Glaisher daily drew weather maps. From 8 August to 11 October 1851, he published them by lithography at the Great Exhibition, selling them to the public at a penny each.

The Outcome of a Conference

The year 1851 was also notable for the appearance of the earliest periodical devoted to meteorology, the weekly *Wind and Weather Journal*, which was published in London and first appeared on 8 July 1851. However, the meteorological developments in 1851 of greatest importance in the history of the Meteorological Office were the schemes proposed by William Reid (1791–1858), a British army officer. He had worked with Redfield on the characteristics of tropical cyclones and published, in 1838, a classic book titled *An Attempt to Develop the Law of Storms by Means of Facts, Arranged According to Place and Time; and Hence to Point Out a Cause for the Variable Winds, With a View to Practical Use in Navigation.*

In the 1840s, Reid persuaded the British Colonial Office to set up meteorological stations in the colonies. Then, in 1851, his persuasiveness was rewarded yet again, when his former commanding officer, Sir John Fox Burgoyne (1782–1871), authorized the establishment of a network of observing stations worldwide. Responsibility for making observations systematically and routinely at foreign stations was given to officers of the Royal Engineers stationed overseas, with overall control of the network given to Henry James (1803–1877) of the Royal Engineers. Reid's persuasiveness did not end there, for in the autumn of 1851, encouraged by Reid, Burgoyne approached the government of the United States via diplomatic channels to suggest expansion of the British network through American collaboration.

In Washington, DC, the recipient of the letter from the British government was Secretary of State Daniel Webster, who wrote to the Secretary of the Navy, William Graham, on 14 November 1851, to ascertain the extent to which the U.S. Navy would be prepared to cooperate with the British. Graham passed the letter to Maury via Maury's superior and, in reply, Maury proposed that a meteorological conference be held. Its purpose would be to review meteorological observing practices, its objective to reach agreement on standard procedures.

Maury pointed out that the greater part of the globe is covered by ocean and argued that no general system of meteorological observations could be considered complete unless it embraced the sea as well as the land. He suggested an amendment to the British proposition:

That England, France, Russia and other nations be invited to cooperate with their ships by causing them to keep an abstract log according to a form to be agreed upon and that authority be given to confer with the most distinguished navigators and meteorologists both at home and abroad, for the purpose of devising, adopting and establishing a universal system of meteorological observations for the sea as well as for the land.

The American government replied to the British government on 6 December 1851, proposing that a conference be held to consider meteorological observing practices on land and at sea. Four days later, Graham gave Maury authority to organize it, after which Maury wrote to a great many diplomats and government officials at home and abroad and to a number of distinguished foreign scientists, among them Glaisher (in his capacity as Secretary of the British Meteorological Society). He suggested the conference be held in Paris, mainly because he was keen that Dominique Arago (1786–1853) should attend. Arago was ailing and not expected to be well enough to travel to another venue.

The replies Maury received were generally favourable, with Humboldt, a close friend of Arago, particularly enthusiastic. The British, too, supported the idea of a conference, except that Sabine opposed inclusion of land meteorology in the conference plans. The Royal Society, to which the response of the American government had been passed for comment and advice, accepted Sabine's objections and strongly recommended to the British government that a conference on marine meteorology be held. The main reason why the Society felt that land meteorology should not be included was that achievement of uniform observing practices in the many countries where meteorological observations were already made was considered too ambitious. Achieving uniformity in marine observing practices was a different matter.

In early 1853, the American government accepted the response of the British and issued invitations to "a conference limited to adopting a universal system for observations at sea". When by the third week of April the British government had failed to respond, Lord Wrottesley took action. A senior Fellow of the Royal Society, Chairman of the Parliamentary Committee of the British Association and a close friend of Burgoyne, he was keenly interested in Maury's *Wind and Current Charts* and convinced of the merit of the proposal for a universal system of meteorological observations at sea. On 26 April 1853, he made a long speech in the House of Lords in which he reviewed Maury's achievements, explained the proposal in great detail, emphasized the potential commercial advantages for British shipping, and urged the British Government to send delegates to the conference.

But still a response from the government did not materialize, so the Americans decided to proceed notwithstanding. In June 1853, they issued invitations to a conference to be held in Brussels, rather than Paris. The reason for the change of venue is not clear, but the poor state of Arago's health may have been a critical factor. He died on 2 October 1853. Whatever the reason, Brussels was an important meteorological centre, where Adolphe Quetelet (1796–1874) was Director of the Royal Observatory and Perpetual Secretary of L'Académie Royale des Sciences, des Lettres et des Beaux-Arts de Belgique.

There were two British delegates, and they were forbidden by the government to make any commitments to spending! Both were Fellows of the Royal Society: Henry James, an expert on land meteorology, and the Arctic explorer Frederick William Beechey (1796–1856), a Captain in the Royal Navy and the professional member of the Marine Department of the Board of Trade. There were twelve delegates in all, representing ten nations, namely, Belgium, Denmark, France, Great Britain, The Netherlands, Norway, Portugal, Russia, Sweden and the United States.[14] Of the twelve, only James and Quetelet were not naval officers. The conference met daily from 23 August to 8 September 1853 and reached agreement on a code of observational practice at sea, including the use of a standard meteorological register for recording observations. The interval between observations was soon changed from two to four hours, but the scheme to which the delegates agreed has otherwise not been modified significantly to this day.

Beechey duly reported to the British government after the conference. At first, there was no response, probably because of ministerial preoccupation with events leading up to the Crimean War. Eventually, however, on 6 February 1854, the First Lord of the Admiralty, Sir James Graham, announced in the House of Commons that a new government department was to be formed, its responsibilities being to collect and digest the meteorological observations already made on board merchant and Royal Navy ships, to increase the number of ships from which observations were

---

[14] In addition to Maury, Quetelet, James and Beechey, the delegates were: Victor Lahure of the Belgian Navy; P Rothe of the Danish Navy; A Delamarche of the French Imperial Navy; Marin H Jansen of the Dutch Navy; Nils Ihlen of the Norwegian Navy; J de Mattos Corrêa of the Portuguese Navy; Alexis Gorkovenko of the Russian Imperial Navy; and Carl Anton Petersson of the Swedish Navy.

obtained, to ensure that observations were made every four hours, and to communicate observations after reduction to Captain Maury.[15]

The original intention was that funds for the new department would be included in the Navy estimates, but, in the event, funds were included in the Board of Trade vote.[16] After the Board's President, Edward Cardwell, submitted the vote to the House of Commons, on 30 June 1854, there was a short debate in which John Ball, the Member for Carlow, "anticipated that in a few years, notwithstanding the variable climate of this country, we might know in this metropolis the condition of the weather 24 hours beforehand". According to Hansard, the reaction to this optimism was "laughter"!

In the Board of Trade vote, the sum of £3200 was included for the first year of the new department's work, plus an additional £1000 for meteorological services to the Admiralty. Regarding who should direct the new department, the government consulted the Royal Society, the outcome being that Captain Robert FitzRoy of the Royal Navy was chosen. He took up his duties on 1 August 1854.

[15] Meanwhile, the Royal Netherlands Meteorological Institute had already been set up. Based in Utrecht, with Christoph Buys Ballot (1817–1890) as its director, it had been founded on 31 January 1854. Its foundation was an outcome of the Brussels conference, and the Institute soon began to publish marine meteorological maps, the first of them in 1856. The first national meteorological institute was that set up in Belgium in 1826. It was based at the Brussels Royal Observatory and directed by Quetelet until 1874. The other institutes which existed before 1854 were those in Russia, Prussia and Austria. Respectively, they were set up at St Petersburg in 1840 (directed by A T Kupffer until 1865), Berlin in 1847 (directed by Dove from 1849 to 1879) and Vienna in 1851 (directed by K Kreil until 1863).

[16] In British usage, a 'vote' is a sum of money granted by a majority of votes, in this case in the House of Commons.

2

Statistics and Storms

Robert FitzRoy had long been interested in meteorology and, in particular, convinced of the barometer's value. In 1829, during his first voyage in command, his ship had nearly capsized in a squall off Uruguay. There had been barometers on board and low pressure indicated, but anchors, topmasts, jib-boom and two men had been lost. He never again ignored the barometer's warnings. Indeed, he interpreted the instrument's readings so skilfully during his second surveying expedition that he weathered the severest gales without the loss of anyone on board or even damage to his ship. The voyage lasted almost five years (December 1831–October 1836) and proved a watershed in FitzRoy's life. It was a personal triumph which enhanced his reputation as a skilled and meticulous surveyor. He was thanked in Parliament, praised by the Hydrographer of the Navy and awarded the gold medal of the Royal Geographical Society.[1] The ship was HMS *Beagle*, and the expedition was the famous circumnavigation of the globe with Charles Darwin on board.

FitzRoy was born on 5 July 1805 and educated first at Rottingdean School near Brighton and later Harrow School. He entered the Royal Naval College at Portsmouth in February 1818 and took only twenty months to complete the course of study, which normally took three years. He went to sea when he was 14, a volunteer aboard HMS *Owen Glendower*, and later served aboard HMS *Superb* and HMS *Hind*. He was promoted to the rank of Lieutenant on 7 September 1824 and subsequently served aboard HMS *Thetis* and HMS *Ganges*. He took command of HMS *Beagle* on 13 November 1828 and advanced to the rank of Captain on 3 December 1834.

After the circumnavigation, FitzRoy wrote his *Narrative of the Surveying Voyages of His Majesty's Ships* Adventure *and* Beagle *Between the Years 1826 and 1836* (published in 1839), and he married and started a family.[2] He was elected Member

---

[1] From 1829 to 1855, the Hydrographer of the Navy was Francis Beaufort (1774–1857), who is best known today for the scales of wind force and weather he advanced in 1806.

[2] He married Mary Henrietta O'Brien in December 1836 and by her had four children. She died in 1852. By his second wife, Maria Isabella Smyth, whom he married in February 1854, he had a daughter, Laura Maria Elizabeth, who was born on 24 January 1858 and lived to a great age, passing away on 6 December 1943.

of Parliament for Durham City in 1841, but he resigned his seat in the spring of 1843 when appointed Governor of New Zealand. There followed an unhappy episode in his career. The settlers considered him arrogant, dictatorial and not at all conciliatory, and he offended the New Zealand Company. He was replaced by George Grey in November 1845. He proposed in the House of Commons on 28 July 1842 that certificates of competency be introduced for the masters and first mates of merchant ships. This move to improve the safety of seafarers bore fruit, for a voluntary system of certification was introduced in 1845, and an Act of Parliament which made the possession of certificates compulsory was passed in 1850. As a result of this Act, the Marine Department of the Board of Trade was formed, with responsibility for the welfare of all employed in shipping and all who entrusted their goods or persons to British ships. The Act rendered the keeping of an official log obligatory and brought about the establishment of schools of navigation.

FitzRoy was appointed Superintendent of Woolwich Dockyard in September 1848 and six months later took charge of the sea trials of the Royal Navy's first screw-driven steamship, HMS *Arrogant*. To the settlers in New Zealand, the name of the ship must have seemed appropriate! In fact, he commanded the ship with great success, showing firmness, tact, compassion, generosity and consideration for others, qualities which were evident during the *Beagle* voyages but not always apparent in New Zealand. It was a surprise and matter of regret to those around him when, in February 1850, he abruptly resigned his command on the grounds of his and his wife's ill health and his need to attend to private affairs. He was placed on half pay and never served at sea again.

After his restoration to full health, he served for a while as a director of the General Screw Steam Shipping Company. He was also, briefly, in 1854, private secretary to an uncle, Lord Hardinge, the Army's Commander-in-Chief. Then, on 1 August 1854, he became Meteorological Statist to the Board of Trade.[3] Though he sometimes used the name 'Meteorological Office' in publications, he did not, in fact, live to see it agreed upon and adopted officially.[4] He was officially head of the Meteorological Department of the Board of Trade.

A New Government Department Is Born

The person who suggested that FitzRoy might be appointed seems to have been Lord Wrottesley, who in early 1854 asked him for his views on the objectives and

---

[3] According to *The Shorter Oxford English Dictionary* (Oxford University Press, 1973), a 'statist' is "one who deals with statistics" or "a person skilled in state affairs, one having political knowledge, power, or influence". For FitzRoy's post in the Board of Trade, it could be argued both definitions were applicable!

[4] He used the name in, for example, the *Report of the Meteorological Department of the Board of Trade*, published in 1855. Occasionally, he referred to the 'Meteorologic Office', as in, for example, his *Weather Book* and the fifth *Report of the Meteorological Department of the Board of Trade*, both published in 1863.

Figure 2.1. Robert FitzRoy, Meteorological Statist to the Board of Trade, the first Director of the Meteorological Office. © Crown Copyright 2010, the Met Office.

operational needs of a meteorological department. FitzRoy was known to be keen to raise the meteorological awareness of seafarers, and his response to Wrottesley, dated 3 February 1854, contained a number of constructive suggestions. He was seeking suitable employment and had made known to Wrottesley, Sabine and other influential members of the Royal Society his interest in the post at the Board of Trade. FitzRoy was himself a Fellow of the Royal Society, elected in 1851.[5] More than that, he was a member of the Society's Council in 1854, but he was not present at the Council meeting on 15 June 1854 when a letter from the Board of Trade was considered.

Dated 3 June 1854 and addressed to the Secretary of the Royal Society, the letter was signed by James Booth, Office of Committee of Privy Council for Trade, Marine Department. It read:

Sir, – I am directed by the Lords of the Committee of Privy Council for Trade, to acquaint you that, with the concurrence of the Lords Commissioners of the Treasury, My Lords have determined to submit to Parliament an estimate for an office for the discussion of the Observations on Meteorology which it is proposed shall be made at sea in all parts of the globe in conformity with the recommendation of the conference held at Brussels last year; and they are about to construct a set of forms for the use of that office, in which it is proposed to publish from time to time, and to circulate such statistical results as may be considered most desirable

---

[5] His thirteen supporters included Darwin and Beaufort. He was also a member of the British Association and served on its Council. He was elected a member of The Athenæum in January 1852 and also joined the Royal Geographical Society, the Royal Astronomical Society and the Ethnological Society. He joined the British Meteorological Society on 27 March 1855 and became a Council member two months later.

by men learned in the Science of Meteorology in addition to such other information as may be required for the purposes of navigation.

Before doing so, however, they are desirous of having the opinion of the Royal Society as to what are the great desiderata in Meteorology, and as to what forms that Society consider the best calculated to exhibit the great atmospheric laws which it may be most desirable to develop.

I herewith inclose a form of Log which will contain all that it is proposed to execute at sea; but it may possibly happen that observations on land upon an extended scale may hereafter be made and discussed in the same office, and in framing your reply it is desirable that such a contingency should be borne in mind and provided for.

In their acknowledgement, dated 24 June 1854, the Royal Society informed the Board of Trade that the President and Council of the Society had "addressed a letter to several of the most eminent meteorologists in foreign countries" to seek their comments and advice. These included Maury, Dove and Quetelet.[6] A considered reply would be sent to the Board in due course.

Meanwhile, at numbers 1 and 2 Parliament Street, Westminster, FitzRoy started work.[7] His approach clearly impressed Sabine, for, on 26 August 1854, he wrote in a letter to FitzRoy: "It appears to me that you are commencing in a very good business-like practical manner". Insight into FitzRoy's views on the objectives, priorities and modus operandi of the Meteorological Department can be found in the Report of the 24th Meeting of the British Association, held in September 1854. Set out in two memoranda which were included in the Presidential Address delivered by the Earl of Harrowby, these views were provided by FitzRoy in response to Harrowby's request for a report on the importance of meteorology to the interests of maritime commerce and navigation.

In Memorandum I, FitzRoy pointed out that it was "a question of the greatest importance to determine the best tracks for ships to follow, in order to make the quickest as well as the safest passages". With reference to Maury's work, he explained how knowledge of prevailing winds and currents could be applied to save not only time and lives but also expense (to the merchant, shipowner and insurer). In Memorandum II, he reported that he had been appointed head of the new Meteorological Department on 1 August, "referring to Dr Lyon Playfair, of the Department of Science and Art, and to Admiral Beechey, of the Marine Department, for such assistance as they could render". As soon as registers and instruments were ready and an office prepared, he would be assisted by four or five persons whose duties he would superintend. He

---

[6] The others were: A T Kupffer (Director of the Russian Meteorological Observatories); K Kreil (Director of the Austrian Meteorological Observatories); J von Lamont (Director of the Bavarian Meteorological Observatories); C H D Buys Ballot (Director of the Dutch Meteorological Observatories); C Hansteen (Director of the Norwegian Meteorological Observatories); A Bravais (President of the French Meteorological Society); N Martins (Editor of *l'Annuaire Météorologique de la France*); L F Kaemtz, Dorpat (author of a treatise on meteorology); G A Erman of Berlin, and E Heis of Aix-la-Chapelle (both eminent meteorologists); G Forchhammer, Copenhagen (investigator of ocean currents and sea-water density); A D Bache and J Henry (in connection with the American Meteorological Observatories).

[7] This was the home of the Meteorological Office until 1869. The Cenotaph now stands very close to the site.

Figure 2.2. The first home of the Meteorological Office at 1 and 2 Parliament Street, London, with wind vane and rain gauge on the roof. © City of Westminster Archives Centre.

expected that several ships would be supplied with 'abstract logs' (meteorological registers) and instruments in October, and that the office would be "in full work" in November. In fact, it was early 1855 before the department was fully staffed and fully operational.

FitzRoy's memorandum of 3 February 1854 to Wrottesley shows that he originally envisaged a meteorological department with a director assisted by only two people, a draftsman and a clerk. The Board of Trade proposed a staff of seven, including the director. In the event, four clerks were appointed, all from within the Board of Trade. The intention was that one of them would serve as FitzRoy's deputy, but the Board's Statistical Department refused to release the man who was seen as FitzRoy's second in command, so the post of deputy was initially left vacant.

The most senior of the clerks who had joined FitzRoy by early 1855 was William Pattrickson, a skilled draftsman. He received a salary of £180 for the year 1855–1856, and £190 for 1856–1857. Of the others, Thomas Henry Babington was paid £81 0s 7d for 1855–1856 and £88 15s 0d for 1856–1857, whereas F R Townsend was paid £81 0s 1d for 1855–1856 and £90 0s 0d for 1856–1857.[8] FitzRoy's salary was £600 per annum in 1856–1857, charged as £300 to the Board of Trade's vote and £300 to the Admiralty's. For the first two years of the Meteorological Department's existence, the

---

[8] In the monetary system used in the United Kingdom until 1971, there were twenty shillings in a pound and twelve pence in a shilling. Thus, £81 0s 7d becomes £81.03 in the system used today.

salaries of the clerks were taken from the Board of Trade's general vote. Thereafter, they were charged to the Meteorological Department's own vote.

In Memorandum II, FitzRoy reported that the Admiralty had ordered all records in the Hydrographical Office be placed at the disposal of the Board of Trade and that all other documents accessible to Government would be made available, too. There would be no want of materials, said FitzRoy. He accepted that Maury's *Instructions* or *Sailing Directions* had brought about real benefits for mariners, but he ventured to think that they "presented too much detail for the seaman's eye" and had "not been adequately condensed". In his opinion, their practical value was not as great as generally supposed. As additional information would not remedy this shortcoming, he proposed to collect ships' meteorological observations, which he would reduce and average.[9] This would enable him to compile diagrams, charts and meteorological records, from which seafarers could obtain information about conditions at particular points on the ocean.

Whilst the Royal Society's reply to the Board of Trade's letter of 3 June 1854 was awaited, FitzRoy had no specific instructions regarding the work he was expected to carry out. His brief was rather general. As stated in a letter he received in September 1854 from James Booth (Office of Committee of Privy Council for Trade, Marine Department), he was expected to improve knowledge of ocean currents and prevailing winds and thereby "diminish the risk and labour of navigation" and "shorten the duration of voyages". He set to work with a will.

Within a month of taking up his appointment, he had prepared a circular to be issued to the largest shipowners, seeking their cooperation over putting into practice the recommendations of the Brussels conference. Suitable instruments would be lent to any master who volunteered to keep a meteorological logbook (then called a 'register'), and masters who were diligent as meteorological observers would receive copies of charts and books as and when these were published. Because the Meteorological Department's financial resources were limited, however, the Board of Trade hoped that the wealthier shipowners would purchase instruments at cost price. This turned out to be a forlorn hope. There was hardly any response from shipowners.

FitzRoy soon began to supply instruments, instructions and registers to the commanders of Royal Navy ships and the masters of British merchantmen.[10] The instruments were loaned free of charge, on condition that masters undertook to return the

---

[9] In the words of the *Meteorological Glossary*, published by the Meteorological Office (Her Majesty's Stationery Office, 1991, Met.O.985, 335 pages): 'Reduction' is "the substitution of computed values for those directly observed, the purpose being to eliminate the effect of some particular factor or factors. The reduction process is most commonly used to eliminate the effects of varying height on observed surface values of air temperature and pressure and is termed reduction to sea level".

[10] The instruments had been tested and calibrated at Kew Observatory and were therefore considered reliable. From the outset, the Meteorological Department cooperated with Kew Observatory, then controlled by a committee of the British Association. This cooperation was important, for the use of reliable and comparable instruments was, and remains, essential for obtaining accurate meteorological data.

observations made on their voyages. Mariners who cooperated with FitzRoy received not only instruments, instructions and registers but also copies of Maury's charts and sailing directions, provided gratis by the American government. In his *Weather Book* (published in 1863), FitzRoy explained that instruments, charts and registers were furnished only to a limited number of carefully selected ships. "Many more ships might have been similarly provided with instruments", he wrote, "had the willingness of their captains alone affected the supply; but as only a certain number of good instruments could be purchased by Government annually, with due regard to the Parliamentary vote of money, and as the agents required instruments to be kept for the purpose of comparison with those sent or returned, besides those wanted for occasional supply at numerous stations, the number was necessarily limited".

Realizing that registers from ships equipped with instruments he had supplied would not immediately be available in sufficient quantities for meaningful statistics to be compiled, and wishing to use his time fully and constructively, FitzRoy began to extract and modify data from Maury's publications for presentation in a graphical form which he considered simpler and more readily understood by seafarers than the presentational method used by Maury. In his opinion, Maury's pilot charts and sailing directions were so detailed and complex as to be almost unintelligible to most mariners. Accordingly, he devised 'wind stars', diagrams that showed at a glance the directions and strengths of the winds which occurred most frequently at different times of year in areas of the ocean 10 degrees 'square' (i.e., 10 degrees of latitude by 10 degrees of longitude). All of the observations available in each 'square' were averaged and referred to the mid-point of the square, which thus became a single 'observatory'.[11]

The first *Report of the Meteorological Department of the Board of Trade* was dated 23 May 1855 and issued in January 1856, but it was not an official publication. It was published privately by FitzRoy, probably at his own expense.[12] Curiously, the Board of Trade rejected it, apparently because the Superintendent of the Marine Department, Admiral Beechey, refused to accept it. The reason for the refusal may have been that FitzRoy had produced the report on his own initiative, without official approval. However, he evidently believed that he was expected to produce the report, for he wrote in the Preface, addressed to the President of the Board of Trade: "In obedience to your Lordship's orders, I have the honour to present the following *Report of the Meteorological Department of the Board of Trade*".

---

[11] By combining observations within ten-degree squares, FitzRoy employed a technique attributed to William Marsden (see Chapter 1). By so doing, however, Maury and FitzRoy both made the same error. If a vessel was moving quickly, a square would be crossed in a comparatively short time, permitting few observations. From a vessel moving more slowly, more observations would be made. Accordingly, observations were not all of equal weight, causing mean values to be biased towards occasions of light winds.

[12] FitzRoy dipped into his own pocket several times in his career. When equipping the *Beagle* in 1831, for example, he spent a considerable amount of his own money (£300) on chronometers, doing so because he considered he needed more of these instruments than the Lords of the Admiralty deemed necessary.

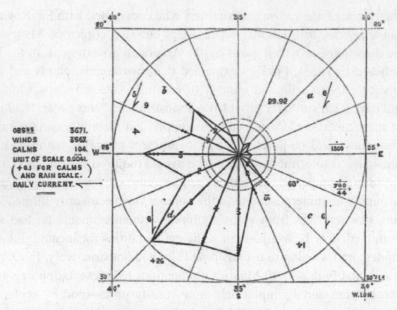

Figure 2.3. Wind star, from FitzRoy's *Weather Book* (1863), showing the directions and strengths of the winds which occurred most frequently between latitudes 20°S and 30°S and longitudes 30°W and 40°W. © Crown Copyright 2010, the Met Office.

In the report, FitzRoy traced the background to the establishment of the Department and reviewed the work he had already carried out. He reported that more than fifty merchant ships and thirty men-of-war had been supplied with instruments since the Department had been set up and said that more ships could have been supplied had there not been a lack of "good marine barometers". With the money available, he commented, the Department could not afford to buy more than a certain number of barometers each year. He praised the Kew Committee of the British Association and wrote approvingly of the accuracy and robustness of the instruments provided by Kew Observatory.

FitzRoy was clearly enthusiastic about the work accomplished by his Department and optimistic about the future. He hoped to make a significant contribution to safety at sea whilst at the same time saving money and time on voyages. The emphasis in the report on the usefulness of the barometer shows his confidence in this instrument for diagnosing and foretelling the weather, but his remarks concerning the availability of funds for the purchase of barometers make odd reading, given his statement that the balance in hand out of the £3200 allowed for 1854–1855 reduced the requirement for 1855–1856 to £700!

The Great Desiderata of Meteorology

The Royal Society replied to the Board of Trade on 22 February 1855, making recommendations that were soon adopted by the Government as a basis for defining the Meteorological Department's objectives and functions. The reply, which took the form of a long letter, was prepared by a committee, among its members Charles Darwin and Heinrich Dove.[13]

The great desiderata of meteorology were considered under eight headings: barometer; dry air and aqueous vapour; temperature of the air; storms or gales; thunderstorms; temperature of the sea and investigations regarding currents; auroras and falling stars; and charts of the magnetic variation. The Royal Society recognized that compass variation did not, strictly, belong in the domain of meteorology but included it because it had been discussed at the Brussels conference.

The recommendations of the committee were that:

- Monthly, quarterly and annual means of barometric pressure, aqueous vapour and air temperature, together with the variability of each, should be ascertained and tabulated for geographical areas delimited by specified meridians and parallels, the objective being to cover, in aggregate, the entire ocean.
- The temperature of the sea's surface in different months of the year should be carefully observed. In addition, the temperature, direction and speed of ocean currents, and their variations in different months and different years, should be considered an important subject of inquiry.
- The varying limits of trade winds and monsoons should be investigated.
- Fluctuations of temperature on a large scale, such as might affect great portions of the globe simultaneously, should be investigated by a comparison of 'five-day means' made at all fixed stations.[14]
- Charts of the magnetic variation should be constructed.[15]
- Observations should be made at British military stations in the Mediterranean and on the coasts of Australia and New Zealand. In addition, hourly observations should be made for at least one year at a station in the West Indies, to supply diurnal corrections for existing observations.

It was further recommended that Her Majesty's Government should "promote and facilitate the mutual interchange of meteorological publications emanating from the governments of different countries". In the view of the Royal Society, "one of the chief impediments to the advancement of meteorology consists in the very slow progress which is made in the transmission from one country to another of the observations and discussions on which, under the fostering aid of different governments, so much labour is bestowed in Europe and America".

Although the Royal Society's reply was concerned mainly with marine meteorology, the Board of Trade's reference to observations on land was not overlooked. Indeed, the part of the reply which dealt with air temperature was more applicable to

---

[13] Dove was Director of the Royal Prussian Meteorological Institute. Detailed comments were also received from Maury and Quetelet, as well as from Heis, Erman and Kreil. The committee consisted of the Officers of the Royal Society, together with the members of the Society's Physics and Meteorology Sub-committee.

[14] Five-day means are the averages of daily values taken on five consecutive days.

[15] This task was subsequently undertaken by the Hydrographic Department of the Admiralty, who published the required charts in 1858.

observations made on land than to those made at sea. The part which dealt with the barometer also contained advice specific to land stations. In the words of the reply, "care should be taken to ascertain by the best possible means (independently of the barometer itself) the height of the station above the level of the sea".

Further comments and recommendations were added by the Royal Society in 1856. It was reiterated that one of the most important objects of the Meteorological Department, for both practical and theoretical purposes, was the procurement of statistics of wind direction and force in different seasons of the year over the parts of the Atlantic Ocean most usually traversed by ships. Atlantic islands such as Madeira, Bermuda, Ascension, St Helena and the Azores were all places where, in the view of the Royal Society, self-recording anemometers might be installed. This recommendation was soon implemented, at least in part, for, in the spring of 1859, Thomas Babington was sent to Bermuda and Nova Scotia to install anemometers.

The work that FitzRoy carried out in his first few months as Meteorological Statist was broadly consistent with the recommendations made in the Royal Society's letter of 22 February 1855. In the longer term, however, he does not appear to have felt himself unduly constrained by the contents of the letter or, come to that, later correspondence. He appears to have considered the letters advisory, rather than binding. Nevertheless, the Royal Society considered their letter a directive, given that the Board of Trade had approved the recommendations it contained.

The Formation of the Scottish Meteorological Society

Organized meteorology in Great Britain received a further boost in 1855 with the formation of the Scottish Meteorological Society. Indeed, its formation was another outcome of the conference held at Brussels, though not as directly as the foundation of FitzRoy's Department.

The establishment of a meteorological society in Scotland was urged by one of Britain's two delegates to the conference, Henry James. To him, such a society was needed to process the mass of observational data which had accumulated at the office of the Northern Lights, and he suggested that members of the Highland and Agricultural Society might be interested in forming the nucleus of a society devoted to meteorology. In the event, Sir John Stuart Forbes, a prominent agriculturalist, and David Milne Home, a barrister with a strong interest in geology and meteorology, convened the meeting at which the decision to form a Scottish Meteorological Society was made.[16]

At that meeting, held in Edinburgh on 11 July 1855, Alexander Keith Johnston was appointed Honorary Secretary, a position he held until his death in 1871. He was

---

[16] Sir John was the elder brother of James David Forbes, who was mentioned in Chapter 1 for the reports on meteorology which he presented at the meetings of the British Association held in 1832 and 1840.

known internationally as a scientific geographer and was, indeed, the Geographer-Royal for Scotland. Thanks largely to him, the Society's membership grew rapidly. Later in the year, James Stark, the Superintendent of Statistics under the Registrar-General of Scotland, became Meteorological Secretary. Initially, Johnston and Stark worked together to organize the collection and tabulation of observations from stations in Scotland, but Stark soon took sole charge of the observational programme, and in 1856 the Council of the Society recognized his efforts by appointing him a paid secretary on a part-time basis.

Stark contacted observers all over Scotland and soon established a network of observing stations. In the *Quarterly Report of the Registrar-General for Scotland* for the quarter ending 31 March 1856, he published a note which stated that he felt the *Quarterly Report* "would lose much in interest and value if the influence which weather exerts on the health of the population was not traced". He undertook to "collect, arrange and reduce on one uniform plan the observations made in different parts of the country". Thenceforth, abstracts of meteorological observations were published in the *Quarterly Report*.[17]

The Council of the Scottish Meteorological Society set aside funds to support Stark's work for a period of three years (1856–1858, inclusive). Towards the end of this period, they approached the Government in London for a subsidy from public funds, arguing that the work Stark was carrying out was of public value. The response of the Treasury was that no direct grant could be made to the Society. Instead, the Treasury suggested, the computing work that was involved in processing data for the *Quarterly Report*, as well as the preparation of each report, should be carried out by the Astronomer-Royal for Scotland, Charles Piazzi Smyth. If this arrangement was not acceptable, the observations might be sent to the Royal Observatory at Greenwich.

Whatever next? Observations made in Scotland processed in England? The Council of the Society could not accept such an outrageous proposal. A compromise was agreed. Observations would continue to be sent to the Society's Meteorological Secretary, who would check them for clerical and other errors. They would then be passed to the Astronomer-Royal for Scotland, who would prepare the *Quarterly Report*. Stark resigned at the end of 1858. The new arrangement came into effect at the beginning of 1859, when his successor took office.[18]

One may ask why the Treasury did not suggest that the observations made in Scotland be processed by FitzRoy's Department. In all probability, the view was that his work was concerned specifically with marine and coastal observations and wholly

---

[17] Weather in relation to public health was also a matter of considerable concern in England at this time.

[18] The arrangement continued until 1893, when the Scottish Meteorological Society again assumed sole responsibility for the reports to the Registrar-General. Stark's successor was J D Everett, who resigned after only six months, to take up an appointment in Nova Scotia. He was succeeded by A H Burgess, who also did not remain in the post for long. Then, in December 1860, the right man for the job was found, a Dunblane schoolmaster called Alexander Buchan. He became one of the great names of meteorology in the nineteenth century and held the post of Meteorological Secretary until he died, aged 78, in 1907.

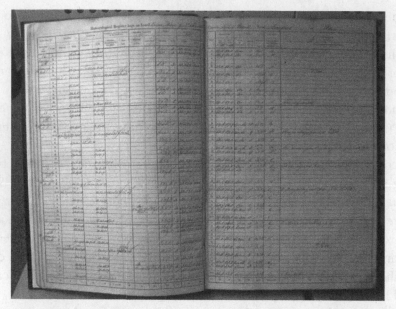

Figure 2.4. Ship's meteorological register (logbook), April 1857. © Crown Copyright 2010, the Met Office.

directed towards the needs of seafarers. It was therefore not appropriate that the work of the Society should be transferred to his Department. The idea of processing the Scottish observations at Greenwich was, on the other hand, reasonable, apart from the political difficulty, because Glaisher had already established his own network of amateur observers and had since 1847 abstracted their observations when preparing the quarterly *Remarks on the Weather* published by the Registrar-General of Marriages, Births and Deaths. Thus the work in Scotland could be considered complementary to Glaisher's. There was little official contact between the Scottish Meteorological Society and the Meteorological Department of the Board of Trade in the 1850s or early 1860s.[19] Such contact came later.

Progress at the Meteorological Department

After the initial burst of activity in FitzRoy's Department, a routine developed. He and his staff amassed data and thereby advanced knowledge of winds, weather and ocean currents. Instruments were obtained, tested at Kew and issued to ships. Logbooks were returned and scrutinized. Observations were recorded and tabulated and the logbooks carefully filed.

Masters of the merchantmen who collaborated with the Meteorological Department received sets of Maury's sailing charts, and the most diligent of them were awarded

---

[19] FitzRoy was elected an Honorary Member of the Scottish Meteorological Society in 1864.

prizes. In the second *Report of the Meteorological Department of the Board of Trade*, dated 11 March 1857, it was reported that five captains had been presented with "valuable telescopes" for returning meteorological registers which were "exemplary in a high degree".[20]

As far as the Meteorological Department's programme of work was concerned, all went smoothly in the early years. But among the staff, there was an atmosphere of unrest, the origins of which lay in a change of government. The coalition of the fourth Earl of Aberdeen collapsed in January 1855, and Lord Palmerston, the next Prime Minister, appointed Edward John Stanley President of the Board of Trade in place of Edward Cardwell.[21] One of Stanley's first actions was to recommend that Lieutenant Simpkinson of the Royal Navy be made FitzRoy's deputy. What ensued illustrates another characteristic of FitzRoy: his loyalty towards subordinates.

The appointment of Simpkinson proved unsatisfactory, so FitzRoy requested his deputy's dismissal. Simpkinson resigned. Piqued at the rejection of his nominee, Stanley refused to appoint a successor. FitzRoy nominated Pattrickson. However, a reorganization within the Board of Trade brought about the promotion of Babington over the head of Pattrickson. FitzRoy thought highly of Babington but nevertheless believed that, on merit, Pattrickson should have gained the promotion, and he felt that Babington had been favoured because of family connections with Lord Macauley (Thomas Babington Macauley). Pattrickson continued to draw a higher salary because of his specialist skills as a draftsman, and FitzRoy continued to consider him his deputy. Babington was willing to accept this arrangement, but Townsend now refused to obey Pattrickson. The ensuing difficulties led to Townsend's resignation and his replacement by a Junior Clerk, Jennings. This man also refused to serve under Pattrickson, so he was removed, at FitzRoy's request. Eventually, thanks to a strenuous effort on the part of FitzRoy, Pattrickson was granted special status as deputy, but without promotion in substantive rank.

Throughout this episode, relations between FitzRoy, Babington and Pattrickson seem to have remained harmonious. Indeed, there were no more outbreaks of unrest during FitzRoy's time at the helm of the Meteorological Department. The leadership qualities he had shown when in command of *Beagle* and *Arrogant* stood him in good stead. He was respected and well-liked by his subordinates.[22]

FitzRoy applied for another job in 1857, that of Chief Naval Officer in the Board of Trade's Marine Department, a post made vacant by the death of Admiral Beechey. However, the job went to Bartholomew James Sulivan (1810–1890), who had served

[20] The telescopes were purchased from Dollond's and cost five guineas (£5.25) each. Copies of the orders for them are held in the UK's National Meteorological Archive.

[21] Edward John Stanley (1802–1869), Baron Eddisbury of Winnington and second Baron Stanley of Alderley served as President of the Board of Trade from 1855 to 1858. Cardwell served from 1852 to 1855.

[22] Pattrickson left FitzRoy's Department at the end of December 1863 but remained an employee of the Board of Trade.

under FitzRoy as a midshipman aboard the *Beagle*. Failure to secure the job does not appear to have unsettled FitzRoy unduly, and he merely asked that the Meteorological Department should not be under the control of Sulivan. Promotion (in 1857) to the rank of Rear-Admiral on the Reserve List by virtue of seniority came as something of a consolation.[23] The loss of his eldest daughter (in 1856) was a severe blow, but he recovered from this and from the preferment of his one-time junior and once again threw himself enthusiastically and energetically into the work of his Department.

In 1842, when giving evidence before a Select Committee of the House of Commons, FitzRoy had urged the use of the barometer not only at sea but also at seaports, to provide warnings of approaching storms. He was particularly keen to help fishermen, who often left port unaware that bad weather was on the way. They suffered from unpredicted storms as much as any seafarers and maybe more than most. In 1857, his wish was granted: the Board of Trade gave him permission to issue barometers to the most exposed and least affluent fishing villages on the coasts of Great Britain and Ireland.

In the third *Report of the Meteorological Department of the Board of Trade* (dated 22 June 1858), FitzRoy stated that he had supplied "ten substantial barometers, durable and easy to observe", adding that "such assistance is urgently needed on the East Coast of England".[24] Commenting on newspaper discussions of disasters at sea, he noted that losses of ships and life might have been avoided by judicious use of barometers. The sturdy, simple and easily read instruments he supplied were known as 'fishery barometers', and instructions for their use were supplied gratis with them. By 1866, barometers and instructions had been supplied to ninety-five of the smaller and less affluent ports of the British Isles.[25]

Prior to 1857, public barometers had already been installed in several places. At Eyemouth near Berwick, for example, David Milne Home had not only set up a barometer but had also distributed printed directions for its beneficial use, and before 1850, there were instruments at Aberdeen, Peterhead and a number of other places in Scotland.

In his efforts to improve the safety of fishermen, FitzRoy was supported by a number of benefactors, among them the fourth Duke of Northumberland, who helped defray the cost of placing barometers at fourteen places on the coast of Northumberland. In the autumn of 1859, the Duke wrote to the President of the British Meteorological Society, Thomas Sopwith, suggesting that barometers would be useful if placed at the fishing villages and harbours on the Northumbrian coast. If the Society was willing

---

[23] Promotion to the rank of Vice-Admiral by virtue of seniority came in 1863.

[24] The villages were Newhaven, Anstruther, Arbroath, Rosehearty, Whitehills, Port Easy (Portessie), Lybster, Dunbeath, Lerwick and St Ives. The loan of a barometer would soon be made to Mount's Bay (Cornwall).

[25] For a detailed discussion of the fishery barometer programme, see 'Safety networks: fishery barometers and the outsourcing of judgement at the early Meteorological Department', by Sarah Dry, published in 2009 in the *British Journal for the History of Science*, Vol. 42, pp. 35–56.

Figure 2.5. FitzRoy display in the National Meteorological Library, showing barometers and model of HMS *Beagle*. Photograph by Malcolm Walker, by kind permission of the Met Office.

to take on the management of the observations, he would pay £100, or half the cost, of purchasing and installing the instruments. The Duke's proposal was supported by the Society's Council, whereupon Glaisher instructed Negretti & Zambra to prepare fourteen barometers. These were compared with the Greenwich Standard and installed by Glaisher himself. The Royal National Lifeboat Institution also cooperated with FitzRoy by placing fishery barometers at many of their stations.

The French Government cooperated, too. Barometers made in Paris were installed at ports on the coasts of France, and translations of FitzRoy's *Barometer Manual* were distributed to French fishermen. FitzRoy produced several books and pamphlets concerned with barometers, the first of them a pamphlet on the marine barometer (published in 1856). His publications contained instructions on the use and care of all types of barometer and provided advice on how to foretell the weather from changes in barometric pressure, shifts of wind, observations of clouds, colour of the sky and behaviour of birds and animals.[26]

Among the other publications that were issued by the Meteorological Department, there was a series of *Meteorological Papers, Published by Authority of the Board of Trade*, compiled or edited by FitzRoy. Of these, the third was the most notable. Published in 1858, it consisted largely of a translation of Dove's *Ueber das Gesetz der Stürme* (mentioned in Chapter 1). The editor was certainly FitzRoy, but the name of the translator was not given. The second edition was a much more substantial work than the first. According to its preface, it had been translated "with the author's sanction and assistance" by Robert Henry Scott, of Trinity College, Dublin. He may

---

[26] Many of FitzRoy's proverbs about wind and weather can be found in *Weather Lore*, 4th edition, compiled and arranged by Richard Inwards, edited, revised and amplified by E L Hawke and published by Rider & Co for the Royal Meteorological Society in 1950. FitzRoy included some in the *Barometer Manual* and some in the *Barometer and Weather Guide*. He had coined and compiled rhymes and maxims since the 1820s.

also have been the translator of the first edition, though confirmation of this has not yet come to light. There was mutual admiration between FitzRoy and Dove, as shown by the dedication of the second edition to FitzRoy, but he did not adopt all Dove's concepts. Nevertheless, the model of mid-latitude weather systems that was taking shape in his mind was very like that proposed by Dove (see Chapter 1).

Operational Storm Warnings

Making use of the observations which flowed into his Department from the ships of the Voluntary Observing Fleet, as well as observations from coastal and inland stations, FitzRoy constructed synoptic charts. These enabled him to study weather patterns and thereby show that changes in the weather at any given station could be foretold (or, as he preferred to call it, 'forecast') by intelligent interpretation of the atmospheric conditions occurring at places upstream. Soon he was making weather forecasts, which he shared with his staff.

The next stage in the development of his Department was obvious to FitzRoy. If warnings of storms were issued to seafarers, the heavy loss of life in shipwrecks might be reduced. In fact, he had suggested as much in 1842, in his evidence before the Select Committee, but it was not within his brief to issue such warnings. His Department's *raison d'être* was to improve ocean climatology and thus help seafarers operate their ships more safely and efficiently. The Department's function was, therefore, largely commercial.

In France, since July 1856, the director of the Paris Observatory, Urbain Le Verrier, had collected weather observations from a network of nineteen stations linked by the electric telegraph to produce daily weather bulletins for seafarers, but he had not attempted to forecast the weather.[27] And in The Netherlands, too, the meteorological institute was exploring ways and means of publishing telegraphic weather reports and issuing storm warnings to seafarers.

A disaster during the Crimean War had brought about the work being carried out by Le Verrier. The French ship of the line *Henri IV* and more than two dozen other vessels had been wrecked in a ferocious storm which occurred on 14 November 1854 during the blockade of Sevastopol. The French government had asked Le Verrier to investigate how such storms developed, and he had concluded that a timely warning of the tempest in question could have been given. The main recommendation of his report had been to establish a network of meteorological reporting stations so that warnings of the existence of storms could be provided in future. The French government had accepted this recommendation.

FitzRoy decided that an operational storm warning service for shipping ought to be provided by his Department, making use of the electric telegraph to gather observations

---

[27] Le Verrier succeeded Arago in 1854.

from weather stations and to communicate warnings to ports and harbours. To this end, he gained some support from the scientific community, notably from Sir John Herschel, whom he consulted over the soundness of his idea. In September 1859, moreover, at the Annual Meeting of the British Association, the Association's Council passed a resolution "praying the Board of Trade to consider the possibility of watching the rise, force and direction of storms and the means for sending, in case of sudden danger, a series of storm warnings along the coast".

Tragically, the attention of the whole nation was soon to be focused on the need for such warnings. In the early hours of 26 October 1859, the auxiliary steam clipper *Royal Charter*, 58 days out from Melbourne and bound for Liverpool, was driven ashore by winds of almost hurricane force and totally destroyed on the north-east coast of Anglesey, near Moelfre. Only 39 of the almost 500 on board survived.[28]

The inquiry could establish no particular cause for the disaster. The *Royal Charter* was a well-found iron ship, launched in 1855. There were excellent instruments on board, including three barometers. Her engines were running when she ran aground, and there appeared to be nothing wrong with her sails. True, there were many other wrecks that night, but ships that were much less seaworthy than the *Royal Charter* weathered the storm safely. FitzRoy was certain that a suitable warning system could have prevented the disaster, and he produced charts to show that the storm could have been tracked and its path predicted had the means existed.[29] By his analyses of this and other storms, FitzRoy amply demonstrated the validity of his cyclone model.

His arguments in favour of barometers, storm warnings, better understanding of weather systems by seafarers and effective use of the electric telegraph received support from many quarters. For example, soon after the *Royal Charter* catastrophe, a correspondent (simply identified as E.G.R.) wrote as follows in *The Athenæum* (31 December 1859):

Many disasters at sea might be prevented if every vessel carried a marine barometer. Had the Commander of the Royal Charter attended to the warnings of the barometers on board, and struck yards, &c., and made all snug aloft, it is possible that that most fearful loss might have been avoided. Yet no coaster or fishing vessel ever carries a barometer! Ought it not to be made compulsory on all vessels to provide themselves with these instruments; and ought not the Board of Trade examinations of captains, mates, &c. to include a knowledge of their indications in various climates? Till this be done, I would suggest that barometers should be erected in public situations on shore, and a signal be devised (to be hoisted as required), signifying that the barometer indicated foul weather. This should be done at the various coastguard stations; and even our vessels of war, especially when in the Channel, should keep it flying, as a signal to craft in sight of them. These barometers should not be entirely donations. Part of the expense

[28] The exact number on board is not known for certain because the passenger list was lost in the wreck, but it was probably 498.

[29] It is widely accepted that the *Royal Charter* Gale was the most violent storm of the century and probably comparable in intensity to the most devastating storm to visit the British Isles in recorded history, the tempest of 26–27 November 1703 (6–7 December New Style).

should be borne by the Board of Trade, the other raised by small (say shilling) subscriptions among the class to be principally benefited and their employers. I believe that the sure way to render any movement unsuccessful is to make it wholly eleemosynary. Beachmen, fishermen, &c. who had contributed to the erection of a barometer would be interested in its preservation, and observant of its indications. A cheap book, explaining its construction and its indications in plain Saxon English, that could be understood by such people, should be published and sold to them.

In his capacity as President of the British Association for the session 1859–1860, the Prince Consort took a close interest in the matter, as did the Astronomer-Royal, Sir George Airy. According to a letter dated 19 December 1859, from the Board of Trade to the Council of the British Association, and forwarded to Airy, the Board of Trade instructed FitzRoy to report on the use of the electric telegraph to warn ports of approaching storms and asked the Association to forward any suggestions to FitzRoy direct. FitzRoy duly put forward proposals, and these were approved by the Association's Council at a meeting held at Buckingham Palace on 25 February 1860, the Prince Consort in the chair. FitzRoy proposed that the UK be divided into three 'weather districts'. In each of the three, officers would be selected, instructed and provided with instruments. There would be three or four officers in each district, required to send "such telegraphic messages to London *occasionally* as their instructions specify". The messages would be posted at Lloyd's and transmitted to other selected stations, where they would "likewise be conspicuously posted".

Justifying his proposal, FitzRoy explained to the Council that storms generally did not arrive unannounced. At Valentia, south-west Ireland, for example, gales were typically preceded by rough seas.[30] Because the weather systems responsible for these seas advanced in particular directions, the time of arrival of gales at other places could be predicted. Regarding the cost of telegrams, he pointed out that agency fees and the cost of supplying instruments to ships had been reduced. Accordingly, he suggested, the money which had been saved from the amount voted for his Department could be used to defray the cost of telegrams. This was approved by the Board of Trade, and the role of the Meteorological Department as a storm-warning centre was officially authorized by the President of the Board of Trade (Thomas Milner-Gibson) under a minute dated 6 June 1860.[31] Authority to issue weather forecasts was not, however, granted. Authority was given for the existence of storms at one place to be announced by telegraph to other places, nothing more. The Board of Trade and the Council of the British Association were not in favour of attempting to foretell anything except the approach of storms known to exist elsewhere. This was the position of Le Verrier, too.

[30] Wind-waves formed in a distant storm generally arrive ahead of the storm in the form of swell. Sometimes, however, swell arrives from a storm which has dissipated over the ocean and never reaches land. The occurrence of rough seas on a coast is therefore not an infallible indication of an approaching storm.

[31] The expenditure of the Meteorological Department per annum had thus far never matched the sum voted, falling short by £959 in 1856–1857, £553 in 1857–1858 and £587 in 1858–1859.

Board of Trade officials were not convinced FitzRoy was able to forecast the weather correctly. As the Board's Assistant Secretary, T H Farrer, put it in a communication to him: "official timidity prompts us to question whether it might not be better to confine ourselves to registering and publishing facts, and leave foretelling the weather for a subsequent stage". In reply, FitzRoy defended his forecasts, saying that "six months' trial would prove their character". "It had been ascertained", he said, "that atmospheric changes on an extensive scale were not sudden, and that premonitions were more than a day in advance, sometimes several days". To avoid compromising those in authority over him, he offered to initial the weather forecasts that he proposed to issue to *The Times* and the *Shipping Gazette*.

The President of the Board of Trade remained sceptical. He declined to authorize the issue of weather forecasts and wrote to FitzRoy as follows: "I don't see any objection to the collection of facts as to weather, posting them at Lloyd's and transmitting them to various ports and to Paris, but it appears to me that the Government cannot take the responsibility of drawing conclusions and foretelling the weather for the practical guidance of merchant shipping". In a letter to FitzRoy dated 6 June 1860, Farrer asked if the warnings of storms that FitzRoy wished to give might not be provided through the medium of, or as a member of, the Council of the British Association. FitzRoy's reply, dated 9 June, suggests that the Council's views on the matter tended to mirror those of the Board of Trade. The Treasury took a different view, showing approval for FitzRoy's scheme by adding £500 to the estimate of the Meteorological Department for the year 1861–1862!

On 1 September 1860, the collection of weather reports by means of the electric telegraph began, with reports of shade temperature, wet-bulb temperature, wind force and direction, barometric pressure and state of the weather transmitted from thirteen coastal stations, these being at Aberdeen, Greenock, Berwick, Hull, Yarmouth, Dover, Portsmouth, Jersey, Plymouth, Penzance, Queenstown (now Cóbh), Galway and Portrush.[32] Observations were made once a day, at about 9 a.m., and transmitted to the Meteorological Department in London, with the telegraph companies charging only for the cost of the telegrams and furthermore rebating them by one-third.

Telegraph clerks made the observations as part of their regular duties. Though they were given no more by way of training than a printed list of instructions, and observing stations were apparently not inspected on any systematic basis, the reporting arrangements worked satisfactorily from the outset. The same arrangements worked satisfactorily in France, too, where telegraph clerks had made weather observations since 1856. Both FitzRoy and Le Verrier believed that the prompt and regular despatch of telegrams which resulted from the use of telegraph clerks to make the observations

[32] The wet-bulb temperature is ascertained by means of a thermometer whose bulb is covered with muslin wetted with pure water. Unless the relative humidity is 100%, the 'wet-bulb' shows a lower temperature than the thermometer whose bulb is exposed to the air. This occurs because heat is required to evaporate the water, and this is supplied by the air which is in contact with the wet muslin.

was an advantage that outweighed any lack of meteorological knowledge on the part of the observers. In America, where the electric telegraph had been used since the mid-1840s to send weather messages, it had been used since 1854 to supply weather reports each morning to the Smithsonian Institution. The reports were used by the Director, Joseph Henry, in collaboration with James Espy (see Chapter 1), to make predictions of the weather, with the first weather forecast in America appearing in the *Washington Evening Star* on 7 May 1857.

In early 1860, through the French Ministry of Marine and later in a letter to Airy, Le Verrier proposed an exchange of weather observations with the British. In return for observations from five stations in the British Isles, the French offered reports from any five places on the Continent. FitzRoy sought permission from the Board of Trade to accept the offer and this was granted, the expense to be covered by the sum of £700, which had been approved for defraying the cost of telegrams. Negotiations for the transmission of data from France began in April 1860, and the first exchange of reports took place five months later. On 14 September, they were received from the original thirteen stations and also from Liverpool, London and Kew. Reports were received that day, too, from Copenhagen, Helder, Brest, Bayonne and Lisbon. Two more British stations were added on 8 October 1860, these being at Valentia and Nairn, though several days elapsed before reports were received from the latter.[33]

For the time being, FitzRoy contented himself with merely receiving observations and passing them to the principal newspapers, which published them daily. Then, on 6 February 1861, he issued the first of his storm warnings for shipping, or, as he called them, 'cautionary signals'.[34] On the Tyne, the warning was disregarded, with disastrous consequences. Many ships were wrecked and many lives lost. However, greater heed was taken of the eight further warnings that FitzRoy issued in the period 6 February to 19 March 1861. In the fourth *Report of the Meteorological Department of the Board of Trade*, published in 1862, he claimed in justification of the warnings that very few ships had been wrecked on the coasts of the British Isles "during the notoriously tempestuous weather" of February and March 1861.

Warnings were sent by telegram to places likely to experience inclement weather, each message containing a list of the coastal stations affected, together with advice on the warning signals to be displayed. Near each of the telegraph stations in receipt

---

[33] Copies of *The Daily Weather Report* are held in the National Meteorological Library and Archive at Exeter. The British stations whose observations were transmitted to France are not identified in this publication. The telegraph station at Valentia was chosen partly because it had been linked to the Atlantic Telegraph Company's transatlantic cable and partly because Valentia Harbour was much used by ships of the British Navy. The cable was completed in August 1858 but remained serviceable for only a few weeks. Not until 1866 was a permanent and successful link between the Old and New Worlds established.

[34] The idea of storm warnings for shipping was first proposed by William Marsden (after whom 'Marsden squares' were named). His scheme failed, though, for lack of support. In the 1830s, Captain A B Becher tried to introduce a system whereby storm warnings were issued to seafarers. His initiative also came to nothing, for lack of means. The Dutch were the first in Europe to operate a storm-warning service. Buys Ballot first published telegraphic weather reports in 1859 and first issued storm warnings on 1 June 1860.

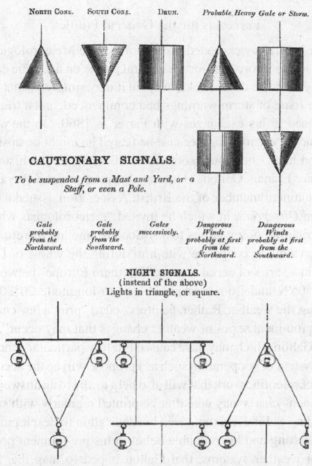

Figure 2.6. Storm-warning signals, from FitzRoy's *Weather Book* (1863), showing cones and drum and the night signals. © Crown Copyright 2010, the Met Office.

of a warning, at a conspicuous point on the coast, a cone or drum or combination of the two was hoisted on a staff, the cones and drums being about three feet in height and made with hoops and canvas, the latter painted black. A cone with its point upwards warned of gale-force winds expected from the north, whereas a cone with its point downwards warned of a gale from a southerly direction. A drum alone showed that stormy winds from more than one quarter could be expected. A drum and cone together warned of dangerous winds, with the point of the cone showing the direction of the wind expected to occur first. At night, signal lanterns were hoisted, the equivalent of a cone being represented by a triangle formed by three lanterns, the equivalent of a drum being represented by a square formed by four. A spherical shape was not used, to avoid confusion with time-balls.

Forecasts for the General Public

On 1 August 1861, FitzRoy exceeded his authority. The Meteorological Department began to publish weather forecasts for the general public on a routine daily basis. Why he acted precisely when he did is not known, but it may simply be that six months had elapsed since the issue of storm warnings had commenced, and a trial of six months had been mentioned in his exchanges with Farrer in 1860. On the other hand, it is possible that he acted when he did because he feared he might be upstaged.

In July 1861, a man who would soon make his mark on the history of the Meteorological Office, Francis Galton, Honorary Secretary to the Royal Geographical Society and a prominent member of the British Association, issued a *Circular Letter to Meteorological Observers*, in which he invited "meteorologists who have been in the habit of contributing observations to any Society, and are therefore familiar with methods of observing", to cooperate with him during the whole of December 1861 "in order to obtain a series of aërial charts of Northern Europe" between the latitudes 42°25'N and 61°00'N and from western Ireland to longitude 20°30'E. He did not propose to forecast the weather. Rather, he proposed to "print a few charts, containing one of the most prominent series of weather changes that may occur".

The words of Galton which may have caused FitzRoy particular concern were: "The result of a wide system of cooperation such as I propose will be the accomplishment of a valuable piece of scientific work that will also help to afford an answer to the question whether synchronous charts may hereafter be printed regularly with success". It may have been unsettling to FitzRoy, too, given his dedication to developing a sound basis for weather forecasting and his probable belief in his pre-eminent position vis-à-vis understanding of weather systems, that Galton hoped to map the "broad eddying currents of air, heat and moisture which determine our climate of whose directions, shapes and mutual relations we are at present in lamentable ignorance".

Formation of *The Daily Weather Map Company* by Glaisher, Sopwith and others in the summer of 1861 may also have been an unwelcome development for FitzRoy. A publication to be called *The Daily Weather Map and Journal of News, Literature and Science* was to be produced by the Company. According to its prospectus, the publication would comprise a map of the British Isles, showing, at a large number of stations, the state of the weather, including wind, at 9 a.m. at each locality by a simple and semi-pictorial combination of symbols. The observations would be transmitted to London by telegraph and the chart would be ready for circulation by 11 a.m. It was surely something of a relief to FitzRoy that this venture was short-lived, for the prospectus stated that the charts would "afford the best data wherefrom to anticipate coming changes and foretell the approach of rain, storms or sunshine".

In February 1862, after a further six months had elapsed "for gaining experience by varied tentative arrangements", as FitzRoy put it in his *Weather Book*, a system was established in the Meteorological Department. Forecasts were compiled from twenty reports received each morning (except Sundays) and ten each afternoon, besides five from the Continent. Forecasts for two days in advance (what he called 'double forecasts') were issued to six (later eight) daily newspapers and to one weekly paper, as well as to Lloyd's, the Admiralty, the Horse Guards, the Board of Trade and the Humane Society. Forecasts were prepared directly from the observations and despatched within an hour of telegrams being received. Operational charts were not drawn.

Though retrospective synoptic charts had been drawn in the Meteorological Department since 1857, the technique of drawing isobars to reveal patterns of barometric pressure was still in its infancy, and its validity was indeed questioned by some, including FitzRoy. In any case, the data available for the construction of operational synoptic charts were scanty. FitzRoy's method was to interpret changes in barometric pressure intelligently in conjunction with observations of wind, temperature, visibility and other indicators. He also employed his own model of extratropical weather systems.

According to FitzRoy, the forecasts added almost nothing to the cost of running his Department, but their usefulness was increasingly being recognized. As an example of the benefits that his work had brought, he noted in his *Weather Book* that a meeting of the shareholders of the Great Western Docks at Stonehouse in Plymouth had been informed that the deficiency in revenue in 1862 "was to be attributed chiefly to the absence of vessels requiring the use of the graving docks for the purpose of repairing the damages occasioned by storms and casualties at sea".

FitzRoy was the originator of the term 'weather forecast' and made plain forecasts were not intended to be predictions. "Prophecies or predictions they are not", he wrote in the *Weather Book*; "the term forecast is strictly applicable to such an *opinion* as is the result of a scientific combination and calculation". However, he added, forecasts may occasionally be marred by unexpected atmospheric behaviour, including "a rapid electrical action not yet sufficiently indicated to our extremely limited perception and feeling".

Officials of the Board of Trade were not entirely happy over FitzRoy's initiative, as shown by the comments of Farrer in a substantial report dated 12 February 1862 which was published as an appendix to the 1862 *Report of the Meteorological Department of the Board of Trade*. However, the Board did not attempt to stop FitzRoy issuing forecasts, but they did monitor the accuracy of the warnings for shipping and the forecasts for the public.

Departmental Growth

By 1862, the staff of the Meteorological Department numbered ten, including FitzRoy. Only two were on the regular establishment of the Board of Trade, namely Babington and Pattrickson. The former was in charge of the scientific duties of the Department and increasingly becoming involved in the work of forecasting, whilst Pattrickson was responsible for the general management of the office and financial business correspondence (though his skills as a draftsman were still employed). The other seven members of staff were all non-established supplementary employees, among them George James Symons, who worked on the extraction and reduction of observations and assisted with the telegraphy. He had joined the Department on 12 March 1860, on the invitation of FitzRoy.

By 1860, Symons had established a network of voluntary rainfall observers, and this continued to grow whilst he was employed in the Meteorological Department, to the extent that FitzRoy became concerned. He recognized the importance of the work but felt, nevertheless, that Symons would soon be unable to combine it with his departmental duties. There were exchanges between the two men, partly because FitzRoy objected to Symons writing to newspapers with regard to rainfall work, and the upshot was that Symons resigned in 1863 to work full-time on the collection and collation of rainfall statistics. Despite the resignation, relations between the two men remained cordial, and, in fact, FitzRoy served as a voluntary observer for the network for a short while.[35]

In 1863, Symons began to publish a monthly rainfall circular, which developed into *Symons's Monthly Meteorological Magazine*, first published in February 1866. This was not, however, the first periodical devoted to meteorology, for Richard Strachan, who joined the Meteorological Department in 1862, founded the *Meteorology Magazine* in 1864. But Strachan's journal was short-lived. Only four issues were ever published, the first in April 1864, the last in July 1864.

According to the 1862 *Report of the Meteorological Department of the Board of Trade*, Strachan was responsible for instruments, along with Frederic Gaster. Of the other staff, Harding and his son (also J S Harding) attended to records, stores, correspondence and translation, while G Harvey Simmonds worked with Symons on the extraction and reduction of observations and assisted with the telegraphy. Two youths served as office juniors, dispatching weather reports or telegrams, and were "otherwise actively employed in searching for papers, extracting and copying". The office hours were 10 a.m. to 6 p.m. for some, 11 a.m. to 5 p.m. for others.

In the fifth *Report of the Meteorological Department of the Board of Trade*, published in 1863, FitzRoy summarized work in progress. Besides attending to "the usual daily duties of correspondence, accounts, registry, records, observation of instruments, and telegraphy", he and his staff had collected, from all navigable regions:

- Reports of storms, gales, fog, rain, lightning and other meteorological phenomena
- Readings of temperature, barometric pressure and wind speed taken aboard ships
- Anemometer readings from Halifax, Bermuda, Orkney, Ascension and elsewhere

---

[35] At first, Symons called his network of observers his 'system' of collecting rainfall records, with himself as the 'centre'. It was renamed the 'British Rainfall Organization' in 1900 after his death. For a profile of Symons, see 'The man behind the British Rainfall Organization', by Malcolm Walker, published in 2010 in *Weather* (Vol. 65, pp. 117–120).

- Measurements of the specific gravity and temperature of sea water
- Observations of sea ice
- Measurements of the speed and direction of ocean currents
- Observations of electrical, magnetic, auroral, meteoric and other such phenomena

In addition, FitzRoy and his staff had extracted from reliable logbooks "remarkable occurrences, or facts of a very noteworthy character, having a scientific connexion". They had also carried out "researches into the nature and periods of atmospheric changes, whether supposed to be derived from central lunisolar action or otherwise". In the 1863 *Report*, FitzRoy also provided a list of the many charts, books and pamphlets published by the Meteorological Department up to 1 April 1863, showing the total number printed for sale and distribution, as well as the number distributed gratis. The list included the number of Maury's publications supplied gratis by the American government.[36]

Financial statements were given in the 1863 *Report*, too, showing the sums voted by Parliament from the foundation of the Department onwards, together with the amounts paid and chargeable to each year's vote. These statements show there was a surplus on the Board of Trade account each year up to and including 1860–1861, though it was pointed out in the *Report* that these surpluses were nominal, because the sums voted were amounts *authorized*, not *drawn*. There was a small deficit on the Admiralty account in each of the years 1857–1858, 1858–1859, 1859–1860 and 1862–1863, offset by surpluses in the other years.

The Admiralty vote remained constant at £1000 per annum throughout the period in question. The Board of Trade vote, which remained constant at £3200 per annum until 1858–1859, decreased to £2400 for the year 1859–1860 and £2300 for 1860–1861 before increasing to £2800 for 1861–1862 and £3800 for 1862–1863. The deficits were £164 8s 4d for the year 1861–1862 and £640 7s 3d for the year 1862–1863, and it was noted in the 1863 *Report* that a deficit of £641 would be provided for in the Board of Trade vote for 1863–1864.

Criticism and Controversy

The *Royal Charter* disaster affected FitzRoy profoundly. The ship had safely negotiated storms and other hazards on its voyage from Australia, only to be dashed to pieces on the rocks of Lligwy Bay in full view of dozens of horrified spectators, none of them able to help those who perished. FitzRoy considered this a disaster which could have been prevented and threw himself into his work to such an extent that

---

[36] Of the *Wind and Current Charts*, 17,100 had been received and 10,150 issued by 1 April 1863. The other publications were Maury's *Sailing Directions* (800 received, 730 issued), *Monographs* (100 received, 50 issued) and *Abstract Logs* (100 received, 50 issued). These were substantial publications, especially the *Sailing Directions*, which contained 800 pages.

he did not report on the activities of his Department for almost four years. The third *Report of the Meteorological Department of the Board of Trade* was published in June 1858, the fourth in May 1862.

In those four years, FitzRoy not only continued to carry out the work he was expected to do but also exceeded his brief. This did not in itself bring official reprimand, but criticism of his forecasting techniques soon came, from various quarters. His storm warnings and weather forecasts, though helpful on the whole, were not always accurate. This came as no surprise to him, for he was aware that his methods were imperfect, as shown by his insistence on the word 'forecast', rather than 'prophecy' or 'prediction'. Unfortunately, he assumed that the views of those such as Galton who considered weather forecasting unscientific were directed against him personally. In the 1860s, meteorology was an emerging science not yet firmly based on absolute laws of physics and, furthermore, believed by many to be incapable of mathematical expression. This was a time of debate over the nature and methodology of science.

Some championed 'practical science', from which there were tangible benefits for society and in which, in the case of meteorology, weather wisdom and amateur observers played important roles. Others believed worthwhile progress could be made only through advances in 'abstract science' or, as some called it, 'philosophic science'. FitzRoy was essentially a man of 'practical science' and never claimed to be "a truly scientific man", as he put it in his *Weather Book*. He was "only", he wrote in *The Athenæum* (24 November 1860, Part 2, p.710), "a superficial follower, however devoted an admirer, of *real* philosophers". Perhaps he was being modest. Perhaps, on the other hand, he was a little unsure of himself.

Whatever the truth of the matter, FitzRoy had no need to react as he did to the views of those who considered weather forecasting unscientific. He was a pioneer of meteorology with intuitive insight into the ways of the atmosphere, a man who introduced well-founded empirical methods that were significantly more scientific than those based entirely on weather lore. In retrospect, it is easy to say that he worried unnecessarily, but he was a sensitive man and, moreover, did not take too kindly to criticism. The eventual outcome of his reaction to the doubts over the scientific respectability of his work was that he all but isolated himself from the scientific community, and problems began to mount for him.

The *Royal Charter* disaster was not the only event in 1859 that affected FitzRoy profoundly. So, too, did the publication of Darwin's book *On the Origin of Species*, which appeared in November of that year. To FitzRoy, a devout Christian with conservative views, Darwin's theory of organic evolution by natural selection was unacceptable. It contradicted Biblical 'truth'. Moreover, to make matters worse, the findings of palæontologists were making nonsense of a literal interpretation of the Bible. FitzRoy believed the explanation of Creation given in the *Book of Genesis* literally true.

Though Darwin's conclusions did not come as a total surprise to FitzRoy, for the two men had discussed Darwin's observations and their implications many times during the *Beagle* voyage, publication of the book was, nevertheless, a disappointment to FitzRoy. Whenever an opportunity arose, he attempted to refute Darwin's theory, sometimes in writing (usually under a pseudonym), sometimes in public debates. He became obsessed with the matter and it preyed on his mind. There was criticism of FitzRoy's forecasting techniques from astro-meteorologists, too, the early 1860s being an especially active time for astro-meteorology.

The 'Astro-Meteorological Society' was founded in 1860 by some of the foremost astrologers of the day, notably Richard James Morrison (pseudonym 'Zadkiel'), William Henry White, Christopher Cooke and Stephen Saxby. The inaugural meeting was held on 29 November 1860, and meetings took place regularly until January 1862, when the Society was dissolved. Two months later, Morrison and White attempted to re-launch it by founding the 'Copernican Meteorological Society'. There was little enthusiasm, though, and the Society survived for only a month or two. But this was long enough for some of its members to take issue with FitzRoy publicly. They were scornful of him and others who did not accept that the moon and the planets could influence weather significantly.

FitzRoy had a tendency to overreact and engage in public controversy that did not help the cause of his Department. In *The Times* of 2 March 1861, for example, he wrote dismissively of "persons who profess to know intuitively more than real philosophers", particularly those who "have 'prognosticated' or, as some say, 'prophesied' changes, or storms, at definite times, upon some vague ideas of 'lunar' influence, or (so-called) 'astro-meteorology'". "Perhaps", he added, "the alchymists and astrologers of old were wiser, in their generation, than these prophets". "Astro-Meteorology is a sham and calculated to mislead the public." This resulted in a challenge from the Astro-Meteorological Society. An advertisement was sent to *The Times* "challenging the gallant admiral to mortal meteorological combat". *The Times* refused the advertisement, but the *Daily Telegraph* published it on 8 March 1861. FitzRoy was invited to disprove the ability of astro-meteorologists to predict the weather, but he wisely did not accept the challenge!

Further criticism came from a number of people with entrepreneurial ambitions. Led by Glaisher, a promoter of *The Daily Weather Map Company*, these were people who sought financial gain from the supply of weather information and considered the supply of free material by the Meteorological Department an obstacle to their ambitions. So far as Glaisher was concerned, there may have been, additionally, a lingering niggle over the decision in 1854 to place the Department within the Board of Trade, rather than the Royal Observatory at Greenwich, where a meteorological department had existed since 1840. Criticism came, too, from shipowners, who were concerned less about the safety of their crews than about the loss of revenue caused by captains keeping their vessels in port when storm warnings were in force.

To set against the criticism, there was support for FitzRoy from some quarters. A survey carried out by the Meteorological Department showed that most seafarers approved of storm warnings, and some expressed astonishment that anyone should question their value. The forecasts were popular also with the general public, as articles and letters in contemporary newspapers and magazines indicate. The Royal Society expressed support, too, though somewhat qualified, and the French showed their approval of FitzRoy's techniques by introducing, in 1863, a storm-warning system that was essentially the same as that operated by the British. Further recognition from the French came the same year, when FitzRoy was elected a Corresponding Member of the Académie des Sciences de Paris (though not entirely for his meteorological work). In addition, FitzRoy enjoyed royal patronage, for Queen Victoria frequently consulted him about weather prospects before she crossed the Solent to the Isle of Wight.

FitzRoy's principal reaction to the criticisms was to work harder than ever, not only during office hours but also when off duty. In fact, he wrote the *Weather Book* as "a holiday task, hastily performed", according to a letter he sent to the Secretary of the Royal Society on 5 March 1863. He apologized for submitting "so ill-digested and obscurely-written a work". "In a week or two", he said, "I hope to lay before you a second edition of this book, which has been carefully revised and, I hope, rendered somewhat less obscure". He included in the book a whole chapter devoted to the effect of the moon and sun on the atmosphere, which he called the 'luni-solar effect'.

He sought the opinion of Sir John Herschel in December 1862 on the ideas he proposed to include in the chapter. Impatiently, though, he did not wait for the reply to arrive before publishing the book, and when it did come, in March 1863, he found that Sir John had rejected all of his ideas. Herschel was much interested in meteorology, and in January 1864 he published a substantial article titled 'The Weather, and Weather Prophets' (published in *Good Words*), in which he wrote dismissively of "lunar prognostics" and explained how weather resulted from physical processes involving solar radiation, evaporation and condensation of water, and density differences between warm and cold air.[37] This was not, however, intended as an attack on FitzRoy. On the contrary, his references to the work of FitzRoy were approving.

Astro-meteorologist Saxby took issue with both FitzRoy and Herschel, devoting a whole chapter of his book (*Saxby's Weather System or Lunar Influence on Weather*) to a discussion of FitzRoy's luni-solar theory, titling it 'The World's Disbelief in the Moon's Influence on Weather as Shown by the 'Weather Book' – 'Luni-Solar' Theory'. In the following chapter, he further criticized FitzRoy's ideas and disputed

---

[37] In *A Manual of Scientific Enquiry; Prepared for the Use of Her Majesty's Navy: And Adapted for Travellers in General* (London: John Murray, 1849, 488pp.), there is a substantial section on meteorology, written by Herschel. In this, he pressed for the introduction of uniform ways of making observations at sea and for the recording of such observations in standardized registers. These recommendations, it will be noted, were published four years before the Brussels Conference considered observing practices at sea.

the views which Herschel had expressed in his article on the weather and weather prophets.

FitzRoy was able to respond philosophically to the views of Herschel, but the views of Saxby and other astro-meteorologists provoked him, to the extent that he included in the sixth report of his Department, published in 1864, a 'Postscript' which contained his comments on "a small volume, just published", almost certainly Saxby's book. FitzRoy was now the one who poured scorn. "There is nothing new, or even modern, in the assertion of lunar influence on weather", he wrote; "some have denied, while others have advocated it, time out of mind". "What is *now* wanted", he said, "is the *modus operandi*; how dynamic effects, which we observe, are caused?"

More and more, FitzRoy felt the need to respond to those who disagreed with him, as the increasing frequency of his letters to *The Times* and other newspapers shows, and in official publications he was defensive, too. But he did, nevertheless, show himself sensitive to the views of the scientific establishment and pointed out in the 1864 *Report* of his Department that the "meteorological communications" of his Department were "highly appreciated on the continent". He pointed out, too, that the work of his Department had increased over the years and requested that the annual vote be increased to £5800. He was not rewarded. The parliamentary vote for the Meteorological Department for 1864–1865 was £4270, made up of £3700 from the Board of Trade and £570 from the Admiralty, this being a reduction of £530 on the vote for 1863–1864, when the Board of Trade vote was £3800 and the Admiralty vote £1000.

In 1864, FitzRoy must have felt beleaguered, for, in addition to the aforementioned pressures on him, he came under attack in the House of Commons. The Member for Truro, Augustus Smith, launched an assault on all forms of government spending and stated that he did not think the Board of Trade should "undertake the functions of Aeolus"! FitzRoy's mode of response was by now familiar. He wrote to *The Times*. In his letter, published on 14 May 1864, he defended his own position and pointed out that the harbours of the Scilly Isles were now less frequented by vessels in distress than in the days before storm warnings were issued. As Smith was Lessee of the Scillies, the implication was that his real concern was a loss of revenue from harbour dues.

The immediate cause of Smith's outburst was the publication of a Parliamentary Paper in April 1864. Entitled *Weather Forecasts*, it contained the results of surveillance carried out by the Board of Trade over a considerable period of time. Since 1 July 1861, the Board's Wreck Department (which was outside FitzRoy's control) had monitored the accuracy of storm warnings, with checks carried out by seventy-four coastal observers, most of them coastguards. Who set up this monitoring system is not known, but some have pointed the finger of suspicion at T H Farrer, Assistant Secretary to the Marine Department of the Board of Trade. Farrer had never appeared wholly enthusiastic about FitzRoy's work, not least the verification process which the Meteorological Department used to assess storm warnings.[38] This was far from systematic and not a little ad hoc at times.

Doubts over FitzRoy's work surfaced in a letter Farrer sent to the Secretary of the Royal Society on 27 February 1863. In this, he referred to the Royal Society's letter to the Board of Trade dated 22 February 1855 (concerning 'the great desiderata of meteorology') and asked the President and Council of the Royal Society to reconsider the advice then given. He asked whether:

- The science of meteorology was now in such a state as to admit of a permanent reliable system of storm signals and daily weather forecasts
- The progress and useful application of meteorological science would be more efficiently promoted by devoting the money voted by Parliament to the original objects contemplated, viz. the collection, tabulation, and discussion of meteorological phenomena, or by devoting it to the system of telegraphy and weather forecasts

In his reply, dated 27 March 1863, Dr W Sharpey, Secretary of the Royal Society, reported that the President and Council had "placed themselves in communication with Admiral FitzRoy" and learned from him "that the original objects for which the Meteorological Department was formed were still kept steadily in view". They were satisfied that the system in place in the Department for collecting and processing what they called 'Ocean Statistics' was of such efficiency that FitzRoy had time to extend his activities to the work he had initiated. They commented that "forewarnings of storms must as yet undoubtedly be viewed as in great measure tentative" but "noticed with great pleasure" that "replies to inquiries circulated by the Board of Trade" as to the importance and success of the warnings were mostly favourable. There was no criticism of FitzRoy in Sharpey's letter, only support.

At this stage, therefore, FitzRoy was vindicated. However, his problems seemed unending, another being that a number of storm warnings issued outside Britain were falsely attributed to him. The worst example occurred in late 1863, when two warnings caused public alarm in Portugal and Gibraltar. Once again, FitzRoy wrote to *The Times*. In a letter published on 18 January 1864, he disclaimed responsibility for "these absurd but injurious predictions", saying that he did not possess the proof necessary to identify the culprit. It is not clear who was responsible, though it may have been Saxby.

Even Maury turned against FitzRoy, publishing in the Paris magazines *Courrier des Sciences* and *Bulletin International* comments of a critical nature concerning meteorological telegraphy and the feasibility of providing accurate storm warnings for the British Isles. FitzRoy countered these in the 1864 *Report* of his Department, in an appendix (dated 25 February 1864). Captain Maury, he said, "like many other really

---

[38] The impression given by Board of Trade papers held in the National Archives at Kew, is, however, that FitzRoy and Farrer were on good terms, in which case Farrer may simply have been acting as the mouthpiece for others who were not enthusiastic about FitzRoy's methods.

Figure 2.7. The *pendule de voyage* presented to FitzRoy by the French government in 1864. © Crown Copyright 2010, the Met Office.

competent and authoritative judges of scientific questions, has not had time, means, or inclination to study the daily complications, and published facts of atmospheric changes. His opinion of the systems in operation cannot be so valuable as the prestige of his name might lead the public to suppose".[39]

In the appendix, FitzRoy explained the system he used and made a curious statement which was ridiculed in an editorial published in *The Times* on 18 June 1864: "Facts are as the ground – telegraphic wires are roots – a central office is the trunk – forecasts are branches – and cautionary signals are as fruits of this youngest tree of knowledge". The ridicule hit FitzRoy hard, for the subject of meteorological telegraphy was very close to his heart. Would nothing go right for him? The presentation of a handsome *pendule de voyage* to him by the French government in October 1864, together with a letter of gratitude from the Ministry of Marine, lifted his spirits a little, but only a little, for the French had previously awarded Maury the Legion of Honour.[40]

By the end of 1864, the Meteorological Department was effectively being run by Babington, with the forecasting work almost entirely in his hands. FitzRoy's health was deteriorating by the day and, to add to his woes, he was losing his hearing. He was absent from the office more and more. By early 1865, his attendance had dwindled to almost nothing. The attacks on his forecasts, his personal financial difficulties, the *Origin of Species* affair and worries over his standing in scientific circles had taken

[39] On 20 April 1861, Maury resigned his U.S. Navy commission and joined the Confederacy. Several days later, he accepted the position of commander in the Confederate States Navy. Because of his fame, he was sent to England to promote the Southern cause, and during the Civil War he was successful in acquiring war vessels for the Confederacy.

[40] FitzRoy wrote to the Secretary to the Admiralty, Lord Clarence Paget, on 7 October 1864 to thank him for forwarding the clock. Laura FitzRoy bequeathed it to the Meteorological Office and it is still in their possession, now on display in the National Meteorological Library at Exeter.

their toll. It was nothing new for him to become a little dispirited, and he had indeed suffered from depression whilst on the *Beagle*, but deep despondency now descended on him all too often.

In January 1865, Mrs FitzRoy consulted doctors over the soundness of her husband's mind. By April 1865, she was even more concerned about his health. In a letter to a relative, she reported that the doctors had united in prescribing total rest, as well as absence from his office for a time. Leave had been granted him, she said, but his active mind and over-sensitive conscience had prevented him from profiting by it. He did not wish to put the work he was paid to do upon others. He was continually restless to be at his post, she added, and hastened back the moment he felt better, only to find himself unable to work satisfactorily when he got there.

On 20 April 1865, FitzRoy inadvertently took too much opium, which caused him to become too ill to leave his bed for two days. On 25 April and subsequent days, he felt well enough to travel from his home in Upper Norwood to central London, though not always to his office. The news of Abraham Lincoln's assassination brought him yet more anguish.[41] On 29 April, he met Maury in London, afterwards returning home "worn out by fatigue and excitement" and in a state of great "nervous restlessness", as Mrs FitzRoy put it in her diary.[42] Just before 8 a.m. on Sunday 30 April, he went to his dressing room and cut his throat. Mr Frederick Hetley, MD, FRCS, was called, but there was nothing he could do. By mid-morning, Admiral FitzRoy was dead.[43]

---

[41] Lincoln was murdered on 15 April 1865. It appears that news of the President's death reached FitzRoy about ten days later.

[42] On this occasion, FitzRoy wished to take leave of Maury, who was about to proceed to the West Indies.

[43] At the inquest (reported in detail in *The Times* on Thursday 4 May 1865), the Reverend Tremlett, a friend of FitzRoy, reported that he had, on 29 April, "urged FitzRoy to resign at once his post as meteorological officer of the Board of Trade". He had urged this because he was greatly worried by the "alarming symptoms" FitzRoy was suffering: noises in the ears, inability to sleep at night and twitching of his hands.

3

Inquiry and Criticism

The speculation began before FitzRoy was cold in his grave.[1] Who would succeed him? Some considered Glaisher a likely successor. Others supposed the post would again be filled by a naval officer. In fact, many months passed before anyone at all filled it on a permanent basis. FitzRoy's second in command, Thomas Babington, was in charge for nineteen months, but his appointment was never more than temporary, and he had to wait a long time for even that to be approved. He waited until 9 June 1866, by which time a government inquiry into the work of the Department had taken place.

A New Beginning

The first moves to set up an inquiry were made little more than a week after FitzRoy's death. Edward Sabine, the President of the Royal Society, reported at the meeting of the Society's Council on 18 May 1865 that he had been consulted by the President of the Board of Trade, Thomas Milner-Gibson, about "arrangements in consequence of the death of FitzRoy". In his response, dated 10 May 1865, he had come straight to the point. "Should it not be desired to fill up the vacancy occasioned by Admiral FitzRoy's death *immediately*, time would be afforded for a reconsideration of the duties of the Office, which might be productive of advantage in many respects." Babington was "competent to conduct and continue the system of storm warnings (with such assistance as he may require and with such moderate increase of his own salary as may be deemed suitable)". Nevertheless, "the time may be viewed as suitable for obtaining and considering evidence and opinions as to the advantages, present and prospective, of continuing the practise of storm warnings". And, he suggested, "it may be unnecessary to continue the publication of the daily forecasts". The ocean statistics work could, he thought, be transferred to the Admiralty's Hydrographic Department.

---

[1] FitzRoy was buried in the grounds of All Saints' Church, Upper Norwood, south London.

Details of procedures in the Meteorological Department were sought by T H Farrer, one of the Secretaries of the Board of Trade. In particular, he asked Babington to provide an explanation of the method used to produce weather forecasts and storm warnings. Babington's response took the form of a statement entitled *Forecasts and Cautions*, which was submitted to the Board of Trade on 11 May 1865. In it, he outlined the routine work and advised that details of the Department's methodology could be found in the chapters of FitzRoy's *Weather Book* concerned with storm warnings, meteorological telegraphy and weather forecasting.

He reported that the routine work was as follows:

- Telegrams were received about ten o'clock each morning (except Sundays) from eighteen places around the coasts of the British Isles, as well as from Heligoland and a few French ports. The observations thus telegraphed were first reduced or corrected for elevation, temperature and scale errors and then entered on prepared forms.
- The first copies of the forms, with all the telegrams, were "passed to the chief of the Department, or the person appointed by him, to be studied for that day's forecasts". At eleven o'clock, weather reports, together with forecasts, were sent to *The Times* (for its second edition), as well as the *Shipping Gazette* and the principal afternoon papers. At about the same time, forecasts of weather expected in the English Channel and on French coasts were telegraphed to Paris for the Ministry of Marine. This work was completed by 11.30 a.m., leaving everyone in the Department "free to turn his attention to other duties". Late in the afternoon, telegrams were received from a few selected stations, whereupon the morning forecasts were modified, if necessary.
- In addition, storm warnings were occasionally sent to British coastal stations and to Paris. They were also, when it appeared advisable, sent to Hamburg, Hanover and Oldenburg "by the request and at the expense of the governments of those states".

Forecasts, Babington said, were "the result of theory and experience combined". They were not *predictions* but *opinions*, though they were probably, he considered, the *best* opinions that could be formed. He reviewed the meteorological indicators that were used when making forecasts and said that he offered "no argument, or opinion, with regard to the advisability or otherwise of the continuance of the present system of forecasting, because none was asked for". He did, however, mention that London forecasts were sent daily to the French Ministry of Marine at their request.

The recommendations concerning the great desiderata of meteorology which had been given by the Royal Society in February 1855 were soon to be in the spotlight again. In a letter to Sabine dated 26 May 1865, Farrer formally asked the President and Council of the Royal Society to review the advice given in 1855 and examine the work of the Meteorological Department since then. On behalf of the Lords of the Committee of Privy Council for Trade, he sought the opinion of the Royal Society on a number of points:

- Were the objects specified in the letter of 22 February 1855 still as important for the interests of science and navigation as they had then been considered?

- To what extent had any of these objects been answered by the work of the Meteorological Department so far?
- What steps should be taken for making use of any observations already collected or any compilations already made by the Department?
- What was the nature of the basis on which the system of daily forecasts and storm warnings established by Admiral FitzRoy rested? Were they founded on scientific principles, so that they, or either of them, could be carried on satisfactorily notwithstanding his decease?
- If they, or either of them, could be carried on satisfactorily, could the Royal Society suggest any improvement in the form and manner of doing it?
- He assumed it "desirable to continue the publication of the daily reports of weather received from various stations" and asked the Royal Society to "make any suggestions as to the extent to which it should be carried out, and the form in which it should be done".
- Finally, he asked for any general suggestions regarding the mode, place or establishment in, at or by which the duties of the Meteorological Department could best be performed.

One of Farrer's questions was surprising. He asked if it was desirable that monitoring of the accuracy of weather forecasts and storm warnings be continued beyond the end of the period which had been covered by the surveillance summarized in the Parliamentary Paper published in April 1864 (i.e., beyond 31 March 1864). The question was curious because Babington had already carried out a survey for the period 1 April 1864 to 31 March 1865, and a manuscript copy of his report had been enclosed with Farrer's letter. Farrer may have been hinting to the Royal Society that the Board of Trade still harboured misgivings over the verification process employed by the Meteorological Department. Be this as it may, the Board of Trade appeared to have confidence in Babington, given that Farrer's letter informed the Royal Society that the Board would "gladly place the knowledge and services of Mr Babington, Admiral FitzRoy's second, at the disposal of the Royal Society for the purpose of the above inquiries".

The Royal Society's Council discussed Farrer's letter at their meeting on 1 June 1865 and finalized their reply two weeks later. Dated 15 June 1865 and signed by Sabine, the reply took up six pages of the Royal Society's *Proceedings* (June 1865, Volume 14, pages 311–317). In the Council's view, the objects specified in the letter of 22 February 1855 were still as important for the interests of science and navigation as when originally formulated, and much had been accomplished "in the collection of facts bearing on marine meteorology". As no "systematic publication of the results" had yet been made, however, the Council were "unable to reply more specifically". They recommended that marine observations "should be placed in the hands of the Hydrographer, with a view to the introduction of the results into the Admiralty Charts". It was "very desirable", they advised, that "further observations be made, especially with reference to oceanic currents and great barometric depressions, and generally on all subjects comprehended under the denomination of 'Ocean Statistics'". The Council recommended that the storm-warning service should continue "under the superintendence" of Babington and declined to express any opinion on the daily forecasts of weather.

Farrer's final question was considered in some detail, the conclusion being that a "central office" should be established, to collect, reduce, co-ordinate and publish observations relating to "the Land Meteorology of the British Isles". As the Council pointed out, offices of this kind had already been established in almost all the "principal States of the European continent". A chain of stations was suggested, with stations "at nearly equal distances in a meridional direction from the south of England to the north of Scotland", each "furnished with self-recording instruments supplied from and duly verified at one of the stations regarded as a central station". The stations, the Council suggested, might be at Kew Observatory, Stonyhurst College, the Falmouth Polytechnic Institute, Armagh Observatory, the Glasgow University Observatory and Aberdeen University. Another fitting location for a station, they thought, might be Valentia, given the hope of a permanent telegraphic link between there and America in the near future. The Council pointed out that the British Association's observatory at Kew already possessed the principal self-recording instruments and recommended, therefore, that the observatory might, "with much propriety and public advantage, be adopted as the central meteorological station".

There was silence from the Board of Trade for several months. Then, on 24 October 1865, Farrer replied to Sabine's letter. He seemed irritated. A number of the Royal Society's answers were not, in the Board's opinion, helpful. For instance, so far as meteorological observations on land were concerned, the Board did not "clearly understand whether the Royal Society think that they should be substituted for, or be in addition to, the meteorological observations at sea". They agreed that "any observations of a scientific nature would be better conducted under the authority and supervision of a scientific body such as the Royal Society, or the British Association, than of a Government Department". However, they did not "see how they could advise the Government to sanction any plan which would involve the establishment of two separate offices for meteorological purposes, one under the Board of Trade at Whitehall, and the other at Kew".

"As regards meteorological observations made at sea", said Farrer, "the Board of Trade were not satisfied that they fully understood the views of the Royal Society". In the letter from the Council, he complained, there was scarcely any reference to meteorological observations at sea. Without expert guidance, the Board of Trade could not "determine what steps should be taken with regard to the Meteorological Department". The Board needed to know the value of the observations collected already by seafarers. They also needed to know "what steps should be taken to make the observations useful" and what "further observations of the same kind should be collected".

"With the view of clearing up these points", advised Farrer, the Board of Trade were "disposed to suggest the appointment of a small Committee, consisting, say, of three or four persons, to examine the whole of the data already collected by the Meteorological Department; to inquire whether any and what steps should be taken for digesting and publishing them; and also to report whether it is desirable that observations of a similar kind shall continue to be collected". If the Royal Society accepted this suggestion, they were asked to appoint, as a member of the committee, "some gentleman whose acquirements would enable him to give valuable advice on the scientific part of the subject". The Admiralty would also be asked to appoint a member. Farrer's letter was considered at the meeting of the Royal Society's Council on 2 November 1865 and a reply sent the same day. The Council would be "quite ready to assist in this inquiry in the manner proposed". A resolution was passed that Francis Galton be nominated by the President to be a member of the committee. He was General Secretary of the British Association and keenly interested in meteorology.[2]

Sabine reported at the Council meeting on 30 November 1865 that he had received a letter dated 20 November 1865 from Sir James Tennent, Permanent Secretary to the Board of Trade. This informed the Royal Society that Farrer had been nominated as the Board's representative on the committee and that Staff Commander Frederick Evans RN, Chief Naval Assistant to the Hydrographer of the Admiralty, had been nominated by the Admiralty.[3] If, said Tennent, the Council of the Royal Society saw no objection, the points the Board of Trade proposed to refer to the committee were as follows:

- What data, especially meteorological observations at sea, had been collected by, and existed in, the Meteorological Department of the Board of Trade?
- Should any steps be taken to arrange, tabulate, publish or otherwise make use of such data? If so, what steps were needed?
- Was it desirable to continue meteorological observations at sea, and, if so, to what extent and in what manner?
- Assuming that the system of weather telegraphy was to be continued, could the mode of carrying it on and publishing the results be improved?
- What staff would be necessary for the above purpose?

Neither the Board of Trade's Meteorological Department nor the Greenwich Royal Observatory's Magnetic and Meteorological Department was represented on the committee. Babington was, however, consulted during the committee's deliberations. The position of the Royal Observatory in British meteorology was potentially a sensitive matter, as the minutes of the Royal Society's Council meeting on 15 February 1866

---

[2] Galton published *Meteorographica, or Methods of Mapping the Weather* (Macmillan, 1863) and also wrote 'Meteorological instructions for the use of inexperienced observers resident abroad' (published in the *Proceedings of the British Meteorological Society*, 1862, Vol. 1, pp. 397–400).

[3] At the time, Captain George Henry Richards was the Hydrographer of the Admiralty. Evans succeeded him in 1874. Farrer succeeded Tennent as Permanent Secretary of the Board of Trade in 1867.

show. At that meeting, a letter from the Astronomer-Royal dated 31 January 1866 was read and the following reply approved:

The President and Council of the Royal Society are much concerned to hear that certain statements occurring in a communication from the President and Council to the Board of Trade on the 2nd of November 1865 are understood by Mr Airy to throw discredit on the Royal Observatory, and unduly to exalt the merits of another institution. The President and Council desire to assure Mr Airy that nothing in their communication to Her Majesty's Government was intended to imply any disparagement of the Meteorological Department of the Royal Observatory; and as little did it enter into their mind to exalt the Observatory at Kew to the disadvantage of Greenwich.

According to the minutes of the Council meeting on 15 March 1866, the matter was concluded satisfactorily through exchange of correspondence.

The Galton Report

The findings of the committee of inquiry were laid before Parliament on 13 April 1866, presented in a document that has come to be known as the Galton Report because he chaired and dominated the committee. The inquiry was not, however, a solo effort, as a note of appreciation in Galton's autobiography indicates.[4] In this, he acknowledged "the singular grasp and thoroughness" of Farrer and commented that the occasional brief notes he received from him, "in the course of the inquiry, were models of clearness combined with cordiality".

Given Galton's views on the scientific approach to meteorology, which had so unsettled FitzRoy, it was not surprising that the report's conclusions were critical of the Meteorological Department's work. Galton had an aversion to anything he did not consider 'scientific', and that included weather forecasting as practised by FitzRoy, which he considered wholly unscientific. To Galton, meteorology needed to be based on absolute laws of physics. His attitudes to science in general and meteorology in particular mirrored those of Sabine, with whom he had been friendly for some considerable time. Indeed, as he stated in his autobiography, the friendship had exercised great influence in shaping his future scientific life.

The Galton Report was a substantial document. Part I was concerned with 'Measures Taken, or to be Taken, for Procuring Meteorological Statistics of the Ocean'; Part II was titled 'Weather Telegraphy, Foretelling Weather, and Observations of Weather Within or Affecting the British Isles'; Part III dealt with costs; and the final part contained answers to the questions which had been put to the committee. There were, in addition, eighteen appendices, which contained a range of supplementary material, including forms, tables, charts, analyses, copies of official correspondence,

---

[4] The autobiography was called *Memories of My Life* (Methuen & Co, 1908, 339 pp.); see, in particular, chapter XVI (pp. 224–243). Galton was born on 16 February 1822 and died on 17 January 1911.

specimens of weather reports and forecasts, opinions of port officers on the efficacy of storm warnings, an "attempted digest of maxims employed by the Office in foretelling weather", and a statement of the Meteorological Department's income and expenditure up to 1 December 1865. These appendices occupied thirty-eight of the eighty-one pages.

In the committee's opinion, the views expressed by the Royal Society in their letter of 22 February 1855 had been adopted by the Government of the day and therefore constituted "the instructions under which the Meteorological Department was to pursue its labours". They pointed out that it was never a "part of the functions of the Department as originally instituted to publish un-discussed observations on the one hand, or to speculate on the theory of meteorology on the other". "Still less", they continued, "can it be considered to have been a part of those functions to attempt the prognostication of weather".

A compliment was paid when the committee noted that a great many ships had been supplied with instruments and registers, but approval soon gave way to criticism. "The number of these registers was steadily increasing", the committee remarked, "and would, no doubt, have been very much greater if the attention of Admiral FitzRoy and of his Department had not become gradually diverted from the objects recommended by the Royal Society to those belonging to a wholly different department of Meteorology, namely, the Prognostications of Weather".

FitzRoy had claimed in the 1862 *Report of the Meteorological Department of the Board of Trade* that he and his staff were close to being overwhelmed by the seafarers' observations they had collected. He considered that the 500,000 sets of observations he had obtained was close to the maximum his staff could cope with, so he thereupon allowed the accumulation of registers to dwindle. The committee accepted there was a limit to the number of observations which could be processed with the resources available to FitzRoy, but the limit they proposed was approximately treble his. They estimated that the number of sets of observations that had been processed by the time of their inquiry was about 550,000, derived from 1298 registers. They considered that the number of observations needed to be around 1,650,000 before the desiderata of the Royal Society were "procured". Accordingly, they recommended that the practice of loaning instruments and issuing registers be re-established on its original basis and "carried on as rapidly and widely as consistent with considerations of expense and convenience".

The committee stressed the need to obtain observations from parts of the ocean not often visited by ships and pointed out the need to avoid overloading the Meteorological Department with observations from the parts most frequented. The registers received by the Department had been "executed with scrupulous care and assiduity". What a pity, therefore, that the Department's analyses of the registers left much to be desired. The same pages had been searched repeatedly for different items, and no register had "ever yet been more than partially examined". Each search had been directed towards some limited object, and a great deal of labour had been spent in "going over and over again the same voluminous records, in order to extract from them different classes of observations".

The collection of observations from seafarers was a function the committee assumed would remain with the Board of Trade, as it had been "well performed by the Meteorological Department before its attention was diverted to the practice of foretelling weather". However, Galton's committee said, the work of processing observations called for considerable knowledge of meteorology and an ability to employ "exact scientific method". In their view, it had "not been satisfactorily performed by the Meteorological Department" and would be better executed "under the direction of a scientific body". Their proposal was that a committee of the Royal Society or the British Association be set up, "furnished with the requisite funds by the government". Alternatively, they suggested, Kew Observatory "might probably be developed so as to carry into effect such a purpose".

As regards the publications of the Meteorological Department, the committee again found fault, stating that they evinced much industry but appeared to have been "selected and published without any plan". For the most part, publications contained compilations of original observations and fragmentary and miscellaneous papers on detached subjects. Where, moreover, observations had been discussed, no uniform method of tabulating results had been adopted. In the committee's view, matters of immediate importance to navigation should be brought to the notice of the Hydrographer for publication, if he thought fit, and works published by the Meteorological Department should not contain original observations, fragmentary papers or "speculations on meteorology". As the Hydrographic Department was "devoting considerable pains to preparation of physical charts, such as Ice, General Ocean Current, and Wind Charts", the committee thought it advisable that the "results collected by the Meteorological Department" be embodied in the charts "in a form available to seamen".[5]

In the part of the committee's report dealing with weather telegraphy, storm warnings and weather forecasting, criticism was heaped on criticism. Even the reason FitzRoy gave for adopting the word 'forecast' was used against him. The committee did not disagree with his explanation, that the word implied less precision and certainty than 'predict' or 'foretell', but they felt that the use of vague phraseology had a tendency to make those who used it satisfied with uncertain conclusions!

The committee pointed out that the resolutions passed by the Council of the British Association in February 1860 had not contained anything that encouraged, let alone authorized, FitzRoy to do more than introduce a service whereby storms already known to exist at one place were "announced by telegraph to other places". There was, they added, "nothing in them upon which to found such an elaborate system of foretelling probable weather as was subsequently adopted". They observed that the approach in France under Le Verrier had been cautious. His view had been that weather forecasting might be brought into disrepute if commenced prematurely; foretelling the weather was too uncertain a business to be practised on an operational basis. FitzRoy had not shared this point of view. He had been determined, the committee said, that Britain "should take an independent course". He had considered that "too much time and labour had been given by meteorologists to registering and publishing facts, and that too little attention had been directed to practical results". Accordingly, he had "persevered in his intention of foretelling, or, to use his own expression, *forecasting*, not only storms announced by telegraph as already existing but weather generally".

Weather forecasting as practised by FitzRoy was, the committee found, based on maxims which had not been derived and established by means of accurate induction from observed facts. From consultations with Babington, they had learned that the methodology used by the Meteorological Department when forecasting the weather was not "capable of being stated in the form of Rules or Laws". Though they did not doubt that many of the conditions and probabilities which formed the basis of FitzRoy's forecasting methodology *could* be expressed in the form of rules or laws that would be accepted by meteorologists generally, they did not find that these fundamentals had been "reduced into any definite or intelligible form of expression", nor were they, as they then existed, "capable of being communicated in the shape of instructions".

It had been stated in the 1863 *Report of the Meteorological Department of the Board of Trade* that storm warnings and daily weather forecasts both rested on the same footing and therefore stood or fell together as part of one system. The committee disputed this, believing that it probably did an injustice to storm warnings, which they considered to have been "to a certain degree successful" and "highly prized". Weather forecasting was a different matter. It was not based on "precise rules" or on "a sufficient induction from facts" and was "not in a satisfactory state".

Patronizingly, the committee reported that daily weather forecasts had proved "popular and interesting" and caused no additional expense. However, they were not "generally correct in point of fact" and there was "no evidence of their utility". There appeared to be no good reason why a government department should continue to undertake the responsibility of issuing them. Galton's dislike of FitzRoy's approach to forecasting had been clear for a number of years. The recommendation of the committee that publication of daily forecasts should cease immediately therefore came as no surprise.[6]

---

[6] This recommendation was accepted without much delay. The last forecast to be published in *The Times* for many years appeared on 28 May 1866.

The committee recommended that the practice of issuing storm warnings should continue but advised that the principles on which they were issued needed to be defined and those principles tested by accurate observation. In the committee's opinion, the system of telegraphing weather reports from distant stations and the publication of telegrams should both continue. However, the committee did not feel that warnings of direction were sufficiently precise or correct to be of practical value. They recommended that the Meteorological Department when issuing warnings of wind force should "make, but not issue or publish, a prediction of the probable direction of the coming gale, endeavouring in so doing to render it as specific as possible". They further recommended that the maxims which had been employed when determining the signal of force or prediction of direction should be noted down at the time. These maxims, they suggested, should be "reduced into a clear and definite shape, and kept in the office ready for reference". For monitoring the accuracy and efficacy of storm warnings, the committee preferred the system of the Board of Trade's Wreck Department to that of the Meteorological Department.

In Part III of their report, the committee turned their attention to cost, dealing first with the work on ocean statistics they were recommending. This involved the completion of tasks currently in progress, along with the issue of instruments and registers to merchant ships, the work of increasing the number of observations extracted (to 1,650,000), and the work of "reducing, digesting and tabulating the observations so extracted". The estimated cost of all this was £3200 annually, composed of £1500 for the issue of instruments and registers and the remaining £1700 for "discussion and publication of results". The committee advised that the expenditure "ought to terminate in about 15 years, as by that time a sufficient number of observations to determine the Meteorological Means will have been collected and discussed". They assumed that the work of publishing "the results of meteorological observations at sea" in a form useful to mariners would be transferred from the Board of Trade's Meteorological Department to the Admiralty's Hydrographic Office.

An additional £7250 would be required annually, the committee estimated. Of this, £3000 would be required for weather telegraphy and the maintenance of a storm-warning service. The other £4250 would be required for the establishment and maintenance of six stations equipped with self-recording instruments, the collection of observations from lighthouses, coastguard stations and other intermediate stations, and the work of digesting, tabulating, charting and publishing results. A further initial outlay of £2500 would be required for additions to the buildings at Kew Observatory, where, the committee envisaged, some of the work they proposed would be carried out under the control of a scientific body. The committee also pointed out that a new home for the Meteorological Department would probably be needed in the near future, as they understood the premises currently occupied would soon be pulled down.

As Dr J M C Burton has commented, the Galton Report paid "nominal tribute" to FitzRoy but "demolished virtually everything he had accomplished".[7] Very little escaped criticism. The supply of barometers to fishing villages and small ports was, for example, noted with approval, but the report's conclusions were, on the whole, as Burton said, "devastating". A number of weaknesses in the committee's conclusions have been pointed out by Burton. In his words: "Galton was preoccupied with the power of statistics, his judgement was erratic and he lacked the imagination necessary to look upon FitzRoy's work as more than an academic exercise being attempted by someone lacking in scientific expertise". There is no evidence that Galton thought in retrospect that the committee had been excessively critical. On the contrary, he indicated some years later that he did not believe they had been critical enough!

Reactions and Consequences

After the Galton Report was laid before Parliament, months passed before there was any official reaction. In the House of Commons on 30 July 1866, W H Sykes asked the President of the Board of Trade "on what footing" the Meteorological Department was to be placed for the future, how far the recommendations contained in Galton's report were to be carried out, and whether the storm signals were to be continued in the manner FitzRoy had used them. In reply, the President of the Board, Sir Stafford Northcote, stated that he had found when he became President that "no step had been taken upon it". He promised that the report would be considered "as soon as possible" and advised the House that it was impossible to state at that juncture what decision would be arrived at.[8]

Farrer wrote to the President of the Royal Society on 30 August 1866. He regretted the loss of so much time and reported that the Board of Trade yet again sought advice! They were, he said, "prepared to adopt and support the course proposed" by Galton's committee and had reason to believe that the Admiralty were of the same opinion. However, the Board considered it necessary to "obtain the consent of the Treasury to the proposed expenditure", and, "before taking steps for that purpose", would "be glad to learn the views of the President and Council on the subject of the measures recommended by the Committee". In particular, they wished to know if those measures were "well calculated to advance meteorological science in the most efficient way" and if "the machinery and establishment suggested by the Committee" would achieve the desired purpose. If the answer was in the affirmative, the Board would be obliged

[7] See Burton's paper on 'Robert FitzRoy and the early history of the Meteorological Office', published in 1986 in the *British Journal for the History of Science*, Vol. 19, pp. 147–176.

[8] On 30 July 1866, Northcote had been President of the Board of Trade for only four weeks. Colonel Sykes was a naturalist and soldier, with many years of military service in India behind him. He was especially interested in meteorology and had made numerous contributions to tropical meteorology.

if the President and Council would provide a detailed statement of the establishment it would be "necessary to provide at Kew, or in connexion with Kew, for the purpose of receiving and discussing meteorological observations". They would also be glad of a similar statement "with respect to local observations in the United Kingdom" and "an estimate of the cost of both". Furthermore, the Board wished to "learn the views of the President and Council with respect to the body under whose management and responsibility the establishments in question should be placed".

Another eight weeks passed. Then, on 27 October 1866, the Secretary of the Royal Society, Dr W Sharpey, replied to Farrer. After first apologizing for the delay, he reported that the considered opinions of the Royal Society were as follows:

- The measures recommended by Galton's committee were "generally well calculated to advance Meteorological Science in a very efficient manner".
- The collection of observations from the masters of merchant ships was probably best performed "through the medium of such agencies as a Government office can command".
- The work of digesting and tabulating results of observations was a function which would be better performed under the direction of a scientific body, furnished with requisite funds, than if left to a government department. Marine observations should be limited to those collected by British observers, and other observations should be limited to those made within the British Isles, including those made at lighthouses and coastguard stations. It was assumed that "the aid afforded by Government would be in the shape of an annual vote, so made as to leave the Royal Society, or other scientific body charged with the duty, perfectly free in their method and in their choice of labour, but upon the condition that an account shall be rendered to Parliament of the money spent, and of the results effected in each year".
- The President and Council did not accept the recommendation that responsibility for issuing storm warnings should be given to the scientific body under whose direction meteorological observations were discussed. These warnings were "founded on rules mainly empirical" and were likely, in a few years, to be "much improved by deductions from the observations in land meteorology" which would have been collected and studied by that time. As their basis was thereby likely to become less empirical and more "strictly scientific" than was currently so, the management of storm warnings might *in due course* be "fitly undertaken by a strictly scientific body". Storm warnings, it was pointed out, "did not originate in any recommendation from the Royal Society". If the government wished to continue issuing them, then they, not the Royal Society, should decide how this should be done.
- Given the recommendation in the Galton Report that publication of the results of observations made by seafarers was a function which properly belonged to the Admiralty's Hydrographic Department, it seemed desirable that the Hydrographer himself should be a member of the committee which superintended the work of the Meteorological Department.
- There was no reason to question the estimates of cost made by Galton's committee, but any detailed statement, either of staff required or of salaries to be paid, was premature.

Finally, Sharpey reported, the President and Council considered that the department responsible for observations, reductions and tabulations should be "under the direction and control of a Superintending Scientific Committee, who should have (subject to the approval of the Board of Trade) the nomination of all appointments, as well as the power of dismissal, of the several officials receiving salaries or remuneration". The committee's members would receive no remuneration, but the assistance of a "competent paid secretary" would be required and the salary of that person would need to be included "in the estimates requested". "Should the nomination of the Superintending Committee be entrusted to the President and Council", Sharpey wrote, "they would be prepared to recommend gentlemen competent to undertake the duties".

The Royal Society now seemed close to achieving what appears to have been their covert ambition for some time. Their opinions on matters meteorological had been sought by politicians and civil servants a number of times since the early 1850s. Now, it appeared, they were close to taking a controlling interest in the Meteorological Department of the Board of Trade.

By the autumn of 1866, many months had passed since the death of FitzRoy. Soon, however, decisions concerning the Meteorological Department came quickly, the first of them in a circular issued by the Board of Trade on 29 November 1866 and signed by Farrer. With effect from 7 December 1866, the storm-warning service would be suspended, but not necessarily permanently, as the Board hoped that the warnings might be resumed by "the new Meteorological Department at no distant time on an improved basis". Meanwhile, daily weather reports would be received and published as before, and if, at any port or place, there was a need for these reports, then they would be communicated by telegraph on the morning they were received, on request and subject to the recipient agreeing to pay the expense of the telegram from London.

A response to Sharpey's letter came on 5 December 1866, when Farrer wrote to the President of the Royal Society to inform him that the Board of Trade, the Admiralty and the Treasury had agreed to the proposals contained in the Galton Report, subject to the modifications suggested by the Royal Society in their letter of 27 October 1866. Furthermore, the Treasury had authorized the preparation of estimates on the basis of the modified proposals. The President and Council were asked to appoint the Superintending Scientific Committee that had been proposed with as little delay as possible.

On 15 December 1866, Sharpey wrote to Farrer to report that the Council of the Royal Society had resolved that a Standing Committee be appointed. Eight Fellows of the Society had been nominated to serve on it. Of these, four were members of the British Association's Kew Committee (Edward Sabine, John Peter Gassiot, William Allen Miller and Warren De la Rue) and two were officers of the British Association (Francis Galton and William Spottiswoode).[9] The others were Captain George Henry Richards (Hydrographer of the Admiralty) and William James Smythe

---

[9] The Kew Committee controlled the meteorological and magnetic work of Kew Observatory, following the Observatory's acquisition by the British Association in 1842. Gassiot was a scientific writer and one of the founders of the Chemical Society. Miller succeeded Daniell as Professor of Chemistry at King's College, London. De la Rue was an inventor and man of science. Spottiswoode was a mathematician.

(a noted meteorologist).[10] Sharpey further reported that the Council wished to be informed if a vacancy occurred on the committee, whereupon they would appoint a new member.

Farrer acknowledged Sharpey's letter on 22 December 1866, saying that the Board of Trade would consider the arrangements proposed by the committee as soon as they were communicated. All was now set for the next phase of the Meteorological Office's existence. The Meteorological Committee of the Royal Society formally came into being on 1 January 1867 and met for the first time two days later.

FitzRoy's Reputation

Meanwhile, there had been a significant repercussion of the decision to suspend the storm-warning service. Babington had resigned as temporary head of the Meteorological Department, doing so on 7 December 1866, the day storm warnings were discontinued. Responsibility for supervising the Department's routine work until the new arrangements were in place had passed to the Chief Clerk, George Harvey Simmonds.

The decision to suspend the storm-warning service was resented by many, which should not have been unexpected, for the surveys carried out by the Board of Trade's Wreck Department, FitzRoy's Department and Galton's committee had all shown that most seafarers considered the warnings beneficial. There was widespread agreement that the system of cautionary signals had helped save the lives of many seafarers, especially fishermen. Given that so many seafarers approved of the service, suspension of it in the middle of winter, the stormiest time of year, surely indicated a lack of judgement on the part of those responsible for the decision.

Strong and vociferous complaints came from seafarers, harbour authorities and many others. Letters were written to *The Times* and other journals. Letters were written to the Board of Trade. Questions were asked in the House of Commons. Even the astro-meteorologists who had so needled FitzRoy spoke out in his defence. And some people wondered why storm-warning services very like that developed by FitzRoy had been established in France and other countries if cautionary signals were as unreliable as critics claimed.

Few sought to belittle FitzRoy's meteorological achievements. Among those who did was Augustus Smith, who had criticized FitzRoy in the House of Commons in 1864 (see Chapter 2). In the House on 8 June 1865, he asked if the Board of Trade intended to continue "prophesying the weather" now that the Meteorological Department had no director. He asserted that the forecasts issued in 1864 had been twice as often wrong as right. He hoped FitzRoy's successor would be "a gentleman of high scientific attainments".

---

[10] Smythe was a colonel in the Royal Artillery. From 1842 to 1847, he was in charge of the observatory at Longwood, St Helena, where he carried out magnetic and meteorological observations under the direction of Sabine. From 12 January to 30 April 1861, he made meteorological observations at Levuka, Fiji.

This was too much for Mrs FitzRoy. "Not content with hunting a good and noble man almost to his grave", she wrote in a letter to Smith, "you endeavour to depreciate and ruin the work of humanity to which he was so earnestly devoted". Her husband had given all for his country, she said, "leaving his family nothing but his reputation". "You can no longer hurt him", she went on, "though I do not suppose you act from personal malice towards one who never injured anyone". "Your motive may be to gain a little popularity by being the economical Member of the House", she wrote, "but you can add one drop of bitterness to the cup of misery already overflowing in the heart of his unfortunate wife who now feels doubly every slur cast on *his* work and most gratefully all the deservedly high terms in which that work is spoken of by thousands and will be long after I am gone and can be no more affected by praise or blame than he now is".

In the debate on 8 June, the President of the Board of Trade defended FitzRoy. He stated that Smith had underrated the reputation enjoyed by FitzRoy in the scientific world and pointed out that the extension of the Meteorological Department's work from the mere collection of statistics into "a more practical application" of the science of meteorology had occurred with the concurrence of the House of Commons and the Royal Society.

The Hydrographer spoke up for FitzRoy, too, and in 1866, after Galton's report had been published, Babington also defended FitzRoy, objecting particularly to the accusation that the statistical work had been neglected. Others pointed out that FitzRoy had dipped into his own pocket a number of times in his career and never been repaid. Indeed, he had died significantly in debt. In the issue of *Punch* published on 31 March 1866, readers were reminded of the fact that he had hired two additional vessels at his own expense when surveying the coasts of South America in the early 1830s. He had "died morally worth millions, fiscally worth less than nothing – in debt £3,000". As the editorial pointed out, "the late First Minister of the Crown" had promised to confer a pension on Mrs FitzRoy but had not honoured this promise. At the very least, thought *Punch*, the Government could have asked the House of Commons to approve a grant "sufficient to liquidate the debt" left by FitzRoy. However, noted *Punch*, the plight of Mrs FitzRoy and her children had not been forgotten: an 'Admiral FitzRoy Testimonial Fund' had been set up by the Liverpool Chamber of Commerce.

Robert FitzRoy was a man who enjoyed popular support for his efforts to improve the lot of seafarers but did not always endear himself to his superiors. In the words of Dr Burton, from his paper on 'Robert FitzRoy and the Early History of the Meteorological Office':[11]

His judgement might have been considered faulty by some, but his sincerity and devotion to duty as he saw it seem not to have been questioned. He laid the foundations of the Meteorological Office in a characteristically controversial manner, and placed the priority of a

---

[11] *Op. cit.* (this chapter, footnote 7).

Figure 3.1. The upper picture, from FitzRoy's *Weather Book* (1863), shows his analysis of the 1859 *Royal Charter* storm, with flows of cold air from the north and flows of warm air from the south. The lower picture is a meteorological satellite image showing cloud patterns over the British Isles at 11:50 GMT on 2 April 2006. The resemblance between the patterns is striking. Upper picture: © Crown Copyright 2010, the Met Office. Lower picture: © NEODAAS/University of Dundee.

Figure 3.2. Robert Henry Scott. © Crown Copyright 2010, the Met Office.

meteorologist's freely-given service to the public firmly ahead of a more academic or commercial rôle in society. He was essentially a man of grand visions, his attention to detail frequently leaving something to be desired, and he could be single-minded to the point of obstinacy. His career was of significant importance to the development of meteorology in Britain.

We may add that FitzRoy's insights in his *Weather Book* are still worth reading today. He provided ideas which guided meteorology into the public arena and meteorology into the twentieth century, and his analysis of the *Royal Charter* storm presaged satellite pictures of the late twentieth century. He pioneered the use of forecasting in a scientific but uncertain discipline, combining not only scientific reasoning and experience but also science, public service and working in the public and political arena, as all directors have had to do.

FitzRoy's Successor

At the first meeting of the Meteorological Committee of the Royal Society, held at Burlington House on 3 January 1867, it was agreed that Robert Henry Scott be appointed Director of the Meteorological Department of the Board of Trade. Born in Dublin on 28 January 1833, he had been educated at Rugby School (from 1845 to 1851) and at Trinity College, Dublin, which he had entered in September 1851. He had gained a Diploma in Engineering in 1856 and also that year become a Bachelor of Arts, gaining First Class Honours in Classics and Science, Chemistry and Geology. Though not a meteorologist, he was no stranger to meteorology. From 1856 to 1858, he had studied chemistry, physics, mineralogy and meteorology in Germany, working under Heinrich Dove in Berlin and the celebrated chemist Justus von Liebig in Munich. He had also, as mentioned in Chapter 2, translated the edition of Dove's *Ueber das Gesetz der Stürme* which had been published in Britain in 1862 as *The Law of Storms*. He had made a living as a teacher in Dublin from 1859 to 1862 and since then served as Keeper of the Minerals and Lecturer in Mineralogy to the Royal Dublin Society.

Scott's appointment as Director of the Meteorological Department was a clear case of personal patronage. He was offered the job by Edward Sabine and appears to have been the only candidate considered seriously. It was not just that both men were natives of Dublin; Scott was a close family friend of Sabine and, furthermore, his executor! Though nearly eighty years of age, Sabine was at the height of his power and influence. He was President of the Royal Society and chairman of the committee which now controlled the Meteorological Department.

The minutes of the meeting of the Meteorological Committee held on 8 January 1867 show that Scott wrote to Sabine on 29 October 1866, naming a salary of £800 per annum and asking that he be allowed a vacation of at least six weeks in each year. His annual salary was indeed £800 when he became Director of the Meteorological Department, and the amount of annual leave he was allowed was indeed six weeks! It is clear from his letter that he was responding to an approach from Sabine, for he began by saying that he had "considered most carefully the proposal" which he had received from him "last Wednesday" (22 October). He went on: "I shall be prepared to undertake the duties of the directorship of such a meteorological office as that which you described to me, in case the Council of the Royal Society think fit to propose my name to H.M.'s Government for that office".

At the meeting on 3 January 1867, it was agreed also that Captain Henry Toynbee be offered the post of Marine Superintendent of the Meteorological Department and that Mr Balfour Stewart, the Director of Kew Observatory, be appointed Secretary to the Committee.[12] Scott assumed command of the Meteorological Department on 7 February 1867, and the title 'Meteorological Office' was adopted formally soon afterwards, at the meeting of the Meteorological Committee held on 25 February 1867.

[12] Toynbee was born on 22 October 1819 and joined the Merchant Navy in 1833 or 1834. It is said that his interest in meteorology was stimulated by his marriage (in 1854) to a daughter of Admiral W H Smyth, a distinguished astronomer and meteorologist. She accompanied him on voyages, and her illustrations supplemented his meteorological logs. Balfour Stewart was a physicist and meteorologist who had been appointed Director of Kew Observatory on 1 July 1859. As Secretary to the Royal Society's Meteorological Committee, he supervised the installation of meteorological stations all over Britain and Ireland. He left Kew in 1871, when he was appointed Professor of Natural Philosophy at Owen's College, Manchester.

Under New Management

The meetings of the Meteorological Committee in the early part of 1867 were concerned mainly with administrative matters such as duties and salaries. At their meeting on 3 January, for example, the Committee agreed that Scott be offered a salary of £800 per annum and resolved that Stewart, as Secretary to the Committee and Director of Kew Observatory, be paid £400 for the current year, commencing on 1 January. At their next meeting, on 8 January, the Committee heard that Scott had accepted the salary offered and noted that the Board of Trade had declined to recommend to the Treasury that officers of the Meteorological Department be civil servants.[13]

Toynbee accepted the post of Marine Superintendent at the salary offered (£350 per annum) but asked in his letter of acceptance, dated 11 January, that he be paid increments of £15 per annum until a salary of £450 per annum was reached. The response of the Committee was to instruct Stewart to write to Toynbee to inform him, as it was put in the minutes of the meeting on 15 January, that they were "only authorized to provide Estimates for the Meteorological Department from year to year, and that to establish an annually increasing salary to the officers would be implying a permanency to the Department which they had no power to do". However, the Committee appears to have had second thoughts, because, on 25 February 1867, they reconsidered Toynbee's request and resolved that they were prepared to "take into consideration his application for a salary increase at the close of the financial year of 1867". This was a promise they kept, for at their meeting on 6 April 1868 they agreed that his salary would be £400 for the financial year beginning that day. And it was still £400 per annum twenty years later![14]

Draft estimates for expenditure in 1867, for submission to the Board of Trade, were agreed at the meeting held on 21 January 1867, these amounting to £9900, with an additional £2900 budgeted under the heading of 'special grant for equipment', composed of £1700 "to complete the instruments of the eight observatories" and £1200 for "buildings and alterations at Kew". The grand total was, therefore, £12,800. The estimated costs given in Part III of the Galton Report totalled £12,950, made up of £10,450 for the routine work of the Meteorological Department and an initial outlay of £2500 for additions to the building at Kew.

Though the distribution of costs agreed by the Committee on 21 January 1867 differed significantly from that proposed in the Galton Report, the total of £12,800 was within the £12,950 recommended in the Report. This being so, it is understandable that the Meteorological Committee believed their estimates would be acceptable, given

---

[13] It was later confirmed that officers appointed by the Meteorological Committee would not be "in any sense in the Civil Service of the Crown", to quote from a letter dated 10 April 1867 from G A Hamilton of the Treasury to T H Farrer of the Board of Trade and forwarded from the Treasury to the Committee on 15 April.

[14] In real terms, however, his salary had increased, for by 1888 the retail price index had fallen to a little over 70% of what it had been in 1868.

that Farrer's letter of 5 December 1866 to Sharpey had indicated acceptance of the Galton Report by the Treasury, including the estimated costs.

It came as something of a surprise to the Meteorological Committee, therefore, that the Treasury objected to the estimates, even though the Board of Trade had accepted them and (on 8 February 1867) forwarded them to the Treasury with a general expression of approval. The Treasury's view, as recorded in the minutes of the Committee's meeting on 22 April 1867, was that the Civil Service Estimates for 1867 (approved in March 1867) included £10,000 for the services formerly performed by the Meteorological Department of the Board of Trade, to be used as the Royal Society saw fit, with no further claims on the public purse and any surplus to be returned to the Exchequer.

The Treasury denied that approval had been given by them for any work required at Kew Observatory, which was embarrassing for the Meteorological Committee, because they had already made spending commitments, believing that the £2900 they had requested as a 'special grant for equipment' had been approved by the Board of Trade as an initial outlay. As noted in the minutes of the Committee's meeting on 18 February 1867, an architect had been consulted, and, as noted in the minutes of their meeting on 1 April 1867, tenders for the work at Kew had been obtained, all within the £1200 earmarked for buildings and alterations at Kew. Instruments had also been ordered for Kew and the other observatories.

Their Lordships of the Treasury were unmoved. As noted in the minutes of the Committee's meeting on 24 June 1867, the Treasury chided the Committee (through the Board of Trade) for anticipating the vote and insisted that the arrangements already made be curtailed. There was, however, a concession. The sum of £570 which had been included in the Civil Service Estimates for meteorological services rendered to the Admiralty (intended for the purchase and verification of instruments) would be considered additional to, rather than part of, the £10,000, but it was made very clear to the Committee that their expenditure in 1867 had to be kept within the £10,570 now sanctioned by the Treasury.

Salaries of the junior staff were agreed at the meeting of the Committee on 25 February 1867. The Chief Clerk would be offered £200 per annum, rising by £10 per annum to £250, and the most junior of the Junior Clerks would be offered £78, rising by £5 10s 0d (£5–50p) per annum to £100.[15]

It was announced at the same meeting that the Chief Clerk, Simmonds, who had taken charge of the Meteorological Department when Babington resigned, himself now wished to resign.[16] The Committee accepted his resignation and chose to advertise

---

[15] It is curious that the Committee could approve increments for clerks but not for Toynbee. The reason may be that clerks were technically employees of the Board of Trade, whereas Toynbee, as an *official* of the Meteorological Office, was technically an employee of the Royal Society.

[16] His resignation was understandable, for his position must have become quite awkward by then. On 21 January 1867, Sabine wrote thus to T H Farrer, Secretary of the Board of Trade: "The Committee would suggest that

for a new Chief Clerk, rather than promote one of the clerks already in post. At the Committee's meeting on 4 March 1867, it was reported that the clerks had accepted the salary offers made to them, and at the meeting on 11 March it was announced that Richard Strachan, Second Senior Clerk, and William Salmon, an instructor in the Royal Navy, had applied for the post of Chief Clerk. The latter was chosen and took up his appointment on 1 April 1867.

Office hours were decided at the Committee's meeting on 14 February 1867 and fixed as 10 o'clock to 4 o'clock daily, except Saturday, when half of the clerks in rotation would be excused attendance after 2 o'clock. Staff were not required to work on Sundays, Christmas Day, Good Friday, Easter Monday, Whit Monday or the Queen's Birthday. Salaries were paid quarterly in 1867. Thereafter, as agreed by the Committee at their meeting on 5 December 1867, they were paid monthly.

During 1867, the workload of the Meteorological Office increased rapidly, and at the meeting of the Committee held on 11 November 1867 Scott was given permission to appoint an additional assistant on a weekly salary. Ten days later, Goodwin White joined the staff, employed as a clerk at £1 18s 5d per week. The total complement of the Office thus rose to ten, with the names and occupations of the staff, as recorded in the *Report of the Meteorological Committee of the Royal Society for the Year Ending 31st December 1867*, being as follows:

Robert H Scott – Director of the Office

Captain Henry Toynbee – Marine Superintendent

William Salmon, RN – Preparation of weather reports. Reduction of logs.

Jas S Harding, Junr – Correspondence – Accounts – Registry of documents.

Richard Strachan – Care and management of the instruments, and correspondence therewith connected.

Frederic Gaster – Discussion of returns from land observatories.

Charles Harding – Preparation of weather reports. Reduction of logs.

Jas S Harding, Senr – Registry of documents and assistance in accounts, &c.

Richard H Curtis – Preparation of weather reports. Reduction of logs.

Goodwin White – Discussion of weather reports and Coast Guard returns.

Balfour Stewart (Secretary to the Committee) and the Rev Thomas Kerr (Director of Valentia Observatory) were also employed by the Committee.[17]

Mr Simmonds, the gentleman now in charge of the Meteorological Department, be so informed and desired to give over the charge to such of those gentlemen as the Committee may appoint to receive it. The Committee would also suggest that Mr Simmonds be informed that the further appointment of clerks on the establishment is vested in them".

[17] Kerr was a former Navigation Lieutenant in the Royal Navy and had trained at Kew for the work at Valentia. The Meteorological Committee engaged him as director of the proposed observatory at a salary of £250 per annum, commencing from 15 August 1867, and stipulated that the rent of the house for the observatory and observer was not to exceed £40 per annum. They also agreed that he be given an allowance of 10 shillings (£0–50p) per week for an assistant. Instruments would be provided by the Committee.

In addition to administrative matters, the Committee also considered aims, objectives and policy issues and turned their attention increasingly to meteorological matters. At their meeting on 14 February, for instance, they received a report by Toynbee on the quality of seafarers' observations. In this, attention was drawn to incorrect observational practices and to shortcomings of registers. Various improvements to 'weather books' (registers) were suggested, and Toynbee reported that a circular letter would be sent to the commanders of ships, explaining their obligations upon taking instruments and giving specific examples of incorrect practices, such as thermometers being kept in warm cabins, rather than the open air, and barometers aboard steamships being mounted too close to boilers!

As recommended by the Royal Society's Council in June 1865, the Committee took steps to establish observatories at Kew, Glasgow, Aberdeen, Stonyhurst, Armagh, Valentia and Falmouth, with the one at Kew designated the 'Central Observatory'. The Committee's original intention, approved by the Board of Trade, was that eight observatories would be established, including one in the north of Scotland. In a letter dated 5 June 1867, however, the Treasury informed the Board of Trade that less money had been allocated for the current year than the Committee had estimated. The consequence was that plans for the observatory in the north of Scotland were abandoned. The planned observatory at Aberdeen might have been abandoned, too, but for the intervention of David Thomson, Professor of Natural Philosophy in the University of Aberdeen, who wrote to the University's Chancellor about the matter in July 1867. The Chancellor was the sixth Duke of Richmond, who was also, coincidentally, President of the Board of Trade! By the end of 1867, the observatories at Kew, Glasgow and Stonyhurst had become fully operational, and the other four became fully operational before the end of 1868.[18]

The Committee also took steps to improve the quality of observations made at telegraph stations and other places which possessed instruments supplied by the former Meteorological Department, including fishing villages and small ports equipped only with fishery barometers. Most of these places were visited by Scott, Toynbee or Stewart, who made arrangements for defective instruments to be recalibrated or replaced, if necessary.

From time to time, the Committee considered applications for information, instruments or funding, as on 25 February 1867, when they approved an application from a well-known mountaineer, Edward Whymper, for thermometers to measure sea temperatures on a forthcoming expedition to Greenland.[19] An application by the Scottish

---

[18] Kew Observatory was superintended by the Kew Committee of the British Association, Glasgow Observatory by Robert Grant, Professor of Practical Astronomy in the University of Glasgow, Stonyhurst Observatory by the Council of Stonyhurst College (observer the Rev W Sidgreaves, SJ), Valentia Observatory by the Rev Thomas Kerr, Falmouth Observatory by the Royal Cornwall Polytechnic Society (observer Mr Squire), Aberdeen Observatory by Professor David Thomson (observer William Boswel) and Armagh Observatory by the Rev T Romney Robinson, the Observatory's astronomer. Robinson is best known to meteorologists as the inventor of the cup anemometer.

[19] Edward Whymper was the first to climb the Matterhorn (in 1865). He visited Greenland in 1867 to ascertain the nature of the interior and investigate the possibility of crossing it.

Meteorological Society to the Board of Trade for government funds to process the observations made at certain Scottish stations was, however, rejected. At their meeting on 24 June 1867, the Committee instructed Scott to point out to the Board of Trade that the Scottish Meteorological Society was a private scientific society which furnished information to the Registrar-General of Scotland. This being so, the Committee considered that the application should be channelled through the Registrar-General, should he require the observations.

Not all administrative matters were dealt with by the Committee, and one in particular raised the possibility of divided loyalty. At their meeting on 21 March 1867, the Royal Society's Council considered a letter from Scott, dated that same day and addressed to the President. It read as follows: "Sir, As I have been directed by the Board of Trade to apply to the Royal Society for permission to use a crown, or other device, on the seal of this Office, in addition to the simple legend, I have to request that you will ask the Council of that body to grant permission to the Meteorological Office to use the Crest of the Royal Society on its seal and die for papers". With Sabine in the chair, as President of the Royal Society, not, on this occasion, Chairman of the Royal Society's Meteorological Committee, the Council refused the request. No reason was given in the minutes of the meeting; these simply state that "the Council did not deem it expedient to comply with Mr Scott's request". But this was a minor matter. The Meteorological Office now seemed to have left behind the problems and uncertainties of recent years, adopting, in accordance with the wishes of the Royal Society, a much more structured and scientific approach than before the untimely death of Admiral FitzRoy.

4

The Fight over Forecasts

When control of the Meteorological Department passed to the Royal Society, the Meteorological Committee and officials of the Board of Trade may have thought that the storm of protest over the suspension of the storm-warning service would soon blow itself out. If so, they were much mistaken. The stream of complaints which began to flow as soon as the suspension was announced continued unabated, and a campaign to restore the service soon developed, with a formidable champion in the person of Colonel W H Sykes, FRS, MP.

The Campaign in Parliament to Restore Storm Warnings

Sykes's offensive began in the House of Commons on 15 February 1867, when he asked the President of the Board of Trade whether the Storm Signals as hitherto practised by the late Admiral FitzRoy were to be continued. If so, in what manner would they be continued, and by whom? If they were to be discontinued, would it not be prudent to invite the Chamber of Commerce of the United Kingdom to express an opinion on the subject? Furthermore, he wondered, was the Meteorological Report which had appeared in *The Times* and other papers to be continued, and, if so, could not observations from the Paris, Brussels and St Petersburg Observatories be added to it?

In reply, Sir Stafford Northcote explained why the storm-warning service had been suspended and informed the House that "as soon as the matter had been placed in the hands of the Royal Society they informed the Government that they were not prepared to continue the storm signals, as they rested upon a mere hypothetical basis". He further reported that he had received a note from Sabine which had informed him that the Committee were of the opinion that no advantage would be gained by receiving communications from Paris and Brussels, as these places were not on a coast and telegrams from St Petersburg would not be received in time for publication in the daily newspapers. The Committee, said Sir Stafford, already received telegrams from six coastal stations on the continent and had dispensed with none since the Meteorological Department had been in their hands. "At present", he added, quoting from Sabine's note, the Committee were "not prepared to recommend any additional expense to be incurred on this head".

Complaints and enquiries about the suspension of the storm-warning service came from far and wide. At the meeting of the Meteorological Committee on 4 March 1867, Scott reported that most of the foreign correspondence received by the Meteorological Office related to this matter, and at the meeting on 18 March 1867 he reported that memorials pressing for a resumption of the warnings had been received from the Leith Chamber of Commerce and from the underwriters and shipowners of Glasgow and Greenock. Many of the complaints and enquiries had been ignored or brusquely brushed aside, including a communication from Le Verrier in February 1867, when he sought information about the suspension of the service.

However, the letter from the Board of Trade at the end of May 1867 could not be disregarded. Dated 31 May 1867, signed by T H Farrer and addressed to "Robert H Scott, Esq., Director, Meteorological Office", it read as follows:

Sir, I am directed by the Lords of the Committee of Privy Council for Trade to state that a large deputation has waited on the Duke of Richmond to urge that some warning should be given of apprehended danger from storms, and I am to ask whether it might not be possible for the Committee appointed by the Royal Society, upon such conditions and under such limitations as they might think necessary, to give effect to a desire which is strongly expressed by many competent and influential bodies and persons.

Notice had to be taken of this letter. The Duke of Richmond was now the President of the Board of Trade and the deputation included members of both Houses of Parliament.

The Committee discussed Farrer's letter at their meetings on 3 and 7 June 1867 and Scott replied on the 8th, saying that the Committee were not willing to "prognosticate weather, or to transmit what have been called 'storm warnings'". However, he said, they were collecting information which they confidently anticipated would enable them to frame rules by which prognostications could be made, one of the main objects being "the advancement of meteorological science in this important practical direction". The observatories, he said, were not yet in practical operation and, indeed, could not be until the necessary funds had been voted by Parliament.

Scott reminded Farrer that the Meteorological Office had issued a circular in March 1867 in which (as it was put in the circular) they had offered to "forward each day, by post, free of charge, to any port which may require it, a copy of the daily weather report, the same as that furnished to the second edition of the London morning papers". Moreover, the Office was prepared, on application, to "furnish, without unnecessary delay, any telegraphic information which it may have received", half the cost of the transmission to be borne by the local authorities that wished to receive the information. The Committee were "willing to communicate information to any accessible place upon the terms laid down in their circular, and to an extent limited only by the sum placed at their disposal for the purpose".

Sykes was impatient to know the outcome of the appeal to the Duke of Richmond. In the House of Commons on 7 June, he asked the Vice-President of the Board of Trade, Stephen Cave, "whether any and what action had been taken to restore 'storm warnings', consequent upon the Deputation of Members of Parliament to the President of the Board of Trade".[1] In reply, Cave told the House that the Duke of Richmond had "communicated personally with General Sabine" and been assured that the Meteorological Committee were willing to provide warnings of storms "as far as practicable".

Cave asked the House to remember that the Committee "consisted of scientific gentlemen named by the Royal Society, at the request of the Government" and "gave their time, labour and talents gratuitously". They had been appointed because "a Government Department felt itself incompetent to deal with a purely scientific matter". They must be left to "decide what could and could not be done in the practical application of the present results of meteorological science". Sykes pressed his point, saying that he wished to know "whether the Committee had expressed their willingness to undertake the subject of storm warnings". Cave replied that "they were willing to do so as far as possible, and that the subject was under their consideration".

Sykes was on his feet again in the House of Commons on 24 June, asking the Vice-President of the Board of Trade whether the Meteorological Committee found it practicable to renew the storm warnings, as formerly practised by the Meteorological Department of the Board of Trade. Cave's answer added nothing. He proceeded to do no more than reiterate the contents of Scott's letter of 8 June.

Sykes was still not satisfied. In the House of Commons on 8 July, he asked Cave what interpretation was to be put on Scott's letter, for it appeared to be contradictory, in that it clearly stated that the Meteorological Committee refused to renew storm warnings but at the same time contained the implication that storm warnings would be sent to local authorities which were willing to pay half the expense of the telegrams. Cave denied there was any inconsistency, saying that the Committee declined to transmit storm warnings but also offered to communicate "information". According to Cave, they declined to prognosticate what the weather would be the following day or the day after but were willing to send telegraphic information of what the weather actually was in any particular place.

Thomas Gray, the Permanent Assistant Secretary in the Marine Department of the Board of Trade, replied to Scott's letter on 11 July, saying that the Board approved of the Committee's proposal that a portion of the cost of sending telegraphic information concerning storms should, as a general rule, be borne by the ports which desired

---

[1] The Vice-President of the Board of Trade answered questions in the House of Commons because the President, the Duke of Richmond, was a peer and therefore a member of the House of Lords.

such information. However, he said, the Board suggested that the Committee might consider sending information gratis to impecunious fishing villages. The Committee considered this suggestion at their meeting on 15 July and authorized Scott to reply that they were prepared to "transmit telegraphic intelligence of weather" free of charge to the poorest fishing villages if the Board of Trade undertook to provide, "from time to time, the names of any stations to be placed on the free list".

In his letter of 15 July, Scott further informed Gray that the Committee had "in contemplation a scheme of signals for weather reports" and suggested that "the Board of Trade should present to each station a complete set of signal shapes and gear". The Committee did not feel, however, that they or the Board should bear the expense of maintaining the gear or paying the wages of those who managed and hoisted the signals. The Board agreed and informed Scott that they were "willing to hand over to the Committee the old signals, gear, &c., supplied under the direction of Admiral FitzRoy", but had no funds at their disposal for the purchase of new signals.

Sykes took the campaign to restore the storm-warning service another step forward on 19 July 1867, when he moved in the House of Commons that it was "inexpedient to continue the present arrangement with the Committee of the Royal Society in consequence of the suspension of storm signals, at the expense of £10,000, the average cost of the Meteorological Department of the Board of Trade having been £4,300 per annum". He held "those gentlemen of the Royal Society who appointed a committee to take charge of the Meteorological Department of the Board of Trade" responsible for the suspension of the storm-warning service and the Meteorological Committee responsible for its continuing suspension. "In spite of the opinion expressed by various public bodies as to the expediency of the storm warnings on the score of humanity", he said, the Committee "still abstained from publishing them". They continually asserted that "they could not take on themselves to predict storms" and they appeared to be obsessed with the view that the weather predictions which were issued hitherto had not been founded on scientific data.

Sykes pointed out that twenty-eight petitions, bearing a total of 1744 signatures, had been presented to the House of Commons in favour of the resumption of storm warnings, with five of the petitions from incorporated bodies and others from different ports. Moreover, he added, a report submitted to the Literary and Philosophical Society of Manchester in April 1867 contained not only the accusation that the Meteorological Committee were utterly regardless of public opinion and feeling on the subject of storm warnings but also an expression of hope that the Board of Trade would again take on the management of the Meteorological Department. There were, he said, 50,000 fishermen along the eastern coast, "and it was a perfect mockery to require them to pay half the expense of telegraphing". "Such an arrangement", he commented, "would be of no more use than the daily meteorological report in the papers, telling what had occurred yesterday". He wondered what the Committee would do prospectively.

Sykes hoped the Board of Trade would insist on the restoration of storm signals and saw no reason why this demand should not be satisfied, "except the fear on the part of the Committee of the Royal Society that their scientific dignity would be compromised". The objection was, he said, "nothing more than a piece of scientific coxcombry and pedantry." In his view, the Board of Trade had no right to spend £10,000 in the manner he had described, the expense formerly having been only £4300.

Stephen Cave stated that the present Government could not be held responsible for the current situation, "as they found the arrangement almost concluded when the change of Government took place".[2] He did not, however, distance himself from the issues. On the contrary, he restated the views of the Galton Committee on storm warnings and defended the cautious attitude of the Meteorological Committee towards an early restoration of a storm-warning service. He reminded the House that the Galton Committee's inquiry into the work of the Meteorological Department had found that FitzRoy, "having obtained permission to organize the system of storm warnings", had diverted almost the whole of the funds voted for the purpose of observations "from the original scientific object to an object deemed more immediately practical".

In the view of the Royal Society, said Cave, storm warnings were based on imperfect data and were therefore liable to be inaccurate. Storm warnings "were stated to be like both Queen's Speeches and ministerial answers, which might be read a hundred different ways. "It was said", he went on, "that they were twice wrong for once right, and that they would be very mischievous but for the fact that no-one paid any attention to them".

The Government's view was, Cave said, that the system of weather telegraphy and foretelling weather was not carried on by precise rules and had not been established by a sufficient induction from facts. The storm warnings had, however, been, to a certain degree, successful and were highly prized. The Government thought that the daily weather forecasts for the general public ought to be discontinued and an endeavour made to improve the storm warnings, by defining the principles on which they were issued and by testing those principles by accurate observation. Above all, he said, the Government thought that "steps should be taken for establishing a full, constant, and accurate system of observing changes of weather in the British Isles".

The next speaker, Thomas Milner-Gibson, a former President of the Board of Trade, agreed with the course of action taken by the Government, at least in principle, and trusted that it would be "possible at no distant time to resume the storm warnings".

---

[2] Cave was a member of the minority government of the fourteenth Earl of Derby, which took office in July 1866 and survived into 1868. The years 1866 and 1867 proved momentous not only for the Meteorological Office but also, economically, socially and politically, for Great Britain as a whole. There was economic disorder, the bank rate reached 10%, there was widespread unemployment, the harvests of 1865 and 1866 were ruined by heavy rains, there were riots and disturbances in London and other cities, the Fenians created alarm in Ireland and England, and political life at Westminster was dominated by the crises over the 1866 and 1867 Reform Bills.

This was a view shared by another former President of the Board of Trade, Edward Cardwell. The response to his speech was a resounding "Hear, hear", after which Sykes withdrew his motion. It was not, however, the end of the matter as far as he was concerned.

Efforts in the Press to Restore Storm Warnings

In August 1867, the House of Commons voted the sum of £10,500 for the work of the Meteorological Office. This was "not too much for the purpose", said *The Daily Telegraph* in an editorial on 23 August 1867, "provided it be wisely spent". It was, they noted, "three times as much as that taken by the same office when under the direction of Admiral FitzRoy", which prompted them to ask what the Office had to show for the increased expense. The question was, they thought, far from easy to answer, for it had "ceased to do several things which it performed during the lifetime of the lamented Admiral". It had once issued daily weather forecasts and, "when needful, caused storm signals to be hoisted round the coast". It now did neither. Why were those services no longer performed? What reason could the Office show for its existence? Suspension of the storm-warning service had been condemned by those most directly concerned, the seafarers. The list of shipwrecks was now enormous. The Galton Report was full of prejudice, and some of the methods used by the Galton Committee to assess the achievements of FitzRoy's department were dubious or fallacious. The Office was "in a very unsatisfactory state; its new career has been anything but brilliant, and one of the first steps towards giving it efficiency must be the restoration of the storm signals". Editorials in, and letters to the editors of, other newspapers were in similar vein.

At the meeting of the British Association in September 1867, Sykes was scathing. In his paper on *Storm Signals – Their Importance and Practicability*, which he delivered before a large audience, he claimed that 305 of 405 storm warnings "given under the system lately in use" had proved correct, and he asked if such results did not "sufficiently justify the continuance of these warnings". It was evident, he complained, that those in authority did not think so. The warnings had been stopped, and he wanted to know the reason why. To laughter and applause, he suggested that the argument employed by the Meteorological Committee was "a pedantic affectation of science – literally, the coxcombry of science"! The Committee considered the reliability of the warnings questionable, he said, "on the ground that Admiral FitzRoy had obtained his conclusions mainly on empirical data". To remedy this, the Committee proposed to establish eight additional observatories and "at the end of fifteen years expected to be able to predict storms on philosophical data, not on empirical data". He was sceptical, doubting they would obtain these results in the next fifteen years if they had not done so in the last fifteen. "Here were we", he went on, to further applause, "the most maritime nation in the world, having set the example to other countries in this matter of storm warnings, and yet we were now dropping them". "We were", he said, to further laughter, "too scientific for the work".[3]

As weather systems typically moved from west to east, Sykes noted, "the advent of storms could be telegraphed from the west of Ireland to England several hours before the wind actually reached those parts". Indeed, he suggested, the Atlantic cable might be used to provide a week's warning of stormy weather, if it was true that storms generally travelled across the Atlantic from North America to the British Isles. This was an interesting idea, though not, we now know, meteorologically reliable.[4]

The only members of the Meteorological Committee who were present to hear Sykes's attack were Gassiot and Stewart. They tried to defend the Committee's position, but what they said was unconvincing and tended to reinforce Sykes's arguments. The tide of opinion ran against them. The chairman of Dundee's Chamber of Commerce thanked Sykes for his "admirable address" and hoped that the British Association would urge "in the strongest manner the necessity of resuming storm signals throughout all the ports of the kingdom". David Milne Home, Chairman of the Scottish Meteorological Society's Council, said he was a personal friend of the late Admiral FitzRoy and "many a time saw the manner in which he made his forecasts". He could assure the meeting that FitzRoy's storm warnings were based on correct data and not mere hypothesis. Even one so committed to science as James Glaisher spoke against the Meteorological Committee. He particularly took issue with Stewart, who had stated that "poor fishermen and others should be left to draw their own deductions from the reports sent them". His reply to that was: "For God's sake, do no such thing; the fishermen would each form different conclusions; experienced men in headquarters should judge for them".

Milne Home seconded the views expressed by Sykes but did not support him unreservedly, for he exonerated the Royal Society from some of the blame that Sykes had placed upon them. The reservation was not, however, sufficient to stop him moving "that this section apply to the Council of the British Association to make a communication to the Board of Trade urging them to institute arrangements for causing the storm-signals to be resumed".[5] The resolution was seconded by the Association's President, the Duke of Buccleuch, who told the meeting he had been one of the deputation who had appealed to the Duke of Richmond. His view was that "we do not ask the Board of Trade or any body of men to volunteer or undertake the duty of being 'weather prophets'; all we want is that correct information be given

---

[3] In the *Report* of the British Association meeting held at Dundee in 1867, there is no mention of Sykes's paper. Summaries of it and the ensuing discussion can be found in *The Times* of Wednesday 11 September 1867 (p. 6, column e) and in Symons's 'Report of the British Association meeting at Dundee on the subject of meteorology; storm warnings' (*Symons's Monthly Meteorological Magazine*, 1867, Vol. 2, pp. 101–105).

[4] The idea of storms travelling all the way across the Atlantic Ocean from eastern North America to the British Isles was a popular one and proved persistent, despite evidence to the contrary. It was refuted only when reliable data from the North Atlantic became available in 'real time'.

[5] The section in question was that concerned with 'Mathematical and Physical Science', i.e., Section A.

that will enable the public (or at least maritime men) to judge what kind of weather they may expect when they leave port". After contributions to the debate from several others, the resolution was amended slightly and carried by acclamation.

The press did not let the matter drop. In an editorial published on 16 September 1867, for example, *The Daily News* pointed out that the British Association had "placed itself in direct opposition to a no less learned body, the Royal Society". They poured scorn on the suggestion that "Admiral FitzRoy made up his warnings by mixing his own seaman's knowledge of weather signs with the telegraphic information he received". And a letter from FitzRoy's erstwhile junior, superior and friend, Rear-Admiral Sulivan, published in *The Daily News* on 24 September 1867, contained a full explanation and justification of the methodology used by FitzRoy and Babington. In Sulivan's opinion, thousands more lives would be lost and many more properties damaged or destroyed if a resumption of the storm-warning service had to wait until the scientific approach favoured by the Meteorological Committee could guarantee warnings more reliable than those issued by FitzRoy.

Resumption of Storm Warnings

The Director of the Meteorological Office never became involved personally in the controversy. The blame for suspension of the storm-warning service was invariably placed on the members of the Galton and Meteorological Committees who were obsessive over the need for scientific rigour. It fell to Scott, however, to reply to Farrer's letter of 30 October 1867, in which he asked if the Meteorological Committee were "prepared to issue cautionary notices during the forthcoming winter, and, if so, of what description, on what basis, and on what terms".

Scott replied on 7 November, informing Farrer that a "system of weather tele-graphy" had been approved by the Committee and that specimen signals would be tested at "one or more of the chief ports before being issued to the public". Each signal consisted of "a semaphore, similar in principle to a railway-signal post". Messages would be sent free and complete sets of signals presented to the stations named by the Board of Trade as 'poor fishing villages'. Other stations would purchase the signals and pay half the telegraphic transmission costs. Until the new system of signals had been tested, it was proposed that the drum be employed as a cautionary signal to indic-ate the existence of a gale. The Board of Trade accepted this proposal, whereupon Farrer wrote to the Committee (on 13 November 1867) to inform them that permission to use the drum had been granted and "arrangements for using it should be made with the least possible delay".

The Meteorological Committee asked the Board of Trade to announce the resump-tion of the storm-warning service by means of a circular sent to such ports and fishing stations as the Board considered expedient, stating in it that the Meteorological Office was prepared to transmit, free of charge, to all parts of the coast accessible by telegraph, "immediate intelligence of all storms of whose existence it may have received information". The intelligence would be "made public by the use of the drum". The Committee's request was made in a letter from Scott to Farrer dated 19 November and Farrer replied on 22 November, agreeing not just to the request but, in fact, to all the proposals contained in the letter, including the recommendation that the Board of Trade's Wreck Department should check the accuracy of the storm warnings, as in FitzRoy's lifetime. The circular was issued by Farrer on behalf of the Board of Trade on 30 November 1867, and the first storm signals were hoisted on 10 January 1868. From that day to this, a storm-warning service for seafarers has continued uninterrupted. However, it was only the storm-warning service that recommenced in the winter of 1867–1868. The decision to cease publishing daily weather forecasts for the general public was not reversed at this time.

Trials of the experimental signals in 1868 showed that the idea of conveying information to ships by means of a system of semaphore telegraphy was not satisfactory, so, in 1870, the idea was dropped.[6] Careful investigations of the weather before and after every gale were carried out by Toynbee, Scott, Galton and other Committee members, the objective being an enhanced understanding of depressions and, therefore, improved storm-warning service. The outcome was a *volte-face* for the Committee, and for Galton in particular, for the investigations led, at least in part, to a reintroduction of the system of storm signals employed by FitzRoy (including indications of wind direction)! On 13 January 1874, Scott informed the Board of Trade that the Committee were prepared to reintroduce FitzRoy's system, though he did not admit that the studies of gales had played any part in their deliberations.

On 14 February 1874, the Board of Trade issued a circular (No.717, entitled *Telegraphic Weather Intelligence*) announcing that the system would be reintroduced on the 15th of the following month, using, as in FitzRoy's time, cones to indicate wind direction and a drum and cone together to indicate a heavy gale, with the cone showing the expected direction. A drum was never to be used without a cone. The system was subsequently used for more than a century. Indeed, signals were displayed at prominent coastal locations around the British Isles until 1 June 1984, when the system of hoisting cones and drums or the equivalent in lights was superseded by electronic means of communicating storm warnings.

Once the storm-warning service had recommenced, the storm over its suspension soon blew itself out, and the Meteorological Committee were then able to devote themselves fully to realizing their ambition of establishing a scientifically respectable Meteorological Office which provided a valuable public service, especially for seafarers.

A routine developed in the Office. Indeed, the day-to-day work became almost humdrum. Instruments were supplied to ships of the Merchant and Royal Navies; the

---

[6] The semaphores were actually devised by Toynbee.

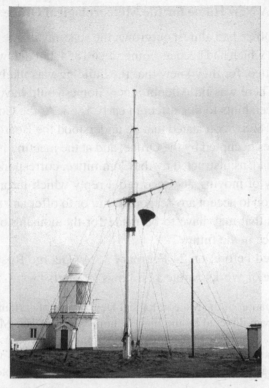

Figure 4.1. South cone, St Ann's Head, Pembrokeshire, 17 September 1965. Photograph by Malcolm Walker.

registers returned by mariners were examined by Toynbee and his assistants; instruments were loaned to navigation schools; the work of establishing observatories and otherwise advancing what the Committee called 'land meteorology' continued; *Daily Weather Reports* were prepared for the afternoon editions of London newspapers; storm-signal drums were issued to coastal stations and storm warnings were issued as and when necessary.[7]

The revived storm-warning service appears to have operated satisfactorily, given that the Committee, who would surely have been its sternest critic, claimed in their report for 1868 that it had proved, on the whole, successful. However, the subject of storm warnings appears to have remained one that was sensitive for the Committee, for the term 'storm warning' was avoided in the report, with 'telegraphic intelligence of storms or atmospherical disturbances' preferred!

[7] It was noted in the *Report of the Meteorological Committee of the Royal Society for the year ending December 31, 1869* that the Committee had supplied instruments to the navigation schools at Aberdeen, Plymouth, Leith and South Shields on the understanding that the instruments would be used for educational purposes. It was also noted that the Board of Trade had allowed a set of instruments to remain at the Church of Scotland Training School in Edinburgh.

A New Home for the Meteorological Office

By early 1869, the Office had almost outgrown the space available to it in the building on Parliament Street which had been its home since 1854, but that was probably immaterial to the Committee, for they knew that the building was likely to be demolished in the near future. There was little doubt a new home would have to be found soon. There had indeed been hints to that effect in early 1868. At the Committee's meeting held on 3 February 1868, Scott stated that he understood the Board of Trade intended to vacate the premises occupied by the Office, and at the meeting held on 17 February he reported that he had, as instructed by the Committee, corresponded with the Board about the possibility of moving and received a reply which informed him that they were "not in a position to accept any responsibility or to offer any advice with respect to the arrangements that may have to be made for the housing or furnishing of the Meteorological Office in the future".

A year then passed before, on 24 February 1869, George Russell, a Secretary in Her Majesty's Office of Works, wrote as follows to Scott:

Referring to the occupation of the Meteorological Department of certain rooms at Nos.1 and 2 Parliament Street, I am directed by the First Commissioner of Her Majesty's Works, &c., to inform you that this Board [the Office of Works] require to resume, as early as practicable, possession of those rooms and of the furniture in them, the property of the Government, and I am to inquire when it will be convenient to the Meteorological Department to vacate the premises.

Scott replied on 25 February, asking whether, in the event of the Committee having to furnish an office, it would be possible to purchase from the Office of Works any of the furniture or other government property used by the Meteorological Office, "more especially certain presses, fifteen in number". Two weeks later, through a letter dated 13 March 1869, Russell informed Scott that the Treasury had valued the presses in question at £111 8s 0d and sanctioned disposal of them to the Office, the sum to be paid to the credit of the Paymaster-General on the Office's account.

At the meeting of the Committee held on 1 March 1869, Scott was instructed to write to the Board of Trade to ask if it was contemplated that the Committee should be expected to find accommodation for the Office out of their annual grant. He did so on 2 March but never received a direct answer. The letter he did receive from the Board, signed by Thomas Gray and dated 8 April 1869, simply informed him that the Office of Works had no spare accommodation. Gray did not say in so many words that the Office would have to pay for its accommodation in the future, but the inference was plain to the Committee.

The search for a new home in central London soon proved successful, and the Office moved to 116 Victoria Street on 31 May 1869.[8] Having occupied the premises

---

[8] At the beginning of 1889, 116 Victoria Street was renumbered 63 Victoria Street. On the issue of the *Daily Weather Report* for 4 January 1889, the address is given as 116 Victoria Street. On the issue for the following day, it is given as 63 Victoria Street. The block containing numbers 1 and 2 Parliament Street was demolished in the early 1870s.

at Parliament Street free of charge, the Committee were not best pleased that rent and other charges now had to be paid, but they had no alternative; the additional expense was unavoidable.

An Important New Publication

The move to new premises was not the only notable event of 1869. Balfour Stewart resigned as Secretary to the Meteorological Committee in October of that year, and the intention of the Office to issue its first official climatological publication, the *Quarterly Weather Report*, was announced that year.

Though Stewart resigned as Secretary, with effect from 31 March 1870, he continued to serve as Superintendent of Kew Observatory until 1871. His duties as Secretary to the Committee and inspector of outlying observatories devolved to Scott, but the Committee did not transfer any of his salary (£400 per annum) to Scott. Instead, they decided to offer the British Association (the body which controlled Kew Observatory) an annual allowance of £400, partly to defray the costs involved in examining the records from other observatories and partly to retain the services of Stewart's assistant, the mechanic and inventor Robert Beckley, who had designed, constructed and maintained instruments at Kew Observatory since 1854. The allowance, which was accepted by the British Association, was in addition to the £250 per annum that Kew already received from the Office for serving as the central observatory.

The main purpose of the *Quarterly Weather Report* was to publish the graphs produced by the self-recording instruments at the seven observatories, which was quite an undertaking, given the volume of material and technical difficulties involved. The use of lithography to reproduce the continuous autographic records turned out to be clumsy and expensive, so an alternative method of copying and combining the graphs by means of a special pantograph was tried.[9] This proved a great deal more successful than lithography and was, accordingly, adopted.

Publication of the *Quarterly Weather Report* began in August 1870 with the appearance of Part I of the report for 1869. This contained not only a detailed introduction and the autographic records of pressure, rainfall, temperature and wind that had been produced at the seven observatories in January, February and March 1869 but also simple weather maps, remarks on the weather and other information. Part II was published in December 1870.

In respect of content and quality, the publication lived up to its early promise, the only major reason for criticism being the lateness of some issues. Parts III and IV of the 1869 report, for example, were so delayed by production difficulties that they did not appear until well into 1871. The problems were overcome, however, and the volumes for 1870 and 1871 were published in 1871 and 1872, respectively.

---

[9] The Office had used lithography to produce the *Daily Weather Report* since 12 January 1869. The pantograph was designed for the purpose by Francis Galton.

Then delays began to mount, to such an extent that the last volume of the *Quarterly Weather Report* ever published, that for 1880, came out in June 1891, by which time two other Meteorological Office publications, the *Weekly Weather Report* and the *Monthly Weather Report*, had begun to prove more viable and popular.[10] The former was first published for the week ending Monday 11 February 1878, the latter in January 1884. The value of the *Weekly Weather Report* to the general public was quickly recognized, for decisions to insert it regularly in *The Times*, *The Daily News*, *The Mark Lane Express* and *The Gardener's Chronicle* were made after only two or three issues of the *Report* had been published.

Weather Charts for the Public

By the middle of the 1870s, weather charts had become available to the general public on a daily basis. They had appeared in the *Daily Weather Report* since 11 March 1872 and in *The Times* since 1 April 1875, while charts furnished by the Meteorological Office showing wind direction and force at 8 a.m. at the telegraphic reporting stations of the British Isles had been published in the *Shipping and Mercantile Gazette* since the beginning of January 1871.[11]

By the end of March 1872, weather charts were being forwarded regularly to Queen Victoria and supplied free of charge to the Royal United Service Institution, the Meteorological Society and Professor Balfour Stewart. Another notable development was the service provided by *The Times* since 1 January 1876: from telegrams reporting observations made at 6 p.m., a special map was drawn by the Office for publication in the morning edition of the newspaper. The expense of this enterprise, about £500 per annum, was borne by *The Times* for several years.

Francis Galton took an active part in the production of the weather charts which were published in *The Times*. From about 1860 onwards, he had developed pictorial methods of presenting weather information. Initially, his weather charts were primitive, because the amount of data available to him was limited and the techniques used for mapping weather were undeveloped. Nevertheless, he had been able, as early as 1863, to identify the anticyclone. From 1867 onwards, he had been able to step up his investigations, as he now had at his disposal all the data received by the Meteorological Office, including the records from the autographic instruments at the observatories. By the early 1870s, he was constructing weather maps very similar to those which appear in newspapers today. In some respects, his investigations complemented and extended FitzRoy's studies of weather systems and the studies of Dove and others. They especially complemented the studies of Alexander Buchan, who, in the 1860s, used weather

---

[10] Copies of the *Quarterly Weather Report*, *Weekly Weather Report* and *Monthly Weather Report* are held in the UK's National Meteorological Library and Archive at Exeter.

[11] In the UK's National Meteorological Library at Exeter, there is, on the open shelves, a full set of *Daily Weather Reports* from 1860 to the present day.

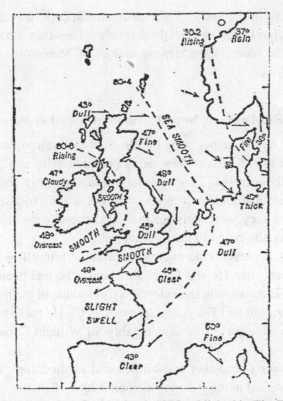

Figure 4.2. The weather chart for 31 March 1875 published in *The Times* on 1 April 1875. The dotted lines indicate the gradations of barometric pressure. Variations of temperature are marked by figures, state of the sea and sky by descriptive words, and direction of the wind by arrows – barbed and feathered according to its force.

charts he (Buchan) constructed to investigate the progress of storms across the North Atlantic.

Besides the investigations of weather systems carried out by Galton, Buchan and others, meteorological case studies were carried out by the Office, one of the most notable being that by Toynbee in connection with the loss of the *City of Boston*, a ship which left Nova Scotia on 28 January 1870 and disappeared. With the help of an assistant, Charles Harding, Toynbee collated a large amount of information from every available source to construct a series of charts of wind, weather and barometric pressure over the North Atlantic for the period of eleven days from 28 January to 8 February 1870, their main conclusion being that "the *City of Boston* must have experienced a very severe gale".[12] They also concluded, as had Charles Meldrum, the Government Observer of Mauritius, from his studies of weather systems in the

[12] The report by Toynbee and Harding was published in 1872 and called *A discussion of the meteorology of the part of the Atlantic lying north of 30°N, for the eleven days ending 8th February 1870, by means of synoptic charts, diagrams and extracts from logs, with remarks and conclusions* (Official Meteorological Office Publication No. 13, 64 pp., plus numerous charts and diagrams).

southern hemisphere in the early 1860s, that "atmospheric waves of high and low pressure follow each other on an easterly course".[13] Thus they anticipated by several decades some of the ideas of the Bergen School of Meteorology on extratropical weather systems.[14]

Owen Rowland – A Notable Amateur Weather Forecaster

By the early 1870s, a practice begun by FitzRoy had been reintroduced in the Meteorological Office. Weather forecasts were being prepared daily and compared with the weather that actually occurred. They were for internal use only, though, not intended for publication. Galton and his colleagues were not willing to issue forecasts to the general public until they were satisfied that predictions were reliable enough and sufficiently scientifically based to justify such a step.

One person was, however, bold enough to promote himself as a respectable and reliable weather forecaster. He was Owen Rowland, who had been, as he said in his prospectus, "Chief Electrician to the Telegraph Committee of the Board of Trade, and Editor of the *Electrician* and *Telegraph Journals*, &c". He had assisted some of the foremost telegraph engineers of the day, notably Sir William Cooke and Sir Charles Wheatstone.[15]

Rowland's weather forecasting capabilities and methodology were summarized thus in a piece published in the *Western Mail* on 7 June 1875:

An association has been formed for the purpose of supplying Mr Owen Rowland's forecasts of meteorological variations and changes of weather, according to the prospectus, from 24 to 72 hours in advance, thus affording timely notice of the approach of stormy squalls, thunder, heat, frost, fog, blight, dry, or wet weather, and electric or other disturbance, terrestrial or atmospheric. The system is the result of an invention several years since by Mr Owen Rowland, late electrician to the Board of Trade, of an instrument by means of which the slightest electric or atmospheric variation is indicated.

The instrument was a 'terrometer', and the association was called 'Owen Rowland's Weather Forecasts'. Its office was at 8 Queen Victoria Street, Mansion House, London, EC, and its observatory at Aled Villa, Torriano Avenue, Camden Road, London, NW.

In the early 1870s, Rowland offered to supply weather forecasts to the Board of Trade.[16] On 14 January 1871, for example, he wrote to Thomas Gray, the Permanent

---

[13] See 'Some early synoptic charts for the Indian Ocean' by S G Cornford (in *Colonial Observatories and Observations: Meteorology and Geophysics*, 1997, Occasional Publication No. 31, Department of Geography, University of Durham, edited by J M Kenworthy and J M Walker, pp. 177–212).

[14] The importance of these ideas is considered in later chapters.

[15] Wheatstone (1802–1875) was Professor of Experimental Physics at King's College London from 1834 until his death. He made many contributions to acoustics, optics and cable telegraphy and gave his name to the Wheatstone Bridge (which he used extensively for measuring electrical resistance but did not, in fact, invent). Cooke (1806–1879) was also a physicist and pioneer of the electric telegraph. He entered into partnership with Wheatstone in 1837, and together they invented and patented telegraphic devices.

[16] There is a file in the National Archives at Kew which contains correspondence between Rowland and the Board of Trade (file MT9/93/M13381/1874).

Assistant Secretary in the Board's Marine Department, saying that he was "prepared to give notice of storms, gales, thunder-storms, rainy, showery or fine weather, fog, mist, frost, snow, thaw, calm, breezy or gusty weather, the appearance of the Aurora Borealis, and the force and duration of each kind of weather". In doing this, he said, he needed to do no more than "read the movements of simple but unerring instruments". He wrote this letter "privately", saying that he would "soon address the Board on the subject", which he did, within a very few days. On 16 and 18 January 1871, he wrote again to Gray, wondering in his letter of the 18th what type of anemometer was used at coastal stations. He claimed that the wind speeds indicated by those instruments did not agree with the reports of his correspondents at ports.

Signs of irritation can be found in a minute dated 26 January 1871 from Gray to a junior clerk in the Board of Trade:

This gentleman who sends these papers has a scheme which he runs in opposition to the Meteorological Office. I have told him that we cannot recognize him in any way: but that if he chooses to send us reports we will receive them and put them away on the condition that he does not expect us to write to him.

A draft of a letter dated 9 August 1871 shows that Gray did, however, write to him again. In reply to a letter from Rowland dated 2 August, Gray informed him that his communication had been forwarded to the Director of the Meteorological Office and suggested that he send any further comments directly to the Office for attention, since the Board of Trade now had "no control over or opinion to offer on matters contained therewith". Rowland did not appear to know that management of the Meteorological Office had passed from the Board of Trade to the Royal Society at the beginning of 1867.

But it seems likely that he did know, for on 11 August he wrote to Gray again, saying that he was aware the Board of Trade was not responsible for the Meteorological Committee. However, he did not intend to address the Committee, for he knew "full well that it would be useless". The implication was that he had already approached the Office and been ignored or rebuffed.

This was not, though, the end of the story. Farrer wrote to Scott in September 1874, asking if he knew anything about Rowland. It appears that Rowland had sent the Board of Trade a forecast for 27 September. In Scott's absence, Toynbee had replied, on 30 September, informing Farrer that the Office had heard nothing of Rowland since his letter of 2 August 1871. Rowland continued to send the Board of Trade his forecasts, at least for a while, for there are copies of them in the National Archives for some days in October 1874, addressed to "The Secretary, Board of Trade, Whitehall". He was still practising as a forecaster in 1875, as shown by the piece in the *Western Mail* and a report in *The Aberdeen Journal* for 15 September 1875, but nothing is known of his activities thereafter, and he died in 1877.

The report in *The Aberdeen Journal* concerned a ship that had been lost and alleged that Rowland's Association had issued a warning of fog. As for Rowland's forecasts as a whole, the report concluded that they were "made upon the results of seven years' experiments" and had been "found to be correct on the leading points". Thus it appears that Rowland had begun his 'experiments' around 1868 and come to be respected by some as a weather forecaster, though not by the staff of the Meteorological Office, who were critical of his forecasts, not to mention his methodology, which depended on an instrument of doubtful reliability, his terrometer.

Trans-Atlantic Weather Warnings

The belief that storms travelled all the way across the Atlantic Ocean from eastern North America to western Europe formed the basis of the storm-warning service that was run by the Weather Bureau of the *New York Herald* in the late 1870s and early 1880s. The essential feature of the service was that telegrams were transmitted to the UK and France via the Atlantic cable to report the departure of weather systems from North America, the idea being that mariners leaving ports in the British Isles and other parts of western Europe would be forewarned of heavy weather they could expect to encounter on their voyages. The first warning was issued on 14 February 1877.

The idea of using weather reports from North America to give advance notice of weather on the other side of the Atlantic was not, in fact, a new one. Attempts had already been made to try it practically. From 1868 to 1871, the Office had received cablegrams daily from Newfoundland through the courtesy of the Anglo-American Telegraph Company and published them in the *Daily Weather Report*. However, these cabled reports had not proved useful for forecasting purposes, so the scheme had been discontinued.

The Weather Bureau of the *New York Herald* gathered information about the weather from stations on land in North and Central America and from ships arriving in the ports of eastern North America after voyages across the North Atlantic. Then, the Bureau's director, Jerome Collins, analysed the information and applied his knowledge of Atlantic storm tracks and other facets of marine meteorology to produce storm warnings which were issued to British and French ports. Originally an engineer, Collins was an enthusiastic meteorologist with much confidence in his own scientific ability, though some considered him a dilettante. The cost of running the Bureau, including the charges for sending cablegrams, was borne by the newspaper's proprietor, James Gordon Bennett Junior.

A survey published in the *Gentleman's Magazine* in 1879 focused on twenty-seven storm warnings issued by the *New York Herald* and rated seventeen "a perfect success", eight a "partial" success and the other two "misses". However, many meteorologists in Europe were not so sure the warnings were that reliable, among them Robert Scott, who performed his own analysis, which he published in the March 1878 issue of the *Nautical Magazine*. He carried out this work in his own time, not as part of his official duties, but may have consulted Toynbee, the Office's Marine Superintendent.

The Weather Bureau of the *New York Herald* was not recognized officially in the UK.

Scott's paper ran to twenty-eight pages. In it, he argued (correctly) that the forecasts issued by the *New York Herald* could not be reliable because information about the scale, evolution, movements and intensities of weather systems that were far from land was simply not available. He pointed out that information from mariners who had encountered bad weather over the Atlantic some days beforehand was of little value, except in very general terms during periods of unsettled weather. Many systems which were active when they left eastern North America dissipated over the Atlantic, and new systems formed over the ocean where none had been expected. Systems that were never suspected in North America could form over the Atlantic far from land and reach the British Isles as vigorous storms. Weather systems went through life cycles of growth to maturity and subsequent decay. His analysis showed that only 45% of the forecasts issued by the *New York Herald* were successful, not 80% as the newspaper's Weather Bureau had claimed.

Unsurprisingly, the *New York Herald* disagreed with Scott's conclusions and included a lengthy rebuttal of his criticisms in a thirty-four–page document which listed the storm warnings the Weather Bureau had cabled to Europe in 1877, 1878 and 1879.[17] The Bureau thanked Scott for transmitting to them daily weather bulletins issued by the Office but were otherwise critical of him, pouring scorn on his analysis.

An opportunity for Collins to explain to an international audience the methodology he used to produce his storm warnings arose at the International Congress of Meteorology which took place in Paris from 24 to 28 August 1878. This meeting was organized by the French Meteorological Society, and fourteen nations were represented.[18] Most of the leading meteorologists of France attended, but only two of the twenty-five meteorologists from outside France were well-known internationally: Buys Ballot, who represented the Dutch government, and George Symons, who represented the Meteorological Society. No one from the Office attended.

The paper Collins presented on the last afternoon of the meeting was published in full in the proceedings of the congress.[19] A summary of that afternoon's session was also published in the proceedings, in which it was noted that "Mr Collins, pressed for time, expounded his ideas rapidly in English, interpreted [in French] by Monsieur

---

[17] List of the storm warnings cabled to Europe by the New York Herald *Weather Bureau since the commencement of operations and the manner of their fulfilment – answers to critics. 1877-'78-'79* (New York, June 1879, 34 pp.). There is a copy of this document in the National Meteorological Archive at Exeter, and the Archive also possesses press cuttings from the *New York Herald* which explain and justify the methodology employed by the newspaper's Weather Bureau. As they were written for publication in the newspaper, they of course present the work of the bureau in a favourable light!

[18] The countries that were represented were France, the United States, the UK, Belgium, The Netherlands, Denmark, Norway, Sweden, Switzerland, Italy, Hungary, Spain, Austria and Alsace. Surprisingly, no one represented Russia or Germany, which were leading nations in meteorology in the 1870s, and no one represented Portugal, Romania or Greece.

[19] See Annexe No. 5 (pp. 109–117) in *Congrès international de météorologie tenu à Paris du 24 au 28 aout 1878* (Paris, Imprimerie Nationale No. 20 de la Série, 1879, 274 pp.).

Angot". And the rushed nature of his presentation is clear again in the next paragraph of the summary, in which it is stated that he gave his explanation of depression movements over the North Atlantic "hurriedly".

In the discussion of Collins's paper, to quote from the proceedings of the congress:

Mr Buys Ballot, while commending the efforts of Mr Collins, pointed out that, according to a statistical study made by Mr Robert Scott, fewer than half of the forecasts issued by the *New York Herald* had proved correct. Though being more favourable to the ideas of Mr Collins than Mr Scott, he (Buys Ballot) considered that some forecasts were unfortunate for maritime commerce in that departures on some voyages had been delayed unnecessarily because of the *New York Herald*'s forecasts. In his opinion, the forecasts of the *New York Herald* should be used with great caution. Mr Collins then presented to Mr Buys Ballot a statistical analysis which he said was more complete than Mr Scott's and showed that the forecasts were more reliable than Scott thought.

Collins's career as director of the Weather Bureau was sadly soon to end, for he perished in the Arctic in 1881, having been sent there in 1879 by Gordon Bennett, a man given to investing funds in newsworthy ventures. He financed the *Jeannette* expedition which sailed from San Francisco in July 1879 with Collins aboard as the expedition's correspondent and meteorologist, the purpose of the expedition being to reach the North Pole via the Bering Strait. However, the *Jeanette* became trapped in ice on the Chukchi Sea in September 1879 and then drifted for twenty-one months before it sank in June 1881.[20] Collins and his shipmates reached the Lena River Delta, but he died of starvation and exposure in October 1881, along with a dozen of his companions. This was not, though, the end of the *New York Herald* story.

The President of the Meteorological Society from January 1880 to January 1882 was George Symons, and in the Presidential Address he delivered on 18 January 1882 he spoke thus of the storm-warning service provided by the *New York Herald*:[21]

The general public, and not a few meteorologists, attach considerable importance to the storm warnings issued by the Weather Bureau of the *New York Herald*. On the other hand, two of my colleagues on the Council [of the Meteorological Society], who have studied the subject independently, have become convinced that they are useless. Whatever may be the ultimate verdict, there is one thing that I should like to see rectified. I do not think that Mr Gordon Bennett has received that recognition which his costly and difficult work deserves. Some years since I had the pleasure of spending several days with the first director of the *New York Herald* Weather Bureau, Mr J J Collins, and I was astonished at the knowledge which he possessed of Atlantic storm tracks, and of the causes which modify them.

Collins had clearly impressed Symons when they met in Paris in 1878.

---

[20] See Guttridge, L F (1987), *Icebound: the* Jeannette *Expedition's Quest for the North Pole* (Airlife Publishing Ltd, Shrewsbury, England, 357 pp.).

[21] See Symons, G J (1882), 'On the present state and future prospects of meteorology', *Quarterly Journal of the Meteorological Society*, Vol. 8, pp. 101–110. The Society became the *Royal* Meteorological Society in 1883.

We do not know what influence, if any, Symons may have had on the meteorological community in the UK in respect of the storm-warning service operated by the *New York Herald*, but the idea of cabling information about weather systems across the Atlantic was clearly still alive in 1884. The Council which now controlled the Meteorological Office heard on 11 June 1884 that the Director of the French Bureau Central Météorologique, Éleuthière Mascart, "was anxious, if possible, to obtain from the United States telegraphic reports of the weather on the coast and over the Atlantic".[22] Scott said that Mascart had asked whether the Council would be willing to cooperate in such an undertaking. He (Scott) had responded to Mascart's enquiry by approaching officials of the Chief Signal Office of the United States and the Hydrographic Office in Washington and been assured these bodies would cooperate. The Council resolved "that Mr Scott be instructed to ascertain the nature of information obtainable from the eastern coast of the United States from arriving ships and from land stations, and the probable cost of its telegraphic transmission to England in a codified form".

These so-called 'American telegrams' were discussed at a meeting of a working group held at the Meteorological Office on 5 September 1884, and the group decided, as it was put in the minutes of the meeting, that they did not consider they were at that time in a position to "speak decidedly on the value of information telegraphed from America" but thought the idea seemed "worthy of a trial". Attention was drawn at the meeting to a paper by the Office's Charles Harding published in the January 1884 issue of the *Quarterly Journal of the Royal Meteorological Society* (Vol. 10, pp. 7–25), in which he had discussed a violent storm that had crossed the British Isles at the beginning of September 1883. In the paper, he had provided details of the storm's progress across the North Atlantic as derived from ships' logs and, in drawing the charts for his paper, he had used observations from 248 ships. This was a tour de force of meteorological analysis and had shown that the warning of the storm for the British Isles that had been issued by the *New York Herald*, though broadly accurate, could have been more precise and issued earlier had more information from the ships that had encountered the storm been taken into account by the Bureau.

For Harding's analysis, there had been more logs available than usual because of a special observational programme carried out by the Office. Owners, captains and officers of steamships and sailing vessels on the North Atlantic had been asked to record meteorological observations on special forms and return the forms to the Office. Using the observations, weather charts of the North Atlantic had been prepared daily for the thirteen months beginning 1 August 1882 and ending 31 August 1883.

---

[22] Control of the Office passed from the Meteorological Committee of the Royal Society to a Meteorological Council in 1877 (see Chapter 5).

In a statement about the programme issued in April 1882,[23] the Meteorological Council said they believed any systematic information which could be obtained regarding the origin, development and laws of motion of atmospheric disturbances which occurred over the Atlantic Ocean would "promote the science of meteorology and be of immediate practical utility". The information, they said, "could not fail to be a benefit to seamen traversing the Atlantic Ocean, and would tend directly to the improvement of the forecasts and storm warnings issued to the British coasts". Reference was made in the statement to the efforts which had been made by the proprietors of the *New York Herald*. However, the Council clearly believed that the newspaper's storm-warning system would benefit from a greater number of weather reports from arriving ships and from a greater understanding of the behaviour of weather systems over the Atlantic Ocean.

We can see from the minutes of the Council's meeting on 2 June 1886 that Scott had doubts about the cablegram service that had been instituted in 1884. "I must say", he reported, "that we are not disposed to value the present ship service very highly". And in a letter to Mascart dated 20 May 1886, reproduced in the minutes, Scott said that "analyses carried out by a member of his staff, Frederic Gaster, had shown the value of the service very dubious".

Nevertheless, the service continued into 1888. Then, as recorded in the minutes of the meeting of the Meteorological Council on 25 January 1888, Scott was instructed to inform Mascart that unless he could "satisfy the Council more completely of the practical utility of the Atlantic telegrams, the Council would consider that it would not be possible to continue to share the expense after April 1st next". In the minutes of the meeting on 11 April 1888, it was recorded that the Council had agreed to pay only the cost of postage of copies of the Atlantic telegrams from Paris to London, and it is clear from the minutes of subsequent meetings that the use of the telegrams was quietly dropped. It seems British politeness had dictated that no one wanted to tell Mascart officially that the telegrams were not useful. He was, after all, Director of the French state meteorological service, and the initiative over obtaining weather information from America by means of the Atlantic cable had been his.

A Mid-Atlantic Observatory – An Idea before Its Time

It occurred to a number of people in America and Europe that a solution to the problem of there being a lack of real-time weather observations from places far from land was to moor ships near the Atlantic cable and attach them to the cable. Two articles published in *Symons's Monthly Meteorological Magazine* in 1885 focused

---

[23] The statement was published in *The Times* newspaper on 14 April 1882 and reproduced verbatim in the May 1882 issue of *Symons's Monthly Meteorological Magazine* (Vol. 17, pp. 49–53).

on this idea.[24] We can assume the author was Symons, for he referred in favourable terms to Gordon Bennett and "his clever meteorological assistant, Mr Collins". No other British meteorologist ever spoke approvingly of the storm-warning service run by the *New York Herald*.

Symons suggested there should be a floating meteorological observatory near 50°N 20°W but acknowledged there were difficulties over this, saying that he was aware the depth of the ocean above the spot in question was about 2000 fathoms and "a mooring chain of that length would be a novelty". However, he pointed out, "broken telegraph cables have been picked up at even greater depths and raised to the surface". He felt sure there was little chance of securing the necessary funding for a floating observatory that was solely for meteorological purposes, so he considered that it "must also be a call station for passing vessels and for those in distress".

The idea of a floating observatory was not a new one. An attempt had been made in the 1860s to establish what Symons said was "a 'call station' for merchantmen entering the Channel". A vessel which had telegraphic connection with the mainland had been placed south of the Lizard, and the crew of the vessel had been asked to make weather observations. However, as Symons noted, "the vessel was not constructed specially for the work, and the connection was so often broken that eventually the scheme was abandoned".

Symons quoted in the first of his two articles from a piece that had been published in the March 1885 issue of the *American Meteorological Journal*. Entitled 'Meteorological Stations in the Atlantic' and written by F S Coburn of North Carolina, it had led Symons to draw attention to the idea of mid-Atlantic observatories in his *Meteorological Magazine*. In the words of Coburn:

The practicability of such stations has been demonstrated by the cable-laying steamship *Faraday*, while laying the cable between France and the United States, and, holding on to the cable, found herself in the course of a cyclone, which passed directly over the vessel without causing her to lose her hold of the cable, and which was at once reported to the European continent, the report giving wind changes and velocity with barometric changes. It is also demonstrated by the light-ship off Frying Pan Shoals, North Carolina, which is anchored about twenty miles from the shore, where it is exposed to the severe storms that occur off the Carolina coast.

Symons considered that a floating observatory with continuously recording instruments could be stationed on the North Atlantic, and he wondered how many years the weather forecasters of Europe would have to "rely upon conjecture, owing to the non-establishment of such an observatory as above suggested".

The Meteorological Office showed no interest in the idea of floating observatories. Perhaps the cost was considered excessive. Perhaps doubts over the existing cablegram

---

[24] 'A floating mid-Atlantic meteorological observatory', in *Symons's Monthly Meteorological Magazine*, Vol. 20, pp. 33–35 and 52–54.

service helped shape the Meteorological Council's attitude. We do not know. There would indeed be weather stations on the oceans eventually, but in the 1880s the idea of them was ahead of its time.

Resumption of Weather Forecasts for the Public

Meanwhile, Francis Galton had satisfied himself that weather forecasts prepared in the Office had become reliable enough and sufficiently scientific to be made available to the public again. On 1 April 1879, the Office published forecasts for the public for the first time since May 1866. In the words of R G K Lempfert, published in 1913, when he was Superintendent of the Office's Forecast Division:[25]

April 1, 1879 witnessed the revival of Admiral FitzRoy's scheme for issuing anticipations of the weather of the immediate future. After twelve years' study of the daily weather charts it was felt that the attempt might again be made. It was made a rule of the Office, which is still observed, that a reason should be given for every forecast issued, that is to say, that the forecaster should set down the general nature of the changes anticipated during the period the forecast covers, and this 'general inference' is published regularly in the *Daily Weather Report*. As regards the period covered by the forecasts, our aims have remained to this day much more modest than those of FitzRoy. Until quite recently, the period was limited in all cases to 24 hours. In 1879, forecasts were prepared three times a day; those issued in the morning and afternoon were placed at the disposal of all newspapers free of charge, but the evening ones were circulated only to *The Times*, *The Standard* and *The Daily News*, the three newspapers which at that time shared the expense of the evening service originated in 1876 by *The Times*. In December 1880, the parliamentary grant to the Office was increased so as to enable the Office funds to bear the cost of the evening service, and from that date the evening forecasts have also been placed at the disposal of all newspapers without charge.

Soon after the regular issue of weather forecasts to the public recommenced, the Office announced its readiness to answer, by telegraph, enquiries regarding the weather likely to occur up to twenty-four hours ahead. The charge for each enquiry was set at three shillings, made up of two shillings for the message and reply, plus one shilling for "the trouble to the postal authorities and the Meteorological Office" (to quote the words used in the *Report of the Meteorological Council for the Year Ending the 31st of March 1880*). According to this *Report*, the number of enquiries received through the Post Office during the year was 340 and the number of personal enquiries at the Meteorological Office during the same period 78. That suggested, the *Report* said, that, "although the public appears to take an interest in the weather information supplied by the newspapers, there is not at present any considerable demand for private information".

---

[25] Taken from 'British weather forecasts: past and present' (*Quarterly Journal of the Royal Meteorological Society*, 1913, Vol. 39, pp. 173–184).

Hay-harvest forecasts also commenced in 1879, in collaboration with the Royal Agricultural Society, the Royal Dublin Society and the Highland Society. That year, weather forecasts were sent gratis during the hay season to about thirty observers selected by the Councils of these Societies, on condition that recipients made the forecasts known as widely as possible and, in addition, assessed and recorded the accuracy of each forecast. The scheme was considered satisfactory, with 76% of the forecasts rated completely or partially successful, as calculated from the reports of observers. Accordingly, the scheme was repeated in the summer of 1880 and in the summers of many years thereafter.

5

Squalls and Settled Spells

With the resumption of a storm-warning service more or less as FitzRoy conceived it and the reintroduction of daily weather forecasts for the public, the fight over forecasts was won. Never again in peacetime would the Meteorological Office withdraw these services. The fight over forecasts was not, however, the only important development in the Office in the years after the Royal Society took charge. A threat to Kew Observatory had to be addressed, too.

Changes in the Role of Kew Observatory

For as long as the British Association had managed Kew Observatory (since 1842), the cost of sponsoring and maintaining the Observatory's work had proved a drain on the Association's finances. The Association's Council decided in 1869 that the time had come to discontinue the Observatory's financial support. Money would be provided for a further two years, and the connection between the Association and Kew would then cease.

The Royal Society's Council discussed this matter at their meeting on 19 January 1871 and considered taking over the Observatory, but hesitated over the costs involved. Then, J P Gassiot came to the rescue. A letter from him was read at the Society's Council meeting on 16 March 1871. He proposed to give to the Society securities amounting to £10,000, on trust, for continuation of the Observatory's work, on condition that both the Observatory and the income of the Trust Fund were to be entirely under the control and management of a Committee appointed by the Society's Council. He specified that those who served on this Committee should receive no remuneration for so doing.

The offer was accepted and possession of the Observatory formally passed to the Royal Society at the meeting of the British Association held in August 1871. The Society's Council decided that its Kew Committee should consist of the members of its existing Meteorological Committee, though the functions of the two committees would remain separate. Thus, on 29 June 1871, the following were charged with managing Kew Observatory, with effect from 2 August 1871: Edward Sabine, Francis Galton, Gassiot himself, Warren De La Rue, William Spottiswoode, Admiral Richards, Colonel Smythe and Sir Charles Wheatstone.[1]

The Meteorological Committee resolved on 3 July 1871 that the Kew Committee be granted the sum of £650 per annum, £250 of it for the maintenance of Kew Observatory, the rest for scientific work. We do not know the response of the Kew Committee to this resolution, but we may assume they raised no objections! It seems, though, that the Meteorological Committee realized eyebrows might be raised over the dual-membership arrangement, for in their report for the year ending 31 December 1871 they felt it necessary to state that the management of the Observatory and the Office were totally independent of each other!

Samuel Jeffery became Superintendent of Kew Observatory on 1 August 1871, succeeding Balfour Stewart.[2] To some, he proved a disappointment, particularly when compared with Stewart and his predecessors, namely Francis Ronalds, who was Honorary Superintendent from 1843 to 1852, and John Welsh, the first paid Superintendent (1852–1859).[3] When reviewing the *Report of the Kew Committee for the Year Ending October 31st, 1875*, the editors of *Symons's Monthly Meteorological Magazine* (May 1876, Vol. 11, p. 60) said that they could not "remember ever seeing the name of the superintendent attached to any scientific communication whatever".

George Mathews Whipple succeeded Jeffery in February 1876. He had spent almost all his life in the Observatory, having been there since 1858.[4] And since Balfour Stewart's resignation, he had borne much of the responsibility of the management. Symons approved of Whipple's appointment and noted in the May 1876 issue of his

---

[1] Wheatstone was involved in the work of Kew Observatory from the early days of the British Association's period of control. In connection with the Observatory, he designed meteorological instruments which could record their observations or transmit them over telegraph wires. Also, along with Gassiot, he had become a member of the Kew Committee in 1849 and, like Gassiot, remained a member in 1871. He joined the Meteorological Committee in April 1871, replacing W A Miller (who had died on 30 September 1870).

[2] Jeffery had been a Senior Clerk in the Meteorological Office since 25 January 1869, paid £3 a week to help with the task of reducing the observations made at the seven observatories. He was no stranger to observatory work, having been employed at the Rossbank Observatory in Tasmania from 1842 to the end of 1854, as Superintendent for the last twenty months. Thereafter, he had been unemployed and impecunious for many years. Sir William Denison, the Lieutenant-Governor of Tasmania, considered him a most impractical man and remarked that "somehow or other Mr Jeffery always managed to disagree with those above him"!

[3] Ronalds was the inventor of the electric telegraph and a keen meteorologist. In 1847, with W R Birt, he devised a method of maintaining a kite at a constant height for purposes of meteorological observation, and whilst at Kew he devised a system of continuous automatic registration for meteorological and other instruments by means of photography. Welsh entered the University of Edinburgh in 1839 with a view to becoming a civil engineer. There, he studied under James David Forbes, author of the reports on meteorology that were presented at the meetings of the British Association held in 1832 and 1840 (see Chapter 1). In 1850, he was appointed assistant to Ronalds at Kew; and in 1852 he made four balloon ascents from Vauxhall Gardens, London, one of them to a height of 22,930 feet.

[4] Whipple became Magnetic Assistant in 1862 and Chief Assistant in November 1863. He helped improve Kew's magnetic instruments, invented optical apparatus, made a series of pendulum experiments for ascertaining the constant of gravitation and undertook studies of wind pressure and speed, including, at the Crystal Palace in 1874, an investigation of the cup anemometer invented by Robinson. He took a University of London Bachelor of Science degree in 1871, after a period of study at King's College.

Monthly Meteorological Magazine that he had earlier "complained of the anonymity of the Kew Committee". "We do so still", he said, commenting that he had no idea who the Committee were, or whether they would allow the new Director to do more than carry out their orders.

Though apparently aimed at the committee in general, these comments may have been a veiled attack on Sabine in particular, who had not attended a meeting of the Meteorological Committee since 7 December 1874. Though kindly, considerate, conscientious, zealous, efficient and dignified, he was also obdurate and scientifically conservative. Symons possibly thought it time for the elder statesman of science (now in his late 80s) to step down as chairman of the Kew and Meteorological Committees. If so, he was unlikely to have known that Sabine would soon do this. He ceased to be chairman of these committees at the end of June 1877, though not because of any comments published in *Symons's Monthly Meteorological Magazine*. He stepped down when a change in management of the Office occurred.

Thanks to Gassiot, Kew Observatory did not have to close, but its income from the endowment was only £250 per annum and its financial position far from secure. To increase income, calibration work was undertaken on a much greater scale than hitherto. Testing of meteorological instruments had taken place at the Observatory since the 1840s, and fees had been charged for carrying out this work. With the expansion of the calibration work in the early 1870s came the testing of other types of instrument, including clinical thermometers, watches, telescopes, binoculars and sextants. The Observatory soon became self-supporting.

A procedural change occurred in 1876, when the Kew and Meteorological Committees decided to transfer the work of examining the records of the various meteorological observatories from Kew to the Office's London headquarters. The reason for this change, as stated in the annual report of the Office published in 1877, was that the work could be "most efficiently and economically performed in connexion with that of the preparation of the plates for the *Quarterly Weather Report*". The financial consequence was that, from November 1876, Kew ceased to receive an allowance of £400 per annum for the examination of records. Instead, as it was put in the 1877 report, "a special sum of £100 a year" was allotted to Kew Observatory to enable it to retain the services of an extra assistant who would be available "to take temporary charge of the other observatories in case of absence of their staff from illness or other causes".

An Unsuccessful Proposal from the Astronomer-Royal

The Astronomer-Royal, Sir George Airy, succeeded Sabine as President of the Royal Society in 1871 and seems to have remained as sensitive over the position of the Royal Observatory in British meteorology as he had been in 1865 and 1866, when there had been, as mentioned in Chapter 3, an exchange of correspondence between him and the Society over the relative merits of Kew Observatory and the Meteorological Department of the Observatory at Greenwich.[5]

At their meeting on 29 January 1872, the Meteorological Committee considered remarks which had been submitted by Airy to the Council of the Royal Society eleven days earlier. Airy had pointed out that a grant was paid to Kew Observatory each year and commented that:

In estimating the propriety of this grant, it is to be remarked that within a few miles of Kew is the Meteorological Department of the Royal Observatory of Greenwich, instituted by Her Majesty's Government several years before the establishment of Kew, maintained by the Government at considerable expense, more complete in its equipment than the Kew Observatory, at least equal to it in the excellence of its instruments, and under the most careful daily superintendence, and perfectly able to furnish to the Meteorological Committee, at insignificant expense, all that is now furnished with an annual expense of £250 to the Government.

He did not think it could be "considered right still to load the Government with this unnecessary expense" and accordingly proposed to the Council that "steps be taken immediately for the transference, to the Royal Observatory, of the observations in conjunction with the Meteorological Committee now taken at Kew". As Superintendent of the Royal Observatory, he could guarantee that everything practicable would be done "to render its observations available to the Meteorological Committee". He concluded by stating that his remarks did not "in any degree apply to other expenses incurred at Kew for the Meteorological Committee", to which, on general grounds, he saw no objection.

The reactions of the Committee to Airy's proposal may be sensed from the minutes of their meeting on 29 January 1872, in which it is recorded that, "after the consideration of various drafts proposed by the several members of the Committee, Mr Scott was directed to prepare a reply in accordance with the general opinions expressed, and to forward it to Mr Airy"! Judging by the terseness of the letter he sent to Airy on 2 February, some indignation had been expressed at the Committee's meeting, which was only to be expected, given that several of those present had been personally involved in the work at Kew over many years and the others would have been inclined to show loyalty to Kew by virtue of their dual membership of the Meteorological and Kew Committees.

Scott informed Airy that the Meteorological Committee did not consider it their province to enter into a discussion with the head of another observatory. They would be willing, though, to advise the Board of Trade, should they be consulted about the matter. In the opinion of the Committee, it was "an essential part of any system, such as theirs, that the normal observatory should be controlled by an independent unpaid

---

[5] Airy was Astronomer-Royal from 1835 to 1881. He was President of the Royal Society for only two years, resigning because he felt he could not devote sufficient time to the Society's affairs.

governing body". The Committee should not become associated with an official paid department of the government over which they had no power.

In his reply, dated 8 February and considered by the Committee at their meeting on 12 February, Airy assured the Committee that he did not wish to interfere with the proceedings of the Meteorological Office. He considered the Committee's opinion "on the necessity of retaining Kew as an active observatory worthy of the most respectful attention", however much it differed from his own. He asked if he might be allowed to read Scott's letter of 2 February at the meeting of the Royal Society's Council on 15 February, and Scott was instructed to reply that the Committee saw no objection.

Airy did not, thereafter, pursue the matter with the Committee. Instead, he attempted to bypass them and put his case directly to the Board of Trade. At the meeting of the Royal Society's Council held on 11 April 1872, he gave notice that he would, at the Council's next meeting, move three resolutions:

1. That it rested within the competency of the President and Council to investigate the necessity of retaining Kew Observatory as the central meteorological station, and to make such representation thereon to the Board of Trade as they may judge expedient.
2. That, in the opinion of the President and Council, it was desirable that charge of the meteorological observations made at Kew be transferred to the Royal Observatory.
3. That a copy of the last resolution be transmitted to the Board of Trade.

He duly put these resolutions to the Council of the Royal Society at their meeting on 18 April but failed to find anyone to second the first resolution and accordingly deemed it inexpedient to move Resolutions 2 and 3!

What machinations there had been behind the scenes, if any, we may never know. Only three of the eighteen present at the meeting were members of the Meteorological Committee (Galton, Spottiswoode and Wheatstone), so there was no question of the meeting being dominated by those who controlled Kew Observatory. Regarding how different the course of the Office's history would have been had Kew's observational work passed to Greenwich, we can only speculate! The Superintendent of the Royal Observatory's Magnetic and Meteorological Department was, of course, James Glaisher. What part he played in Airy's approach to the Meteorological Committee and the Royal Society's Council, we do not know, but Airy would surely have consulted him.

London and Scotland Disagree

Another dispute which occurred in the early 1870s may at the time have seemed petty and little more than a matter of honour. In fact, though, it turned out to have significant consequences for meteorology in Scotland in the longer term. The nub of the problem was that the Meteorological Committee had persuaded one of the Scottish Meteorological Society's most experienced observers to make observations for the Committee. In the view of the Society, this breached an agreement under which the

Committee undertook not to communicate with their observers except through the Council of the Society.

The Committee believed they had right on their side, because, on 9 November 1872, Scott had written to the Secretary of the Society to inform him that he had "received instructions from the Meteorological Committee to establish a telegraphic reporting station at Stornoway" and, furthermore, that the "local authorities" had submitted to the Committee the name of John Smith, the gardener at Lews Castle, as "a probable observer for that station". According to Scott, Mr Smith had given a written assurance "that his new duties would not in any way interfere with his relations to the Society" and had stated that his employers were "well pleased that he should undertake the reports". The Council of the Society considered the appointment of Mr Smith poaching, but the Committee did not accept that anything unreasonable had occurred.

The Council of the Society considered the matter at their meeting on 20 January 1873 and forwarded to the Committee a long excerpt from the minutes of that meeting, whereupon there ensued an exchange of correspondence between the Council and Committee in which the positions of the two sides became polarized. By March 1873, what had begun as a minor difficulty had greatly escalated.

There is nothing in the minutes of the Committee's meetings to indicate that the Board of Trade showed any immediate interest in the matter. On 20 March 1873, however, Thomas Gray, the Permanent Assistant Secretary in the Board's Marine Department, wrote to Scott, enclosing a letter dated 12 March 1873 which he had received from the Marquess of Tweeddale, the President of the Scottish Meteorological Society. Gray asked for the Committee's observations on the letter in order that the Board might consider them before deciding whether or not to receive a deputation from the Society.

The letter began with an expression of regret that the Society had been "compelled to trouble the Board of Trade with their affairs" and went on to say that the proceedings of the Meteorological Committee had left them no alternative. The Marquess reminded the Board that the Society derived its income "wholly from the voluntary contributions of persons interested in science". Moreover, it received "no pecuniary aid from Government", even though the "results of their labours were adopted by Government and regularly published as an integral part of the Registrar-General's Reports for Scotland". Indeed, the Society had to pay the Government rent for the rooms they occupied in Edinburgh. This being so, the Society had applied to the Board of Trade on 5 March 1870 for "a small grant to assist them in carrying on and extending their numerous investigations". The application, he admitted, had proved unsuccessful.[6]

---

[6] This was not the first application by the Scottish Meteorological Society for a share of the annual government grant voted for meteorological purposes. It was reported in a newspaper published on 7 May 1869 that a request from the Society in early 1869 had been refused by the Chancellor of the Exchequer. A cutting containing this report is held in the National Meteorological Archive at Exeter, but the newspaper in question has not been identified.

The Marquess then turned to the matter of the disputed agreement, which, he advised, was a written one, entered into in May 1870 and signed by the then Chairman of the Meteorological Committee (Sabine). The Committee had agreed, he said, that, "with four specific exceptions, they were not to communicate directly with the Society's observers and that all communications were to be conducted exclusively through the Society's Council". In the Council's view, this agreement had been violated by the Committee, and he provided documentary evidence to support the Council's claims. The documents showed, he said, that the agreement "was subsequently acted upon", yet the Committee now asserted, "in justification of their recent proceedings, that no such agreement was ever confirmed".

The situation was clear. The Meteorological Committee had broken a written agreement, and the Council felt that they now had no option but to "request the assistance and advice of the Board of Trade". The Council felt they had been unfairly treated, in that a portion of the Government grant to the Office was now being employed in obstructing and undermining the Society by offering payment to their necessarily unpaid observers. The Council were at a loss to understand why Government money should be employed to counteract the efforts of individual societies.

In his reply to Gray, Scott informed the Board of Trade that, "in establishing a telegraphic reporting station at Stornoway, nothing could have been further from their desire than to interfere with or cause any inconvenience to the Scottish Meteorological Society". The observations from Stornoway were, he remarked, "not only different from those supplied by the observer at that station to the Scottish Meteorological Society but of such a character that they might readily be supplied by any observer to a number of societies or individuals without prejudice to any of them". Moreover, he pointed out, "the entire special outfit of the station had been supplied direct from London". As the Council of the Society did not want their observers to be engaged without the Council's prior agreement, however, the Committee would in future "most readily conform to this wish on the part of the Society". The Committee, Scott added, were "desirous to continue their friendly relations with the Society in the interests of science which they cultivated in common".

The Board of Trade appears to have considered the whole affair a trivial matter which had been allowed to grow out of proportion. They showed no inclination to become involved, their hope being, as expressed in the letter from Gray to Scott which was read at the meeting of the Meteorological Committee held on 5 May 1873, that the Committee and the Society, notwithstanding what had occurred, would be "able to agree to some common plan whereby they could work in harmony in future".

For the time being, the matter rested there. The President and Council of the Scottish Meteorological Society had made their point, and the Meteorological Committee had promised they would not poach the Society's observers in future. The Council continued to press for government support, however, and achieved some success in 1877, when two awards were made: a grant of £1000 for services rendered to the

Registrar-General for Scotland during the previous twenty years and an allowance of £150 per annum to the Society's secretary to help fund the work of inspecting the stations from which vital statistics were compiled.

The Birth of the International Meteorological Organization

Though the Meteorological Committee's exchanges with Airy and the Scottish Meteorological Society were not insignificant in the history of the Meteorological Office, they took up time which the Committee could have better used for considering other matters, such as the developments in international meteorology which unfolded in 1872 and 1873.

Since the Brussels Conference (see Chapter 1), meteorologists and other scientists had come to recognize the need to develop trans-national networks of meteorological stations and to standardize techniques of making, recording, analysing and disseminating meteorological observations. Accordingly, they had attempted a number of times to organize conferences which might bring about formal cooperation between national meteorological services, but all had come to naught for one reason or another, the unstable political situation across Europe in the 1860s being one of them. Eventually, an attempt proved successful, thanks to the efforts of Heinrich Wild, Carl Bruhns and Karl Jelinek,[7] who were backed by the governments of Russia, Germany and Austria, respectively. The outcome was a conference which took place at Leipzig from 14 to 16 August 1872. Scott and Buchan were among the fifty-two who attended, representing, respectively, the Meteorological Office and the Scottish Meteorological Society. A number of German scientists attended, too, but no one from France took part, the reason possibly being that international relations remained delicate after the Franco-Prussian War of 1870–1871. The only participant from outside Europe was an American, Dr E H Sell of New York.

Another who attended was Buys Ballot, and he played an important part, for a paper he published in January 1872, *Suggestions on a Uniform System of Meteorological Observations*, provided one of the incentives for convening the conference. Indeed, the letter of invitation which was sent to the world's meteorologists encouraged delegates to peruse the paper before attending the conference. The letter contained a list of twenty-six questions for discussion relating to the most pressing meteorological problems of the day, the first of them concerning a matter that has still not been fully settled, the need to introduce the metric system into meteorology (which Scott opposed vehemently). Given the pioneering, preparatory and consultative nature of the conference, it comes as no surprise that many differences of opinion were revealed

---

[7] Wild was a physicist, climatologist, inventor of meteorological instruments and director of the Central Geophysical Observatory at St Petersburg. Bruhns was an astronomer and geodesist, a professor at Leipzig University and director of the Royal Saxon Meteorological Institut. Jelinek was a mathematician and the director of Austria's Central Institute for Meteorology and Terrestrial Magnetism.

at it, particularly with respect to the best instruments to use, the correct ways to mount them, the most appropriate times to make observations and the most suitable units to adopt for measurement.

Most of the world's foremost meteorologists attended the conference, and they reached broad agreement on standardized methods of observation and analysis. Moreover, they prepared the way for holding an international meteorological congress at which, it was proposed, delegates would take steps to establish a permanent body to deal with meteorological problems that were common to the international community.

The Congress was held in Vienna from 2 to 16 September 1873, attended by thirty-two representatives of twenty governments, among them Scott and Buchan but, again, no one from France.[8] There was also no one present in a personal capacity because of the decision made at Leipzig that participation in the Congress would be limited to representatives of governments. The Congress acknowledged this was an error and expressed regret that some distinguished meteorologists had thus been excluded. Like the British delegates to the Brussels Conference, Scott and Buchan were given strict instructions that they must "abstain from pledging Her Majesty's Government in any way". In the event, however, no decisions involving expenditure were taken.

The agenda of the Vienna Congress, as agreed at Leipzig, consisted mainly of items relating to practical matters, such as the calibration and checking of instruments, scales and units to be adopted, and times of observations. There was a lack of unanimity on a few matters, such as the use of the metric system, but agreement was reached on many, including definitions of certain meteorological phenomena, a classification of weather stations and a list of the symbols to be used to denote precipitation types on weather charts and in climatological tables.[9] The idea of establishing an 'international meteorological institute' for collecting and publishing observations for climatological purposes was dropped, as the establishment of one was not considered feasible, mainly because of a lack of funds to create and maintain it.

As proposed at Leipzig, steps were taken at Vienna to establish a permanent body which would deal with meteorological problems common to the international community. A committee was formed. Called the Permanent Meteorological Committee, it initially had seven members, among them Buys Ballot (President) and Scott (Secretary), but the number increased to eight in 1878, when a representative of France was appointed. The Committee's main tasks were to:

• Publish the proceedings of the Congress and communicate them through diplomatic channels to the governments represented at the Congress and to the French government

[8] Scott represented not only the British government but also the Meteorological Society. Buchan represented both the Scottish Meteorological Society and the British government, but the government refused to pay his expenses! The Council of the Scottish Meteorological Society raised the necessary funds privately from the Society's own members.

[9] The Vienna Congress recommended the continued use of Luke Howard's cloud classification, and the First Conference of Directors of Meteorological Services (at Munich in 1891) adopted a developed form of it.

Figure 5.1. Second-order weather station at The Hollies, Hastings, 12 September 1884. Photograph in the National Meteorological Archive reproduced by kind permission of the Royal Meteorological Society.

- Ensure that the decisions of the Congress were implemented
- Prepare the agenda of the next Congress
- Draft and approve the statutes and rules of an international meteorological organization

The inaugural meeting of the Committee was held in Vienna on 16 September 1873, immediately after the Congress closed, and further meetings were held in September 1874 (in Utrecht), April 1876 (in London) and October 1878 (in Utrecht again).

At the meeting in 1874, the Committee resolved that a weather station producing continuous automatic records or making hourly readings be designated a 'first-order station' and a station observing at least twice a day and making certain supplementary observations be designated a 'second-order station'. From Great Britain, observations from fifteen second-order stations were requested, which proved an embarrassment for the Meteorological Office because their stations did not, for the most part, meet the specifications for this type of station. The Office did not run a network of climatological stations. In England and Wales, this was left to the Meteorological (from 1883 Royal Meteorological) Society, while in Scotland it was left to the Scottish Meteorological Society. There followed negotiations with the Meteorological Society and, subsequently, with the Scottish Meteorological Society, as a result of which observations from stations maintained by these societies were used. In recognition of the cooperation, both societies were invited to send representatives to the Congress held at Rome in 1879.

The rules and statutes of the 'International Meteorological Organization' (IMO) were drafted at Utrecht in 1878, and the Organization formally came into being at the Rome Congress, when an International Meteorological Committee (IMC) was established, with terms of reference similar to those of the Permanent Meteorological Committee, which it replaced. At Rome, Scott was appointed Secretary to the IMC, and he held this post until 1900. His commitment to international meteorology is shown not only by the length of time he remained the Committee's Secretary but also by the fact that he served as Foreign Secretary of the Meteorological Society from 1873 until his death (in 1916), except in 1880–1881 (when he was a Secretary of the Society) and 1884–1885 (when he was President).

Another Change in Management of the Meteorological Office

The minutes of the Royal Society's Meteorological Committee in the 1870s do not radiate dynamism. For the most part, they make dull reading and suggest that the Committee lacked inspiration. Without new members to provide fresh ideas, this was likely. Changes in Committee membership were few and far between.

Besides the change in 1871, when Wheatstone succeeded Miller, there were only four changes in the period 1866 to 1876: Spottiswoode resigned in June 1873; Frederick Evans, a member of the Galton Inquiry in 1865–1866, joined the Committee in March 1874 (in his official capacity as the new Hydrographer to the Admiralty); Gassiot resigned in October 1875; and Wheatstone died on 19 October 1875. Lieutenant-General Richard Strachey replaced Spottiswoode, and the Royal Society accepted the suggestion of the Committee that the fourth Earl of Rosse be appointed an additional member.[10] When Admiral Richards ceased to be Hydrographer to the Admiralty, the Royal Society assented to the request of the Committee that he be allowed to remain one of their number, as they were, as it was put in the minutes of their meeting on 16 February 1874, "most anxious to retain the benefit of his valued experience". Gassiot and Wheatstone were not replaced.

An inquiry into the work of the Office was carried out in 1874 and 1875 by the 'Royal Commission on Scientific Instruction and the Advancement of Science', chaired by the eighth Duke of Devonshire. The Office was but one of a number of bodies investigated, and the inquiry reviewed staffing, financial arrangements, expenditure, costs and stocks of instruments, arrangements for supplying instruments, observations made for the Office at sea and on land, meteorological services for the public and scientific aspects of meteorology. Evidence was collected from Glaisher, Buchan, Balfour Stewart, Strachey and others, among them the Rev Robert Main, who was in charge of the Radcliffe Observatory at Oxford.[11]

---

[10] Strachey was an army engineer in India for many years. He invented a number of devices that were useful to meteorologists and also helped set up a meteorological service in India. The Earl of Rosse was an astronomer who was particularly interested in the radiation of heat from the moon and in the magnetic observations made at Valentia Observatory.

[11] Main was an astronomer, mathematician and Fellow of the Royal Society. He was Chief Assistant at the Royal Observatory under Airy from 1835 to 1860 and thereafter Radcliffe Observer.

Scott was questioned, too, in April 1874, and his answers provide some insight into the operations, organization and tensions of the Office at the time. In early 1874, for example, there were twenty-nine telegraphic reporting stations in the UK for morning reports and seven for afternoon ones, and volunteer observers were also used. The observers, Scott informed the Commission, all held their offices directly under the Office and were, in this respect, different from the observers of all other meteorological organizations, except the federal weather service in the United States. Asked if the idea of using Greenwich as the Central Observatory had ever been entertained, Scott responded by reviewing the correspondence that had passed between the Astronomer-Royal and the Meteorological Committee in 1872. When asked if the Office received a large number of returns from the Merchant Service, he replied "not very many" and reported that the average number of ships sending in registers each year over the past seven years had been about seventy-five. It was "necessary to become personal friends of captains", he said, for if their companies *ordered* them to observe for the Office they became unhelpful and ceased to make observations! The quality of the registers received was, he added, "very high indeed".

Scott's chief complaint concerned what he and others considered the anomalous position of the Office vis-à-vis the Royal Society and the Board of Trade. In his words: "The Government considers us under the Royal Society, and only lately the senior Secretary of the Royal Society has told me that he always considers us under the Board of Trade; but the Government distinctly disclaims all connexion with us, whilst the Royal Society equally disclaims all control over us, except merely the nominations of the members of the Committee". The chief advantage of being under the control of the Royal Society was, Scott considered, the "perfect freedom from political management" which it provided. In his opinion, there would be a loss of prestige if there were no connection between the Office and the Royal Society.

Another inquiry into the work of the Office was carried out in 1876, this time by a Treasury Committee, its remit being to review the workings and achievements of the Office and recommend the best method of administering the funds supplied by the government for the provision of a national meteorological service. This Treasury Committee was appointed in November 1875 and had as its members Farrer, Galton, Strachey, T Brassey, J D Hooker, R R W Lingen and D Milne Home, with Sir William Stirling Maxwell as Chairman.[12]

The need for the inquiry was explained in a Treasury Minute dated 2 November 1875, in which the Lords Commissioners of the Treasury expressed the view that the annual grant for the work of the Office (still £10,000 per annum) was "so considerable" they should satisfy themselves "the results obtained were such as to warrant the

---

[12] Brassey was the Member of Parliament for Hastings and took a keen interest in maritime affairs. Farrer had long taken a close interest in the work of the Meteorological Office (see Chapters 2 and 3). Hooker was President of the Royal Society from 1873 to 1878 and Lingen was the Treasury's Permanent Secretary. David Milne Home represented the Scottish Meteorological Society.

application of so large a sum of public money". Therefore, the inquiry should "be directed to the following points":

- How far had the statistics hitherto collected led to the discovery or confirmation of any meteorological laws?
- How far had the principles on which storm warnings were given been justified by results?
- How far was the appropriation of a large sum of public money in aid of meteorology justified, bearing in mind the fact that it was not the policy of British Government to give direct assistance to the study of any science, except with a view to the more immediate application of scientific theories to practical purposes in which the public rather than individuals had a direct interest?
- Should the Committee decide to recommend that public expenditure for meteorological purposes be continued, they should proceed further to consider on what system it may be best administered. With this object, full information should be obtained with regard to the mode in which the present grant was applied, and in connexion with this part of the inquiry their Lordships of the Treasury wished that the representations of the Scottish Meteorological Society would receive the consideration of the Committee.

There was a sense of déjà vu about this, except that the attention paid to the Scottish Meteorological Society was a new departure and suggested that the government had, at long last, acknowledged that the Scots possibly had a case to be heard. The Minute mentioned a letter dated 21 April 1874 from the Marquess of Tweeddale to the Treasury "submitting the claims of the Society for aid from the State, also previous applications of similar purport from the same Society". It mentioned, too, that Treasury ministers had been "unwilling to propose to Parliament any grant in aid of Meteorological Science beyond that made to the Meteorological Committee of the Royal Society". Now, however, "recognising the value of the labours of the Scottish Society", their Lordships of the Treasury said they would be "glad if an arrangement could be made by which it should participate in the Parliamentary grant in proportion to the services rendered by it in furtherance of the objects of that grant". This all seemed very encouraging for the Scots.

The Treasury Committee's report, with supporting evidence and other documentation, was presented to Parliament in February 1877, and its main recommendation, that the future management of the Office be vested in a paid Council, was quickly adopted by the government. The reaction of *Symons's Monthly Meteorological Magazine* to the report was more than a little cynical. Galton and Farrer were "enquiring into the success of the plans which they had recommended", while other members of the Treasury Committee (especially Strachey, who was a member of the Meteorological Committee) were "enquiring into the success of what they themselves had done". Furthermore, "one-third of the witnesses were connected with the present Meteorological Committee and gave more than half the evidence"! And the opinions of Glaisher, Balfour Stewart and other prominent meteorologists had not been canvassed.

A detailed and critical review of the report appeared in the March and April 1877 issues of *Symons's Monthly Meteorological Magazine* (Vol. 12, pp. 19–25 and 33–37), in which attention was drawn both to contradictory evidence and to recommendations which appeared sensible to Symons. For example, the Treasury Committee noted that "the want of communication by telegraph on Sundays caused a serious defect in the [storm-warning] system which ought to be remedied". On this, Symons commented, "either the storm warnings are useless or it is as criminal to stop them on Sundays as it would be to extinguish the lamps in all the lighthouses at 12 o'clock on Saturday night"![13]

Symons was pleased that the Treasury Committee considered every effort should be made to cooperate with and assist "the different societies or other local bodies engaged in meteorology". However, he did not find much evidence in the report that a significant amount of financial support for societies was being recommended. The Committee considered, not unreasonably, that no expenditure should be incurred that was not essential and saw no reason why observers should necessarily be paid for their services. "There is evidence to show", they said, "that a large and trustworthy amount of co-operation may be obtained in all parts of the UK from observers who do not require remuneration for their services, and it seems very important that such co-operation should be fostered to the utmost".

The Treasury Committee recommended that payments to those who observed for the Scottish Meteorological Society be made "from the grant placed at the disposal of the Meteorological Council" but only "as necessary for obtaining observations at stations required for the purposes of the Council" or for inspecting stations, compiling and checking registers or "special researches conducted by the Society with the approval of the Council". Grants should not be made to "ordinary observers", they said, or "for any general purposes of the Society which lie beyond the scope of the observations to be placed under the Council". To Symons, this left the "distinct impression that the Scottish Meteorological Society closely represented the woman mentioned in the New Testament, who, by her continual coming, obtained that which she wished". The Scottish Society, he said, must be treated in the same way as similar bodies, such as the 'English' Meteorological Society and his own network of rainfall observers.

Commenting on the Treasury Committee's statement that "there is evidence of a connection between weather and health, but it does not appear that any special meteorological observations are wanted at present, or are likely to be wanted in future for this special purpose, other than the observations which the Council should collect for general purposes", Symons drew attention to the work carried out by Glaisher for more than a quarter of a century. "If this is a specimen of the support to be given to original investigators", he wrote, "we do not think that the Council will receive many

---

[13] Weather telegraphy on Sunday mornings and the issuing of storm warnings on Sundays began in 1877.

valuable offers". Glaisher, he said, had been "very harshly treated lately". His tables were "far from perfect", but he had "not had £10,000 a year to spend upon them, and latterly even the miserable pittance of £150 per annum, which he used to receive for compiling them, had been withdrawn". Symons remarked that the pages of the report dealing with Glaisher's work adequately showed the fate of an English man of science who was not in favour with the powers that be.

The response of the Meteorological Committee to the Treasury Committee's report was to resign en bloc, with effect from 31 March 1877. They were, however, persuaded to remain in office for a further two months, and, indeed, their last *Report* covered the period of seventeen months ending on 31 May 1877. Two meetings of the Committee were held after 31 May, one on 18 June 1877, the other on 9 July, the business of the latter being merely to read the minutes of the meeting held on 18 June and hand over the reins to the new Council, whose members attended the meeting.

Disappointment and Discontent

With the change in management of the Office, there came no change in its administration. Scott remained in charge and Toynbee remained Marine Superintendent. Of those who had been members of the old Committee, four became members of the new Council, namely Evans, De La Rue, Galton and Strachey. All three members of the committee who investigated the work of the Meteorological Department of the Board of Trade after the death of FitzRoy (Galton, Evans and Farrer) were still taking a close interest in the Office.

The original intention was that the Council would consist of a chairman and three nominated members, as well as the Hydrographer (Evans), who would be an ex officio member. In the event, the proposal of the Royal Society that there should be a fourth nominated member was accepted, as a letter from Treasury official Lingen to the Royal Society shows. In this letter, dated 28 June 1877, he stated that the Treasury did not object to the proposal that an additional member be appointed, provided that no more than £1000 would be required for remuneration of the new Council! George Stokes became the fourth member, and Henry Smith, who had served as a member of the Devonshire Commission, was appointed chairman.[14] Scott became 'Secretary of the Meteorological Council' and ceased to enjoy the title 'Director of the Meteorological Office'. Buchan was appointed 'Inspector of the Stations in Scotland' and 'Agent for the Meteorological Council in Scotland', while Scott continued to be responsible for inspecting Irish stations.

The change in management of the Office was generally welcomed, but an obvious shortcoming was soon pointed out. The new Council resembled the old Committee

---

[14] Stokes was Lucasian Professor of Mathematics at Cambridge from 1849 until his death (in 1903) and Secretary of the Royal Society from 1854 to 1885 (President from 1885 to 1890). He had broad scientific interests. Smith was Savilian Professor of Geometry at Oxford and a specialist in the theory of numbers.

in having no meteorologist among its members. This was, to quote from an article published in the journal *Nature* on 19 July 1877, "a matter of surprise and regret among meteorologists".[15] The appointments of Smith and Stokes were, however, well received.

In the view of *Nature*, scientists would be justified in looking for both research and experiment from the new Council, "in addition to the dreary piles of observations which have cumbered all scientific libraries for the last half century". Would the members of the new Council conduct researches, *Nature* wondered? They did. In the two decades that followed the Council's formation, its members, particularly Stokes and Strachey, contributed about twenty papers to Meteorological Office publications or learned journals, and members of the Office's staff published more than thirty papers in the *Quarterly Journal of the Royal Meteorological Society*. However, the majority of the publications by Council members simply described instruments or discussed observations. Few advanced meteorological science to any extent. Many of the so-called researches by staff of the Office were really no more than case studies of significant weather events or statistical analyses of observations, and most were carried out by staff on their own initiative.

Those who expected the change in management of the Office to bring about renewed dynamism were surely disappointed. Inspiration and innovation were as lacking in the 1880s and 1890s as they had been before the Council assumed control. Indeed, as Alston Kennerley has mentioned in his biography of Frank Thomas Bullen, a clerk in the Office from 1882 to 1899, initiative appears to have been discouraged.[16] In Kennerley's words, Bullen "thoroughly disliked his experience in the Office, where he encountered the petty tyranny of senior clerks over their juniors, the demand for absolute conformity, the denial of any initiative, and the total lack of any variety". Several of these senior clerks had served in the Office longer than Scott. They were set in their ways and resistant to change.

Further evidence of unhappiness can be found in a letter dated 29 January 1895 sent by 'Anomine' to Richard Strachey, then Chairman of the Meteorological Council.

Sir, I beg to say that there is a lot of private work done in this Office, and which is all done in Office time. It must take each one more than two hours each to do the work. This is unfair to us as underpaid clerks. They earn *three* or *four* pounds weekly, at this business. Yours most respectfully, 'Anomine'.

What came of this anonymous complaint, we do not know.[17]

In the late 1870s, the Council had been forced to focus on matters that had been sources of discontent for years, salaries and superannuation. In FitzRoy's day, staff

---

[15] The article was reproduced verbatim in the August 1877 issue of *Symons's Monthly Meteorological Magazine*.

[16] 'Frank Thomas Bullen, 1857–1915: whaling and nonfiction maritime writing' by Alston Kennerley (*The American Neptune*, 1996, Vol. 56, pp. 353–370).

[17] When Bullen became a clerk in the Office, he augmented his income by running simultaneously a picture-framing business and a haberdashery shop. He left the Office when appointed columnist on the *Morning Leader* and subsequently made a living as a writer and lecturer.

had been government (Board of Trade) employees and entitled to civil service salaries and pensions. Under the Royal Society, as noted in Chapter 3, staff of the Meteorological Office were not considered public servants. Their salaries were lower than for comparable jobs in the civil service and private sector, and they were not entitled to superannuation. Dissatisfaction over salary levels and the lack of pensions was voiced by Office staff on a number of occasions.

Salaries were reviewed annually by the Meteorological Council and individual increases awarded as considered appropriate. In 1879, dissatisfied with their salary awards, three members of staff applied for, and received, larger increases. However, the Council ruled that this method of adjusting salary awards would not be permitted in future and turned down applications for pay rises the following year. The junior and temporary clerks then wrote jointly to the Council to press for an application to be made to the Treasury for an increase in the Office's annual grant that would allow larger salary increases to be paid. The request was turned down, as a result of which a number of temporary clerks resigned, probably further disenchanted by the Treasury's flat refusal to consider a proposal of the Council's then Chairman, Professor Smith, for provision of superannuation allowances for the staff, despite strong Royal Society support. There was sympathy for the clerks amongst the Council's members, to such an extent that, despite the earlier ruling, applications for revised salary awards by several members of staff were given special consideration by the Council's Chairman and Secretary, but no awards were in fact made. Thenceforth, annual salary reviews by the Council continued as before 1879.

A couple of incidents in 1880 provide further illustrations of the intransigence of the Treasury. The Office's engraver died that year after a long illness. Because he had served the Office well for ten years, the Council made a special award of £20 to his widow. Some months previously, a similar gratuity had been awarded to the widow of a temporary clerk. When the Treasury came to know of these gratuities, Lingen sent the Council a letter which stated firmly that any such payments were forbidden in future. However, the Council did not fully comply, and their intention to continue using ad hoc methods of dealing with staff matters was made clear in a note attached to the Office's accounts for 1884–1885:

The Council think it necessary to state that the Treasury having declined to recognize claims to pension of persons employed in the Meteorological Office, it must be understood that no such claims can be recognized by the Council as arising of right, and that the scale of salaries now fixed must be regarded as discharging all liabilities in relation to the services performed in the Office. This will not preclude the Council from considering whether in any special case some retiring allowance may not be made on account of prolonged and meritorious service, within the amount available from the grant placed at their disposal.

The problem of superannuation had to be faced in 1882, when a long-serving clerk retired, J S Harding senior. He had joined the Meteorological Department in 1860 and transferred from the Board of Trade to the Meteorological Office in 1867. Thus he had been a government employee for seven years and therefore believed he qualified for a civil service pension. However, the Treasury refused any payment, whereupon, regardless of the possible displeasure of the Treasury, the Meteorological Council decided to pay a pension out of their own funds. Harding was awarded £42 16s 5d per annum, the amount he would have received had he been employed in the civil service for twenty-two years. The question of a pension arose again six years later, when Toynbee retired, and the Council then dealt with the matter in a similar manner, granting a pension of £144 per annum.

The Treasury exercised a strict control over expenditure, and the Council controlled expenditure carefully too, but its members were able, at the same time, to show compassion and consideration, being closer than the Treasury to the staff on a day-to-day and personal basis. As Jim Burton has noted in his study of the Meteorological Office to 1905:[18]

The role of the Council in its staff relations does seem to have been a largely beneficent one and, despite the low salaries, a degree of loyalty to the Office appears to have been generated. Several members of the same families were sufficiently attracted to its life to follow one another on to its staff, and men with alternative careers and seemingly better prospects available to them outside would sometimes opt instead to work within the Office.

Science at Last

The lack of scientific vitality in the Office was not in itself a cause for criticism. Apart from Scott, the staff were not trained scientists, and the remit of the Office was to provide a reliable service to seafarers and others, not to spend a disproportionate amount of the funds received from the public purse on advancing the science of meteorology beyond what was required to improve the service. Scott was conscientious and methodical and a capable administrator, but he never provided scientific leadership. Research was left to the universities, the meteorological societies and those who conducted the 'special researches' commissioned by the Meteorological Council. To fund these, the Council were authorized to dip into the Office's annual grant, which had been increased to £14,500 in 1877 in accordance with the recommendation of the Treasury Committee. One of the first of the special researches was a study of the methods of hygrometry carried out by a young physicist called William Napier Shaw, who was destined to become Director of the Office. Other special researches included a study of the heights and motions of clouds by means of photographic techniques and an investigation into the chemical and physical properties of London fog.

[18] Burton, J M C (1988), *The History of the Meteorological Office to 1905*, Doctor of Philosophy dissertation submitted to The Open University, 339 pp. Copy held in the National Meteorological Library, Exeter.

There was also a programme of observations from manned balloons to study vertical distributions of temperature and other meteorological conditions, but it ended in disaster. On 10 December 1881, Captain James Templer of the King's Royal Rifle Corps ascended from Bath (Somerset) with two assistants in a balloon lent by the War Office to ascertain for the Office the vertical distribution of temperature and the amount of snow in the air. Near Bridport, the balloon touched the ground. Templer was thrown out, and so, too, was one of his assistants, James Agg-Gardner (who fell heavily and broke a leg). The other assistant, Walter Powell, a Member of Parliament and an experienced and enthusiastic aeronaut, was still in the car when, without warning, the balloon suddenly rose. He was carried away over the English Channel and never seen again.

An investigation into the destruction of the Tay Bridge in a gale on 28 December 1879 led to the invention of a new type of anemometer. The bridge collapsed while a train was crossing it, and all seventy-eight people on board perished. In the ensuing inquiry, questions were raised regarding the allowances engineers made for wind force when designing bridges and other structures, and doubts were expressed, too, about the abilities of anemometers to measure gusts accurately. The inquiry showed that the current state of knowledge and understanding of these matters left much to be desired. Accordingly, in June 1885, the Royal Meteorological Society appointed a Wind-Force Committee:

- To investigate the relation between Beaufort's notation of wind force and the equivalent speed in miles per hour, as well as the corresponding pressure in pounds per square foot
- To inquire whether any existing scale could be adopted or modified, and, if not, to determine such equivalents as may be recommended for general and international use
- To report on the best mode available for the attainment of a satisfactory solution of the entire question of wind force

Though the Society took the initiative, the Office became involved too, principally through Richard Curtis, who was in charge of the section of the Office responsible for collecting and analysing records produced by self-recording instruments at observatories. With the help of the Office, anemometers of different types were set up at exposed sites on Anglesey, the Scilly Isles and elsewhere. Doubts over the ability of these instruments to respond fully to gusts because of friction in the bearings of their rotating cups led the British meteorologist William Henry Dines to invent, in 1890, his 'pressure-tube anemometer', an instrument that depended on a device called a Pitot tube.[19]

In 1883, a new observatory was built, this one on the summit of Ben Nevis (1344 m), the highest point in the British Isles. Hourly observations commenced on 28 November

---

[19] A Pitot tube is an open-ended tube in which, when the open end faces into wind, the pressure exceeds that in the surrounding atmosphere. It was invented by a Frenchman, Henri Pitot, in the 1730s.

Figure 5.2. Ben Nevis Observatory, 1890. © Crown Copyright 2010, the Met Office.

1883 and continued for almost twenty-one years. The observations were made by a superintendent and two assistants, all resident, and the observatory was managed by a committee which consisted of the Council of the Scottish Meteorological Society and representatives of the Royal Societies of London and Edinburgh.[20] An appeal for money to build the observatory raised more than £4000, and most of the funds to maintain it came from private subscriptions. The Scottish Meteorological Society's application for a grant of £400 from the Government's Scientific Research Fund to help build and equip the observatory was refused, but a similar application to the Meteorological Council was successful, with a contribution of £100 per annum for the maintenance of the observatory promised, on condition that records of the observations made on the summit of Ben Nevis would be supplied to the Meteorological Office. In the event, observations were sent not only to the Office but also to meteorological services throughout Europe.

By the end of the nineteenth century, the Office was well-respected internationally, thanks partly to the professional way its operational activities were conducted but mainly to Scott's efficiency and effectiveness as Secretary of the International Meteorological Committee. In the 1880s and 1890s, he and others from the UK participated in a number of international meteorological gatherings, among them meetings of the IMO's Commission for the Polar Year.

This Commission was established at the Rome Congress in 1879, its main function being to plan the International Polar Year of 1882–1883. The importance of setting

---

[20] See *The Weathermen of Ben Nevis* by Marjory Roy (Royal Meteorological Society, 2004, 62 pp.). Remarkably, Clement Wragge climbed Ben Nevis every day from 1 June to 14 October 1881, making meteorological measurements and observations at various places as he ascended and descended. Simultaneously, his wife made observations near sea level at Fort William.

up meteorological stations in polar regions had been pointed out by Buys Ballot at Leipzig in 1872, but the inspiration for the enterprise of 1882–1883 was in fact provided by Lieutenant Karl Weyprecht of the Austrian Navy. His ideas bore fruit in the form of an observational programme which ran from 1 August 1882 to 31 August 1883 and focused on meteorology, geomagnetism and auroral phenomena. Sadly, he did not live to take part in the enterprise. He died on 29 March 1881.

Twelve nations took part, and expeditions were sent to the Arctic and the South Atlantic, among them a British expedition to Fort Rae, Canada. However, the participation of the British in the Polar Year was reluctant and did not involve anyone from the Office, even though standard instruments were provided by the Office and by Kew Observatory. The reluctance of the British to take part is difficult to explain, for British explorers had over the years undertaken expeditions to the Arctic and the waters around Antarctica, and Scott, as Secretary of the IMC, was aware of developments as they unfolded.

There was, nonetheless, a substantial contribution to the Polar Year by the Office. Daily synoptic charts for the North Atlantic Ocean were constructed from all available information that could be collected, not only from British ships but also from ships of other nations. These charts were drawn and published for parts of the Polar Year period, and the work was undertaken because many scientists believed that meteorological events in high latitudes affected the behaviour of the atmosphere in middle and lower latitudes. Parallel efforts to construct synoptic charts for the South Atlantic were made by German meteorologists.

By the 1880s, there was already controversy about the climatic effects of carbon dioxide emitted from factories. However, measurements of concentrations of this gas in the atmosphere were made by only one Polar Year expedition, the French one to Cape Horn, where the mean concentration of the gas was found to be 256 parts per million (ppm), compared with a mean of 284 ppm in France. Those who contended that human activities were causing concentrations of carbon dioxide in the atmosphere to increase thus gained a shred of supporting evidence.

The Value of Weather Forecasts

By the early 1890s, the annual grant to the Office had increased to £15,000 per annum, which seems woefully inadequate today. There were some, though, who considered the amount too much and doubted that the British taxpayer received value for money, while others sprang to the defence of the Office. Hugh Clements, for example, in the July 1891 issue of *Tinsley's Magazine* (pp. 167–172), thought the sum of £15,000 "paltry". He put it thus:

It is really a disgrace to a great maritime and industrial country like ours to starve the meteorological service for the sake of £100,000 or £200,000. However, it is always the case in this country that if money is wanted for any really useful home purpose, with the object of benefiting the people at large, it cannot be obtained, but if the sinews of war are required for naval or military aggressiveness millions are squandered.

There ought to be more stations along the western coasts of Scotland and Ireland, Clements said, as well as stations in the Faroe Islands and Iceland "connected to the mainland electrically".[21]

Anticipating, to some extent, a long-range forecasting technique that would be practised operationally several decades later, Clements thought that the weather would one day be forecast with considerable accuracy a long time in advance from the weather records of the past. Accordingly, he considered it "now essential that inland stations be very largely increased". He concluded that the Office was indeed worth the £15,000 it received, but he was, nevertheless, critical of the Office's forecasts.

The official view of the Office was that the weather forecasts issued to the general public were, on the whole, reliable. In the early 1890s, they claimed that over 80% of forecasts were either completely or partially successful, and the success rate they claimed for storm warnings rose steadily from 79.3% in 1885 to 92.0% in 1894, judged by the occurrence of gales or strong winds after warnings had been issued. The definitions of 'complete success' and 'partial success' were, however, open to question. The criteria used by the general public were not necessarily those used by forecasters! Under Scott, the Office never provided rigorous definitions.

To the general public, weather forecasting left much to be desired, but those for whom storm warnings were intended were broadly satisfied. There had undoubtedly been improvements in forecasting since the early 1860s. Nonetheless, forecasting remained an art that was far from perfect. The techniques used by forecasters in the 1890s had developed little since the days of FitzRoy.[22] Indeed, the Office had taken a step backward in the 1880s by adopting an uninspired model advanced by Ralph Abercromby. In contrast to the models of Dove, Jinman and FitzRoy (see Chapters 1 and 2) and an innovative three-dimensional model put forward in the 1870s by Clement Ley, Abercromby's model amounted to nothing more than a classification of isobars by their shape.[23]

The Office probably had little option but to use Abercromby's model, for money had been spent on it. His work had been commissioned by the Meteorological Council, and he had been paid an honorarium of £60 in 1885, in addition to an earlier interim payment of £15 15s. But that was not the whole story. Ley had also received an honorarium

---

[21] A submarine telegraph cable was laid from Shetland to the Faroe Islands and Iceland in the autumn of 1906 and used almost immediately for the transmission of meteorological observations. These observations were included in the Meteorological Office's *Daily Weather Report* from the beginning of 1907.

[22] See Gaster, F (1896), 'Weather forecasts and storm warnings: how they are prepared and disseminated', *Quarterly Journal of the Royal Meteorological Society*, Vol. 22, pp. 212–228.

[23] See Ley, W C (1880), *Aids to the study and forecast of weather*, Meteorological Office, HMSO, Official No. 40, 38 pp. See also Abercromby, R (1885), *Principles of forecasting by means of weather charts*, Meteorological Office, HMSO, Official No. 60, 123 pp.

of £60 from the Council (in 1880), the payment taken from the money earmarked for 'special researches'. He had been asked by the Council to revise FitzRoy's *Barometer Manual*, and his draft of it, which included his cyclone model, had been accepted by the Council, after some revision. The Council had asked Abercromby to add to Ley's book sections on weather telegraphy, storm warnings, types of gales and the use of the barometer by seafarers, but his first draft had turned out to be incompatible with the part written by Ley. Accordingly, the Council had decided to publish the two parts as separate texts. Ley's appeared in August 1880, Abercromby's in 1885.

Ley's model was ahead of its time. In the 1880s, though, his work was criticized, even ridiculed, not least because it indicated that the axes of extratropical depressions were inclined to the vertical, which some considered would cause the depressions to topple over! Ley supported his idea of inclined axes with barometer records from mountain observatories, but to no avail; his model was discarded, and the concept of a depression as a revolving cylinder of air with an upright axis persisted for another forty years. Instead, Abercromby's model was adopted by forecasters in the Office, and his book came to exert a considerable influence on weather study in the UK. Indeed, extracts from the book were still being quoted verbatim in works published after the Second World War. Far from advancing weather forecasting, his model proved a hindrance. To be fair to Abercromby and the Office, though, his book was generally well received when published and only later came to be considered retrogressive.

Advances in Theoretical Meteorology

An approach to forecasting that was completely scientific was impracticable in the latter years of the nineteenth century, not least because observations were inadequate in both quantity and quality. Without sufficient data from stations on land and at sea, and also from instruments carried aloft, forecasters could not ascertain the current state of the atmosphere. Kites and balloons could be used, but they could not ascend to any great height (typically no more than a few hundred metres).[24] Without apparatus for measuring temperature and humidity in the atmosphere above the ground, and without devices for transmitting upper-air data to stations on the ground, knowledge of the atmospheric state aloft was virtually non-existent. Only from studies of cloud types and cloud movements could information about conditions aloft be derived, and in developing his cyclone model Ley had in fact done exactly that.

Without expressions of physical and dynamical processes in the form of mathematical equations, methods of weather analysis and forecasting which eliminated personal judgement were not available. Nevertheless, foundations of the objective

---

[24] In the years 1883–1885, whilst resident at Tunbridge Wells, Edmund Douglas Archibald carried out a number of kite ascents with suspended anemometers. He used a special form of kite which he himself had devised, and on some ascents an altitude of 400 metres was achieved.

weather prediction techniques that are used routinely by meteorologists today had been laid by the 1890s. Sir William Thomson and James Joule in England, Karl Reye in Germany and Henri Peslin in France had produced important advances in atmospheric thermodynamics, and William Ferrel in the United States had formulated equations of motion for a body moving on a rotating earth.[25] On these equations he had constructed a mathematical model of the general circulations of the atmosphere and ocean. As Gisela Kutzbach put it, in her book on *The Thermal Theory of Cyclones*, he made "the first great contribution to the field of geophysical hydrodynamics after Laplace".[26]

Other notable contributions to theoretical meteorology before 1890 were the kinematic model of a cyclone proposed by the Norwegians Henrik Mohn and Cato Guldberg and the insights into hydrodynamical conditions in atmospheric vortices provided by Thomson and Helmholtz. Thomson's classic contribution, his circulation theorem, was published in 1867 and governed how vortices develop in the atmosphere.[27] It formed the basis of the theoretical treatment of atmospheric motions by the Norwegian hydrodynamicist Vilhelm Bjerknes in the 1890s. Helmholtz explored in theoretical terms equilibrium conditions along the surfaces of discontinuity that separate air masses of different properties and constructed a model of the general circulation in which these surfaces were important elements. For several months of 1879, Napier Shaw studied under Helmholtz in the University of Berlin, attending his lectures on hydrodynamics and working in his laboratory.

During Scott's time in charge, the administrative and operational activities of the Meteorological Office were carried out competently and professionally, and the Office played an important role in international meteorological affairs. Moreover, staff of the Office advanced knowledge of the behaviour of weather systems and helped improve meteorological instruments. Scientifically, though, the Office lacked flair and vitality. None of the 'special researches' commissioned by the Meteorological Council had been carried out by staff of the Office, nor had any contributions to theoretical meteorology been made by staff of the Office. But the wind of change was about to blow, as we will see in the next chapter.

[25] Thomson was raised to the peerage in 1892, thus becoming Baron Kelvin of Largs. Together with the German physicist and physiologist Hermann von Helmholtz, he shaped nineteenth-century physics.

[26] *The Thermal Theory of Cyclones: A History of Meteorological Thought in the Nineteenth Century* by Gisela Kutzbach (American Meteorological Society, 1979, 255 pp.).

[27] He published his work in the *Proceedings of the Royal Society of Edinburgh* (Vol. 6, pp. 94–105).

6

The Emergence of Science

Some of the most distinguished scientists in the land were members of the Meteorological Council, but to little effect. As a scientific institution, the Meteorological Office was moribund in the 1890s. A scientist with vision was required.

Changes in membership were rare, and those that did occur were forced. Sir John Lefroy deputized for Richard Strachey from April 1878 to April 1879 (whilst Strachey was in India, advising the India Meteorological Department).[1] Professor Smith died in February 1883, and the vacancy that resulted from Strachey succeeding Smith as Chairman was filled by the Radcliffe Observer, Edward Stone. William Wharton succeeded Frederick Evans as Hydrographer in July 1884, and George Darwin joined the Council in February 1885, when Warren De La Rue resigned because of failing health. George Stokes resigned in November 1887, when elected Member of Parliament for the University of Cambridge, and his place on the Council went to Alexander Buchan, to represent the Scottish Meteorological Society. Thereafter, there were no more changes in membership for nearly ten years.

Within the Office itself, changes of personnel were also few and far between, the most notable being the retirement of Henry Toynbee on 30 June 1888 and the appointment of Charles Baillie in his place.[2] A change in the Office's status occurred in October 1891, when the Meteorological Council became incorporated under Section 23 of the Companies Act 1867, but this merely secured for the Office a legal status as an association under the 'no profits clauses' of the Act and did not make it any more or less scientific than hitherto.

---

[1] The India Meteorological Department was formed in 1875, its first Director H F Blanford. He gained a second-in-command in 1882, when a clerk in the Meteorological Office, W L Dallas, was appointed 'Scientific Assistant'. Dallas had joined the Office in July 1872. He served the India Meteorological Department until November 1906, when he retired.

[2] Baillie had been Assistant Marine Superintendent in the Meteorological Office since October 1879. He had previously been Director of Nautical Studies at the Imperial Naval College, Tokyo, Japan.

A Scientist with Vision

The injection of enthusiasm, vision, inspiration and innovation the Office needed was eventually supplied by Dr William Napier Shaw, who joined the Meteorological Council on 27 May 1897, taking the place of Edward Stone, who had died earlier that month. Shaw was an academic with a strong interest in meteorology. He had not only carried out the study of hygrometry mentioned in Chapter 5 but had also delivered courses on meteorological physics at the University of Cambridge. He had, furthermore, published a substantial paper on cloud formation in the 1895 volume of the *Quarterly Journal of the Royal Meteorological Society* and served on the Kew Committee of the Royal Society since 1894.

Born in Birmingham on 4 March 1854, Napier Shaw had been educated at King Edward's School in that city, after which, in 1872, he had proceeded to Emmanuel College, Cambridge, where he had read mathematics. He had graduated sixteenth wrangler in 1875 and gained a distinction in physics in the Natural Sciences Tripos the following year. He had been elected a Fellow of Emmanuel College in 1877 and soon afterwards employed in the Cavendish Laboratory, where he had worked under James Clerk Maxwell.[3] In 1880, after the death of Maxwell, the new Cavendish Professor, Lord Rayleigh, had appointed him a Demonstrator.[4] A University Lectureship in Experimental Physics had followed in 1887, and he had been elected a Fellow of the Royal Society in 1891.

The manner of Shaw's appointment to the Council was unconventional. He explained it thus in a set of reminiscences published in 1934 to mark his eightieth birthday, 'The March of Meteorology: Random Recollections' (hereafter called 'Random Recollections'):[5]

In May 1897, there was great uproar in Cambridge at the voting on the report of the Women's Degrees Syndicate of which I was one of the secretaries. The Senate House square was filled with dons voting and being pelted with missiles of all kinds from the windows overlooking the square. In the middle of the uproar, Michael Foster, Secretary of the Royal Society, asked me if I would accept nomination by the President and Council [of the Royal Society] as a member of the Meteorological Council, where there was a vacancy on account of the death of E J Stone, the Radcliffe Observer and previously Astronomer Royal at the Cape. I consented, and the first meeting of the Council that I attended began its business by recording my appointment with a subdued protest on the part of the Council that they had not been consulted before the appointment was offered to me.

---

[3] Maxwell was the greatest physicist since Newton. He was the first Cavendish Professor of Experimental Physics at Cambridge (from 1871 until his death in 1879).

[4] John William Strutt, the third Baron Rayleigh, was Cavendish Professor of Experimental Physics from 1879 to 1884. He was awarded the Nobel Prize for physics in 1904 for his work on atmospheric gases which led to the discovery of argon. He was also a great hydrodynamicist whose work on convection and instability was important for meteorology.

[5] *Quarterly Journal of the Royal Meteorological Society*, 1934, Vol. 60, pp. 101–120.

Figure 6.1. William Napier Shaw. © Crown Copyright 2010, the Met Office.

Shaw was taken aback at the state of the Office. He noted in his 'Random Recollections' that there was a lack of enthusiasm, even though the Council included a number of distinguished scientists among its members. The average age of the Council was high. Strachey was 80, Galton 75 and Buchan 68. Moreover, Scott was 64 and approaching retirement. He had been at the helm of the Office for thirty years. He was respected as head of the Office and as Secretary of the IMC, but he had never been someone with the vision and boldness of FitzRoy. He was solid and dependable and, in Shaw's words, "more of a critic than a creator, an excellent disciplinarian and correspondent, a very regular inspector of the stations in Ireland, his native country".

Furthermore, Scott was not the only long-serving member of the Office. The senior clerks had all been employed by the Office for many years. Indeed, most had served under FitzRoy, namely James Staughton Harding of the Secretariat, his brother Charles of the Marine Branch, Frederic Gaster of the Forecast Division, Richard Strachan in charge of instruments and Richard Curtis in charge of observatories.[6] They were all approaching retirement. Their chief concern now, Shaw said in his 'Random Recollections', was whether they were still entitled to pensions, as they had been in FitzRoy's time, when they ranked as civil servants. Among the senior staff, only one was a comparative newcomer, Charles Baillie, who had joined the Office in the autumn of 1879.

Shaw was astonished to find so little interest in weather forecasting. At meetings of the Council, he recalled, no one ever raised any question about forecasts. The Council's chairman told him that it was Gaster's responsibility to produce them, but

---

[6] J S Harding had been appointed in 1855, C Harding in 1861, Gaster in 1859, Strachan in 1858 and Curtis in 1861.

he (Strachey) "understood that F J Brodie [one of the forecasters] had a very keen sense of when it was going to rain"! The London observations which were used by the forecasters, Shaw was told, were those made by Gaster at his home in Brixton.[7] The scientific basis of weather forecasting had advanced little since the days of FitzRoy.

Meetings of the Council were, said Shaw, "peculiarly interesting". As he recalled in his 'Random Recollections':

The business consisted mainly of letters and draft replies, reports of work, or inspections, or analysis of forecasts and storm warnings. All these were set out in the Agenda book; but the Council had not seen them. The first business was the Minutes, which were last meeting's Agenda paper printed and circulated. So we considered all the items 'on the report stage' and the Chairman noted our conclusions in quite few words. When the minutes of the last meeting were disposed of, we adjourned.

Shaw found, nevertheless, that the Council responded to suggestions. When, for example, he reported, in 1898, that C T R Wilson, a young physicist in the Cavendish Laboratory, had made fundamental discoveries about the ionization of cloud droplets, the Council supported his suggestion that Wilson might continue his researches on atmospheric electricity and awarded a grant of £200 for the purpose.[8] There were, however, frustrations, as we see in the *Annual Report of the Meteorological Council for the Year Ending 31 March 1900*. A note states that the Council were "saddened" that they could not make the balloon or kite investigations recommended by Wilson. This was partly because of the expense involved and partly because of practical difficulties, the most significant being "the disposition of the land and its intersection by inclosures and roads which tends to aggravate the possible consequences of any misadventure with the apparatus".

The need to solve the problem of superannuation became urgent in 1899, as the retirements of Scott and other long-serving staff approached. The Council appealed to the Royal Society to apply to the Treasury for an increase in the parliamentary grant to cover the additional cost, arguing that all of the available funds were committed for maintaining the current work of the Office. The application was refused. The Treasury's response to the Council, through the Royal Society, was that provision for superannuation was indeed essential but had to be paid for out of the current grant. In other words, pensions had to be provided, but afforded by reducing the costs of the Office's operations.

After the death of Baillie, which occurred unexpectedly on 24 June 1899, some members of the Council suggested that financial economies might be achieved by appointing a Marine Superintendent who would also act as Secretary to the Council, but nothing came of this idea. The new Marine Superintendent, Captain Hepworth,

---

[7] Gaster retired early on 31 December 1903 because of ill health. An official weather station for London was established the following year in St James's Park and came into operation on 8 November 1904.

[8] Wilson became a Nobel laureate in 1927 for his work on cloud physics.

did not become Secretary.[9] Instead, Shaw succeeded Scott. Financial savings were made by closing some telegraphic reporting stations and reducing costs in other ways, as a result of which the Office was able to afford retirement pensions.

Questions were asked in the House of Commons about the Office's pension position. On 4 July 1899, for example, Charles Beresford (the Member for York) asked the First Lord of the Treasury, Arthur Balfour, whether he was aware that, "owing to the necessity of providing pensions for old servants", the Office was "about to cut down its expenses by withdrawing the small payments made for observations at York and other stations". He asked too, "in view of the important national work done by the Meteorological Office", if the First Lord could see his way to increasing the Government grant. Balfour avoided the question, saying that it should be addressed to the Chancellor of Exchequer.

On 7 August 1899, Duncan Pirie asked the Financial Secretary to the Treasury (Robert Hanbury) whether the accounts of the Meteorological Council showed "a sum set aside to provide for a superannuation fund from the Annual Grant". If so, what was its amount? A condition of the Grant was, he said, that it "should be applied exclusively to the scientific work of meteorology". This being so, the payment of superannuation out of it would surely be a misappropriation of public money. Hanbury pointed out that the sum of £144 had been expended on pensions in the year that ended 31 March 1898, but there was no provision for a general superannuation fund.

A retirement age of 65 was introduced by the Council in 1899, along with the rule that staff would normally retire on half pay if their approved service amounted to at least thirty years. Scott and Strachan retired in 1900, Scott on half pay, with a pension of £400 per annum, Strachan with an annuity of £150 on his life.[10] The Chairman of the Council (Strachey) wrote to the President of the Royal Society on 16 November 1899 to advise him that the Council had approved of the resignation of Scott as from 1 January 1900 and that Shaw had been chosen to succeed him.[11]

Shaw reported for duty on New Year's Day, only to find that official sanction for his appointment had not yet arrived. He believed, therefore, that he was unemployed, because arrangements for others to deliver his lectures and take his tutorials at Cambridge had been made after an announcement of the appointment had appeared prematurely in several newspapers. To his surprise, he received, in early January,

---

[9] Captain Melville Willis Campbell Hepworth RNR, an officer in the Merchant Navy who had contributed a number of excellent weather logs to the Meteorological Office, was appointed Marine Superintendent on 24 July 1899, on probation for one year, at a salary of £350 per annum, no pension guaranteed. He took up his appointment on 2 October 1899 on his return to England from the Pacific.

[10] In accordance with the new retirement rule, Strachan was asked to retire, as he was already 65. He declined and was thereupon technically dismissed. His official retirement date was 31 March 1900, but he fought for several months thereafter against the decision to dispense with his services, without success.

[11] A letter in the National Archives (BJ 1/199) shows that Strachey had offered Shaw the job of Secretary as early as July 1898. In the letter to Strachey dated 26 July 1898, Shaw referred to "the question which you put to me at the Meteorological Council on Wednesday last" and suggested that the two of them might meet "to talk the matter over". Strachey had, incidentally, been knighted in 1897.

formal notice from the Master of Emmanuel College that he must either go into residence at Cambridge or resign his emoluments, which he believed he had already done! He was unemployed for two months, "learning wisdom from the outgoing Secretary and something about the Office Library", as he put it in his 'Random Recollections'.

We do not know how the Council came to believe Scott's last day in office would be 31 December 1899, for he does not appear to have announced his retirement by then. In a letter published in *The Times* on 19 February 1900, he stated that he had "already passed the age of retirement from Her Majesty's Civil Service" and had, therefore, tendered his resignation, "to take effect as soon as the necessary arrangements can be completed – i.e. February 28".

Scott announced in the letter that his successor would be Napier Shaw and commented that he had "every confidence in saying" that his meteorological colleagues would find him "in every way a most valuable addition to their ranks". The American journal *Monthly Weather Review* expressed approval thus (1900, Vol. 28, p. 68):

The appointment of Mr Shaw as Secretary of the Meteorological Council and Superintendent of the Meteorological Office at London will be recognized by everyone as demonstrating the high position meteorology has at last attained among the sciences cultivated in England. For a long time it has, we fear, been at the bottom of the list. Many a time we have been assured that its problems were too difficult for the analyst, and its relations with agriculture and mercantile affairs too intimate to free it from the sordid everyday relations that characterize business rather than science. But when now we see the finest physical laboratory of the nation relinquish its distinguished assistant director and encourage him to devote his energies to this most difficult branch of experimental and theoretical physics, we at once realize that under his guidance meteorology is certain to assume no second rank in England.

Scott retired on 28 February and Shaw became Secretary the following day, his salary £750 per annum, which was £50 per annum less than Scott had received throughout his tenure of the Office headship.[12] Retirement did not, in fact, bring to an end Scott's formal association with the Office, and Shaw did not, as it turned out, sever his connection with Emmanuel, for the College re-elected him a Fellow, on the understanding that it would be a supernumerary position without emolument and that he would give each year to the University on behalf of his College a course of four lectures on the physics of the atmosphere.

Shaw's appointment left the Meteorological Council a member short of the minimum number specified in the Council's Articles of Association. He retained his place on the Council when he became Secretary, thereby leaving it with only six members. This difficulty was overcome by altering the Council's constitution (with the sanction of the Royal Society and the Treasury). Fresh Articles were adopted, under which the members of the Association (the Council) numbered at least seven but not more

[12] Given that the retail price index in 1900 was only about 75% of what it had been in 1867, Shaw's salary was worth more in real terms than Scott's had been when appointed.

than ten, with five of them designated 'Directors'. Of these, the Hydrographer was a director ex officio, and the other four were, under the Association's rules, nominated by the President and Council of the Royal Society.

Galton retired from the Council in March 1901, whereupon the remaining five members (Buchan, Darwin, Shaw, Strachey and Wharton) were appointed Directors, remunerated for their services and made responsible for controlling the business of the Council.[13] Five additional members were appointed: John Young Buchanan, William Henry Dines, Arthur Schuster, the fourth Earl of Rosse, and Shaw's predecessor, Robert Scott.[14] Rosse and Buchanan were appointed to serve for five years, the others for three. The role of the additional members was to assist with the management of the Office in general and to help prepare the annual report and estimates in particular. They were not expected to attend more than two or three Council meetings a year, and they received neither remuneration nor honorarium. The new Council formally took over control of the Office on 10 April 1901 and met for the first time two weeks later.

Tributes to Galton were generous, and deservedly so, for he had taken a close interest in the Office over a great many years and contributed much to its development. An era had ended. His influence on the Office had been profound and far-reaching. He had always wanted the Office to be a scientific institution of the highest order. Though he had achieved some success, there was still much to be done, but his time had now passed.

An era had ended, too, with the death of George Symons, on 10 March 1900. This remarkable meteorologist had continued to run his nationwide network of rainfall observers and edit *Symons's Monthly Meteorological Magazine* until his death, and he had, moreover, for four decades, keenly supported the British Association and the Royal Meteorological Society. So highly regarded was he that the Society opened a Symons Memorial Fund and used some of the proceeds to institute the Symons Gold Medal, an award to be made biennially "for distinguished work done in connection with meteorological science". It was first awarded in 1901, when the recipient was Alexander Buchan. Today, it is the most prestigious of the Society's prizes, and a considerable number of Meteorological Office staff have received it. The work of the rainfall organization was continued by Herbert Sowerby Wallis, and the editorship of *Symons's Monthly Meteorological Magazine* passed to Hugh Robert Mill.[15]

---

[13] The total sum allotted for remuneration of directors was limited to £1000 per annum, as authorized by the Treasury.

[14] Buchanan was a chemist and oceanographer who had taken part in the HMS *Challenger* Expedition of 1872–1876. Schuster was from 1888 to 1907 Professor of Experimental Physics at the Victoria University, Manchester, in succession to Balfour Stewart.

[15] The organization became the British Rainfall Organization (BRO) in 1900. Wallis became an unindentured apprentice to Symons in 1872. He edited *British Rainfall* with Symons from 1890 until 1900 and then became, with Mill, joint director of the BRO. Mill, a distinguished geographer and meteorologist, was sole director of the BRO from 1903, when Wallis resigned on grounds of ill health, until 1919, when responsibility for the BRO

Three years after Symons died, meteorology lost another remarkable meteorologist when, on 7 February 1903, James Glaisher passed away, aged 93. For sixty years, he had remained at the forefront of meteorology, continuing to take an active interest in the subject since his retirement from the Greenwich Royal Observatory in 1874. He had not only published many meteorological papers after his retirement, the last in 1902, but had also continued to supplement the official records and tables supplied by the Royal Observatory for the Registrar-General's weekly, quarterly and annual reports and summaries, using for the purpose the weather reports of volunteer observers in different parts of the UK. He had given up this work in March 1902, the reason being, he said, "advancing age"!

The response of the Registrar-General to Glaisher's resignation was to ask the Meteorological Council to assume responsibility for returning to him periodical reports and statistics of the weather. This they agreed to do, having first ascertained that no special reorganization of the Office would be required. Steps were taken to continue the weekly reports from 1 April 1902 and the quarterly and annual reports as soon as possible thereafter. The Council had for some years past performed a similar service for the Registrar-General for Ireland, while the Scottish Registrar-General's reports had been, and continued to be, supplied by the Scottish Meteorological Society. The change in arrangements of the Registrar-General for England provided an opportunity for bringing the returns published by the three Registrar-Generals into similar form, and it was noted in the *Report of the Meteorological Council for the Year Ending 31 March 1903* that the three had reached an understanding, an aspect of it being that the Council would incorporate "in the general climatological organization of the Office the climatological stations comprised in the late Mr Glaisher's organization".

Further Developments at Kew Observatory

Kew Observatory became part of the newly formed National Physical Laboratory on 1 January 1900, thus ending the long association of Francis Galton with Kew, for he had become a member of the British Association's Kew Committee in 1858 and served on it and its successor, the Royal Society's Kew Committee, ever since, becoming chairman in 1889. In fact, Galton had played a part in bringing about the change in control of the observatory. In the early part of 1895, his cousin, Sir Douglas Galton, had visited the Physikalisch-Technische Reichsanstalt near Berlin and come away so impressed he had campaigned for the establishment of a similar body at Kew.

The response of the Association was to appoint a committee, with Sir Douglas in the chair and his cousin Francis another of the fourteen members, the committee's remit

passed to the Meteorological Office. See 'A short history of the British Rainfall Organization', by D E Pedgley (*Occasional Papers in Meteorological History*, No. 5, Royal Meteorological Society, 2002, 19 pp.). See also 'The British Rainfall Organization, 1859–1919', by D E Pedgley, published in *Weather* in 2010 (Vol. 65, pp. 115–117).

being, as it was put in the Association's resolution, "to report on the establishment of a National Physical Laboratory for the more accurate determination of physical constants and for other quantitative research".[16] The report which the committee prepared for the Association's annual meeting in 1896 contained estimates of the monies required for building, equipping and maintaining the proposed Laboratory, along with the recommendation that the Royal Society should be responsible for defining and managing the Laboratory.

On 16 February 1897, the Prime Minister, the Marquess of Salisbury, received a deputation. Scott and Shaw were among the twenty-nine who attended the meeting, along with Francis Galton, Sir Douglas Galton, Lord Rayleigh and the President of the British Association, Lord Lister. The case for a National Physical Laboratory (NPL) was introduced by Lord Lister and presented in detail by Arthur Rücker (Professor of Physics in the Royal College of Science, London). When summing up, Lord Rayleigh pointed out that enormous sums were being devoted in Germany in aid of science.

There appears to have been a positive outcome of the meeting with Lord Salisbury, because, later in 1897, the Government appointed a committee of inquiry into the possible establishment of an NPL, with Lord Rayleigh as chairman. The committee heard evidence from Francis Galton (Chairman of the Kew Committee) and Charles Chree (Superintendent of Kew Observatory), and members of Rayleigh's committee visited the observatory in the early part of 1898.[17]

The main conclusion of the inquiry, which was contained in a report presented to the Treasury in July 1898, was that an NPL should be established under the control of the Royal Society. Four recommendations stemmed from this conclusion, the two of greatest significance for the Office being that:

- A public institution should be founded for standardizing and verifying instruments, for testing materials and for the determination of physical constants
- The institution should be established by extending Kew Observatory, and the scheme should include the improvement of the existing buildings and the erection of new buildings at some distance from that Observatory

The Royal Society accepted the recommendations and drew up a plan whereby Kew Observatory would be incorporated with the NPL, with the management vested in an Executive Committee that would replace the Royal Society's Kew Committee. The latter committee would, accordingly, cease to exist and was, indeed, wound up on 2 November 1899. The day-to-day work at Kew would, nevertheless, proceed as

---

[16] The idea of a National Physical Laboratory had first been put forward in the 1870s by the Devonshire Commission (see Chapter 5) and proposed again in 1885 by J A Fleming when presenting to the Society of Telegraph Engineers and Electricians a paper 'On the necessity for a national standardising laboratory for electrical instruments'. Yet another case for such a Laboratory had been made in 1891 by Sir Oliver Lodge in that year's Presidential Address to the British Association.

[17] Chree served as Superintendent from 1893 until he retired from the Meteorological Office in 1925. He was an authority on terrestrial magnetism.

hitherto, carried on by the existing staff of nineteen. Official approval for the plan was forthcoming, financial aid for the NPL was voted by Parliament, and Dr Richard Glazebrook, who had been a Demonstrator in the Cavendish Laboratory with Shaw in the early 1880s, became Director of the new Laboratory on 1 January 1900.[18]

Rayleigh's committee reviewed the formal long-standing association of the Meteorological Office with Kew Observatory and agreed that the Office's contribution of £400 per annum to the Observatory would continue. In return, the Office would receive specified autographic records and tabulations and also employ members of staff at Kew whose role would be to inspect self-recording instruments at observatories. It appears that the Scottish Meteorological Society did not know of this agreement, for they assumed the Office's contribution to Kew Observatory would no longer be necessary and accordingly suggested to the Treasury, in October 1898, that the £400 they thought would be saved by the transfer of the facility at Kew to the NPL might be made available to them to help maintain their Fort William and Ben Nevis observatories. They estimated that an additional £1000 was needed to keep their observatories open for two more years.

The Meteorological Council not only turned down the Scottish Society's application for that amount but also decided that the existing grant of £350 per annum to the Society for maintenance of the observatories would cease in the autumn of 1901. The Office needed the money, they said, to meet superannuation obligations. Further applications to the Council and the Treasury for funds to keep open the observatory on Ben Nevis were made by the Society in the period 1899 to 1902, and the importance of maintaining it was the subject of questions asked in the House of Commons in 1899 and 1902, but all to no avail. Extension of the grant of £350 per annum to 31 December 1902 merely delayed the inevitable. Closure was unavoidable if money to sustain the observatory was not forthcoming. It was not, and the observatory closed on 1 October 1904.

Kew Observatory was the home of the NPL from 1900 to 1902 but was never big enough or otherwise suitable for purpose. Bushy House at Teddington was converted and became, in March 1902, the NPL's headquarters. Thereafter, Kew Observatory became the 'Observatory Department' of the NPL, concentrating on the calibration, repair and design of meteorological instruments and continuing to make routine meteorological observations.

Besides there being no active representative of meteorology or terrestrial magnetism on the NPL's committee of management, there was also a technical difficulty. Since 1898, proposals for electric tramways in Kew and Richmond had threatened the continued accurate measurement of the earth's magnetic field at Kew Observatory. To overcome this problem, the NPL announced in 1903 that a new observatory would be built at Eskdalemuir (southern Scotland), and to help pay for it there would be

---

[18] Glazebrook had been one of those at the meeting with the Prime Minister on 16 February 1897.

a sizeable compensatory donation (£10,000) from the tramway company.[19] The first
sod was cut on 19 July 1904, and Eskdalemuir Observatory came into operation on
11 May 1908 for studies of geomagnetism, atmospheric electricity and seismology
and for making weather observations.

Glimpses of Everyday Life

Shaw soon learned that "the chief business was letters" (as he put it in his 'Random
Recollections'). There was confusion if he arrived before eleven, because the letters
were not ready for him, and the letters missed the country post if he stayed after
four![20] To Shaw, this was not satisfactory. To him, his job involved a great deal more
than correspondence. He was a scientist and believed the Office should aspire to be a
scientific institution of the highest calibre. As he said in his 'Random Recollections',
the Office was, in fact, considered in official circles a scientific establishment, yet no
member of staff had received any formal scientific training, and the science was all
concentrated in the Fellows of the Royal Society who governed the Office. He pressed
for the appointment of a personal assistant with adequate scientific qualifications,
and his wish was granted. At the meeting of the Meteorological Council on 14
March 1902, he and the chairman of the Council were given authority to select
someone who would undertake duties that were, as it was put in that minute, "mainly
in connection with scientific investigation". The person appointed would not take
any part in the administration of the Office. When the Secretary was absent, the
Marine Superintendent would be in charge, and when both were absent, the Chief
Clerk.

The Council heard on 25 April 1902 that Rudolf Gustav Karl Lempfert, an assistant
master at Rugby School, was to be employed as 'Secretary's Scientific Assistant' from
1 May 1902, the appointment being for one year, on probation, at a salary of £200 per
annum. If re-appointed, his salary for the second year would be £300. The Council
undertook to provide facilities for him to undertake experimental investigations, also
to allow him to give lectures, provided they did not interfere with his work for the
Office. His holiday entitlement would be six weeks, plus public holidays. Lempfert had
been a pupil and colleague of Shaw. Born in Manchester in 1875, the son of German
immigrants, he had been educated at Manchester Grammar School and Emmanuel
College, Cambridge. He had gained First Class Honours in Parts I and II of the Natural
Sciences Tripos and subsequently been employed as an Assistant Demonstrator in the

---

[19] The South Western Railway Company's line along the southern boundary of the Old Deer Park was electrified in
1916, further affecting the work of Kew Observatory adversely, and by 1926 all of Kew's geomagnetic work had
been transferred to Eskdalemuir.

[20] In 1900, the Meteorological Office was open for general enquiries from 10 a.m. to 4 p.m. on weekdays and 10
a.m. to 1 p.m. on Saturdays. It was open for telegraphic enquiries from 8.30 a.m. to 8 p.m. on weekdays and 6 to
8 p.m. on Sundays.

Figure 6.2. The Meteorological Office headquarters at 63 Victoria Street, London, in the late nineteenth century, showing the display of notice-boards on the first-floor balcony and the pianoforte maker's shop on the ground floor. © Crown Copyright 2010, the Met Office.

Cavendish Laboratory. He had taken up his appointment at Rugby in 1900. Apart from Scott and Shaw, he was the first graduate employed in the Office, and the first member of staff other than them to possess any formal scientific training.

A fascinating glimpse of life in the Office in 1902 was provided by Lempfert in an article published half a century later.[21] From this, we learn that the Office possessed no telephone and only one typewriter! Outgoing letters were "written autograph in copying ink and press-copied before dispatch", with "copy for the printer supplied in long-hand". The passer-by could recognize the Office by the display of notice-boards on the first-floor balcony which provided reports of the sea-state at Dover and the weather at Valentia, Scilly, Holyhead, Yarmouth and The Needles. At 63 Victoria Street, the home of the Office since 1869, the ground floor was occupied by a piano shop, and the Office's premises were on the four floors above and also in the basement, which was used for storage.

---

[21] 'Some reminiscences of the Meteorological Office of the year 1902' by R G K Lempfert (*Meteorological Magazine*, 1954, Vol. 83, pp. 161–166).

What Lempfert called "the office-keeper's den" was on the first floor, and so, too, were the instrument store and Statistics Division, the latter in a "large but extremely low room which ran from front to back of the building". Shaw's room and the Marine Superintendent's were on the second floor, the former used for meetings of the Council and, later, for Monday evening *colloquia* modelled on those held in the Cavendish Laboratory. His room also housed most of the library, which had "begun to overflow into other parts of the building".[22]

Two rooms on the third floor, Lempfert said, were allocated to the Forecast Division, while the large front room was occupied by the Observatory Division and the remaining small room housed the Marine Division. On the top floor, "the front room was occupied by four ladies of the Marine Division, the only women in an otherwise all-male staff". Lempfert's room was also on the top floor, but the best room on this floor was, he considered, that tenanted by Mr and Mrs Drane, the resident caretaker and his wife. She provided, as he put it, "for those members of the staff who wanted it (and most of them did) a substantial lunch for 1s 2d". The lunches "were sent to the recipients on trays and eaten at their desks". The Instrument Division was, Lempfert recalled, part of the Marine Division in 1902 and was, for the most part, "an instrument store for the supply of instruments to the navy, the mercantile marine and the telegraphic reporting stations".

When Lempfert joined the Office, John Curtis was in charge of the Statistics Division, with Duncan Bell under him as Librarian. In this capacity, Bell was personally responsible for the meteorological section of the Royal Society's *International Catalogue of Scientific Literature*. In 1906, when Curtis succeeded James Harding as Chief Clerk and Cashier, Lempfert became Superintendent of the Statistics Division, with Duncan Bell his principal assistant. Lempfert recalled that both Curtis and Bell gave much of their time to municipal politics, with Curtis becoming Mayor of Fulham and Bell an alderman of his borough.

The publication called *Meteorological Observations at Stations of the Second Order* was perhaps, Lempfert said, the Statistics Division's "most important task". In compliance with a resolution of the International Meteorological Committee, this publication contained observations made daily at 9 a.m. and 9 p.m. at a number of selected stations and monthly summaries for a larger number. "No doubt", said Lempfert, "the Committee hoped that the data would be published promptly, but in most countries there was a lag of several years". The Office was no exception, but arrears were, he said, cleared during the following five years.

Another of the Statistics Division's responsibilities was the *Weekly Weather Report*. For this, he recalled, the schedule was tight. When he joined the Office, the *Monthly*

---

[22] It was stated in the *Annual Report of the Meteorological Council for the year ending 31 March 1905* that there were then 18,400 volumes and pamphlets in the library. It was noted that the library was available to students and others from 10 a.m. to 4 p.m.

Figure 6.3. Afternoon tea in the Forecast Room of the Meteorological Office at 63 Victoria Street, *c*. 1904. From left to right: R Sargeant, F Snell, W Hayes. © Crown Copyright 2010, the Met Office.

Weather Report was, as he put it, "a very slender affair, confined almost entirely to summaries for the telegraphic stations". However, it expanded considerably when observations from the stations formerly run by Glaisher were included. Indeed, the enlargement of the *Monthly Weather Report* rendered *Meteorological Observations at Stations of the Second Order* redundant as regards monthly summaries, as a result of which publication of the latter ceased with the volume for 1907. The main function of the Observatories Division was to tabulate and scrutinize the records from Kew, Aberdeen, Stonyhurst, Oxford, Falmouth and Valentia Observatories. Only at Valentia did the Office now have full-time staff of its own. Lempfert recalled that the copper plates used for producing the long-defunct *Quarterly Weather Report* had been kept, but were requisitioned for scrap in 1915.

Lempfert recalled that Frederic Gaster was the principal forecaster in 1902 and had been for many years. However, he was now in failing health and took little part in the daily routine. Nevertheless, the London observations included in the *Daily Weather Report* were those taken in his garden at Brixton and telegraphed to Victoria Street twice a day. The active forecasters in 1902 were, Lempfert said, F J Brodie and R Sargeant, assisted by G G Francis and A R Simpkins, all of whom "had years of experience in plotting synoptic charts and watching the vagaries of British weather". How they coped was something Lempfert wondered about, for there did not appear to be any "reserve of experienced man-power in the [Forecast] Division to provide for sickness or leave or public holidays". After about a year in the Office, Lempfert himself "was called on to take a share in forecasting". He said that the staff he had met when he joined the Office had all "climbed the ladder" of assistant to telegraphist to forecaster over a period of fifteen, twenty or more years.

In the Forecast Division, all information was received by Post Office telegram, and the messages had no priority over those of the general public. The forecasters had to operate the telegraphic equipment themselves, which meant they had to be proficient in sending and receiving Morse. Weather reports were exchanged in an international code which had been approved by the Permanent Meteorological Committee in 1874. There was no upper-air information, as the balloon and kite investigations which had been made up to that time had been solely for experimental purposes. Because there were no typewriters, material for the *Daily Weather Report* had to be written and drawn in lithographic ink, which meant staff had to be not only meteorologists and expert telegraphists but also draughtsmen and calligraphists.

Further insight into life and conditions in the Office at the beginning of the twentieth century can be found in the reminiscences of A T Bench.[23] He included not only his own recollections of the period 1898 to 1910 but also those of colleagues H E Carter, W Hayes and H L B Tarrant. These reminiscences provide additional information not mentioned by Lempfert.

Bench commented that the boy clerks who joined the staff around the turn of the century might almost have thought they were joining a family party, for, in the staff of forty persons, there were four or five sets of brothers and a pair of sisters. The latter, Miss Rose and Miss Beatrice Smith, Bench said, were "ensconced", with Miss E A Anderson in charge, "in a veritable 'purdah' on the top floor of the building behind double doors and were allowed to arrive ten minutes later and leave ten minutes earlier than the male staff in order (ostensibly) to avoid contact with the men"! "They comprised", he went on, "a section of the Marine Branch and such was their strict seclusion that permission to interview them had to be obtained from their chief".

Around the corner from the Office, there was, Bench recalled, a public house. In his words, "the invariable practice of the two senior forecasters" was to pay a visit to this pub "immediately before settling down with the synoptic chart to dictate the forecast". Street musicians often played outside the Office's premises and thus disturbed Shaw, who would, as Bench put it, "send Mr Snell down to request them to soften their music, as his room overlooked the side street". Whether or not other members of the Office's staff objected to the music is not recorded by Bench, but he did mention that several were musical. These included Lempfert, an accomplished viola player, W G James, a flautist and organist, Rose Smith, a member of the London Choral Society, and L H Powers, who played the violin at staff dinners. Another notable achievement of Powers was, Bench said, that he discovered, in 1900, that observers at certain seaside resorts joined up intermittent burns on sunshine cards by a hot wire in order to increase their daily totals![24]

---

[23] 'The Meteorological Office: 63 Victoria Street, Westminster, London, S.W., January 1898–November 1910' (*The Marine Observer*, 1963, Vol. 23, pp. 79–84); 'Reminiscences of the Meteorological Office, 1898–1910' (*Meteorological Magazine*, 1981, Vol. 110, pp. 323–329). Bench entered the Office on 24 January 1898 as a Boy Clerk in the Autographic Records Branch and became a Probationer in the Forecast Division in September 1902. He retired from the Office on 12 March 1947, having reached the age of 65.

[24] This is a reference to the Campbell-Stokes sunshine recorder, in which a glass sphere focuses the sun's rays on a graduated card. The card does not burn when the intensity of solar radiation is insufficient, as, for example, when the sun is low in the sky or behind cloud.

Bench said that Gaster had lunch in his room and then practised "the curious habit of removing his dentures before his meal, placing them on a table napkin beside his plate"! At one period, furthermore, Gaster "had a cask of ale in the basement for his own consumption and for anyone else who would pay for a drink". This arrangement was short-lived, however, as "he found that the ale disappeared and he lost money"! As one of his official duties, Gaster was Meteorological Correspondent for *The Times* and almost every day prepared for this newspaper a special account of the day's weather experienced over the country.

Time-keeping and discipline in the Office were strict. A red line was drawn in the attendance book at 9.15 a.m., and anyone who had to sign below this line twice in a week had to face Shaw. Talking was not permitted, unless it concerned an official matter, and no one smoked whilst on duty. Despite the strictness of the official discipline, senior clerks sent boy clerks on errands almost daily, often to the Army and Navy Stores for tobacco and whisky! As Bench noted, office hours were originally 10 a.m. to 4 p.m., but demands for better pay were met by raising salaries and at the same time adding two hours to the working day, thus extending it from 9 a.m. to 5 p.m. The practice of working full days on alternate Saturdays was dropped in 1901, with staff thereafter working half a day every Saturday.

Because the line to the Central Telegraph Office was closed on Sunday mornings, Bench mentioned, staff went there instead to work. "The junior of shift", he said, "went from his home to take the morning observations at Westminster and then walked to Charing Cross Post Office to send them by telegram to the Central Telegraph Office". Forecasters worked to a shift pattern, the hours of duty being as follows: on weekdays, 8 a.m. to 4 p.m. or 1.30 p.m. to 8.30 p.m.; on Saturdays, 1 p.m. to 8.30 p.m.; on Sundays, 8 a.m. to noon at the Central Telegraph Office and 6 p.m. to 8.30 p.m. at Victoria Street. A day off in lieu of Sunday duty was introduced during the First World War. The annual leave of four weeks was, Bench said, generous compared with that of other offices.

Those classed as 'staff' received their salaries monthly by cheque drawn on the Western Branch of the Bank of England, while boy clerks and temporary clerks were paid weekly in cash. Bench recalled that boys received 12s 6d a week when they were 15 years of age, and their wages increased by 1s 6d a year to a maximum of £1 a week. The maximum received by a Junior Clerk was £150 per annum, whilst the pay scales of Senior Clerks and the Chief Clerk rose to £275 and £333 per annum, respectively. Overtime payments were made to members of the Forecast Division as compensation for time worked before 9 a.m. and after 5 p.m. and for Sunday duties.

Before there was a telephone in the Office, communication with the outside world was by letter, telegram or 'by hand', with messages conveyed by boy clerks. Journeys could often be made wholly or in part by the horse-drawn buses which passed 63 Victoria Street three times every hour, and occasionally boy clerks were sent to the London Docks to deliver barometers, such journeys being made partly by the

Underground Railway. In a case of urgency, the use of a hansom cab would be authorized.

Installation of the Office's first telephone, in 1903, caused so much excitement, Bench said, that "discipline was temporarily relaxed and a crowd of seniors and juniors stood around as the Post Office engineers fitted it to the wall in the office-keeper's little room". When the first call was made, by the Chief Clerk, to the Royal Meteorological Society, there was a hushed silence as he informed them that the Office was now 'on the telephone', with the number 'Victoria 153'. Thereafter, Bench recalled, members of staff rushed to use the telephone "to ring up friends, and a few expressed surprise and indignation when asked to pay for private calls"! Before there were telephones, communication between rooms was carried out by means of speaking-tubes, with a whistle at each end for attracting attention.

The first person to use a typewriter was Frank Snell, an expert telegraphist, who had been transferred to the Meteorological Office from the General Post Office. Until September 1905, the forecasts and remarks on the weather that were distributed to newspapers and press agencies were written and duplicated by hectograph process (i.e., by means of a gelatin plate that received an impression of a master copy). There-after, they were typed on stencil sheets for reproduction on a Gestetner Duplicator.

Bench stated that the Meteorological Office was subject to the attention of cranks. "There was", he said, "an occasion one evening when the forecast staff had a visit from a man who declared he was a 'weather chart'". Bench recalled that Brodie asked the man "if he had come so that he could be filed away with the official charts". This appears to have provoked some argument, but the man was eventually "persuaded to go away and come another day"! Rather more welcome were the visits paid by notable meteorologists such as Alexander Buchan, who was, said Bench, "a tall venerable figure, with a long patriarchal beard".

A Fresh Approach

From the outset, Napier Shaw made his research intentions clear. He published several papers each year, and the quality of his work was such that some of the papers were reproduced in journals abroad.[25] He led by example, and a research ethos soon developed in the Office.

One of the first projects initiated by Shaw was an inquiry into the occurrence and distribution of fog in London. A grant of £250 for the project was agreed by the London County Council on 22 October 1901, and the Meteorological Council moved quickly to choose someone to conduct the inquiry, approving at their meeting

---

[25] A complete list of the works of Napier Shaw can be found on pages 249 to 263 of a book containing and entitled *Selected papers of Sir Napier Shaw FRS* (Macdonald & Co., London, 1955, 275 pp.). The magazine founded by Symons changed its name in 1901, from *Symons's Monthly Meteorological Magazine* to *Symons's Meteorological Magazine*. 'Monthly' appeared in the title in January 1901 but not in February or thereafter.

on 6 November the appointment of Captain Alfred Carpenter, a retired Royal Navy officer. Observations of the occurrence and duration of fog began forthwith at many of the stations of the London Fire Brigade and were supplemented soon afterwards by observations made by private individuals and others across London. Observations were made from the middle of December 1901 until 5 April 1902, when they were discontinued for the summer.

Carpenter prepared a report which embodied the results of the inquiry and included an account of the meteorological conditions associated with the occurrence and distribution of fog in London. In this, he pointed out that smoke produced by the inhabitants of London aggravated fog problems. He recommended that observations be continued through another winter and that additional readings of temperature be taken at 5 a.m., to test the possibility of forecasting the persistence of fog during the day from meteorological conditions present in the early morning. In the event, he could not carry out the work himself, as his health had suffered as a result of exposure to the atmosphere of London during foggy weather. Lempfert took his place as director of the inquiry. The programme of observations began again on 1 September 1902 and continued until 28 March 1903. When it recommenced, however, only £15 of the grant remained. Because this was not sufficient to cover expenses during the winter, the Meteorological Council applied to the London County Council for an extension of the grant. This was refused, so the inquiry could not be continued after the winter of 1902–1903. Observations were made until 28 March 1903, and the apparatus was then withdrawn.

To help with the investigation of fog in the London area, Mr P Y Alexander of Bath offered to lend the Meteorological Council a signal balloon, along with recording instruments. The offer was accepted, and trial ascents were made in early 1903 from the grounds of the NPL at Teddington. The weather, though, was less than cooperative! Early 1903 was unusually windy, so the winter was unusually free of fog, and, furthermore, as it was put in the *Report of the Meteorological Council for the Year Ending 31 March 1903*, "an attempt to utilize the first approximately calm day resulted in the balloon breaking free from its moorings and carrying away some of the instruments". Somewhat damaged, the balloon was recovered from the north coast of France!

A notable meteorological occurrence in early 1903 prompted investigations by Lempfert and Shaw. On 21 and 22 February, across southern England and much of Wales, visibility was reduced by a thick haze which an observer in London considered remarkable because a strong wind had been blowing at the same time. From many places south of a line from Suffolk to Anglesey, there came similar reports, with observers mentioning deposits of yellow or reddish-brown dust. In collaboration with Dr Mill of the British Rainfall Organization, Lempfert studied the dust-fall, making use of reports supplied by observers and chemical analyses of the dust carried out by Dr John Flett of His Majesty's Geological Survey. Using all available meteorological information, Mill and Lempfert traced the course of the dust and concluded that its source had been in the northwest of North Africa.[26]

Subsequently, Lempfert and Shaw traced trajectories of air in and around several other weather systems and published their findings in a *magnum opus* of 107 pages that has been recognized as a classic work in meteorology. It was published by the Meteorological Office in 1906 and bore the title *The Life History of Surface Air Currents. A Study of the Surface Trajectories of Moving Air.* The cyclone model which they produced focused on trajectories of air rather than patterns of isobars and suggested the physical processes responsible for rainfall. It was a major advance but was largely ignored at the time, and Shaw himself was dismissive, considering the model a superficial representation of a cyclone. This was a pity, because it was not only a worthy successor to those of Dove, FitzRoy and Jinman but also anticipated by some thirteen years the model of the Bergen School of Meteorology, which was widely adopted by forecasters and other meteorologists in the 1920s and 1930s. However, Shaw did come to realize the significance of the work he and Lempfert had carried out, for he noted in his 'Random Recollections' that their study of the great dust-fall and a notable storm five days later had begun "the analysis of the motion of the air of a cyclonic depression into distinct currents which had been so fruitful in the hands of the Norwegian meteorologists".

Another project supported by the Meteorological Council was an investigation of the upper atmosphere by means of diamond-shaped box kites. This was initiated and partly funded by the Royal Meteorological Society and carried out off the west coast of Scotland in the summer of 1902 by W H Dines, then President of the Society. Observations were made partly from a small island in Crinan Bay but chiefly from the deck of a steam tug in the Sounds of Jura and Scarba and on the open sea south of Mull.

In all, seventy-one ascents were made, and kites reached 10,000 feet on several occasions. Self-recording instruments were used to measure vertical profiles of temperature and humidity. The work showed that observations could be made from aboard ships without undue difficulty or undue risk of loss, and the Council considered, as they said in their report for the year ending 31 March 1903, that it was "more desirable" to investigate the upper air over the sea than to do so over land. However, the expense involved made it "clearly improbable that a vessel could be specially retained for such experiments".[27]

[26] See Mill, H R and Lempfert, R G K (1904), 'The great dust-fall of February 1903, and its origin', *Quarterly Journal of the Royal Meteorological Society*, Vol. 30, pp. 57–91.

[27] For details of the work carried out by Dines, see his paper on 'The method of kite-flying from a steam vessel, and meteorological observations obtained thereby off the west coast of Scotland' (*Quarterly Journal of the Royal Meteorological Society*, 1903, Vol. 29, pp. 65–85) and his paper with Napier Shaw on 'Meteorological observations obtained by the use of kites off the west coast of Scotland, 1902' (*Philosophical Transactions of the Royal Society of London*, Series A, 1903, Vol. 202, pp. 123–141).

Figure 6.4. W H Dines and companion launch a box kite in Crinan Bay, Scotland, 1902. Photograph in the National Meteorological Archive reproduced by kind permission of the Royal Meteorological Society.

Convinced of the value of investigating the atmosphere by means of kites, Shaw appealed to the owners of steam yachts to become involved in such work, doing so through a letter published in *The Times* on 2 June 1903. Only one response was received, from Mr C J P Cave, of Ditcham Park in Hampshire, who subsequently used kites and winding gear he provided at his own expense to make measurements over Barbados on nine occasions in April and May 1904.[28] He later made measurements over Barbados and southern England by means of sounding balloons.

Further investigations of the upper air by means of kites were made by Dines near Crinan in the summers of 1903 and 1904 and at Oxshott, his residence near Leatherhead in Surrey, in early 1904 and between October 1904 and September 1905. As before, he was supported by the Meteorological Council, the British Association, the Royal Meteorological Society and the Government Grant Committee of the Royal Society. He continued to carry out upper-air investigations for the Meteorological Office for some years after 1905, though from Pyrton Hill in Oxfordshire from November 1906 onwards. Oxshott was rapidly becoming a London suburb and was by then, he considered, too thickly inhabited for kite work.

Another who made meteorological observations over the sea by means of kites was George Simpson, a future Director of the Meteorological Office. In July and August 1905, from a trawler, he measured temperature and humidity profiles above the North

[28] Charles John Philip Cave was a gentleman of leisure who wrote a classic work entitled *The Structure of the Atmosphere in Clear Weather*, published by Cambridge University Press in 1912. In this, he presented results from 200 pilot-balloon ascents launched from Ditcham Park.

Sea to heights of several thousand feet. Eight ascents were made, and one reached a height of 5800 feet.[29] The study was initiated by the Kite Committee of the Royal Meteorological Society, and the observations were made by Simpson because Dines was otherwise engaged, as mentioned previously. Simpson had been working for Napier Shaw since 1 March 1905 as a scientific assistant, recommended for this post by Arthur Schuster, a member of the Meteorological Council. When, in the autumn of 1905, Schuster founded a meteorological department in the Victoria University at Manchester, Simpson was placed in charge, thus becoming the first lecturer in meteorology in any British university.[30]

Whilst on the North Sea, Simpson took the opportunity to investigate the effects of wind on sea state for the Meteorological Office. The annual reports of the Meteorological Council for the years ending 31 March 1902 and 31 March 1903 state that the Office had not been able to respond to a request by the Mersey Docks and Harbour Board for a revision of the table of speed equivalents of the Beaufort wind scale because the Office's staff had been fully occupied with other work. As noted in Chapter 2, scales of wind force and weather had been advanced by Admiral Beaufort in the early nineteenth century. His wind scale had been formulated in terms of the canvas that could be carried by a fully-rigged frigate. This had been revised in 1874 to take account of changes in the rig of warships and expanded two decades later to include particulars of the sail required by fishing smacks, but it had become clear by the early twentieth century that the increasing use of steam-powered propulsion was rendering a specification based on the canvas carried by a sailing ship impracticable.

Simpson proposed an alternative scale of wind force, based on the sea's appearance. Published in 1906, it was soon accepted by mariners and meteorologists and in 1939 adopted officially by the IMO. Also in 1906, Simpson devised a wind scale for observers on land. Like a scale used by the Palatine Meteorological Society in the 1780s, it expressed wind speeds in terms of: the rustling of leaves; movements of smoke, flags and branches; the whistling heard in telegraph wires; and other effects of wind.[31]

Simpson would later spend time in the Antarctic as the meteorologist for the 1910–1913 *Terra Nova* expedition of Captain Robert Falcon Scott, which ended in Scott's death. Whilst in the Antarctic, he measured vertical profiles of temperature and humidity by means of meteorographs attached to small balloons.[32] For this expedition, as for Scott's *Discovery* expedition of 1901–1904, the Meteorological Office supplied

---

[29] Simpson discussed his findings in a paper called 'An attempt to fly kites for meteorological purposes from the Mission Ship attached to a deep-sea fishing fleet in the North Sea', published in the *Quarterly Journal of the Royal Meteorological Society* in 1906 (Vol. 32, pp. 15–28).

[30] Simpson remained at Manchester for only one year. He joined the staff of the India Meteorological Department in 1906.

[31] For information about the Palatine Meteorological Society, see Chapter 1.

[32] Meteorographs are self-recording instruments that are attached to kites or small balloons and sent aloft to measure pressure, temperature and humidity.

instruments. They also supplied them for the *Discovery* expedition's support ship, the SS *Morning*, and for the Scottish National Antarctic Expedition of 1902–1904, led by William Spiers Bruce. In addition, they employed temporary staff to reduce observations of the *Discovery* expedition, paid for by a Royal Geographical Society grant of £500.

A New Form of Communication

An important invention in the closing years of the nineteenth century was wireless telegraphy. We do not know who first pointed out the meteorological opportunities offered by this form of communication, but it was surely inconceivable they were not recognized by forecasters soon after Marconi succeeded in transmitting messages across the Atlantic, which he achieved in December 1901. Forecasters had long dreamt of current weather observations being available from points far from land.

In fact, a meteorological use of wireless telegraphy had been suggested by the Italian physicist Thomas Tommasina prior to December 1901. According to the July 1901 issue of the *Quarterly Journal of the Royal Meteorological Society* (Vol. 27, p. 198), he had adapted receiving apparatus to follow the course of distant thunderstorms and "even to forecast rainy weather twelve hours in advance". He felt sure his apparatus would be of great use on ships at sea for locating storms and believed it much more reliable than the instruments currently used for predicting the weather!

The distinguished physicist Norman Lockyer reported in a letter published in *The Times* on 14 January 1903 that he had discussed with Napier Shaw "the desirability of obtaining information regarding barometric pressures from ships crossing the Atlantic by utilizing Marconi's marvellous system". Moreover, he had taken the matter further and contacted Marconi, who had said in an ethergram he had sent Lockyer that he hoped to be able to help. Lockyer commented in his letter in *The Times* that he was "sure that all friends of science will be grateful to Mr Marconi for such generous and invaluable assistance, which will undoubtedly be of enormous advantage to British meteorology".

The Meteorological Council acted quickly. It is stated in the annual report for the year ending 31 March 1903 that they had been "in correspondence with Lloyd's respecting proposals for an arrangement by which outgoing and incoming Atlantic liners should report readings of barometric pressure and wind to Lloyd's stations in the south and north of Ireland by wireless telegraphy". The Council regretted, however, that they had not found it possible to "conclude any arrangements". Several years were to pass before the use of wireless telegraphy to collect weather reports from ships on the high seas became routine, but ships did occasionally transmit weather reports by wireless telegrams.

The first reports published in a British newspaper appeared in the *Daily Telegraph* on 6 August 1904, from the SS *Tunisian* of the Allan Line. As the author of the article in the *Daily Telegraph* commented, these wireless messages represented "the first step in the development of a system" by which the newspaper hoped to supply the data which weather forecasters required most in order to predict the weather over the British Isles. He had every reason to hope, he said, that they inaugurated "a system destined to be of the utmost value to the people of the UK, and ultimately to the whole of Western Europe".

The first systematic attempt to collect and publish meteorological reports from Atlantic liners was made by the *Daily Telegraph* in 1906, and the first official effort to collect them from ships was made by the Office in January 1907, when arrangements were made with the Admiralty for the occasional transmission of observations from ships of the Royal Navy. The establishment of a parallel service from Atlantic liners was considered at the time prohibitively expensive. However, reports were already being received from some liners and providing useful information for forecasters (if received in time to be used by them). From early 1907, space was allocated on the last page of the *Daily Weather Report* for meteorological reports received by wireless telegraphy.

By late 1908, the Office was in a position to contemplate the experimental collection of weather reports from Atlantic liners. But first, as usual, Treasury approval had to be obtained, which on this occasion was forthcoming, though reluctant. The Treasury wrote to the Office on 31 December 1908 with their agreement to add £400 to the vote for the next financial year for the purpose in question, and the Office subsequently collaborated with the German Central Institute for the Promotion of Maritime Meteorology, the Deutsche Seewarte, in a scheme proposed by the Germans in October 1908 whereby weather reports would be collected regularly from liners crossing the Atlantic. The scheme called for ships to report twice a day when between the longitudes of 10°W and 45°W and transmit weather messages to one or other of two signal stations in Ireland, those at Malin Head and Crookhaven. The Office and the Deutsche Seewarte would each pay a subscription of £52 10s, plus the charges for transmission of messages from the receiving station to their offices. Observations would be made at the same hours as at land stations, 7.00 a.m. and 6.00 p.m. Greenwich Time, and would also be made at 1.00 p.m. from ships then east of longitude 20°W.

Provisional arrangements were in place by the end of December 1908, with the Marconi International Marine Communication Company Limited agreeing special rates with the Office for the receipt of weather messages. However, the Germans were not able to start in January, so the British went ahead alone. The first report was received at 2.15 p.m. on 10 January 1909, from the SS *Corsican* of the Allan Line, situated at 51°N 15°W when the observation was made. By the end of January, seventy-eight reports had been received from eighteen ships of eight shipping lines, but only fourteen reports had arrived in time to be of use to weather forecasters. Another 469 reports from forty-three ships were received in February and 672 from fifty-four ships in March. In all, therefore, 1219 messages were received in the first three months of 1909. Of these, 223 arrived within twenty-four hours of the time at which the observations had been made and only sixty-seven within two hours.

Collaboration between the British and the Germans extended from February to April 1909, in accordance with the agreement that the initial trial would be for three months, but the Germans then withdrew from the scheme for lack of funds, whereupon the Office continued the service without interruption (and also continued to receive weather reports from ships of the Royal Navy). There was, however, further collaboration with the Deutsche Seewarte in August and September 1909, with revised instructions, but the Germans thereafter withdrew from the scheme completely, after which, once again, the Office continued the service without interruption.

At last, there existed a means of obtaining weather reports from seafarers far from land in time for them to be useful to forecasters. Lempfert reported in 1913 that some ninety-five ships were then on the Office's books as regular contributors, with the number of reports from ships varying between 400 and 600 per month.[33] Potentially, and indeed in practice, the observations from ships were of great value to forecasters, but the quality of the observations was sometimes questionable. As noted in the Office's annual report for the year ending 31 March 1910, it could be difficult to reconcile barometer readings between ships and shore stations. The problem of measuring pressure at sea was, however, addressed by the Office, with guidance on this and other aspects of making weather observations at sea published in 1915 in the *Marine Observer's Handbook*.

Shaw's Early Impact on the Meteorological Office

By 1897, when Shaw joined the Meteorological Council, "physicists in universities had lost the active interest in atmospheric problems shown by their mid-century predecessors". So said the distinguished meteorologist Sir David Brunt in a lecture delivered in 1955.[34] Interest in meteorology had then been, he said, "at an all time low", and Shaw had been "distressed by the hostility in high quarters to any effort at progress". Brunt mentioned that he still had in his possession a letter dated 7 April 1934, in which Shaw had "expressed resentment at the indifference of some, and the hostility of others, during the latter years of the nineteenth century".

Within a very few years of becoming the Council's Secretary, Shaw brought to the Office a research ethos, and he greatly improved the efficiency of the Office's administration. As well as the projects described in this chapter, Shaw took a great interest in agriculture and, in particular, relations between autumn rainfall and wheat

---

[33] Lempfert, R G K (1913), 'British weather forecasts: past and present', *Quarterly Journal of the Royal Meteorological Society*, Vol. 39, pp. 173–184.

[34] Brunt, Sir David (1956), 'The centenary of the Meteorological Office: retrospect and prospect', *Science Progress*, Vol. 44, pp. 193–207.

yields. He also became an influential figure in the IMO and began to develop the interest in meteorological education which remained with him for many years. He was at all times, Brunt said, "an inspiring leader, ready not only to express his views on any subject but also to listen sympathetically to views expressed by others".

The lack of scientific vitality which had characterized the closing decades of the nineteenth century soon seemed long ago, and the Office's services developed, too. Storm warnings were prepared for seafarers. Weather forecasts were produced for the general public, and tailored services were provided.[35] The needs of farmers were addressed. Instruments were supplied to ship owners, and fishery barometers were occasionally installed in needy ports, as in FitzRoy's day. Instruments were loaned to colonial governments and other bodies, sometimes for special projects, such as Antarctic expeditions. Cooperation with the Royal and Scottish Meteorological Societies continued, and so, too, did international cooperation.[36]

The Office now had at the helm a scientist with vision, but change was in the air. There were important developments during the remainder of Shaw's time at the helm, some of them dramatic.

---

[35] In 1904, for example, weather forecasts were provided for owners of carrier pigeons.
[36] Napier Shaw was elected a member of the International Meteorological Committee in 1900.

7

A Decade of Change

To mark the fiftieth anniversary of the Office, Laura FitzRoy presented a portrait of her father. This was reported at the meeting of the Meteorological Council on 19 October 1904, but the anniversary otherwise passed largely unnoticed. FitzRoy's era had long since passed. However, another reason for the passing of the anniversary with so little attention paid to it could have been preoccupation with the report of an inquiry published in 1904.

Yet Another Inquiry

The impending closure of the Ben Nevis and Fort William observatories was the subject of questions asked in the House of Commons in 1902. For example, John Dewar asked the First Lord of the Treasury on 28 July whether the Government, in giving an annual grant of £15,300 to the Office, would make it a condition that these observatories "be maintained in a state of efficiency, or consider the advisability of making an additional contribution to the Meteorological Council towards the expense of properly maintaining them". The Prime Minister, Arthur Balfour, replied, saying that he had been advised "it would not be desirable to impose conditions on the Meteorological Council or to inquire into this or that particular observatory".[1] He was not prepared, he said, to answer the part of the question relating to an additional contribution.

A week later, Mr Dewar was on his feet again. He asked Mr Balfour on 4 August if he had been made aware of the "dissatisfaction in scientific circles" at the impending closure of the two observatories. Would he "order an inquiry into the distribution by the Meteorological Council of the annual grant of £15,300, so as to secure that an adequate allowance be made to these stations of scientific observation"?

---

[1] Mr Balfour had been First Lord of the Treasury since 1895 and also Prime Minister since 11 July 1902.

Mr Balfour did not avoid the question this time. He said he was aware of the interest the question had excited and had looked into the history of it. He noted that the Treasury Committee which had carried out an inquiry into the work of the Office in 1876 (see Chapter 5) had recommended that similar inquiries be carried out from time to time. However, this recommendation had not been followed, and he considered that it would now be right to have an investigation. "This would involve no slur or slight on the Scientific Committee who allocate the funds". It was, he said, "a purely scientific matter and ought so to be treated".

Time passed, and the matter appeared to have been forgotten. Then, in the House of Commons on 21 November 1902, Sir John Stirling-Maxwell asked the First Lord of the Treasury what progress had been made "in the appointment of a committee to inquire into the subject of the meteorological grant". Mr Balfour replied that he hoped "at no distant date to be able to make a statement on the subject".

Action soon followed. A Treasury Minute was published on 9 December 1902. A committee was to be appointed. It would be chaired by Sir Herbert Maxwell MP, while another member of the committee would be John Dewar. The other members would be Sir William Abney (a specialist in the chemistry of photography), Sir Francis Hopwood (Permanent Secretary to the Board of Trade), Sir Thomas Elliott (Permanent Secretary to the Board of Agriculture), Mr Thomas Heath (a Treasury official), Dr Richard Glazebrook (Director of the National Physical Laboratory) and Mr Joseph Larmor (a physicist and mathematician who in 1903 became Lucasian Professor of Mathematics at Cambridge). The Secretary to the Committee would be George Barstow of the Treasury.

A long time passed before the committee's report appeared. It was dated 16 May 1904 and laid before Parliament on 9 June. It was a substantial work which comprised two volumes, one containing the committee's findings and recommendations (twenty pages), the other evidence and appendices (122 pages). It revealed that the Committee had focused on five main points:

- The constitution of the Meteorological Council
- The Council's administration of the annual grant of £15,300
- The utility to the public of the Council's work in general and of weather forecasts and storm warnings in particular
- A memorandum submitted by the Council which set out their views on the present requirements of the Office
- The withdrawal of the annual payment towards the upkeep of the Ben Nevis and Fort William observatories

The report revealed, too, that evidence had been received from various scientific bodies and government departments, among them the Royal Society, the Meteorological Council, the Royal Meteorological Society, the Scottish Meteorological Society, the NPL, the Royal Agricultural Society, the Hydrographical Department of the

Admiralty, the Highland and Agricultural Society (Scotland), the Lancashire Sea Fisheries, Ben Nevis Observatory, and the Solar Physics Observatory (South Kensington, London).

The investigations of the Maxwell Committee seem to have made little impression on Napier Shaw, for he said in his 'Random Recollections' that he was not sure he had been a witness! This was a curious lapse of memory, for the Committee stated in their report that they had visited the Office and made themselves "acquainted with the general system maintained in that establishment and the physical conditions under which the staff conducted their work". However, Shaw was indeed a witness, the second to be called, and the report contains references to evidence he gave. The first witness was the Chairman of the Council, Sir Richard Strachey.

Shaw's lapse of memory was all the more curious because he said in his 'Random Recollections' that he had heard rumour there had been a difference of opinion on the Maxwell Committee. As he put it: "One side demurred most strongly to any increase in the grant as being outside the terms of reference, while the other side urged that in consideration of the continuance of the responsibility of the Royal Society, the grant must be increased, though the increase should not be handed over to the Council but paid to the Royal Society for them to use in the interests of meteorology at the discretion of their President and Council." In this respect, Shaw's memory did not fail him. The report of the Committee did indeed contain this and other expressions of disagreement. One such expression came from Mr Dewar, who disagreed with Clause 48 of the report, which read: "We consider that credit is due to the Council for the comparative success with which they have dealt with the superannuation difficulty". In Dewar's view, the Council should not have adopted a superannuation scheme "without the direct sanction of the Treasury and an additional grant towards the cost of it".

Maxwell's Committee noted that the Office had, in 1891, become incorporated under Section 23 of the Companies Act of 1867. They recommended that the registration of the Office as a company under the Joint Stock Companies' Acts should be cancelled, the company wound up and the Office reconstituted as a department under the control of the Board of Agriculture and Fisheries, with the fixed Parliamentary grant of £15,300 transferred to the vote for that Board.

The second recommendation was that "the Office be placed under the control of a man of science as Director of Meteorology, appointed after consultation with the Royal Society, but responsible to the Board of Agriculture and Fisheries, and making his annual report to that Department". The Committee further recommended that there should be a consultative advisory board consisting of the Hydrographer to the Admiralty, a representative of the Board of Trade, a representative of the Board of Agriculture and Fisheries, and two members nominated by the Royal Society. They also recommended that a second officer be appointed as scientific assistant to the Director, to assist him in the general management of the Office and discuss with him such scientific problems as may arise.

Maxwell's Committee judged the premises rented by the Council "neither suitable in character nor adequate in space" and concluded that a new home for the Office needed to be found. Furthermore, accommodation for the library needed to be provided. It contained in 1902, they said, 17,000 books, pamphlets, maps and charts and was growing at the rate of about 1500 books and pamphlets each year. There was little point in "accumulating treatises, statistics, reports, returns and registers without the means of ready access to them". And the librarian, the Committee pointed out, was a clerk who had to perform other duties as a member of the Statistical Branch of the Office. The workload of this Branch, they noted, had increased significantly since 1902, when the Office had taken on from James Glaisher the preparation of returns for the Registrar-General.

So far as the Ben Nevis and Fort William observatories were concerned, the Committee were "not surprised that the Meteorological Council, oppressed with financial responsibilities, and desirous of effecting economies, should have come to the conclusion that their annual grant of £350 to the observatories might reasonably be withdrawn without detriment to meteorological science". However, they thought it a matter of regret that weather reports from these observatories had not been used in the preparation of warnings and forecasts and believed it would "constitute a bad bargain to allow the observatories to disappear". In their view, £350 per annum was a small amount to ensure the continued maintenance of the observatories. At the same time, there ought to be a reorganization of the management of them and consideration given to the scientific purposes for which they might be used. Moreover, they said, the Office should have the right to publish telegraphic reports from Ben Nevis.

Dissent came from Maxwell and Abney, whose view was that the evidence put before the Committee contained no definite results of research carried out on Ben Nevis. They thought it "probable that the Ben Nevis Observatory might be profitably turned to account in the service of other branches of science besides meteorology" but considered, nevertheless, that the Council should not have to bear the burden of contributing £350 a year "to an establishment which they had not found to have been of distinct service to their work".

The recommendations of the Maxwell Committee were generally welcomed by the scientific community, with the editors of *Symons's Meteorological Magazine* especially positive.[2] They were "particularly pleased" with one point made in the report. That was, they said, "the frank manner in which the Post Office has been condemned for the obstruction it has systematically thrown in the way of the work of the Meteorological Office". The Committee criticized, in particular, the charges levied for postal and telegraphic services and the frequent failure of warning telegrams to reach their destinations in time for adequate precautions against approaching storms to be taken.

[2] 'The report of the Meteorological Committee', *Symons's Meteorological Magazine*, July 1904, Vol. 39, pp. 101–109.

Yet Another Reorganization

After Maxwell's Report was published, much correspondence passed between the Treasury, the Royal Society, the Meteorological Council, the Board of Trade and the Board of Agriculture and Fisheries. Eventually, at their meeting on 8 March 1905, the Council learned that control of the Office was to pass to a body appointed by the Treasury from 1 April 1905 or as soon as possible thereafter.

Details of the Office's new constitution were contained in a Treasury Minute which was dated 20 May 1905 and published on 21 June as a Parliamentary Paper [Cd. 2559]. The Minute revealed that moves to end the anomalous position of the Office vis-à-vis the Companies' Acts had been made and also that the Office would not become once more an integral part of the Civil Service, as it had been in FitzRoy's day. Rather, the Council would be abolished and the overall management of the Office vested in a Meteorological Committee which consisted of the Director of the Office as Chairman, the Hydrographer of the Navy ex officio, two members nominated by the Royal Society, and one member each nominated by the Treasury, the Board of Trade and the Board of Agriculture and Fisheries. Members of the new committee would be appointed by the Treasury and hold office for a period not exceeding five years, though they would be eligible for reappointment. Committee members would be unpaid and meet at least four times a year. Their travelling and subsistence expenses would be allowed if they did not reside in the Metropolis.

The Office's Director would be appointed by the Treasury and receive out of the Office's grant-in-aid a salary of £800 per annum, rising after five years to £1000 per annum. He would not be entitled to a pension. Like other members of the committee, he would hold office for five years and be eligible for reappointment until he reached the age of 65. As well as serving as chairman of the committee, he would be responsible for the administration of the Office. The person currently Secretary to the Meteorological Council would become Director and receive the maximum of the scale (i.e., £1000 per annum, from 1 April 1905). Thus Dr Shaw would thenceforth be the 'Director of the Meteorological Office'. In his absence, the committee could appoint one of its members to act as interim Director.

It was stated in the Treasury Minute that the grant-in-aid would be fixed at £15,300 for the time being and administered by the new Committee, who could, with the consent of the Treasury, delegate to the Director such powers of expenditure as they considered proper. A statement of accounts had to be submitted to the Treasury each year, and the Committee had to submit to the Treasury annually each December a statement of how they proposed to apply the grant for the ensuing financial year. A further requirement was that an annual report be prepared for presentation to Parliament.

The members of the new Committee were named in the Treasury Minute. They were, in addition to Shaw: Captain Arthur Field (Hydrographer to the Navy), Captain Alfred Chalmers (Professional Officer of the Marine Department of the Board

of Trade), Dr William Somerville (Assistant Secretary of the Board of Agriculture and Fisheries), Professor George Darwin (University of Cambridge), Professor
Arthur Schuster (University of Manchester) and Mr George Barstow (nominated by
the Treasury). In his 'Random Recollections', Shaw mentioned that Barstow had
been a Classical Scholar of his own College (Emmanuel) and given us *isanaka-
tabars*.[3]

Barstow was responsible for winding up the Meteorological Council. The minutes
of the Meteorological Committee show that an Extraordinary General Meeting of the
Council was held on 25 May 1906 and a resolution carried that the Council be wound
up voluntarily, with Barstow appointed Liquidator. The resolution was confirmed at
a second Extraordinary General Meeting of the Council held on 13 June 1906. The
first meeting of the new Committee was held on 31 May 1905 and the final ordinary
meeting of the old Council on 12 July 1905. The Committee heard at their meeting
on 3 July 1907 that all the property of the Council had been transferred to trustees
for the Meteorological Committee, viz. Dr Shaw and the Permanent Administrative
Secretary of the Treasury. The necessary transfers of the annuities on the lives of
Strachan and Gaster had also been executed.

The editors of *Symons's Meteorological Magazine* commented on the new constitution of the Office in their usual forthright manner.[4] They thought it "a blot on
the reorganization" that the Royal Meteorological Society would not be represented
on the new Committee and noted that the Office was not being "placed under the
Board of Agriculture and Fisheries as was recommended". Thus, they said, the staff
were to be "denied the status and privileges of civil servants, though accorded the
civil service liability to compulsory retirement at an age limit (without the pension
that reconciles the privileged official to conclude his active service)". They were
pleased that the "working of the Meteorological Office" would be greatly simplified
and approved of the decision to "do away with the anomalous and undignified constitution of the Council as a limited company with dummy shareholders". However, they
were disappointed that the Treasury Minute of 20 May 1905 contained no expression
of thanks to Sir Richard Strachey or other members of the old Council who had stood
down.

On the final day of their last year in control, 31 March 1905, the Council hosted a
reception at 63 Victoria Street, the home of the Office. Sir Richard and Lady Strachey
received the many guests and ladies, and most of the Council's members were present.
The fourteen principal rooms of the Office were open for inspection, and staff of the
Office were on hand to exhibit and explain charts, diagrams and instruments. In the
words of a report in *Symons's Meteorological Magazine*, Shaw and his staff deserved

[3] An *isanakatabar* is a line on a chart showing equal atmosphere-pressure range during a specified time interval.
Isanakatabar literally means 'equal ups and downs of pressure'.
[4] See *Symons's Meteorological Magazine*, 1905, Vol. 40, pp. 97–99.

to be "congratulated on the happy result of their efforts in showing the work of the Meteorological Council at its best".[5]

A New Home for the Meteorological Office

The first annual report of the Meteorological Committee showed that economies had been achieved by abolition of the paid Council, revised financial arrangements with the Admiralty and abandonment of the Ben Nevis Observatory. The savings which had resulted had allowed £500 to be granted to W H Dines for his investigations of the upper atmosphere by means of kites and unmanned balloons (including an honorarium of £100). The report showed, too, that the question of collecting meteorological data from local authorities and private individuals was being addressed, partly to supply material for official publications but also for the purposes of scientific investigations or the requirements of agriculture, sanitation or other weather-dependent industries. Grateful thanks were recorded in the report to Strachey, Wharton, Galton, Scott and Buchan for their long service to the Office through their membership of the Council.

In their inimitable way, the editors of *Symons's Meteorological Magazine* summarized the main points of the report in a readable and sometimes sarcastic form, as the following extract from their review shows:[6]

The report refers in guarded and diplomatic language to the difficulties and disabilities under which the Meteorological Office labours with regard to the telegraphic service, but we who are untrammelled by the considerations which compel a public body to walk delicately in the presence of the Postmaster General can say that the British public loses vastly more in the value of property damaged by storms – the warnings of which have been tied up and kept back by Post Office red tape – than it saves in preserving traditions of working which have grown dear to the official heart. The fishing interests in particular would do well in this matter to impress upon Parliament that the public is lord also of the Post Office. The humour of the situation is that the Post Office at present cheerfully receives and accepts payment for telegrams which it delivers after twelve, twenty-four, or even more hours; and we are told that 'the representative of the Postmaster General, who went into this matter very carefully, was only able to suggest that the meteorological service should be fixed for such hours as to allow an adequate margin for the occasional pressure of ordinary business and other causes of delay'. Might he not have suggested with equal appropriateness that cyclonic storms approaching the British Isles should time their arrival so that the first indications of their proximity might be observed during the hours when it is convenient for telegraph operators to transmit and receive messages?

The minutes of the Committee show that the matter of a new home for the Office had been raised at their first meeting, on 31 May 1905, and the minutes of the second meeting, on 5 July 1905, show that specific requirements had been identified. For example, a room of about 60 feet by 35 feet was considered necessary for a library

---

[5] See *Symons's Meteorological Magazine*, 1905, Vol. 40, pp. 44–45.
[6] See *Symons's Meteorological Magazine*, 1906, Vol. 41, pp. 164–166.

and museum, with an ante-room of about 35 feet by 25 feet on the same floor for exhibiting current charts and other material and for dealing with the press and other enquiries. Proximity to a Post Office was also considered important.

Shaw was requested at the meeting on 5 July "to ascertain whether facilities would be given for suitable accommodation by the Commissioners for the Exhibition of 1851, and if so upon what conditions". These Commissioners owned 86 acres of land in South Kensington on which cultural and educational institutions were being established. The minutes of the Committee's meeting on 1 November 1905 show, however, that other possible locations were being considered, for Shaw reported that the erection of a Post Office near London Bridge was being contemplated. However, this possibility was not mentioned again in the minutes and appears to have been quietly dropped. The Committee seems to have preferred the idea of moving to South Kensington and asked Shaw to write to the Treasury to that effect.

A draft letter from Shaw to the Secretary of the Treasury was discussed and approved at the Committee's meeting on 7 February 1906. He stated that the necessity of being "in close connexion with" the Telegraph Service had become more urgent and suggested that the Office could "see their way to some economy in the expense of working, as well as increased efficiency" if accommodated in the same building as a Post Office. He further stated that he had learned from the Secretary of the Office of Works that the erection of a Post Office on a site at South Kensington was under consideration. Accommodation for the Meteorological Office might be provided, he suggested, on the upper floors of the Post Office's building.

Shaw signalled a desire to widen the Meteorological Office's interests, saying that "further progress of the science of meteorology would be greatly helped by the development of interest in the subject among students". He went on to say that such development required facilities for study that did not exist in the UK. The Office had, he said, an abundance of material that was essential for meteorological research but not readily accessible to students. To develop effectively the facilities it could offer, the Office should be, he suggested, close to a university, so that the subject might have, as he put it, "an opportunity of claiming the attention of students". In South Kensington, the Office would be close to the Science and Art Museums and the headquarters of the University of London (then in the Imperial Institute). He asked permission, therefore, to consult the Commissioners of the 1851 Exhibition and the authorities of the University of London before proceeding further.

The Treasury's reply was dated 9 April 1906 and signed by George Murray, a Joint Permanent Secretary. It stated that the Lords Commissioners of His Majesty's Treasury were favourably disposed to the proposal that premises be constructed for the Meteorological Office at South Kensington in a block of buildings on Exhibition Road that would contain a Post Office. However, the building in which the Meteorological Office might be housed had to be an integral part of an overall development plan for the area. Therefore, the Treasury could not be committed definitely at that time to the provision of the proposed new building but hoped, nevertheless, that erection of the building might prove feasible in due course.

Months passed. Then, Shaw received a letter from Sir Edward Hamilton, a Joint Permanent Secretary of the Treasury, together with a copy of a letter to the Treasury from Sir Schomberg McDonnell, Permanent Secretary of the Office of Works. McDonnell's letter was dated 14 December 1906, Hamilton's 27 February 1907. The letter from McDonnell gave the location of the proposed new building as the corner of Exhibition Road and Imperial Institute Road and went on to say that plans had been drawn up for a building that would be used jointly as a Post Office and a Meteorological Office. Officials of the Post Office had been consulted and so, too, had Shaw. In addition to accommodation for a Branch Post Office, Sorting Office and Meteorological Office, there would be an upper floor which the Board of Education or the Inland Revenue might use. McDonnell referred to "the urgent need for better Post Office accommodation" and informed the Treasury that the Office of Works had made "some provision for commencing the building" in the estimates for 1907–1908.

The Committee discussed the letters from McDonnell and Hamilton at their meeting on 25 April 1907 and approved the reply which Shaw sent the following day. The Committee noted with satisfaction that provision of a flat roof was included and assumed that the Meteorological Office would have access to it and the use of it for meteorological purposes. The Committee approved of the plans for the building, subject to a number of points of detail which needed to be discussed with the Office of Works but were unlikely to delay progress. There was a question over heating of the building, for example, and points of financial detail were raised, but the Committee did not think consideration of them would hinder progress.

The Committee begged their Lordships of the Treasury not to allocate the top floor of the new building in such a way as to prevent it being used for purposes of meteorology or the allied sciences. They said they would be "unwilling to commit themselves and their successors for 21 years to the occupation of a building which had no possibility of enlargement". The space would be needed if there was any expansion of meteorological services. They admitted they did not at that time need any more space than was currently provided but pointed out there were potential developments which might alter circumstances. For example, residential quarters for someone of the rank of superintendent of department would be required as soon as the weather forecast service attained its "normal development as an element in the economics of the country". When the work of weather forecasting passed from the current experimental stage to be of "regular practical service to the agricultural and seafaring communities", it would be "necessary to place the management of the details of the work in the charge of a resident official". This work went on from 8 a.m. to 8.30 p.m., and "under a real working system", a responsible official could be called for at any hour when the Meteorological Office was open.

Shaw received a reply from the Treasury at the end of May. He was asked to refer to the two letters which were enclosed. One was addressed to the Postmaster-General, the other to the First Commissioner of Works. All three letters were written by Sir Edward Hamilton, and all were dated 30 May 1907. The Committee discussed them on 3 July.

The letter to the Postmaster-General was brief. It was partly a covering note for the copy of the letter that was addressed to the First Commissioner of Works and partly called to his attention "the question of financial adjustments" between the Post Office and the Meteorological Committee in regard to heating and the employment of a night watchman. It also expressed the hope that the accommodation of the Meteorological Office under the same roof as a Post Office would "facilitate the delivery of telegraphic news and render the provision of special wires no longer necessary". The letter to the First Commissioner of Works advised him that the Treasury had approved in principle the plans for both the Post Office and the Meteorological Office. As regards the vacant top floor, the letter stated that the Committee should have the first claim and that "no permanent allocation of it should be made without previous communication with them". The Committee decided at their meeting on 3 July 1907 that Shaw should "express the thanks of the Committee for their Lordships' consideration of the representations made to them and to represent the wish of the Committee for the speedy carrying out of the proposals".

A complication then arose, as we see from a letter Shaw sent to the Secretary of the Treasury on 12 November 1907. In it, he stated that the Committee understood an implication of a proposal for a revised scheme of building at South Kensington was a change of site for the Meteorological Office. The Committee considered that such a change would be "detrimental to the interests of the Office" and wished it to be known that if a reconsideration of the arrangements previously agreed became necessary they requested the Lords of the Treasury to provide an opportunity for their views to be heard before any decisive step was taken.

The root of the problem seems to have been that the Royal Commissioners of the Exhibition of 1851 had not been fully consulted over the proposed building. They pointed out that the latest proposal for a new building was a departure from the one made in a letter from the Office of Works dated 17 February 1905, viz. to improve the accommodation of the existing Post Office at South Kensington on the site it then occupied. However, the Commissioners were willing to grant consent to the suggested new plan but first wished to be sure the building would harmonize with those already existing on the South Kensington Estate.

Ruffled feathers were soon smoothed and plans for the new building on the corner of Exhibition Road and Imperial Institute Road went ahead. There were, however, modifications of detail in order to optimize usage of rooms. As we see from the minutes of the meeting of the Committee on 1 July 1908, much thought was given to the dimensions and positioning of bookcases in the room which would serve as a library and museum, especially in respect of floor loading. And the Meteorological Office certainly now felt the top floor needed to be used, such had been the expansion of the Office's work in the year which had elapsed since detailed plans for the building had first been prepared.

However, the Lords of the Treasury objected, saying in a letter dated 30 June 1908 that they were unable "to contemplate the assignment of the third floor" to the Committee and "not prepared to allow any part of the premises to be assigned as quarters for a resident officer". The Committee agreed at their meeting on 1 July that assignment of the third floor was entirely a question for the Treasury to decide but chose to press the point about quarters for a resident officer. They resolved that the Director should ask for an interview with Sir George Murray, the Treasury's Permanent Secretary, "to represent the special circumstances of the Forecast Branch of the Office in respect of resident quarters for an official". Alas, as we see from the minutes of the Committee's meeting on 4 November 1908, the Treasury stood firm. They were not prepared to change their minds. They did, though, allow a room on the top floor to be used as a Physical Laboratory.

Thereafter, plans for the move to new premises at South Kensington proceeded smoothly, and the Committee heard at their meeting on 5 May 1909 that the new building was scheduled for completion in November 1910. This being so, and the tenancy of the premises currently occupied by the Office being a yearly one which expired on 31 December, the time available for the move to take place would be only six weeks. The Committee decided that this question should be raised again later in 1909, when there might be a better idea of whether the building would be completed earlier. There was good news for the Committee at their meeting on 1 December 1909. The contract date for completion of the building was now 22 August 1910, and there was a possibility of an even earlier date. The Committee decided to vacate 63 Victoria Street by the end of 1910 and, as a contingency measure, to arrange for continuation of the occupancy beyond the term of the actual tenancy for a sufficient period to cover the transfer of books and documents.

At their meeting on 6 July 1910, the Committee heard that plans for the move to South Kensington were well advanced. Indeed, the Director reported that he had received a draft lease from Treasury solicitors. Its term would be twenty-one years from 25 December 1910 and the rent £640 per annum, with an additional £40 for heating. But there was a difficulty. Committee members were appointed for five years and therefore not in a position to make a contract for twenty-one years on their own account! They could do so only on the instruction of the Treasury. Accordingly, the Committee drafted a letter to the Treasury, to seek clarification, and the matter was duly resolved.

Other matters which concerned the move to the new building were considered at the meeting on 6 July, among them the opening date, removal arrangements, installation of telephones, and procedures for the printing of the *Daily Weather Report*. The latter publication was printed by lithographers at Denmark Hill, five miles from South Kensington. Shaw argued in favour of the printing being carried out in the new premises. He had been in communication with the Superintendent of Printing at the Stationery Office who was, he said, prepared to "take up the question of supplying a lithographic press to be worked electrically". He had been advised that machinery of a size sufficient to print the *Daily Weather Report*, *Monthly Meteorological Charts of the North Atlantic and Mediterranean* and *Monthly Meteorological Charts of the Indian Ocean* would not cost more than £400, and this expense would be offset by savings that would result from there being no need for journeys to Denmark Hill. There would be savings of not only expense but also staff time. Shaw pointed out that printing on the premises was not a new departure. A hand press for printing the first issues of the morning forecasts and addressing envelopes had been part of the Office's equipment for many years. Moreover, as he further pointed out, the printing of daily weather reports was carried out successfully in the offices of various national weather services. A draft letter from Shaw to the Secretary of the Treasury was approved. In this, he sought permission for the installation of a lithographic printing machine in the basement of the new building.

From the minutes of the Committee's meeting on 1 February 1911, we learn that a tender by Messrs Wyman & Sons Ltd had been accepted for the printing of the *Daily Weather Report* and the monthly meteorological charts of the Atlantic and Indian Oceans, as well as for other lithographic work for the Office. Conditions had been laid down in a contract dated 9 December 1910. Printing of the *Daily Weather Report* on the premises had commenced on 2 January 1911, and production of the monthly charts was in hand. The printing was taking place in a lithographic facility installed in the basement of the Office's new building.

The move from Victoria Street to South Kensington began in September and took several weeks. In the words of an article published in *The Times* on Saturday 29 October 1910, "the transfer proved a more serious business than was anticipated". The old premises were then, said *The Times*, still occupied by the Marine Branch, the Observatory Branch and the section which dealt with accounts and correspondence. The Statistical Branch had been transferred on 27 October, and the Forecasting Branch had started work in the new building the following day. The author of the article in *The Times* thought closure of the building in Victoria Street would "doubtless be regretted", saying that many had been in the habit of "consulting the weather maps and reports" which had been "exhibited at the office doors". The new premises were less accessible to most people, but, said *The Times*, enabled the Office "to display a much larger number of exhibits in a more effective way than was possible in the old cramped quarters". The move was eventually completed on 15 November.

The meeting of the Committee on 6 July 1910 took place at 63 Victoria Street, and the next routine meeting, on 2 November, took place at South Kensington. However,

Figure 7.1. The headquarters of the Meteorological Office at Exhibition Road, London, 1910. © Crown Copyright 2010, the Met Office.

the first meeting of the Committee in the new premises appears to have been an informal one on 19 October. An 'At Home' was held in the new premises on 1 December, attended by 300 guests, mostly from government departments. Many exhibits were on show, including current and historical instruments, climatic charts, ships' logbooks, photographs of clouds, lantern slides, a zoetrope showing the motion of air in travelling storms, and diagrams showing upper-air temperatures obtained from self-registering instruments attached to balloons.[7]

The Post Office occupied the ground floor of the new building and greater part of the basement, while the remainder of the space in the basement was allocated to the Office for the lithographic press and a workshop. Science Museum staff occupied offices on the top floor, this being a temporary arrangement requested by the Office of Works and agreed by the Meteorological Committee. In the remaining space on the third floor, the Office had a small Physical Laboratory and a Photographic Room. The library and museum were in the main hall, which was on the first floor. Adjoining the hall, there were offices for the Director, the Director's Secretary and the Marine Superintendent. The Forecast Room, on the second floor, was connected with the instrument room of the Post Office below by means of a pneumatic tube. Instruments used in the ordinary work of the Meteorological Office were exposed on the roof of the building, and the recording parts of these instruments were displayed in places accessible to the public.

[7] A full list of exhibits can be found in an article by Shaw entitled 'The Meteorological Office', published in the December 1910 issue of *Symons's Meteorological Magazine*, Vol. 45, pp. 201–205.

Advances in Research and Education

Whilst the story of the move to a new home was unfolding, the routine work of the Office evolved, and the Meteorological Committee supported Shaw's efforts to recruit well-qualified graduates. The Committee resolved at their meeting on 5 July 1905 that two posts be created as soon as practicable, viz. a Superintendent of Instruments and a Superintendent of the Statistics and Library Branch. There and then, the Committee decided that Lempfert would act *pro tempore* as the Superintendent of Instruments.

Seven months later, at the Committee's meeting on 7 February 1906, Lempfert was appointed Superintendent of Statistics, and permission to proceed with the appointment of a new Superintendent of Instruments was granted. The post was soon filled, as we see from the minutes of the Committee's meeting on 25 May 1906, in which it was recorded that, "on the recommendation of the Director, Ernest Gold, BA, of St John's College, Cambridge, third Wrangler, 1903, and Natural Sciences Tripos, 1904, had been appointed Assistant to the Director at a salary of £250 *per annum*, dating from 25th June, with a view to his becoming Superintendent of Instruments".

Gold had been born in 1881 and educated first at local primary schools, then at Coleshill Grammar School and thereafter at Mason College, Birmingham. He had entered St John's College in 1900 and become a Fellow of his College in 1906. Nearly thirty years later, he provided some glimpses of life in the Office when he joined the staff:[8]

When I entered the Meteorological Office in 1906, at 63 Victoria Street, the Office was distinguished to the public by a number of boards displayed outside at the level of the first or second floor showing the weather at different places along the coast. In the churchyard opposite was the meteorological station with its Stevenson screen and rain-gauge. There was also a Halliwell recording rain-gauge in St James's Park and when, later on, the Office moved to South Kensington the rest of the instruments joined this rather refractory self-recording rain-gauge.

He mentioned that he was "assigned an upper room which at about half-past one was converted into the luncheon room" for Shaw, Lempfert and himself.

Gold went on to say that before he had been in the Office many months he had been invited to open a Monday evening colloquium. He had taken for his subject a topic Shaw had drawn to his attention, "the results of the observations of upper-air temperature at Lindenberg". These observations had been made, he explained, by Richard Assmann, Director of the Royal Prussian Aeronautical Observatory at Lindenberg.[9] Gold said that he understood little at that time about the significance of upper-air temperatures and was therefore "much surprised" when Shaw thanked him

---

[8] See 'Incidents in the march, 1906–1914', *Quarterly Journal of the Royal Meteorological Society*, 1934, Vol. 60, pp. 121–125. Gold's article was appended to Shaw's 'Random recollections' (*op. cit.*).

[9] Assmann was Director from 1905 to 1914.

for the "lucid way" he had opened the discussion. "That was characteristic of Shaw", he said. Gold joined the Office at a time when knowledge of the upper atmosphere was increasing rapidly. In many parts of the world, meteorologists were launching kites and balloons with instruments attached. No longer were meteorological investigations surface-based. The need to extend knowledge of the atmosphere upwards had been realized.

The sending aloft of instruments attached to kites and balloons had become almost fashionable by the time Gold joined the Office, the reasons for most of the upper-air studies being basic scientific curiosity and the technical challenge of obtaining data from greater and greater heights. However, meteorologists were increasingly coming to appreciate that better understanding of weather systems depended on greater knowledge of the upper atmosphere, and other practical reasons for studying the behaviour of the upper air had begun to emerge. The age of aviation had arrived.

Gold's work in the Office was concerned not just with heading the Instruments Branch but also with pursuing research, much of it concerned with theoretical aspects of meteorology. This research led to the publication of two substantial works in 1908. One was a paper in the *Proceedings of the Royal Society* on 'The Relation Between Wind Velocity at 1000 Metres Altitude and the Surface Pressure Distribution'.[10] The other was a Meteorological Office publication called *Barometric Gradient and Wind Force*. We learn, too, from the *Annual Report of the Meteorological Committee to the Treasury for the Year Ending 31 March 1907* that Gold also designed a scale for reading the wind speed from charts by measuring the distance between isobars. However, his research was not entirely concerned with meteorological theory. In one investigation, he compared readings of barometric pressure made on board ship with readings made on land and found that the readings made on ships could not be relied on when winds were strong.[11] He ceased to be Superintendent of Instruments at the end of September 1907 and then returned to the University of Cambridge, but he did not forsake meteorology. He became Reader in Dynamical Meteorology, appointed, indeed, by the Meteorological Committee.

It was mentioned in Chapter 6 that Arthur Schuster had founded a meteorological department in Manchester's Victoria University in 1905. He was a mathematical physicist of exceptional ability and also a wealthy man from a family of merchants and bankers. At the Committee's meeting on 5 December 1906, he made a generous offer which was accepted. As recorded in the minutes, he offered to fund a Readership in Dynamical Meteorology "for the promotion of scientific study in connexion with official meteorological work and the application of mathematics to the free atmosphere".

[10] Gold, E (1908), *Proceedings of the Royal Society*, Series A, Vol. 80, pp. 436–443.

[11] Gold published this work in a paper entitled 'Comparison of ships' barometer readings with those deduced from land observations; with notes on the effect of oscillatory motion on barometer readings', which appeared in the *Quarterly Journal of the Royal Meteorological Society* in 1908 (Vol. 34, pp. 97–111).

The stipend would be £350 per annum for three years and the appointment made by the Committee "in accordance with regulations making provision for":

- A course of lectures at the University of Manchester, or some other university, one term's residence per year being required at the university
- A report to the Office on methods of conducting or reducing meteorological observations
- Researches to be undertaken by the Reader and the publication of results

Shaw, Schuster and Sir George Darwin agreed to be the members of a sub-committee that would draw up regulations and recommend someone for appointment.

This sub-committee acted without delay and submitted a set of proposed regulations to the Committee on 6 February 1907. The person appointed would hold office from 1 October 1907, or such earlier date as may be arranged, and that person's duty would be "primarily to promote the science of meteorology by mathematical investigation" and by delivering "annually a short course of about twelve lectures". The other regulations specified by Schuster at the Committee's meeting on 5 December were also adopted.

At their meeting on 25 April 1907, the Committee heard from the sub-committee that there had been applications from twelve candidates. Of these, three had "special qualifications for the appointment", viz. Dr C V Burton, Mr E Gold and Dr J W Nicholson. "After full consideration", the sub-committee had chosen Gold. The full Committee approved the recommendation and offered the Readership to Gold. He duly accepted it and held it at Cambridge, with the assent of the University. When he stepped down as Superintendent of Instruments, the observatories and instruments branches of the Office were combined into a single branch under Richard Curtis, a long-serving member of staff who had served under FitzRoy. He had joined the Office in 1861, when only 15 years of age.

The Committee agreed at their meeting on 3 July 1907 that another Special Assistant would be appointed, and Shaw was authorized to recommend a suitable candidate. The appointment would date from 1 October 1907 and in the first instance be for a year on probation at a salary of £250. The Committee heard at their meeting on 6 November 1907 that the successful candidate was Richard Corless, of Sidney Sussex College, Cambridge. He had been ranked eighteenth Wrangler in 1906 and placed in the First Class of the Natural Sciences Tripos in 1907. The Committee also heard at that meeting that John Somers Dines, a graduate in mathematics in 1906 from Shaw's old college, Emmanuel, had been appointed Student Assistant for one year from 1 October 1907 at a salary of £50. In the year since his graduation he had been helping his father, W H Dines, carry out investigations of the upper air at Pyrton Hill, Oxfordshire. A third appointment which dated from 1 October 1907 was that of Charles Ernest Pelham Brooks, who joined as a Probationer on leaving school and became a renowned climatologist and statistician. All three men served the Office for many years.

Meanwhile, Shaw had forged a personal link with the University of London. He had given a course of four lectures in May 1906, taking as his subject 'The atmospheric circulation and its relation to weather'. Then, to quote from the minutes of the meeting of the Meteorological Committee on 5 December 1906, he "had received a communication from the University of London offering to establish a Readership in Meteorology in connexion with the University if he would accept the post". He reported at the Committee's meeting on 6 February 1907 that he had accepted the appointment and proposed to "pay over the amount received by way of stipend into a Meteorological Office fund for experiments and lectures". It was noted in the minutes of the Committee's meeting on 4 December 1907 that a course of twelve lectures given by Shaw on 'Meteorological organization and methods of dealing with meteorological observations' had commenced on 21 October 1907 at the Royal College of Science in South Kensington and was being continued on alternate Mondays.

Gold returned to the Office in 1910, finding it in the throes of moving to South Kensington. Whilst at Cambridge he had been highly productive and had greatly enhanced his scientific reputation, not least by explaining the existence of the newly discovered stratosphere.[12] He had published a number of important works on the upper atmosphere, among them, with W A Harwood of Manchester's Victoria University, a substantial report to the British Association on 'The present state of our knowledge of the upper atmosphere as obtained by the use of kites, balloons, and pilot balloons'.[13] He had become much involved in the work of the Association, taking on, in 1907, the position of Secretary of the Association's Section A (Mathematics and Physics).

Some have wondered why Gold chose to return to the Office when his Cambridge years had been so brilliantly productive that an academic career would surely have followed.[14] Perhaps there were economic reasons, for he was already a husband and father by 1910. Perhaps the opportunities in practical and theoretical meteorology offered by the Office under Shaw's direction appealed to him. We do not know. When he returned to the Office, on 1 October 1910, he was appointed Superintendent of the Statistical and Library Division, and Lempfert became Superintendent of the Forecast Division.[15] Corless remained Special Assistant to the Director but with additional duties as Secretary to the Director and Clerk of Publications.

---

[12] In the closing years of the nineteenth century, to the surprise of meteorologists, measurements made by means of balloon-borne meteorographs showed that temperature ceased to fall with height above an altitude of about ten kilometres. The lower atmosphere became known as the 'troposphere' and the atmosphere above the 'stratosphere'.

[13] Gold, E and Harwood, W A (1909), 'The present state of our knowledge of the upper atmosphere as obtained by the use of kites, balloons, and pilot balloons', in *Reports of the British Association for the Advancement of Science*, Section A (Winnipeg, 1909), pp. 71–124.

[14] See, for example, the obituary of Gold by R C Sutcliffe and A C Best in the *Biographical Memoirs of Fellows of the Royal Society* (1977, Vol. 23, pp. 115–131).

[15] 'Branch' was dropped in the summer of 1910 and replaced by 'Division'. The former term was used in the minutes of the Meteorological Committee's meeting on 4 May 1910, the latter in the minutes of the meeting held on 6 July 1910. No reason for the change was given in the minutes.

Figure 7.2. Ernest Gold. © Crown Copyright 2010, the Met Office.

Figure 7.3. Rudolf Gustav Karl Lempfert. © Crown Copyright 2010, the Met Office.

Schuster renewed the Readership for a further three years, saying in a letter to Shaw dated 26 June 1910 that he hoped "an equally good successor to Mr Gold" would be found. By November 1910, no appointment had been made, and some time was to pass before the vacancy was filled. Eventually, as recorded in the minutes of the Meteorological Committee's meeting held on 1 November 1911, the position was offered to Geoffrey Ingram Taylor of the University of Cambridge, a Fellow of Trinity College who had been ranked twenty-second Wrangler in Part I of the Mathematical Tripos in 1907 and gained First Class Honours in Part II of the Natural Sciences Tripos in 1908. Though offered the post in August 1911, he was not able to undertake any duties before the end of the year because he was suffering from enteric fever. He duly took up the appointment on 1 January 1912 and held it at Cambridge, focusing his attention on turbulent processes in the layer of the atmosphere next to the ground (known as the 'boundary layer').

Shaw was certainly recruiting graduates of high calibre, but the state of meteorological education in schools concerned him. He said in the December 1910 issue of *Symons's Meteorological Magazine* (pp. 202–203) that meteorology had been allowed to fall out of school curricula and its place taken by sciences with which teachers were more familiar. Characteristically, he played a leading role as ambassador for the cause of meteorological education and encouraged his colleagues to follow his lead. He gave talks and wrote articles for science teachers and put the case for meteorological education at meetings of the British Association.

He published a book called *Forecasting Weather* in 1911, which soon became the standard British textbook for the public on the science of weather forecasting, and Lempfert published that same year a small book for the scientific layman entitled *Weather Science*. Another contribution to meteorological education in schools, as reported in the *Annual Report of the Meteorological Committee to the Treasury for the Year Ending 31 March 1910*, was a syllabus for a course of instruction in weather study in elementary schools prepared by the Office for the President of the Board of Education, Walter Runciman, at the request of Sir John Brigg MP. Surplus copies of the *Daily Weather Report* were sent to schools gratis and some schools chose to subscribe to this publication. To assist in the Office's educational work, a collection of lantern slides was accumulated, and a memorandum was drawn up which set out the arrangements made by the Office for educational purposes.

The Budget of the Meteorological Office

By today's standards, the budget of the Office in the first decade of the twentieth century was modest.

The Parliamentary Vote increased from £15,300 per annum in 1900 to £15,500 in 1907 and £15,900 in 1909. The reasons for these increases were given in the minutes of the Committee's meeting on 1 December 1909, in two draft documents which

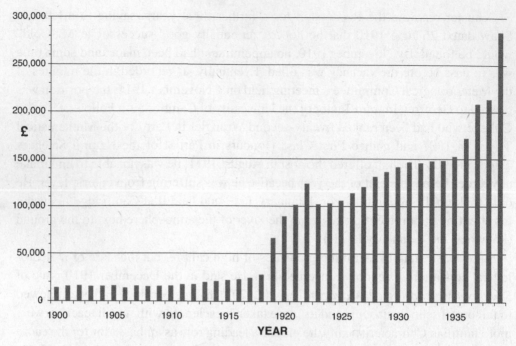

Figure 7.4. Meteorological Office annual expenditure in £ from the year ending 31 March 1900 to the year ending 31 March 1939.

accompanied the estimates for 1910–1911. Both were addressed to the Treasury. One was a statement of the financial position of the Office at the close of the first period of appointment of the Committee (1905–1910). The other was a covering letter. The Committee noted in the statement that the increase in 1907 had been offered for five years on the initiative of the Lords of the Treasury as a contribution towards a subscription of £240 for telegrams from Iceland, and the increase of £400 in 1909 had been made to meet the estimated cost of reports by radiotelegraphy from Atlantic liners.

Shaw pointed out in the statement that no application for an increase had yet been made by the Committee on their own initiative during their term in office. However, they now wished to apply for an increase for a number of purposes, these being:

- To meet the deficiency in the Superannuation Fund (£350)
- To strengthen the Forecast Branch (£500)
- Secretarial work for the Committee and relief in the Statistical Branch (£100)
- Additional expenses of Kew Observatory (£550)
- Capital expenditure for structural alterations to Kew Observatory (£1000)

The Committee pointed out that various services rendered by the Office generated income and quoted figures which provided "strong evidence of the increase in the public appreciation of the work of the Office". The sums received by the Office in return for services rendered and instruments supplied were, the Committee said: £1811 in 1903–1904, £1894 in 1904–1905, £3066 in 1905–1906, £2543 in 1906–1907, £3771 in 1907–1908 and £3397 in 1908–1909.

A reply was received quite soon from Sir Thomas Heath, Assistant Secretary to the Treasury. In his letter, dated 12 January 1910 and addressed to Shaw, he informed the Committee that the Lords of the Treasury would be prepared to ask Parliament to vote an increased grant of £16,500 for 1910–1911 and subsequent years, "apart from any further increase that may be necessary if arrangements were made for transferring [to the Office] the observatories of Kew and Eskdalemuir". Their Lordships would in due course "communicate with the departments concerned and with the Meteorological Office as to the reappointment of the Committee for a further period of five years from 1 April 1910". As regards the sums required for expenditure at Kew, the Lords of the Treasury advised the Committee that no decision could be made until discussions which involved the NPL had been concluded.

In November 1907, Shaw had drawn attention to unsatisfactory financial and operational arrangements between the Office and the NPL in respect of Kew Observatory. In January 1910, discussions and negotiations between the Office, the NPL and the Treasury were still in train. In a letter to the Secretaries of the Royal Society dated 10 January 1910, Sir George Murray, the Treasury's Permanent Secretary, set out proposed arrangements for the transfer of Kew Observatory's meteorological and geophysical work to the Committee and advised that, if arrangements along those lines were made, their Lordships would be prepared (1) "to agree to the observatories at Kew and Eskdalemuir being handed over with their work (except testing) to the Meteorological Committee", and (2) "to provide (a) any necessary sum required to bring the amount available for the staff and maintenance of the observatory at Kew up to £1,400 *per annum*, and (b) any small expenditure required for adapting to the requirements of the Meteorological Committee the approximate amount of accommodation at present devoted to the observatory as distinct from testing work".

At a meeting at Kew Observatory on 14 January, Shaw and Glazebrook agreed on revised estimates for management and maintenance of the building, as well as for caretaking and Dr Chree's salary. Shaw advised the Committee on 2 February that he and Glazebrook had agreed that responsibility for the management of the building and maintenance of the establishment should be transferred to the Office. At the same time, eleven of the fifteen staff, exclusive of the caretaker, would be chargeable to the NPL and the other four to the Office, with all fifteen under Dr Chree's superintendence. The Committee accepted these proposals and agreed that Glazebrook should go ahead and put them to Sir Joseph Larmor, the Royal Society's Secretary.

The Royal Society accepted the proposals, and so, on 21 February 1910, on behalf of the Committee, Shaw wrote to the Treasury, setting out in his letter the proposed arrangements regarding Kew and Eskdalemuir Observatories. A reply was received

Figure 7.5. The distinctive *KO* mark used on instruments tested at Kew Observatory.
© Crown Copyright 2010, the Met Office.

from the Treasury on 26 February. This approved the proposals and fixed the Committee's grant at £16,750 for the year 1910–1911.

Responsibility for the two Observatories passed to the Office on 1 July 1910, with Kew remaining the Office's central observatory. In the administration of the two (and also Valentia Observatory) the Director of the Office was given the assistance of an advisory body, the Gassiot Committee, which had been set up in 1871 to control the funds of the Gassiot Trust. Under the new arrangements, the testing of instruments which had been taking place at Kew prior to July 1910 would be moved to the NPL's Teddington headquarters as soon as the necessary provision could be made, and the Laboratory would retain the distinctive *KO* mark for use on instruments that had been tested at Kew Observatory. Dr Chree remained Kew's Superintendent.

The minutes of the Committee's meeting on 2 November 1910 show that joint occupation of Kew Observatory by the Office and the NPL was not working well. It was making effective coordination of the Office and Observatory staff difficult. As recorded in the minutes:

The practice prevails of using overtime, not exclusively to meet unforeseen contingencies but as a regular supplement to the routine of the observatory, and the laboratory staff and observatory staff help one another out with overtime in a way that makes the application of any ordinary principles of organization inappropriate in this particular case. It is desirable that we should urge the speedy separation of the staffs. The building is not really adequate for the accommodation of the number of persons employed there.

There was no such problem at Eskdalemuir.

The question of the Committee's reappointment for a further period of five years was resolved by a Treasury Minute dated 31 March 1910, in which it was stated that the Committee would be "reconstituted from 1 April 1910 for a further period of five years in the same manner and with the same powers as in the term just expired", and also that Shaw would be "reappointed for the same term at the same salary and with the same powers and responsibilities as heretofore". In addition to Shaw, the members of the Committee would be Rear-Admiral Herbert Purey-Cust (Hydrographer of the Navy), Captain Joseph Harvey (Principal Examiner of Masters and Mates, Board of Trade) and Mr Thomas Middleton (Assistant Secretary, Board of Agriculture and Fisheries), together with Sir George Darwin, Professor Arthur Schuster and Mr George Barstow.

A New Meteorological Branch in Scotland

An important aspect of the work of the Office was the collection, analysis and publication of observations from climatological stations in the UK and the colonies. However, three other bodies collected observations, too, these being the Royal Meteorological Society, the Scottish Meteorological Society and the British Rainfall Organization (BRO). All of these bodies published monthly summaries of climatological data which supplemented, or overlapped with, those of the Office. No publication presented results for all stations, and results for some stations were often published two or three times over in the same or modified form. Shaw therefore suggested, in a memorandum discussed at the Committee's meeting on 5 July 1905, that "the proper course would be for the four organizations to cooperate in a single comprehensive and effective publication", provided funds permitted, which they did not at that time. The Committee decided at the meeting to invite representatives of the Royal and Scottish Meteorological Societies to a conference with the Director and Professor Schuster "to consider upon what terms and conditions the payments hitherto made to the Societies should be continued".

The invitation was issued on 11 October 1905 and a conference took place at the Office on 25 January 1906. Shaw and Schuster represented the Office, Richard Bentley and Francis Campbell Bayard the Royal Meteorological Society and Alexander Buchan and Robert Traill Omond the Scottish Meteorological Society. No one from the BRO was invited. The Committee agreed on 6 December 1905 to cover the travelling expenses of the representatives from Scotland and to "offer a sum of £150 a year to Dr Buchan for three years for the discussion of the Ben Nevis observations in lieu of salary of the same amount which he received as Inspector for Scotland"; and they heard on 7 February 1906 that Buchan had accepted this proposal and resolved that, if he wished to continue the work and the Committee's Parliamentary Grant also continued, "the like sum be assigned for a similar purpose in the two succeeding years". He did wish to continue the work, and indeed did so for a while, but sadly died in 1907.

The minutes of the conference on 25 January 1906 show that various letters and supporting documents were discussed at the meeting. These included Shaw's invitation to the conference, replies from the two societies, a long and detailed memorandum from Shaw dated 16 December 1905 and an equally long and detailed set of remarks on the memorandum from the Scottish Meteorological Society. So multifaceted was the matter and so many were the possible sticking points that agreement at the conference seemed unlikely. In the event, the differences between the various parties did not seem to be as great as might have been anticipated, and the minutes of the conference show that Shaw was "unanimously requested to draw up and circulate for consideration a detailed proposal with regard to the cooperation of the two societies with the Office in a single publication".

Alas, the initial optimism soon faded. There were difficulties and disagreements, and both societies were reluctant to give up their work, some of which had gone on for decades. By early 1907, the BRO had become involved, and at the meeting of the Committee on 6 February 1907 Shaw was able to report an agreement with Dr Mill, Director of the BRO, for a monthly isohyetal map of rainfall for the Monthly Weather Report.[16] As far as cooperation with the two societies was concerned, Shaw reported to the Committee on 6 February that no scheme of joint monthly publication could be carried through in the current year. Accordingly, "negotiations must therefore have reference to 1908 at the earliest". That year came and went without agreement, as, too, did 1909 and 1910.

At last, agreement was reached with the Royal Meteorological Society. The Committee heard at their meeting on 5 July 1911 that the grant of £25 a year for inspections paid by the Office to the Society would not be continued beyond the end of 1911. As a result, the Society decided to discontinue publication of the *Meteorological Record*, which was the Society's equivalent of the Office's *Monthly Weather Report*.[17] The Society would now, as it was put in the minutes of the meeting on 5 July, "refer its observers to the Meteorological Office for the supervision and reduction of the observations and the publication of the data which they afford". In lieu of issuing *The Meteorological Record*, the Society would purchase copies of the *Monthly Weather Report* for distribution to its Fellows, the cost being only that of paper and stitching.

A number of financial questions needed to be resolved, and indeed soon were, allowing agreement to be reached that the Office and the Society would collaborate to prepare a series of normal values of some climatological elements for the British Isles. To consider how this work would be carried out, a joint committee was appointed, its members Lempfert and Gold for the Office, Henry Dickson and Henry Mellish for the Society.[18] The outcome was that the clerical work would be carried out by Society staff under the supervision of the joint committee. And as a contribution towards the cost of the work, the Office would place at the disposal of the Society a sum of £50 per annum for two years.

A critical development in respect of work carried out for, and information supplied to, the Office by the Scottish Meteorological Society came in 1912. The President of the Society, Alexander Crum Brown, wrote to the Registrar-General for Scotland on 7 March. He said that the present position of the Society was causing grave anxiety because subscription income had decreased and the Council of the Society had found it impossible to make good the shrinkage. The Council's position was that "no part of

[16] *Isohyets* are lines on maps joining places which have the same amount of rainfall in a given period.
[17] Publication of *The Meteorological Record* began in 1881. The introduction to this annual volume stated that the Royal Meteorological Society commenced the organization of a series of 'Second Order Stations' in 1874 and another class of stations called 'Climatological' in 1880. Prior to 1881, records from stations were printed as an appendix to the *Quarterly Journal of the Royal Meteorological Society*.
[18] Dickson was the President of the Royal Meteorological Society for 1911 and 1912. Mellish served as President for 1909 and 1910.

the income derived from subscriptions should be diverted from its primary purposes of giving encouragement and direction to scientific research and of meeting the cost of preparing and publishing research papers and papers of educational value". In the view of the Council, "the extra cost of supplying reports to a government department should be met out of public funds". At present, he said, the Society received an annual payment of £100 for reports sent to the Registrar-General for Scotland and an additional £95 per annum from the Office for work done and information supplied. Respectfully, the Council now applied for the annual grant of £100 to be increased to £300.

The Registrar-General for Scotland, James MacDougall, forwarded the President's letter to the Scottish Office in London on 12 April 1912. He pointed out that the grant of £100 had fallen far short of the cost of collecting the necessary material and had always been supplemented out of the general funds of the Society. He trusted that the Secretary for Scotland would see fit to "press the claims of the Society upon the Treasury" for the increased grant that was requested, particularly in view of the fact that "public health statistics were becoming more and more important in their bearing upon the social conditions of the population" and were "incomplete without a reasonably full statement of existing meteorological conditions". He also suggested that the annual grant to the Meteorological Office from the Treasury might be increased "in order to extend and add to the number of centres in Scotland from which meteorological data were supplied". John Lamb of the Scottish Office passed MacDougall's letter to the Treasury on 8 May 1912, saying in his covering letter that the Scottish Office suggested that the annual grant to the Society be increased by the amount desired.

Months passed. Then, on 1 November 1912, Sir Thomas Heath, Assistant Secretary to the Treasury, wrote to the Under Secretary for Scotland. He did not disagree that the Scottish Meteorological Society needed an increased grant. However, the Lords of the Treasury thought a grant by Government which formed so large a proportion of the Society's annual expenditure could be justified only by taking steps to "bring the work of the Society into much closer cooperation with that of the Meteorological Committee, which was the Government authority for meteorology throughout the United Kingdom". He therefore suggested that the Secretary for Scotland should request the Society to "place itself in communication with the Meteorological Office in order that a scheme of readjustment of the activities of the Society and of the Office may be prepared".

Further correspondence passed between the Scottish Office, the Registrar-General for Scotland and the Scottish Meteorological Society. Then, on 25 February 1913, the Permanent Secretary of the Treasury, Sir Robert Chalmers, wrote to Shaw and copied to him all of the correspondence there had been during the previous year in respect of the Scottish Meteorological Society's case for additional funds. He said that the Treasury. proposed to ask Parliament to increase the Office's grant to

£20,000 for the year 1913–1914 and asked Shaw to confer with representatives of the Society to "consider whether from the increased grant placed at the disposal of the Meteorological Committee he could not find the funds necessary to place the finances of the Society on a satisfactory footing and arrange for a closer cooperation between the work of the Society and that of the Meteorological Office".

The Committee considered the matter at their meeting on 12 March 1913 and resolved that Shaw should do as the Treasury requested, on the understanding that the total charge on the Office should not exceed £350. Shaw duly conferred with members of the Society, the outcome being agreement that the best way to proceed would be to establish a branch of the Office in Edinburgh. This was endorsed by the Committee at their meeting on 30 April 1913, with the objects of the branch agreed to be:

- The collection of trustworthy meteorological statistics from municipal and voluntary stations in Scotland and the preparation of summaries for the meteorological reports issued by the Government, and for the statistical reports of the Registrar of Births, Deaths and Marriages for Scotland
- The supply of meteorological information in reply to inquiries
- The promotion by all available means of public technical instruction in meteorology and of the applications of meteorological science in the interests of public health, of agriculture, fisheries and other industries
- The promotion of meteorological researches, including researches on an international basis, which depend on the organization and compilation of observations

The Committee would contribute £350 per annum from their Parliamentary Vote, and in addition, the Treasury would continue to pay the Society £100 on account of the Registrar-General for Scotland. The salary of the branch's Superintendent would be £250 per annum. Rooms would be provided by the Society, and the Society's paid Secretary, Andrew Watt, would be appointed Superintendent. A committee would administer the grant, and the Committee heard at their meeting on 25 June 1913 that its members would be the Director of the Meteorological Office (Chairman), R B Greig (Board of Agriculture for Scotland), James MacDougall (Registrar-General for Scotland), and three representatives of the Scottish Meteorological Society, these being Mackay Bernard (President), Professor R A Sampson (Astronomer-Royal for Scotland) and E M Wedderburn (Honorary Secretary). This committee met for the first time on 19 June 1913.

Observations from all of the climatological stations in Scotland were included in the *Monthly Weather Report* from January 1914, and the Meteorological Committee stated in their annual report to the Treasury for the year ending 31 March 1914 that "for the first time since the collection of meteorological information began to be organized some fifty years ago the available climatological information for all parts of the British Isles is included in a single publication". At last! More than eight years had elapsed since Shaw's efforts to unify climatological data in one publication had begun. There still remained, however, the problem of the BRO, which collected rainfall records from about 5000 observers and published summaries of them in an annual volume called *British Rainfall*.

Communications and Standardization

Meanwhile, the Meteorological Committee sought to improve the distribution of weather forecasts by means of the electric telegraph. They agreed at their meeting on 2 November 1910 that an application be made to the Development Commission for a grant of £5000 a year for five years for the purpose of distributing forecasts by telegraph to all postal-telegraph offices in the UK. The sum requested would be wholly expended in defraying the charges made by the Post Office for transmitting messages by telegraph. Even though they were now sharing a building, there were still disagreements between the Post Office and the Meteorological Office over charges for telegrams. As stated in the minutes of the meeting held on 2 November, the grounds for making the application were that:

• Trustworthy information as to the weather prospects for the ensuing twenty-four hours would be of material benefit to agricultural communities as well as to fishing communities and longshoremen.
• The forecasts issued by the Meteorological Office were sufficiently trustworthy for the purpose.

The draft letter of application for the grant pointed out that forecasts which were prepared in the evening were issued without charge to newspapers and appeared in nearly all morning papers. By this method of distribution, however, the information took about twelve hours to reach readers and failed to reach people who did not take a daily paper. But if the telegraphic method were used, the information would reach all village post offices within an hour of its issue from the Meteorological Office and would be "practically by repetition within reach of everybody in a few hours".

The application was unsuccessful. Indeed, by making it, the Committee made themselves look ridiculous. The reply from the Post Office, dated 30 March 1911, pointed out that the number of their country sub-offices in the UK was so great – 8218 – that the cost of sending a telegram daily to each of them would be prohibitive. Taking the average number of words as fifteen for the text and two for the address, the annual cost for the telegrams alone would be £106,234 for a daily service and £91,099 for a service on weekdays only. Therefore, the application for a grant of £5000 was, to say the least, unrealistic. Attempts were made by the Post Office and the Meteorological Office to settle their differences over charges for telegrams, with financial compromises made by both parties, but their differences had still not been fully resolved when, in 1914, the First World War broke out.

Another issue that had been unresolved for years was the lack of a uniform system of units in meteorology. No consensus was reached at the meeting of the IMC held at Southport in 1903, nor was it at the meeting of that Committee held at Berlin in 1910. However, a decision to adopt the CGS (centimetre-gram-second) system of units for theoretical and thermodynamical studies in meteorology and for studies of the upper levels of the atmosphere was made at the Berlin meeting. A matter which was in fact resolved at Southport, though, was that of the time of the morning observation. The British agreed to alter their standard observation time to 7 a.m. GMT, thus synchronizing their observations with those made at 8 a.m. in countries which used Central European Time.

Daylight Saving Time was a contentious issue which occupied a considerable amount of Parliamentary time in 1908 and 1909, and the Office became involved. Shaw gave evidence to a Select Committee of the House of Commons in May 1908 and also wrote letters to *The Times* on the subject. As far as he was concerned, as he said in a letter published on 6 June 1908, meteorologists were "as much dependent on the sun as the navigator is", and "a separate system of time-keeping would have to be used". He would rather not have to alter the clocks. He was not best pleased that his time had been taken up with appearing before a Select Committee and considered that "the official guardians of the standards" should have been consulted formally before legislators became involved in a matter of such worldwide importance as time.

Though Shaw did indeed object to the introduction of Daylight Saving Time, the statement made by George Courthope in the House of Commons on 5 March 1909, during the debate on the Daylight Saving Bill, was a gross exaggeration. He claimed that "the whole force of the Meteorological Office" was against the Bill and "the whole weight of the Royal Meteorological Society as well". No evidence has been found to support the contention that the Society was against the Bill, and the overall attitude of the Office seems to have been fairly neutral. Shaw certainly testified against the Bill, saying that there might be uncertainty regarding the hours at which observations were taken, but he indicated that any problems which might arise were not insurmountable. The Bill was defeated, but not many years were to pass before Daylight Saving Time was indeed introduced.

The Continuing Impact of Shaw

The first decade of the twentieth century was one of great change for the Office, with the main contributory factors being Shaw's great emphasis on science, the improvements he made to internal administrative procedures, the change in management structure that resulted from the Maxwell Inquiry, the move to South Kensington, the advent of wireless telegraphy, the return of Kew Observatory to Meteorological Office control, and the acquisition of the observatory at Eskdalemuir. Shaw steered the Office through these many changes with great skill and dedication, but there was a toll on him, as we see from a letter he wrote on 8 February 1911 to a personal friend and confidant, C J P Cave (who was mentioned in Chapter 6).[19] In the letter, he said that he was "still very much worried about things at the Office", adding:

I have the sort of feeling that in a properly organized Office everything ought to go of its own accord and the Director might look in occasionally so as to be able to take the credit for the working with some assurance – but it is not that just now at all – I cannot keep up with the wants of various kinds and I get further into arrear as the days go on.

Shaw was a man whose capacity for work was enormous, but he appeared to be driving himself too hard, not just at home but also abroad. He had been since September 1907 President of the IMC, President of the IMO's Commission for Weather Telegraphy and President of the Commission for Storm Warnings and Maritime Meteorology.

At the beginning of the second decade of the twentieth century, meteorologists in Britain and elsewhere must have thought that meteorology would continue to develop along the same general lines as during the previous decade. Forecasters could be expected to continue improving their models of weather systems, experimentalists to continue exploring the upper atmosphere, physicists to extend understanding of atmospheric processes and theoreticians to formulate mathematical expressions of the atmosphere's behaviour. Moreover, the appointment of well-qualified science graduates to the Office could be expected to continue. Indeed, Gordon Miller Bourne Dobson was appointed a Graduate Assistant in 1910 and Francis John Welsh Whipple became Superintendent of Instruments in 1912.[20] But little did Shaw, or probably anyone else, realize how much the world of meteorology was about to change. Clouds were gathering and new challenges lay ahead.

---

[19] This letter can be found among other Shaw correspondence held in the Cambridge University Library.

[20] By the spring of 1912, the last of those who had served under FitzRoy had retired, Charles Harding in September 1911 and Richard Curtis in March 1912.

8

The Great War

The first 'flying machines' were flimsy and the motions of the atmosphere capricious. Many an early aeroplane was damaged or wrecked when dashed to the ground by an unexpected gust of wind. The pioneers of aviation certainly needed meteorological assistance, and the French pilot Louis Paulhan did indeed gain some when, in April 1910, he won the prize of £10,000 offered by the *Daily Mail* for being the first to fly between London and Manchester in under twenty-four hours. The Meteorological Office made elaborate provision to keep him supplied with information about the prospective weather en route.

The Emergence of Aeronautics

The Prime Minister, Herbert Asquith, informed the House of Commons on 5 May 1909 that the Government wished to apply the "highest scientific talent" to problems of aerial navigation. He had appointed a special committee which would be chaired by Dr Glazebrook, Director of the NPL, and include among its members Dr Shaw, Director of the Office.

Shaw played a full part in the work of the Advisory Committee for Aeronautics, as shown by its first report, which was presented to Parliament in 1910. His extensive contribution included meteorological information of importance in respect of aerial navigation, particularly about wind structure and vertical and rotary motions of the atmosphere. Impressed by his contribution, the Advisory Committee proposed that these motions be further investigated, as a result of which the Meteorological Committee included in their estimates for 1910–1911 a sum specifically for work to be carried out at Pyrton Hill by J S Dines and his father (W H Dines). The Committee noted that the Office received from aeronauts many enquiries for forecasts and other information. They also recognized that the Office possessed meteorological records of relevance to aerial navigation which were not being brought effectively to the knowledge of practical aeronauts. Items of interest would be displayed in the museum in the new building at South Kensington.

The minutes of the Committee's meeting on 2 November 1910 contain an announcement of considerable significance for the Office. A proposal to establish a meteorological station at Farnborough under the control of the Office had been made by the Army Council. The purpose of the station would be "to carry out the meteorological observations and investigations necessary for the Balloon School". What was being proposed was the Office's first outstation, and the minutes of the Committee's meeting on 5 April 1911 tell the story of what happened next.

Mr Reginald Brade, Assistant Secretary in the War Office, wrote to the Treasury on 3 February 1911, saying that the Army Council had consulted Shaw and come to the conclusion that it was desirable that a branch of the Office be established at South Farnborough within the precincts of the Balloon Factory. The Council proposed that a trained observer be appointed, paid and directed by the Office, and they offered to provide from Army Funds working space at the Factory equipped with the necessary furniture and laboratory apparatus. They would also provide a person to assist the observer.

Sir George Murray, the Treasury's Permanent Secretary, forwarded the letter to Shaw on 10 February, asking for his views. Shaw had a number of reservations, saying in his reply, dated 21 February, that the Meteorological Committee thought it "better for this work to be done by a civilian, looking to the meteorological service for his pay and prospects of promotion, rather than by an Army officer with special training in meteorology". However, the Committee considered that a Branch Office at Farnborough which was concerned with the practical application of meteorology to aeronautics provided an opportunity to extend the work of the Office "with a promise of usefulness, both to the science of meteorology and to all those interested in aeronautics".

Shaw drew attention to financial aspects which needed to be clarified, noting that no provision for a Branch Office at South Farnborough had been made in the estimates of the Committee. He also pointed out that the combination of scientific training, experimental skill and meteorological experience required by the person who would need to be appointed was not easily found. Moreover, he did not consider £250, the sum available, sufficient to retain the services of a person with the necessary qualifications. And furthermore, the Committee could not "regard as permanently satisfactory an arrangement by which a branch of the Office was maintained on the premises of another Government department by means of funds controlled by a third body". Taking all things into consideration, though, he was prepared to proceed on the terms indicated.

The Treasury sanctioned the establishment of a branch of the Office at the Balloon Factory, "as an experimental measure", whereupon Shaw consulted Glazebrook and thereafter informed the Committee, on 5 April, that he proposed to ask J S Dines to undertake the work of the Branch Office for the year 1911–1912 at a salary of £200. A year later, Dines was reappointed for one year at a salary of £225. Meanwhile, the Balloon Factory was renamed. It became first, on 26 April 1911, His Majesty's Aircraft Factory, then, on 11 April 1912, the Royal Aircraft Factory.

Accommodation for the new Branch Office did not become available for quite some time, but this did not prevent Dines and Dobson from carrying out a preliminary investigation with pilot balloons and preparing daily weather maps for airmen based on telegraphic information from the Meteorological Office in South Kensington. This work was carried out in October and November 1912, using a temporary facility at South Farnborough. Otherwise, in the absence of a permanent facility, J S Dines carried out his pilot-balloon studies at Pyrton Hill.

A potential difficulty arose in the autumn of 1912. The owners of the Pyrton Hill property put it on the market and offered it "for sale for the purpose of an aerological observatory". The Committee did not wish to purchase the site for that or any other purpose, so W H Dines had to seek accommodation elsewhere to continue his upper-air work for the Office. With the assent of the Office, he acquired a property at Benson, six miles west of Pyrton Hill, and moved there early in 1914.

The minutes of the meeting on 6 November 1912 state that the work carried out by J S Dines was being paid for by the NPL out of the grant for the Advisory Committee for Aeronautics, with £400 assigned for the current year. The Meteorological Committee did not consider this arrangement satisfactory and agreed that as soon as work began regularly the grant for it should be included in the Meteorological Office vote.

The accommodation at the Royal Aircraft Factory was completed in November 1913 and J S Dines transferred his work from Pyrton Hill to the new facility at the beginning of the following month, making his first pilot-balloon ascent on the morning of 9 December. His assistant, Harold Billett, joined him on 1 January 1914. A physics graduate of the Royal College of Science and a Demonstrator at the Imperial College of Science and Technology, he had been appointed Assistant Meteorologist at the Branch Office with effect from 1 July 1913. Pending completion of the building at the Royal Aircraft Factory, he had spent the rest of that year at the headquarters of the Office in South Kensington, assisting Shaw and training for the work he would be carrying out at South Farnborough.[1] During December, Dines had only one companion, a Royal Aircraft Factory mechanic-computer, H Allnutt, who was employed as an instrument technician and general assistant. Working conditions were not ideal for several months. There was a shortage of the necessary furniture, fittings and equipment, carpenters and fitters were completing unfinished work, and there

---

[1] The term 'Central Office' was used for several years during the period 1913–1918 but appears to have been an unofficial designation, presumably to distinguish the headquarters of the Meteorological Office at South Kensington from the branches at Edinburgh and South Farnborough and the various observatories.

was no direct outside communication until a telephone was installed in October 1914.

Another development recorded in the minutes of the Meteorological Committee's meeting on 6 November 1912 was the proposal that a meteorologist should join the staff of the Central Flying School of the Royal Flying Corps (henceforth RFC) at Upavon in Wiltshire. The appointment of a meteorologist had been proposed by the War Office and had, in fact, been discussed at the Committee's previous meeting, held on 2 October 1912. At that meeting, the Committee had been informed that arrangements for a course of instruction in meteorology at the School had been discussed with the War Office and a syllabus drawn up. The role of the meteorologist would be to provide courses of instruction for airmen, with experimental work also contemplated. Shaw and J S Dines had visited Upavon on 3 October and the Royal Aircraft Factory the following day.

The Committee suggested that lectures should be given by a member of the Office's staff until the meteorologist appointed had acquired the necessary experience. Lempfert commenced the first course of instruction at Upavon on 17 January and delivered ten lectures on the subject of meteorology in relation to aeronautics. Dobson, who had been in temporary charge of the magnetic work at Eskdalemuir since 1 January 1913, became the instructor and meteorological observer at Upavon with effect from 7 February. Though employed by the War Office, he retained his civilian status and was, indeed, the only civilian on the staff.

It was recorded in the minutes of the Committee's meeting on 12 March 1913 that Ernest Gold had won a competition run by the Deutsche Meteorologische Gesellschaft. The first prize, of 2000 marks, had been awarded for his essay titled 'The International Kite and Balloon Ascents', in which he had discussed upper-air work that had mostly been carried out from 1907 to 1909 but continued at a few stations to 1911. The essay was subsequently published in 1913 as Meteorological Office Geophysical Memoir No. 5.[2] The efforts of Shaw to raise the scientific standing of the Office were certainly bearing fruit. He was gathering around him staff of high calibre.

The Outcome of a Disaster

The first paper Dobson published bore the title 'Pilot Balloon Ascents at the Central Flying School, Upavon, During the Year 1913'.[3] His findings were of interest to G I Taylor, the Schuster Reader, who was by now an authority on air motions in the

---

[2] This series began in 1912 with a paper by the Office's Marine Superintendent, Captain Hepworth, on 'The effect of the Labrador Current upon the surface temperature of the North Atlantic; and of the latter upon air temperature and pressure over the British Isles, Part I'. This work proved important half a century later, as the basis of an experimental method of forecasting weather patterns a month ahead.

[3] See Dobson, G M B (1914), 'Pilot balloon ascents at the Central Flying School, Upavon, during the year 1913', published in the *Quarterly Journal of the Royal Meteorological Society* (Vol. 40, pp. 123–135).

lowest layer of the atmosphere. He had recently returned to England after leave of absence from Cambridge, during which he had carried out meteorological research from a ship on the North Atlantic Ocean. This voyage had not been undertaken for any reason connected with aviation. It was an outcome of a maritime disaster, the loss of the RMS *Titanic* on 15 April 1912 after a collision with an iceberg.

A committee of the Board of Trade recommended that a vessel be stationed between Labrador and Greenland during the iceberg season to shed light on the occurrence of ice in the North Atlantic, and the minutes of the Meteorological Committee's meeting on 6 November 1912 stated that it was probable there would be "opportunity for special meteorological work including the investigation of the upper air over the sea". The minutes of the Committee's meeting on 12 March 1913 show that Taylor had been appointed meteorologist for the voyage but allowed to take the post on the understanding that he would make up the time spent at sea by extending his period of tenure of the Schuster Readership for the corresponding amount of time. In the event, he returned to Cambridge on 14 September 1913.

Details of arrangements for the voyage were given in the reply to a question asked in the House of Commons on 13 March 1913. The President of the Board of Trade, Sydney Buxton, stated that the *Scotia*, a whaler, had been chartered to cruise off the coasts of Newfoundland and Labrador to observe and report sea ice and icebergs which might pose problems for Atlantic steamships. She was equipped with long-range wireless telegraphy, provided free of charge by the Marconi Company, and would be able to keep in touch with wireless stations in Newfoundland and Labrador. She had sailed from Dundee on 8 March 1913 and would be away for three to four months. The cost of the vessel was being shared equally between the Board of Trade and the steamship lines which principally used the North Atlantic.

Mr Buxton went on to say that the vessel carried a staff of three scientific observers, viz. a hydrographer, a meteorologist and a biologist, and he also informed the House that the Royal Prussian Aeronautical Observatory had provided a number of kites for meteorological work, as well as instruments to be attached to these kites for recording air pressure, temperature, relative humidity and wind speed. The minutes of the Committee's meeting on 12 March 1913 show that kites and winding gear were also lent by C J P Cave. They show, too, that the programme of work to be undertaken by Taylor included the maintenance of a four-hourly meteorological log "as taken in the Mercantile Marine", along with investigations of humidity, investigations of atmospheric conditions above fog, comparisons of wind speed and direction at different heights by means of pilot balloons, measurements of temperature in the upper air, measurements of rain at sea, studies of ice drift in relation to wind speeds and directions, comparisons of barometers with different mercury column constrictions, and measurements of sea-water salinity, density, temperature and electrical conductivity in the neighbourhood of ice and elsewhere.

Taylor made good use of his time aboard *Scotia*. He carried out the planned observational programme and measured profiles of temperature, humidity, wind speed and wind direction to a height of about 2500 metres on several occasions. In so doing, he displayed considerable ingenuity. For example, he tethered a kite to the mast-head, so that turbulence from masts and rigging would not cause the kite to plunge into the sea. He described his work in a substantial report for the Board of Trade and published his theoretical analyses in a paper which is today considered a 'classic'. Titled 'Eddy Motion in the Atmosphere', it was published in 1915 in the *Philosophical Transactions of the Royal Society* (A, Vol. 215, pp. 1–26). Though his pioneering contributions to knowledge and understanding of turbulence in the boundary layer were not inspired by the needs of aviation, his work did subsequently prove important in aeronautics.

Meanwhile, the annual Parliamentary grant to the Office had been increased to £20,000 in 1913. Shaw's justification for this was set out in detail in the minutes of the Meteorological Committee's meeting on 4 December 1912. The Office's work in aeronautical meteorology was proving more expensive than anticipated, and the costs of running the observatories had increased. Moreover, a branch was being established at Edinburgh (see Chapter 7), and the work that was resulting from the increasing involvement of the Office in international meteorology was proving an extra burden for him and other members of staff. Meanwhile, too, another person who would become a giant of meteorology had joined the Office. He was a mathematician, Lewis Fry Richardson, who became Superintendent of Eskdalemuir Observatory on 1 August 1913.

War Breaks Out

The Great War brought to an end the old order for meteorologists and almost everyone else. Meteorology had for decades advanced at a fairly leisurely pace. Friendships, cooperation and collaboration between meteorologists and meteorological institutions had developed. In 1914, the priorities and purposes of research and operations changed abruptly, and official relations between institutions changed, too. Research did continue in neutral countries, but the amount of international cooperation and collaboration that was possible was relatively small. The activities of the IMO were suspended when the war began.

Some military techniques were age-old. Smoke-screens were utilized on land and at sea. Appropriate forecasts of winds were therefore required. Mud had long been a hindrance for foot soldiers and cavalry and proved to be a major problem in the trenches of the Great War. Weather forecasts were needed. Amusingly, the official attitude of the Army at the beginning of the war was that British soldiers did not go into action carrying umbrellas![4] But this attitude was soon to change, and a Meteorological Section of the Army was formed in the summer of 1915, its *raison d'être*

---

[4] See 'Some notes on meteorology in war time', by C J P Cave, published in the *Quarterly Journal of the Royal Meteorological Society* in 1921 (Vol. 48, pp. 7–10).

being to provide information to the RFC, which had not been taking meteorological information sufficiently seriously. Advances in aeronautical meteorology before the war had been such that many aviators now considered aeroplanes could fly in any weather, rendering forecasts superfluous! Moreover, some aviators argued that in wartime flights had to be made in spite of the weather. Fortunately, these attitudes of aviators changed quickly as the war developed, and the crucial importance of cloud and fog forecasts and the need for reliable information about upper winds soon became apparent. Zeppelin bombing raids on towns in eastern England in the spring of 1915 further showed the relevance of meteorology. The target of the airships was London, but the wind had taken them far to the north-east of the capital.

Successful deployment of observation balloons called for information about upper winds, too. Moreover, the need to know temperature, humidity, wind speed and wind direction in the upper air came to be appreciated in respect of artillery needs, for the accuracy of gunfire depended on winds and air density in the layer of the atmosphere through which shells travelled. With the high-angle fire that developed during the Great War, knowledge of the atmosphere to ever-greater heights was needed. And the war brought a new reason for understanding air motions in the atmospheric boundary layer. Poison gas was used. For aviators, there was a particular need to forecast thunderstorms, which for them were dangerous weather formations. These storms produced lightning and hail and contained inside their towering cloud systems strong upcurrents and downdraughts. They were also accompanied by vigorous gusts and downbursts near the ground.

The Great War brought new challenges for the staff of the Office. There were changed or new priorities in the workplace. Staff were called up and served abroad in uniform. Some sadly died in active service. The trained scientists Shaw had gathered around him brought great credit to the Office through their capabilities and expertise, and the process of recruiting trained scientists continued during the war. Some were to serve the Office with distinction for many years after the war ended. Many of the men who were called up were replaced by women, and the pre-war employment conditions for women were relaxed somewhat. When Miss Dorothy Chambers, then aged 20, was interviewed by Shaw in the autumn of 1913 for a post as a shorthand typist, there were two requirements other than ability to type: she had to be the daughter of a professional man, and she must not wear a low-cut neckline![5]

The first meeting of the Meteorological Committee during the war took place on 28 October 1914. Some of the business was as mundane as ever. For example, the Office wished to establish a service whereby weather forecasts would be distributed by telephone, but the Postmaster General regretted that he could not see his way to assisting in a scheme of the kind suggested "on the grounds of the accounting work involved and the need for observance of special instructions in the various telephone

[5] Miss Chambers remained with the Meteorological Office for forty years, retiring in 1954.

exchanges". The Post Office and the Meteorological Office were sharing a building but did not seem able to coexist in complete harmony. There was, however, much in the minutes that was war-related, and Minute 478 was headed 'Emergency action during the war'.

The minute in question began with a report that the Office's wireless telegraphy sets at Kew and South Kensington had been removed by order of the Postmaster-General on the day war was declared, 3 August. This was followed by the information that a demand had been received by telegram the same day "to make up the establishment of meteorological instruments at Devonport". The instruments had duly been despatched and the establishments at all other dockyards also made up. Instruments had been lost when ships had been sunk by enemy action, and a theodolite Gold had taken with him when he went to a meeting of the British Association in Australia had been left with the chief officer of the ship, the *Euripides*, on which he had made the outward voyage, to be brought back to England on the return voyage. However, the ship had since been requisitioned by the Admiralty as a transport, and the instrument was presumably still on board. He had taken the theodolite with him to track the pilot balloons he launched from the ship.

The minutes show that telegrams were being censored, with use of the address 'Weather London' banned when censorship began. As a consequence, the flow of weather reports from foreign countries had almost ceased, but reports from the Azores and Madeira had been received in plain language. However, use of 'Weather London' had soon been allowed again, as a result of which the flow of reports had recommenced from all countries except Germany, Belgium and Iceland. Reports from the Atlantic by wireless telegraphy had practically ceased by October 1914.

No meteorological reports were being sent from the UK to foreign countries, but representations asking for restitution of the service had been received from Sweden, Denmark and Holland, and a request for a new service had come from Russia. Reports had been sent in plain language to France from 14 August to 20 October, Brussels from 14 August until the occupation of that city and Antwerp from 7 September until 11 October. The reports to France had ceased because they had been made public in the *Paris Bulletin* and broadcast from the radio station on the Eiffel Tower.

Issue of the *Daily Weather Report* had been unaffected at first, apart from a number of modifications that had seemed desirable. After a meeting between Shaw and the Secretary of the Admiralty on 1 October, however, the Meteorological Committee decided to discontinue the daily issue of lithographed copies of the *Daily Weather Report* to all but certain official addresses, which included the War Room at the Admiralty. Enciphered weather maps and forecasts were being supplied morning and evening to the Naval Air Service, prepared for transmission by radio.

The minutes contain the names of twelve members of staff who had applied for leave to join the Forces. Five had been given permission to enlist and two refused, one on health grounds, the other on account of eyesight. Instructions were awaited for the others. In addition, the minutes state that the Eskdalemuir Superintendent, L F Richardson, a Quaker, had repeatedly asked leave to join a Red Cross unit or the ambulance corps, but permission had so far been declined.

A Professor of Meteorology at South Farnborough

The Meteorological Committee received a substantial report entitled 'Central Observatory for the Investigation of the Upper Air' at their meeting on 28 October 1914. This contained the suggestion that Kew Observatory should be "the central establishment and weather station for London", with Eskdalemuir "the magnetic observatory and meteorological establishment for moorland conditions at 800 feet", Valentia Observatory the establishment "for the study of atmospheric conditions on the Atlantic seaboard" and Aberdeen Observatory the establishment "for the study of atmospheric conditions in the North Sea". The suggested purpose of South Farnborough would be "the study of the turbulence of atmospheric motion, and the application of meteorology to the needs of aircraft". Benson, the report suggested, would be a good location not only for studying upper layers of the atmosphere but also for the application of meteorology to agriculture. The Committee agreed that the subject of Benson's role should be considered further at their next meeting.

This meeting took place on 25 November but devoted very little time to discussion of facilities at Benson. It was concerned mainly with the Office's overall programme of work for 1915–1916, especially meteorological work in connection with aircraft. The minutes show that Shaw was not satisfied with arrangements at South Farnborough or with the experimental programme being pursued there. The grant of £250 had been continued for 1914–1915, but he was "not able to say that any very satisfactory result had so far been produced".

He referred to "a discouraged meteorologist with an inexperienced assistant". He did not think the Branch Office would make "effective progress on novel lines of research without a flying member whose duty it should be to carry out any experimental work required by the meteorologist". The minutes of the Committee's meeting on 28 October 1914 mention that J S Dines was by then working with Lempfert in the Forecast Division at South Kensington, spending nearly half of his time at the Central Office, working night duties to prepare the early morning forecasts for the Royal Naval Air Service. The official story was that assistance was required at South Kensington because the Central Office was short-staffed, but it appears that Shaw did not think Dines was up to the job at the Branch Office. This was probably unreasonable on Shaw's part, for the working conditions at South Farnborough were not conducive to a focused approach to research in aviation meteorology or anything else. Billett was in charge at the Branch Office for long periods and ensured the observing routine was not broken. He asked the Committee for an increase in salary in recognition of his greater responsibility and promptly resigned when his request was met with an offer of books in lieu of payment.[6]

The experience at Upavon, Shaw said, was somewhat similar to that at South Farnborough. The meteorologist there, who had "proved himself well fitted for the post", was responsible not only for meteorological observations and reports but also for the repair of instruments carried by aircraft. He had little time for scientific study. Dobson remained at Upavon until October 1916, after which he transferred to the Royal Aircraft Factory at South Farnborough, becoming Director of the Experimental Department.[7]

Agricultural meteorology was discussed at the meeting of the Committee on 27 January 1915. A request from the French Ambassador had been received in December through the Foreign Office. In this, he had asked to be informed about the organization of the Meteorological Office, with special reference to agricultural meteorology. Shaw's response came in the form of a lengthy memorandum, in which he summarized the whole range of the Office's activities. He stated that the application of meteorology to agriculture was "one of the objects of the Office", but there was no special allocation of funds for this application as such. He went on to say that it was really an open question whether the responsibility for this application belonged to the Office or to the Board of Agriculture and Fisheries in England and the corresponding departments in Scotland and Ireland. The Office issued weather forecasts and published temperature, pressure, sunshine and wind data in the *Weekly Weather Report* and the *Monthly Weather Report*. However, he said, these provisions were in practice very little used by agriculturists. The practical farmer "made his own study of weather and used it in his own way without committing the results to writing". Relations between the Office and the Board of Agriculture were "of the happiest", but, he said, neither side knew "exactly how or where to begin". No mention was made of Benson.

The Branch Office was the subject of another memorandum discussed at the meeting on 27 January 1915. In it, Shaw pressed the points he had made at the Committee's meeting on 25 November. A "man of exceptional scientific ability" was required, someone whose duty would be to give "advanced instruction in meteorology – especially the dynamics of the atmosphere – to the members of the RFC". This person should also discuss with aviators their "experiences in relation to aerodynamics and physics" and carry out research which could be expected to prove valuable to pilots "by enabling them to take advantage of the most complete knowledge and structure of the atmosphere and the dynamical conditions". He pointed out that no such office

---

[6] Billett left the Meteorological Office in February 1915 to take up an appointment in the Meteorological Branch of the Irrigation Department of the Union of South Africa. He enlisted in the South African Infantry in December 1915 and subsequently joined the British Expeditionary Force in France. He died in action in October 1916.

[7] Dobson was appointed a University Lecturer in Meteorology at the University of Oxford in 1919 and went on to become an authority on atmospheric ozone.

existed at present and suggested that the duties were "more nearly described by the title of Professor than any other". Besides the Professor, the staff of the Branch Office should include, he proposed, a meteorologist, an assistant meteorologist (or flying assistant), a clerical assistant and a laboratory assistant. The Committee agreed that Shaw's proposal be submitted first to Colonel Sefton Brancker at the War Office and to Mervyn O'Gorman, Superintendent of the Royal Aircraft Factory, then to the Treasury.

The minutes of the Committee's meeting on 24 March 1915 show that Shaw's proposal, with a few amendments, had been approved by the War Office and O'Gorman. They also show that G I Taylor had expressed his willingness to undertake the duties and suggested that he should undertake a course of training as an officer of the RFC. He had joined the staff of the Royal Aircraft Factory soon after the war began to carry out work in theoretical and experimental aeronautics. However, the Advisory Committee thought it would be inadvisable for the Professor to be a flying officer and resolved that the opinion of the War Office be sought on the question of the preliminary training. They also agreed that enquiries should be made regarding whether the Schuster Readership would be tenable with the Professorship. The Meteorological Committee later heard that it would not, but this was in fact irrelevant, because Taylor had given up the Readership on the outbreak of war.

The Treasury turned down the Committee's request for a Parliamentary grant of £24,000 for the year 1915–1916, proposing instead that it be £22,500. Shaw wrote to the Treasury on 30 March 1915, saying that the Committee regretted that the grant had been set at a lower level than requested but accepted that Their Lordships of the Treasury were not prepared to regard seismology and terrestrial magnetism as official subjects for the Office to pursue. In his letter of 30 March, he submitted revised estimates for the year 1915–1916 and included in it a case for Taylor to be appointed Professor of Meteorology at South Farnborough. The Committee heard at their meeting on 21 April 1915 that Brancker had expressed the opinion that if the Professorship was to continue after the war, funds should be sought from the Treasury by the Meteorological Office, not by the War Office.

Months passed, with very little progress made towards the appointment of a Professor. Meanwhile, C J P Cave was appointed Honorary Special Inspector at the Branch Office, with effect from 17 February 1915, his duty being, as it was put in the minutes of the Committee's meeting on 24 March, "to report from time to time on methods of developing the use of meteorological information in connexion with the Army and the Air Services at South Farnborough and elsewhere, and in particular upon the means of applying the pilot balloon observations to the immediate purposes of the War Services".[8]

---

[8] Cave was officially appointed Meteorologist-in-Charge of the Branch Office on 23 October 1915, when J S Dines was transferred on a permanent basis to South Kensington.

The minutes of the Committee's meeting on 24 November 1915 note that Taylor had received a commission in the RFC and obtained a pilot's certificate. He was "now engaged again at the Royal Aircraft Factory". Eventually, as the minutes of the Committee's meeting on 22 March 1916 show, he was appointed Professor of Meteorology. These minutes state that he had taken up his duties as Professor from 14 February 1916 and been gazetted temporary Major from that date. A letter dated 16 March 1916 from Captain B C H Drew of the General Staff set out the conditions of Taylor's appointment. Sent to Shaw and to various senior officers in the RFC, including the Commandant of the Central Flying School at Upavon, it stated that Taylor was required to carry out research into the properties of the atmosphere in relation to aviation. His office would be located in the Office at South Kensington. He would be assisted by a flying officer whenever required and also by a mechanic based at South Farnborough. His duties were to advise and cooperate with the Office on all questions concerned with atmospheric motions in relation to aviation, but he was mainly to "conduct research or investigations into the structure and properties of the atmosphere in relation to aerial navigation or the use of aircraft in military operations".

Taylor's reports for March and April 1916 show that he had written for a manual a chapter on eddy motions in the atmosphere and begun a chapter on the propagation of sound in the atmosphere. He had also carried out experimental work concerned with the design of bombs and the cooling fins of aeroplane engines. As Professor of Meteorology, he could not be considered a great success, though he did publish an important paper in 1917 on the formation of fog and mist at sea and on land, based partly on work he had carried out on the Atlantic and partly on investigations he had made whilst Professor.[9] His work at South Farnborough turned out to be more concerned with aircraft performance than aviation meteorology. He left the Office in 1917 and became a meteorological adviser to the RFC.[10]

Wartime Field Units of the Meteorological Office

The Meteorological Section of the Army that was formed in 1915 as a unit of the RFC was called the 'Meteorological Field Service', and the first reference to it in the minutes of the Meteorological Committee's meetings occurs in the minutes of the meeting on 2 June 1915. Here, it is stated that "the War Office had forwarded on 5 May copy of a telegram from the General Officer Commanding, RFC in the Field, asking for information in technical form regarding the meteorological situation and

---

[9] Taylor, G I (1917), 'The formation of fog and mist', *Quarterly Journal of the Royal Meteorological Society*, Vol. 43, pp. 241–268.

[10] Taylor returned to the University of Cambridge in October 1919 and remained there for the rest of his career, making outstanding theoretical contributions to fluid and solid mechanics and becoming a world authority on turbulence. He was knighted in 1944.

for forecasts of wind, fog, &c. in connexion with aircraft work". In his reply, dated 7 May, Shaw had "outlined for approval a scheme of daily messages to the War Office for communication to the RFC, and asked for observations direct from some point in Flanders". He had also suggested that an expert meteorologist could be sent out "with the necessary apparatus, in order to facilitate effective communications".

The War Office concurred and agreed that the service should consist of two Super-intendents of the Meteorological Office and two Professional Assistants, and also that the men selected would be granted commissions on the General List of Infantry. The superintendents would be given the temporary rank of Captain and the profes-sional assistants the temporary rank of Lieutenant, but no pay or allowances to these officers would be sanctioned from the Army Funds beyond that to which they were entitled by the ranks held. Shaw reported that three people had been recommen-ded for appointment: A E M Geddes, E H Chapman and F Entwistle. Geddes had gained First Class Honours in Mathematics and Natural Philosophy at the University of Aberdeen in 1906 and pursued an academic career in that university since 1908. He had carried out upper-air research at Aberdeen using pilot balloons and been recommended for appointment as a Senior Professional Assistant from 24 May 1915. Chapman and Entwistle, both science graduates, had been recommended for appoint-ment as Junior Professional Assistants from 22 May. After further consultation with the War Office, Ernest Gold, the Meteorological Office's Superintendent of Statistics, had been recommended for a Captain's Commission and Geddes for a Lieutenant's Commission.

These appointments were all approved by the Meteorological Committee on 2 June, and a further decision was made that Major H G Lyons should proceed to the Headquar-ters of the RFC "in connexion with the initiation of the Field Service". Lyons had been a member of the Committee since 26 February 1913, having replaced Sir George Darwin, who had died on 7 December 1912. When the Committee for the five years beginning 1 April 1915 had been appointed by the Treasury, he had again been named a member, along with Shaw, Schuster, Barstow, Harvey and Middleton, who had all served previously (see Chapter 7). The other member was Captain John Parry, who had succeeded Admiral Purey-Cust as Hydrographer of the Navy in August 1914 and become a member of the Committee ex officio on 1 September.

Lyons had been born in London on 11 October 1864 and educated at Wellington College, after which, in 1882, he had become a cadet at the Royal Military Academy, Woolwich. From there, in 1884, he had entered the Royal Engineers (RE). A posting to Cairo in 1890 had provided him with an opportunity to pursue his long-standing interest in geology, as a result of which, in 1896, he had transferred to the Egyptian Ministry of Public Works to organize a geological survey of Egypt. Two years later, he had retired from the British Army with the rank of Captain and taken up a permanent position with the Egyptian Government as director of a joint geological and cadastral survey department, a post he had held until 1909. This work had covered a range of

Figure 8.1. Henry George Lyons. © Crown Copyright 2010, the Met Office.

environmental sciences, among them meteorology, and Lyons had not only established an observatory and a meteorological office but had also explored the upper air with instruments attached to kites. Furthermore, he had attended the International Congress of Directors of Meteorological Services held at Innsbrück in September 1909. He had returned to the UK the following month to take up a lectureship in geography at the University of Glasgow but moved to London in early 1912 to become an assistant to the Director of the Science Museum at South Kensington. The respect for him in meteorological circles was evident by his election as President of the Royal Meteorological Society in January 1915, an office he held for three years.

When the war began, Lyons returned to the RE and took charge of recruiting at Chatham. However, he was relieved of these duties in the spring of 1915 and was transferred to London, lent by the War Office to the Meteorological Office to head a service whereby weather reports from the Mediterranean would be collected, and weather maps and other meteorological information of value for naval and military operations in the Mediterranean would be supplied. He took charge of this work on 17 May 1915 and became Commandant of the Meteorological Field Service a few weeks later. Now a Major (promoted on 17 April 1915), he travelled to the Headquarters of the RFC at Saint-Omer in the Pas-de-Calais department of France on 8 June to check arrangements which had already been made and to make other necessary arrangements. Gold and Geddes arrived on 11 June.

The need for weather forecasts and other meteorological support in the eastern Mediterranean was stressed by Shaw in the autumn of 1915 when preparations were being made for a winter campaign at the Dardanelles. He advised the Committee of Imperial Defence that it was important to have a meteorologist on the spot, and the War Office heeded his advice. Dr E M Wedderburn departed for the Mediterranean in September 1915 to inaugurate a Meteorological Field Service for the Mediterranean Expeditionary Force.[11] Meanwhile, the Admiralty had recommended a practice which had been discontinued earlier in the year when the fair weather of the Mediterranean summer had set in, this being the transmission of meteorological reports to the Vice-Admiral. It was thought at first that this practice would be sufficient for the Army also, but it was soon realized that it would not. As the Vice-Admiral was constantly moving about, the Army needed their own meteorological service, which Dr (now Captain) Wedderburn would provide.

The duty of the Meteorological Field Service which supported the British Expeditionary Force in France was set out in a memorandum approved by the Meteorological Committee at their meeting on 2 June and forwarded to the War Office two days later. Its first paragraph read as follows:

The duty of the Meteorological Field Service will be to keep Headquarters supplied with the meteorological information that may be required, and to keep the Meteorological Office advised as to information which should be sent to Headquarters from London and as to the means of transmitting it, to carry out the local meteorological observations that are required, and to transmit meteorological reports to the Office daily.

The general organization would be placed in the hands of Lyons, while the staff on field service would comprise a meteorologist with Captain's commission in charge of the service and a professional assistant with Lieutenant's commission for observations of the upper air. Technical or clerical assistance would be provided by the military authorities. Instruments, charts and forms would be provided by the Office.

A further request was received from the General Headquarters in the Field in France for the services of an expert meteorologist to give advice regarding probable changes in the force and direction of the surface wind. The reason for the request was that the Germans had started to use poison gas. Chlorine had been discharged near Ypres on several occasions in April and May 1915. However, agreement was reached that a single service would satisfy the requirements of both General Headquarters and the Headquarters of the RFC.

The weather experienced by Lyons just after he arrived at Headquarters was far from pleasant. There was a severe thunderstorm, accompanied by strong gusts of

[11] Ernest Maclagan Wedderburn was appointed a Senior Professional Assistant (Temporary) on 16 August 1915 to serve with the Office's Forecast Division and the Meteorological Field Service. A lawyer and enthusiastic amateur meteorologist, he was the Honorary Secretary of the Scottish Meteorological Society and had volunteered for war service. He was knighted in June 1942.

wind and very heavy rain. This led him to comment in his report on the visit that such phenomena were "likely to be dangerous to aircraft, and to the temporary hangars, &c. at the aerodromes," so that the RFC were "anxious to have as much notice as possible of their arrival". He said in his report that he had discussed general and specific arrangements with senior officers and identified a number of shortcomings and requirements. Regarding the addressing of telegrams, the original proposal, 'Military Aeronautics in the Field W', had proved unworkable. Accordingly, he had insisted on 'METEOR G.H.Q.' as the code address of the Meteorological Service in the Field, and the service was known informally as 'Meteor' for the rest of the war.

As regards wind observations along the front, Lyons had told the Gas Adviser in the Field (Lieutenant-Colonel Charles Foulkes) that the Meteorological Field Service could only provide "general statements". It would be necessary for Foulkes to "arrange for full and regular observations of wind direction and force at such points along the front as were of importance to him". However, Lyons promised, "these observations would be collated and discussed by the Field Service in order that they might increase the precision of their forecasts". He further said in his report that he had asked Foulkes to "take Gold along the portion of the front concerned in order that he might have a better idea of the relief and features of the country".

Meteor grew and grew as its value came to be realized by the Army in general and the RFC in particular. It was transferred from the RFC to the RE in early 1916 and thereafter known as the Meteorological Section of the RE. A further development came in August 1916, when the Meteorological Committee proposed that a Meteorological Section of the RE (Home) be formed in England to provide a reserve of men who could replace or supplement those abroad. However, the War Office refused the request, and a Home Unit of Meteor was not formed until March 1918. After 1916, Meteor's structure and organization changed little, but the service developed and expanded in France, Italy, the Mediterranean and the UK. At the end of the war, the establishment of Meteor included thirty-two officers and 200 other ranks.[12]

Meteor staff sent information about upper winds to artillery units by priority telegram. These winds were ascertained by means of pilot balloons, and measurements of winds aloft were also made from kite balloons, along with measurements of temperature, by meteorological observers who had been allowed to accompany Balloon Officers when engaged in artillery spotting. Techniques for measuring temperature by means of instruments attached to aircraft were developed in England in 1915 and 1916, at Upavon and (later) South Farnborough, principally by Dobson, who designed an instrument called a 'barothermograph', which provided a continuous temperature trace during a flight.[13] The forwarding of upper-air temperature measurements made

---

[12] One of the officers was C J P Cave, who was commissioned a Captain on 18 June 1915.
[13] See 'G M B Dobson during World War I – his barothermograph and 'bomb'', by B J Booth, published in *Weather* in 2009 (Vol. 64, pp. 212–219).

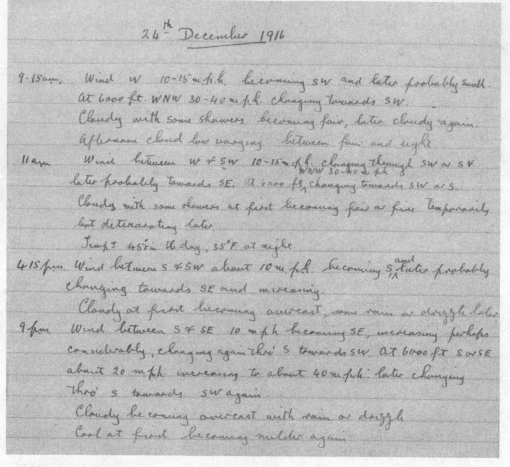

Figure 8.2. Ernest Gold's weather diary, France, 24 December 1916. © Crown Copyright 2010, the Met Office.

over southern England to the headquarters of Meteor in France began in October 1916, and a dedicated Meteorological Flight was formed in France in February 1918, based at Berck-sur-Mer, twenty miles south of Boulogne.

To take command of the new 'Meteor Flight', Gold was keen to appoint Captain Charles Kenneth MacKinnon Douglas of the RFC, a man who had impressed him with ground-breaking papers he had published on meteorological observations made from aeroplanes.[14] Douglas had made numerous flights over northern France during which, from altitudes as high as 10,000 feet, he had observed clouds, measured temperatures and inferred upper winds. He had also published photographs of clouds taken from aeroplanes over England but had not yet been able to take photographs from aeroplanes

[14]   See 'Weather observation from an aeroplane', *Journal of the Scottish Meteorological Society*, 1916, Vol. 3, No. 17, pp. 65–73, and 'Some causes of the formation of anticyclonic stratus as observed from aeroplanes', *Proceedings of the Royal Society of Edinburgh*, 1917, Vol. 37, No. 9, pp. 137–148.

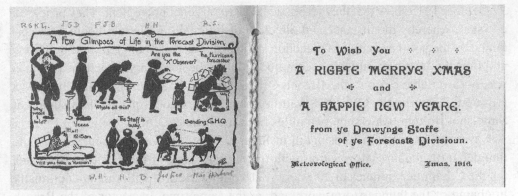

Figure 8.3. Wartime humour. Meteorological Office Forecast Division Christmas Card, 1916, with staff identified by their initials added in pencil, including RGKL (Lempfert), JSD (Dines), FJB (Brodie), HH (Harries), RS (Sargeant), WH (Hayes) and 'Jestico' (C F J Jestico). Card reproduced by kind permission of Steve Poole, great-nephew of J S Dines.

when flying on the line. Furthermore, he had given thought to the physical processes responsible for the conditions he encountered whilst in flight. He seemed to Gold the ideal man to head the Meteor Flight. Initially, though, he could not be released, but his services were soon secured. He assumed command of Meteor Flight on 12 May 1918, so beginning a long association with Gold and the Meteorological Office.

Further Wartime Work of Charles Cave

The transfer of Cave to Farnborough in February 1915 proved productive in a number of ways. Besides his observational and other duties, he worked on techniques for detecting thunderstorms by means of radio-telegraphy, doing so with an engineer called Robert Watson Watt, a graduate of the University College at Dundee. The minutes of the Committee's meeting on 11 August 1915 report that Watt had offered his services for war work and the minutes of the meeting on 27 October state that he had been posted to the Branch Office on 16 September as a Senior Professional Assistant to work with Cave on "the investigation of thunderstorm recording". Watt had received personal tuition in wireless telegraphy from Professor William Peddie at Dundee and hoped for a job during the war that would allow him to use his knowledge of this subject. The post he was given by the Office allowed him to realize his ambition and also to show his flair for scientific research.

For the work on thunderstorm detection, Cave used his own lightning recorder, which he had acquired in 1912 for use at Ditcham Park, his home near Petersfield. This had no directional facility and could therefore be used only for recording the occurrence of lightning discharges. Nevertheless, Shaw encouraged Cave and Watt to use it as an aid for locating storms, and the minutes of the Committee's meeting on

27 October 1915 show that they responded well to this encouragement. They proposed a system whereby the distance and direction of lightning discharges could be ascertained by means of a triangle of detection stations situated at South Farnborough, Kew and Benson. However, suitable accommodation for the necessary detection apparatus was not available close to the Branch Office. Moreover, an aerial 65 feet high was required. Accordingly, Cave and Watt used an abandoned Army wireless station at Smallshot Bottom, Aldershot, two miles from the Branch Office.

Cave moved to South Kensington in June 1917 to "bring the organization of the thunderstorm inquiry into relation with the regular work of the Forecast Division" (as the Committee minutes of 20 March 1918 put it). He was still, though, nominally in charge of the thunderstorm work and, at the same time, Inspector of the Branch Office, but Watt became officially Meteorologist-in-Charge at South Farnborough on 1 June 1917. By then, the supply of trained physicists had become exhausted, and Watt's only assistants were a lady clerk and a corporal mechanic, until December, when E L Hawke joined him. A graduate of Trinity College, Cambridge, Hawke had been a voluntary assistant at South Farnborough from 6 May to 16 August 1915 and thereafter employed in the Central Office.

Cave awarded the credit for the thunderstorm detection work to Watt, saying in 'Some notes on meteorology in wartime', published in 1922 in the *Quarterly Journal of the Royal Meteorological Society* (Vol. 48, pp. 7–10), that his own time had been taken up with lecturing and forecasting at South Farnborough. By early 1918, Watt was in charge of the thunderstorm location work at Aldershot, and he continued to direct it after the war ended. The work was transferred to the Department of Scientific and Industrial Research (DSIR) at the beginning of 1921, and Watt was seconded to DSIR at the same time.[15] He transferred to DSIR in 1925, never to return to the Meteorological Office.[16]

Cave was in hospital for six weeks in the autumn of 1917 and advised on discharge not to return to work at the Central Office. Instead, he was appointed, in January 1918, as the Officer Commanding, Meteorological Section of the RE (Home). His new command consisted of four meteorological offices: Stonehenge Aerodrome, Butler's Cross (nine miles north-west of Stonehenge), Filton (near Bristol) and Shoeburyness (Essex). Stonehenge was a day and night bomber-pilot training establishment, and the pilots required reliable upper-wind data for navigation exercises. From February 1918 to early 1919, Cave and his staff provided this information by measuring the upper winds at regular intervals both day and night. During hours of darkness, this was done by tracking small balloons which carried below them paper lanterns containing lighted candles. Though well accustomed to launching and tracking kites and balloons, as we

[15] The Department of Scientific and Industrial Research was a body formed under the Privy Council in 1915 to enhance scientific and industrial research by means of government funding.

[16] Apparatus and methods developed by Watt in the 1920s and 1930s were used to good effect by the Meteorological Office in the Second World War. When knighted, in 1942, he hyphenated his surname and became Sir Robert Watson-Watt.

have seen earlier in the book, he was not a pilot. However, he had been a passenger on several occasions, as the Flight Log of the Royal Aircraft Factory shows, and he had long been a keen photographer. The archive of the Royal Meteorological Society contains a number of photographs taken by Cave from aeroplanes during the time he was at South Farnborough. The one taken on 15 September 1915 is one of the earliest extant photographs taken from an aeroplane over the British Isles.

David Brunt succeeded Cave as Officer Commanding on 9 February 1919, whereupon Cave returned to the life of a gentleman of leisure he had enjoyed before the war. Brunt was a mathematician who had gained Firsts in Parts I and II of the Cambridge Mathematical Tripos, the latter in 1910. He had been a lecturer in mathematics for a number of years, first at the University of Birmingham and later at the Monmouthshire Training College for Teachers. Then, in 1916, he had enlisted in the Meteorological Section of the RE and served in France, mainly as a forecaster for the RFC but also contributing to the application of meteorology to gunnery. He succeeded Cave when demobilized after the war and was subsequently employed by the Office for a number of years before moving on to an academic career in the Imperial College of Science and Technology.

Serving the Needs of Wartime

The Meteorological Committee discussed the establishments of the Office and Meteor at their meeting on 24 November 1915, having before them a roll of the Office's staff which had been compiled by Shaw and Lyons in response to an appeal made by the Government's Director of Recruiting. He was seeking to identify men who might enlist in the Army. The establishment of Meteor, supplied by Gold, showed that the establishment on 13 October made provision for one Commandant (himself), one Captain, six Subalterns, two Sergeants, sixteen Corporals, one Clerk, six Batmen and one Driver, of whom the Captain and one Subaltern were normally stationed in England. The transport available consisted of one motor car, three motor cycles and two bicycles. And the minutes of the meeting on 24 November further show that Gold had been promoted to the rank of Major as from 28 September 1915.[17]

The roll of the Office's staff showed that most of those who maintained the services at South Kensington and the seven auxiliary institutions were civilians, while the others were members of naval or military units.[18] There were five groups of civilians: scientific staff, clerical and technical staff, auxiliary staff, probationers and boy

---

[17] Lyons was promoted to the rank of Lieutenant Colonel on 29 March 1918 and Gold to that rank on 1 October 1918. Gold was further honoured on 1 January 1919, when it was announced that he had been awarded an honour only recently instituted, Officer of the Order of the British Empire (Military Division).

[18] The Office's Administrative Centre, Marine Division, Forecast Division and Instruments Division were at South Kensington, and the seven auxiliary institutions were the Central Observatory at Kew, Eskdalemuir Observatory, Aberdeen Observatory, Benson Observatory, the Branch Office at South Farnborough, the Weather Station of the Meteorological Office at Falmouth, and the Western Observatory at Cahirciveen (Valentia Observatory). There were also, in various parts of the British Isles, twenty-five observing stations which all reported daily by telegraph.

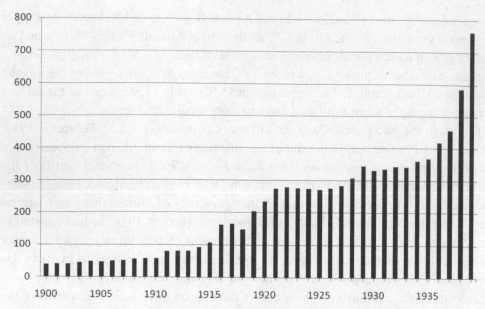

Figure 8.4. Meteorological Office staff numbers, 1900–1939.

clerks, and boy messengers. In the naval and military units, there were eight commissioned and two non-commissioned officers. They included Gold, the Commandant in France, and Wedderburn, in charge of the Meteorological Office for the Mediterranean Force (based in Greece).

Shaw and Lyons considered each group of civilians in turn and concluded, "after careful consideration", that they were able to spare only two of the staff at the present time. Both had just reached the age of 19, then the lower age for war service. A number of men were forty years of age or more and therefore too old. Others had been rejected for war service because they were physically unfit, and those who had already enlisted had been replaced by women and girls. The cases of probationers and boy clerks who were under 19 would be considered further at the appropriate time, with the object of releasing in the course of the following twelve months those who had reached enlistment age and were medically fit.

As far as the civilian scientific staff were concerned, there were few professional assistants of recruitable age. However, as Shaw and Lyons put it:

In view of the fact that this is the only training school for meteorologists in the country, and that during the past year eleven professional meteorologists have been selected to receive commissions on account of their special knowledge, nine for the Army, taken from the professional staff of the Office, and two for the Naval Air Service, taken from the clerical and technical staff, it is not desirable to have less than the present number, six, in training to replace casualties or to meet new demands.

The staff of the lithographic press were employed, Shaw and Lyons pointed out, by Messrs Wyman & Sons Ltd, a company "under contract with His Majesty's Stationery Office to execute certain lithographic work on the premises of the Meteorological Office". In view of the importance of the work the staff in question were doing, however, Shaw and Lyons hoped the work would continue without delay or difficulty.

We do not know the reason behind a question that was asked in the House of Commons on 31 December 1916, but concern over the nationality of Lempfert may have had something to do with it. Major Rowland Hunt asked the Secretary of State for the Home Department whether there were any civil servants on the staff of the Office who had been born in enemy countries. If so, what were their names and when were they naturalized? The Chancellor of the Exchequer, Andrew Bonar Law, said that the Home Secretary had asked him to reply. The answer to the first part of the question was "in the negative", so the second part did not arise. Major Hunt did not appear to have made basic enquiries before asking the question. Had he done so, he would have discovered that Lempfert had been born in Manchester, admittedly the son of German immigrants. Lempfert would soon be honoured for his services to the Office. He was appointed a Commander of the Order of the British Empire (CBE) at the beginning of January 1918.

Another matter which was discussed at the meeting of the Committee on 24 November 1915 was the need for a Press Bureau, with the case put in a long memorandum written by Shaw. The chain of events which led him to write it began on 11 August 1915, when the work of C V Diehl was discussed at the Committee's meeting. He was a newspaper journalist and Director of the Newspaper Weather Bureau whose work had been, as it was put in the minutes, "interrupted on account of the restrictions imposed by Government in connexion with the distribution of weather reports". He had been offered employment in the Office "for the discussion of the climatology of the British Empire, at a salary of £15 per month, with the usual proviso that communications to the Press must be submitted to the Director for approval". But he had rejected this offer, saying the salary was not sufficient for the requirements of himself and his family, also that he could not accept the conditions laid down for the communication of articles to the Press because he was expecting remuneration for his literary work to supplement his pay from the Office. He subsequently changed his mind, though, as we see from the minutes of the Committee's meeting on 27 October. He was now prepared to accept an appointment in the Office under the conditions previously set out, but this did not impress the Committee. They decided to inform him there was now no vacancy on the staff nor any accommodation for him in the Office!

Shaw's memorandum showed, however, that Diehl's renewed application had drawn to the attention of the Committee the need to consider the supply of information to the Press. Shaw did not accept that a journalist could improve on the reports and forecasts that were prepared and issued by his own staff. They were trained. The vast majority of journalists were not. "Without a long period of preparation", he said, the Office could not "make use of the qualifications of a journalist whose work has been of the same kind as that already done efficiently by the Forecast Division". At that time, the work of this branch was confidential and carefully organized, so it would be "highly inadvisable to bring in an anonymous expert". "If employment is to be found, he said, "it must be in some new form of occupation".

At this point in the memorandum, Shaw revealed his continuing desire to improve weather education and the communication of meteorological information to the public. He asserted that very few who were interested in weather had sufficient knowledge of meteorology to understand a weather report. Moreover, he said, "the use of the commonest meteorological instruments is unknown, for example, to many officers of the Flying Corps". Furthermore, weather reports in the metropolitan or provincial press showed, with few exceptions, "a deplorable lack of skill and touch, often a mere wooden recital of technical meteorological figures".

Shaw believed there was a need for a 'Press Bureau', but now was not the time to "embark upon a new enterprise". Such a development could not come before the war ended, and there would even then, of course, be funding considerations. The Committee decided accordingly not to proceed with the matter at that time. Meanwhile, though, production and publication of manuals and handbooks by the Office continued during the war, including *The Seaman's Handbook of Meteorology*, *The Marine Observer's Handbook*, *The Weather Map*, *The Computer's Handbook* and *The Meteorological Glossary*. Meanwhile, too, Shaw had received a knighthood, conferred on him at Buckingham Palace on 12 July 1915. He was now Sir Napier Shaw.

Whither the Meteorological Office?

By 1917, the volume of meteorological matters of national importance which required Shaw's personal attention had grown to such an extent that he needed to be relieved of as much routine work as possible. This difficulty had been addressed by handing to Lyons overall responsibility for general correspondence and for supplies of instruments and stores. At least the source of the problem was nothing more than an excessive workload. Shaw's health was robust. However, several staff of the Office suffered nervous breakdowns during the war, most of them in the Forecast Division. The majority were able to return to work in due course, but one was not. The Office Keeper was admitted to the London County Asylum in the summer of 1917 and pensioned off with effect from 1 January 1918. He remained on full pay but sadly died before the superannuation arrangement could take effect.

The summer of 1917 brought a development of great moment for the Office. The story began in August, when the Government created a Ministry of Reconstruction, the duty of its Minister being to consider and advise on the problems which may arise out of the war and have to be dealt with when the war ended. To this effect, the Minister (Dr Christopher Addison) was required to institute and conduct such inquiries, prepare such schemes and make such recommendations as he thought fit. Attention was soon focused on the Office.

The memoranda Shaw wrote were often long and detailed, and those dated 28 December 1917 and 12 January 1918 were no exception. The one he wrote in December was a response to an enquiry received in October from the Ministry of Reconstruction, in which the Committee had been asked whether any steps had been taken with regard to problems that may occur after the end of the war. The memorandum dated 12 January had been drafted for the Committee to consider and would be sent to the Treasury once the Committee had approved it. Both memoranda were discussed at the Committee's meeting on 23 January 1918.

The memorandum to the Ministry was called *Reconstruction: Memorandum on the Meteorological Services* and contained fourteen sections. The first dealt with "the universal importance of organized knowledge of weather" and provided "the ultimate justification of a public weather service". The second section made the point that "in all civilized countries there was a government meteorological institute which acted in cooperation with the corresponding institutes of other countries". The third section was concerned with "the basis of the meteorological services" and contained a summary of the differences between the meteorological services of the UK, Germany and other countries. Germany's service was fragmented, Shaw said. Each German state had its own meteorological establishment and the Imperial Navy had a separate service for maritime meteorology. Services in other countries had been "built up on the requirements of agriculture and public welfare in respect mainly of rainfall". In contrast, Britain's meteorological service had been based on the requirements of shipping and subsequently modified and extended to form the present organization.

There followed a section headed 'The British Meteorological Services', in which Shaw outlined the Office's links with the Royal Meteorological Society, Scottish Meteorological Society and BRO. The Office, he explained, had long been managed by the Royal Society and financed by Government through an annual grant-in-aid, whereas the other three institutions had always depended on the subscriptions of their members. The head of the Office was the official representative of the country at international meetings and responsible for the discharge of international meteorological obligations. Furthermore, the Meteorological Office acted in an informal advisory capacity to the Colonial Office for the meteorology of the colonies and protectorates which had no meteorological service of their own. In the UK, the Meteorological Office had recently taken over from the two Societies responsibility for the general climatological work of the home countries. It had also, since 1910, controlled eight observatories.

The fifth section of the memorandum contained a review of the voluntary system of data collection and a brief discussion of the value of such observations to meteorological science. The sixth made the point that one of the most important functions of a central meteorological office was the formation and maintenance of a library of meteorological information for all parts of the world.

The seventh section focused on the operational changes that had resulted from the war. As regards these changes, Shaw pointed out that when the war began the Office had made no specific arrangements to anticipate the requirements of the naval and military forces. "Special compilations of data had therefore to be improvised when the demands of the Navy and the Army were made known." In Germany, he said, "some of the compilations which we have found necessary had already formed the subjects of university exercises". It is a matter of regret, he went on, "that in this country there is practically no academic provision for the teaching of meteorology, and the study of weather is not an avenue to a degree at any university of the British Empire". Moreover, he continued, "when war broke out many compilations of data for the services were made at the Office, and two establishments for the collation of data were set up by the Admiralty in charge of amateur meteorologists". As regards the requirements of services for aviation, the Office had been better prepared, largely through the work of the branch at Farnborough.

However, Shaw said, the wartime requirements of the armed services had gone beyond what the Office had been able to supply from its own establishments. New meteorological services had been developed or introduced to meet the needs of war. The provision for the Admiralty that had been made in 1912 when the RFC was formed had since been modified and developed. A Naval Meteorological Service had been established "to satisfy the special and peculiar needs of the Royal Naval Air Services and the Fleet", with headquarters at the Admiralty and the whole service under the Hydrographer of the Navy. A Meteorological Section of the RE had been established at the headquarters of the Army in France, with reserves and auxiliary establishments at home. But the lack of an adequate number of trained meteorologists was one of the main difficulties in connection with the establishment of these new services.

Scientific or technical training had been necessary for most of the personnel appointed to Meteor. Initially, Shaw said, the only meteorologists in Meteor with any rigorous scientific training had been superintendents or professional assistants of the Office who had been appointed since the Office's reorganization in 1905. For the Naval Meteorological Service, the Admiralty had also drawn to some extent on the technical staff of the Office, while others had received professional training at the Office. In consequence, the Office had been forced to meet "a large increase of obligations with its trained staff very much reduced, and supplemented by various improvised expedients". These circumstances, he said, drew attention to the fact that the Office had been the only school of professional training in meteorology. The urgency of the need for more adequate provision of meteorological training had thus been emphasized, not only for the meteorological services of this country but also "of the Dominions and Colonies, who ought to be able to find adequately trained men should they require them". Shaw clearly envisaged a central role for the Office for the provision of meteorological training throughout the British Empire.

At last, in the eighth section of the memorandum, Shaw addressed the matter which the Ministry of Reconstruction had asked the Committee to consider, the future of the Office. He was extraordinarily brief and showed that he believed the Office should continue to function more or less as it had done before the war, as an institution which promoted meteorological science and served maritime, agricultural, public health and various other civilian needs. Forecasts would continue to be issued and climatological data collected, processed and published, and scientific investigation would continue to be an important part of the Office's work. His only novel suggestion was that an information bureau be set up.

He made no mention of meteorological support for civil aviation, even though the Committee, at their meeting on 24 October 1917, had received a report that Lyons and Taylor had been appointed members of sub-committees of the Civil Aerial Transport Committee, a Government body formed in May 1917 to discuss post-war civil aviation possibilities. Shaw did not appear to have recognized that civil aviation would surely after the war bring new priorities for the Office. So far as the Navy, Army and RFC were concerned, he had acknowledged the need to provide meteorological training but not suggested any specific role for the Office in respect of that training. However, he did make a rather general statement: "a special organization will almost certainly be called for to meet the special requirements of the Air Services, as has already been found to be the case for the service of the Navy and Naval Air Services".

In the section of the memorandum concerned with reconstruction, Shaw did not convey any significant sense of foresight or proactivity. In 1917, he was 63 years of age. Retirement was approaching. Perhaps there were signs of him not appreciating, or wanting to appreciate, the extent to which the world had changed in the war. The scientific flame still burned in him, and his desire to improve meteorological education and training was as strong as ever, but he either overlooked the coming of civil aviation or ignored it, even though a member of the Committee was serving on a sub-committee of a civil aviation body set up by the Government, and Shaw himself was close to the Government. He also did not appear to notice an opportunity for the Office. Not just civil aviation in the form of aeroplanes and airships would need meteorological support; so, too, would military aviation. The obvious institution to provide that support was the Office. Why did he not suggest this?

As regards staffing arrangements, Shaw suggested there should be both a Director-General of Meteorological Services and a Controller of the Office. The duties of the former should be, he suggested, to take charge of scientific investigations, organize the disposition of the staff of the various establishments, coordinate their work, and also "act personally as adviser on meteorological questions when called upon by any of the departments of the home Government or by colonial governments in connexion with meteorological services". The function of the Controller would be to take charge of administration and the Office's supply services. He thus proposed that the administrative duties hitherto carried out by the Director be devolved to a Controller.

Shaw justified the suggested division of duties between a Director-General and a Controller in his memorandum of 12 January 1918, saying that the work of the Office would have ceased to function during the war without the informal arrangement whereby Lyons had taken on much of the administration and accordingly allowed him to concentrate on meteorological matters of national importance. Now, he said, further developments were in prospect that made it desirable to adopt forthwith an arrangement on the lines of that he had proposed in his memorandum on reconstruction. There were three reasons.

First, Meteor was about to be expanded. Second, there was now an Air Ministry, which would, he said, require for its own purposes a meteorological organization more extensive than any meteorological unit which existed at the present time within the Office or outside of it.[19] Third, an authoritative book was needed, one which set out the facts and principles of meteorology, including the more recent results of upper-air research. It would be intended for the instruction of the large number of new recruits to meteorology for whom at present the only publications available were introductory books of an insufficient level and "innumerable official publications and scattered papers in scientific periodicals". Shaw considered the book "the most urgent meteorological need at the moment". He wished to write it himself, but in order to do so needed a "greater sense of freedom from the details of administration and official duties" than was "possible with the present informal arrangements".

He thought it desirable that permission be obtained to introduce at once the arrangement he had suggested and proposed that Lyons be appointed Controller. As the latter represented the War Office on the Committee, however, the permission of that department would have to be obtained for him to be so employed, and permission from the Treasury would also be required "for a suitable salary to be assigned to that post". In addition, Shaw considered it essential that Lyons should continue to supervise the Meteorological Section of the RE. He suggested that the dual arrangement of Director-General and Controller should operate in the first instance until 31 March 1920, the day his own five-year appointment as Director was due to end.

His suggestions and proposals were approved by the Committee at their meeting on 23 January 1918, and agreement was reached that an application be made to the Treasury to proceed. During the discussion of his memorandum, however, reference was made to an important development. The Treasury's representative on the Committee, George Barstow, reported, as it was put in the minutes, that:

A proposal was before the Chancellor of the Exchequer for the transfer of the Office as a whole to the new Air Ministry with the least possible delay. It was based upon a memorandum by

---

[19]   The Air Ministry came into existence officially on 2 January 1918.

General Sir David Henderson setting out the necessity for a large meteorological organization in connexion with the combined Air Services.

Henderson was Vice-President of the Air Council, the body which had been given executive powers to organize and plan the amalgamation of the RFC and the Royal Naval Air Service. This union took place on 1 April 1918, thus forming a third component of the UK's armed services, the Royal Air Force (henceforth abbreviated to RAF).[20] As yet, there was no central body to oversee civil aviation.

Following the Committee's meeting, Shaw wrote two letters, both to the Treasury and both reproduced in the minutes of the Committee's meeting on 20 March 1918. The first of them, dated 25 January, was addressed to the Secretary of the Treasury, and he enclosed with it a copy of his memorandum of 12 January, which included the proposal that Lyons be appointed Controller of the Office. He outlined the work of Lyons during the war and emphasized his "quite extraordinary organizing capacity". On behalf of the Committee, he asked for Lyons to be "advanced in rank so as to mark him out as the officer in charge of the whole unit" and expressed the wish that Their Lordships of the Treasury would "see their way to move the War Office in the matter". A salary of £800 per annum was suggested. The Committee agreed that Shaw should become 'Scientific Adviser to the Government on Meteorology' and retain the chairmanship of the Committee in that capacity. Lyons, they agreed, should become Acting Director "during the continuance of the war, retaining his seat on the Committee as representative of the Royal Society".

The other letter to the Treasury was dated 1 February 1918 and addressed to Barstow. It concerned the proposal that the Office be transferred to the Air Ministry. Shaw was clearly not in favour and reiterated objections voiced by the Committee on 23 January, when Barstow had been present. One way or another, the Office served *inter alia* the Admiralty, the Board of Trade, railway companies, travellers by land and sea, the Board of Agriculture, the War Office, public health and water supply authorities, and the Colonial Office. As regards the Air Ministry and what he called "the interests of the general public in relation to aerial navigation", he and the Committee recognized that aviation developments were "novel and most impressive in their present importance and in the rapidity of their certain growth, and especially in the urgency of their requirements in wartime". However, they did not consider the "older interests which are also in the long run of far-reaching national and imperial importance as being permanently subordinate to the interests of the Air Ministry or the public interests which that Ministry stands for".

Shaw was especially concerned for the future of scientific investigation if the Office came under the control of the Air Ministry and hinted that he was not sure free public discussion and free scientific criticism could be guaranteed under such control. He put forward an alternative. He was not against close links with the new ministry and

---

[20] To avoid confusion over the initials *RAF*, the Royal Aircraft Factory at Farnborough was on 1 April 1918 renamed the Royal Aircraft Establishment.

indeed recognized that it would have access to "an immense amount of information that would be useful to meteorologists", but he proposed that the present Committee should continue to control the Office as an advisory body to the Air Minister, with a representative of the Air Ministry on the Committee.

The response of the Treasury to Shaw's letter of 25 January came from Sir Thomas Heath, in a letter dated 22 May which was read at the Committee's meeting that same day. Shaw's proposals had been accepted. He would become for the period of the war 'Scientific Adviser to His Majesty's Government on Meteorology', retaining his existing emoluments and the chairmanship of the Committee. Lyons would become 'Accounting Officer for the Meteorological Office grant' and continue to represent the interests of the Army Council on the Committee. Heath's letter also informed the Committee that Lyons had been promoted to the rank of Lieutenant-Colonel with effect from 29 March 1918 and appointed Commandant of the Army Meteorological Services (Home Section). He became Acting Director of the Office on 23 May 1918, paid as such from 1 May 1918.

As regards the proposed transfer to the Air Ministry, the Committee learned at their meeting on 22 May that the matter was still under consideration. A week later, the Committee met again. There was now a development to report. A letter had been received from Sir Thomas Heath. Dated 24 May, it stated that the Treasury had received a letter from the Air Ministry which suggested that control of the Office should be transferred to the Air Council (the body which administered the Air Ministry). Treasury officials thought it desirable before a decision was taken that a conference be held between representatives of the departments principally concerned. There would be a meeting on 31 May in the office of Stanley Baldwin, Financial Secretary to the Treasury. The Committee nominated as their representatives Shaw and Lyons and agreed, as it was put in the minutes of the meeting on 29 May, that there should be "an independent Central Meteorological Institute for meteorological information and scientific investigation". Each of the fighting services or other departments that required specialized meteorological work would have its own sub-service, "in preference to the idea of one of the fighting services undertaking the administration of an office intended to serve not only the special department to which it was attached but all the other departments and the general public, as well as the general duties of scientific investigation in meteorology and the cognate subjects". There was clearly no support for a transfer to the Air Ministry.

The meeting on 31 May was attended by representatives of the Meteorological Office, Royal Society, Treasury, Admiralty, War Office, Air Ministry, Board of Trade and Board of Agriculture & Fisheries, but the minutes of the Committee's meeting on 26 June 1918 suggest that little progress was made. After this, for several months, the Committee's minutes contain comparatively little mention of the idea of the Office being transferred to the Air Ministry. There was, however, a development worthy of report at the Committee's meeting on 21 August, as shown by a letter from

W A Robinson of the Air Ministry that was reproduced in the minutes. Addressed to the Secretary of the Treasury and dated 2 August, it conveyed the information that the Air Council had agreed that Captain R W Glennie would represent the Air Ministry on the Meteorological Committee. He was Director of the Naval Meteorological Service, which was to be transferred, the letter further stated, from the Admiralty to the Air Ministry on 15 August.

The minutes of the meeting on 21 August show that Shaw had begun to write his *Manual of Meteorology* and the Committee had given him permission to make an agreement with Cambridge University Press for publishing it, on the understanding that copies would be supplied to the Office for the use of the services at approximately cost price and also that the Office should render assistance as regards the use of blocks. The minutes also report that Miss E E Austin, a Cambridge graduate in mathematics and natural sciences, had been appointed Secretary to Shaw from 8 July 1918. Thus began a long association between Shaw and Miss Austin which continued until his death in 1945, after which she served as an executor (with Lempfert).

Meanwhile, Lyons was finding plenty of scope for his energies. Additional staff of all grades needed to be found for the Forecast and Statistical Divisions. Demands for forecasts and climatological information had increased rapidly as the war progressed. Furthermore, supplies of instruments were needed. Lyons could see that service requirements for meteorological equipment would continue to grow and realized that instrument makers would be unable to meet demand at short notice. Accordingly, he decided to place large orders, regardless of the fact that the Office's annual grant-in-aid was inadequate to meet their cost. Accommodation was another problem. The new premises on Exhibition Road were already too cramped, even though the whole of the top floor was now occupied by the Office. Lyons alleviated the problem by using a vacant house on nearby Cromwell Road to serve as an instrument store.

The Prospect of a National Meteorological Institute

Robinson's letter of 2 August showed that the Air Council still wished to transfer the Office to the Air Ministry. In the Council's view, he said, the importance of meteorology in the development of aviation was increasing and could hardly be over-estimated. The Council considered it essential that the RAF should be furnished with a Department of Meteorology in immediate contact with the needs of the Flying Service and staffed by persons who were directly responsible to the Council. The Treasury concurred in principle with the proposals of the Council and trusted that all possible steps would be taken to prevent unnecessary duplication of work with the Office. The minutes of the Committee's meeting on 4 December 1918 show, however, that Shaw was still resisting a transfer to the Air Ministry. He had written yet another long memorandum, dated 20 November 1918, and he had also drafted a letter for the Committee to consider, this letter to be sent to the Secretary of the Treasury.

By now, the war had ended. Perhaps Shaw felt that circumstances had changed. Perhaps he thought there would now be less pressure for the Office to be controlled by a section of the armed services. If so, this was not apparent in the memorandum or letter. But he still did not show any signs of warming to the idea of Air Ministry control. He had in mind a national meteorological institute and continued to press for the adoption of his plan to create one.

The memorandum followed a now-familiar Shaw pattern. First, with great clarity, he reviewed the history, organization and activities of the Office from 1854 onwards, with particular reference to the work of the Office during the previous twenty years or so. This information was undoubtedly intended to provide context and justification, which it most certainly did. Then, commendably briefly, he listed what he called "some such practical programme" which it would be desirable to follow.

The letter to the Treasury pointed out that three meteorological establishments had been set up during the war, viz. the Meteorological Section of the RE, the Naval Meteorological Service, and the Meteorological Service of the Air Ministry, with the latter being developed from the Meteorological Service of the Naval Air Stations which had been transferred to the Air Ministry when the RAF had been formed. "For the efficient working of all these stations", Shaw said, "there should be an organized cooperation between them". During the war, arrangements to meet the requirements of the various services had been improvised. Now that the war was over, and the requirements of each individual service could be disclosed without prejudice to national interests, arrangements could be made to bring about organized cooperation between the services.

Shaw reiterated in his letter to the Treasury the proposals made in his response to the Ministry of Reconstruction the previous December and, for good measure, enclosed a copy of his memorandum dated 28 December 1917, adding that "adequate coordination between the Office and the special meteorological services of the departments of State could only be secured by housing all the services in one building". His desire to develop the Office in its current form was shown by the following statement in his letter:

The Committee are reluctant to contemplate any fundamental change in the organization of the operations of the Office and its observatories. Its scientific work in recent years has been successful beyond expectation. While the meteorologists of other countries have confined their attention to the usual matters of detail, our own have worked towards the establishment of general principles and have achieved astonishing success. When the contributions of the Office are put together they form a coherent and organized advance in meteorological science which has no parallel and which has made it possible to approach the technical questions of the Artillery and Air Services with an insight which has been recognized in communications from officers concerned, and which so far as is known no other country possesses.

He considered that any alteration of the Office's arrangements that "interfered with these fortunate conditions would be a most serious misfortune for British meteorology".

At the same time, the possibility of the Office being controlled by the Department of Scientific and Industrial Research (DSIR) was being contemplated. Shaw wrote to DSIR on 6 December 1918 to ask what form of administration was being proposed for the Office. Sir Frank Heath, DSIR's Secretary, replied on 20 January 1919, saying that, in the first place, it was proposed that a new Advisory Council should be established to deal with the Geophysical Sciences. Attached to this Council, there would be a Meteorological Board, which "would include amongst its members technical officers nominated by the Government Departments directly interested in meteorological work". The number of members of this Board would probably be about the same as for the present Meteorological Committee, and the function of the Board would be similar to that of the Committee (i.e., to administer the Office, prepare and oversee the Office's programme of work, assume responsibility for preparing estimates and budgets, control expenditure, and prepare an annual report). Once the Office was part of DSIR, Heath further stated, the expectation was that it would be established in the neighbourhood of Whitehall.

Yet another lengthy memorandum from Shaw ensued, for the Committee's consideration on 20 January. He proposed that the work of the Office be divided into two, with a section in Westminster concerned with operational and climatological matters and another in South Kensington called a 'School of Meteorology'. He was clearly enthusiastic about the formation of a national meteorological institute as a component of DSIR. What he proposed, he said, was similar to that which existed in Norway, Sweden, Denmark, Holland, Italy and Japan.

In his reply to Heath, dated 12 February 1919, Shaw listed the many civilian and military "requirements the Office had already been called upon to satisfy" and mentioned that the BRO was likely to become a department of the Office in the near future. What he called "local depots for the distribution of meteorological information" were needed in Scotland, Ireland and the great seaports. As he put it:

The proposed development of aviation and aerial transport includes the establishment of aerodromes at intervals along the main air routes, and for these it is certain that trained meteorological assistance will be required which must be represented by a local establishment of the kind indicated for Scotland and Ireland. This assistance, by whatever means the cost is provided, must be available for private adventurers as well as for Government officials. Of these new establishments, some may or may not be directly under the control of the Meteorological Office, but all of them will be dependent upon the Office for some part of the training of the officers, the selection and supply of books of reference and other details of organization.

In order to provide for the work of the Office, Shaw said, expenditure of £61,189 was envisaged for the financial year 1919–1920. Of this, £48,500 would fall on Parliamentary votes, made up of £47,000 directly to the Office, £1000 for the Royal Society and £500 for DSIR. The remaining £12,639 would be provided by 'Appropriations in Aid', of which £9700 was the estimated payment for instruments supplied to the Services or private observers, and £400 would be provided by subventions from the Rosse Trust of the Royal Society and from the Gassiot Trust. The rest represented charges made by the Office for services rendered to departments or to the public, or incidental expenses incurred on their behalf and refunded.

The Meteorological Office Transfers to the Air Ministry

Just when it seemed Shaw might succeed in his ambition of making the Office a national meteorological institute, there came disappointment. Sir Frank Heath replied on 24 February 1919, saying that it was not possible in the present circumstances to reply in detail to the points raised in Shaw's letter. There were no objections in principle to any of the proposals. The problem was that the Government had not yet determined to make any change in the constitution of the Office, nor, indeed, had they decided whether the work of the Office should be transferred to DSIR or to some other department. Heath's letter was but one of the papers considered by the Committee at their meeting on 12 March. The others included, as the minutes state, a confidential communication from Heath, "enclosing a secret memorandum prepared by the Air Ministry and the Chairman's reply thereto". The future of the Office had been discussed by the War Cabinet.

A letter from Heath to the Secretary of the Air Ministry dated 10 January 1919 shows that the Home Affairs Committee of the Cabinet had asked him to consult the Air Ministry about the future of the Office. The Committee of the Privy Council for Scientific and Industrial Research were broadly in favour of the Office being transferred to DSIR but did not consider that all meteorological work should be concentrated in the Office thus transferred. They believed the Air Ministry would have an interest in the matter. A memorandum from the Air Ministry dated 12 February 1919 advised the War Cabinet that a transfer of the Office to DSIR would perpetuate a duplication of state meteorological services, which, for financial and administrative reasons, it appeared desirable to terminate. This duplication could be avoided if the Office were placed under the control of the Air Council.

Other responses to the proposed transfer of the Office to DSIR were broadly favourable. The Board of Trade and the Board of Agriculture saw no objection. The Army also did not object but did however ask that the Office's meteorological provision for military requirements and personnel should now continue in peacetime. The Admiralty concurred in the proposed scheme for the Office's transfer to DSIR but observed that it would be necessary to "define with care the exact relationship existing between the Meteorological Office and the various branch services as regards the execution of their work and composition of staff both of numbers and technical qualifications". Nevertheless, the Air Council pressed for the transfer of the Office to the Air Ministry.

Shaw complained, saying that he had not been fully consulted. In a letter to Heath dated 4 March 1919, he wrote thus:[21]

For some reason which I do not understand I have no opportunity of exchanging views upon meteorological matters with the Air Ministry. I have a good deal of experience of national and international organization which no-one else possesses and I am quite willing to place it at the disposal of the Air Ministry without any *parti pris*, in order that the country may secure all the objects for which meteorological services are intended.

He then went on to dissect the arguments of the Air Ministry which had been set out in their memorandum to Heath. He did not understand why the Office should be transferred to the Air Ministry and did not see much logic in such a course of action.

Shaw was fighting a losing battle. The Committee of Home Affairs decided on 8 May 1919 that the Office would be attached to the Air Ministry. Meanwhile, though, hopes that the Office would become part of DSIR remained alive, and much of the Meteorological Committee's meeting on 4 April 1919 was given over to a discussion of detailed arrangements for effecting this transfer. Meanwhile, too, Lyons had decided the time had come to resign as Acting Director. Shaw resumed the duties of Director on 28 April 1919.[22] The decision to transfer the Office to the Air Ministry was discussed by the Committee on 28 May, but there was nothing they could do to alter the decision. The transfer to the Air Ministry happened formally on 1 July 1919.

Shaw remained unhappy about this decision, as is clear from the *Annual Report of the Meteorological Committee to the Treasury for the Year Ending 31 March 1920*, but he voiced his feelings more forcefully in a paper found among the possessions of Miss Austin after her death (which occurred in 1987). With the paper, there was a copy of a letter dated 17 August 1945 from the then Director of the Office, Nelson Johnson, which shows that Miss Austin had drawn the paper to his attention soon after Shaw's death.[23] The paper is headed 'Note of a meeting of the Home Affairs Committee of the War Cabinet on May 8 1919'. There is nothing in it that indicates when it was written, to whom it was addressed or why it was written, other than to record Shaw's feelings at the time.

The paper begins with the following:

On May 6 or 7, I received as a double-enclosed document the reference of the Treasury to the Home Affairs Committee on the proposal for the transfer of the Office to the Department of Scientific and Industrial Research with certain annexes. There was no covering note nor

---

[21] The letter can be found among the documentation on the coordination of meteorological services laid before the War Cabinet at their meeting on 22 March 1919 (CAB/24/77, G.T.-7018).

[22] A diary written by Lyons noted for 12 April 1919: "Wrote to Sir N Shaw advising termination of arrangement of Meteorological Committee March 20 1918 which was approved by Treasury May 22 1918 to take effect May 1 1918. This was done in consequence of his repeated interference in matters of an administrative character thereby far exceeding the functions of either the Adviser or the Chairman of the Meteorological Committee". On 28 April 1919, he wrote: "Ceased Directorship of Meteorological Office. No report of any kind sent in by Shaw". Relations between Shaw and Lyons appeared to be strained.

[23] Johnson's letter and the typescript of Shaw's note are in the National Meteorological Archive at Exeter.

any intimation of the reason for the communication. I concluded it was 'for information', acknowledged its receipt and put it away safely.

On May 8, in conversation on the telephone, Mr Barstow said he supposed he would see me in the afternoon at the meeting of the Home Affairs Committee, and when I explained that although I had received a copy of the reference I had no notice of any meeting he said I had better be in attendance at 4.15 and that he would communicate with Sir Frank Heath on the subject. I received no further communication and accordingly attended at the hour Mr Barstow mentioned. The duty that I anticipated having to perform was to supply Sir Frank Heath with any material information about the Office and its working which might be found necessary for the conduct of his case and which was not explicit in the papers submitted. I had no expectation of taking part in the discussion of the Committee except by way of cross examination.

Shaw went on to say that the chairman had begun the meeting by stating that the idea of the Office being transferred to DSIR had been approved by all departments except the Air Ministry. After that introduction, General J E B Seely had been called to speak. He was the Under-Secretary of State for Air and the Air Ministry's representative.

Shaw said that Seely "misrepresented the contributions which the Meteorological Office and the Meteorological Service of the Air Ministry had made to the practical requirements of the Air Services during the war". He insisted, Shaw went on, "that the meteorological requirements of the Air Ministry, which were of the most imperative character, could not possibly be satisfied except by the absorption of the Meteorological Office". Shaw then listed the fundamental errors of the speech and commented that it was "ostensibly intended to show that the Air Ministry could not do its duty to the State in guarding the lives of the pilots and passengers of aircraft unless they had control of all the meteorological forces of the country". This showed, Shaw said, that the Air Ministry did not have sufficient knowledge of meteorology to organize a service for their own requirements. They needed to depend on the staff and experience of the Office.

Seely reiterated arguments contained in a memorandum dated 1 May 1919 which he had submitted to the War Cabinet. In this he had contended that the Air Ministry should be responsible for all meteorology and should therefore absorb the Office. The memorandum included comments extracted from a report by Brigadier-General F V Holt, RAF, comments which were critical of the wartime meteorological service provided by the Office for aviators. Seely acknowledged that the work of the Office for farmers and seafarers had long been excellent but argued that the successful development of the Office for the benefit of civil and military aviation depended on the Office being controlled by the Air Ministry.[24]

[24] There is a copy of this memorandum in the National Archives at Kew (CAB 24/79, paper GT7220). It is inconceivable that Seely would insist on the absorption of the Meteorological Office by the Air Ministry without the approval of the Secretary of State for Air, who was, at the time, Winston Churchill. As the memorandum bears the hand-written initials W.S.C., it appears likely that Churchill was the person who ultimately decided that the Office would be attached to the Air Ministry.

To his surprise, Shaw was called on to follow Seely. Not expecting to be called, and therefore unprepared, he said that he could only endeavour to suggest that Seely's statement was exaggerated and did not take account of the more general requirements of a meteorological office. Furthermore, he said, the statement misrepresented the capacity of the existing Office to meet the requirements of the Air Force in the field.

Shaw reported in his paper that the representatives of DSIR had said nothing during the meeting, but the representatives of the Treasury had explained that the other departments which were interested had all expressed their dissent for the proposal to transfer the Office to the Air Ministry. Thereupon, Shaw said, the Home Affairs Committee had peremptorily decided that the Office should be attached to the Air Ministry, not DSIR. This, Shaw commented, "would seem to be a travesty of deliberative process".

Terms for amalgamation of the Office with the Air Ministry's Meteorological Service were agreed at a meeting in the Air Ministry on 9 July 1919, and Shaw undertook informal responsibility for the Meteorological Service. Transfer of the Office from the Treasury to the Air Ministry for purposes of accounting occurred on 1 October 1919, and all expenses of the Office were chargeable to the Air Estimates from the beginning of 1920.

Formal discussions with the Director of the BRO, Dr Mill, about the incorporation of the BRO into the Office began in October 1915. He wished to give up the BRO work, as his eyesight was failing. Conditions attached to such a transfer were discussed over the coming months and years, but the transfer did not happen until 25 July 1919, when a legal agreement was signed and Carle Salter succeeded Dr Mill.[25] *Symons's Meteorological Magazine* was combined with the *Monthly Circular of the Meteorological Office* and became the *Meteorological Magazine* as from February 1920.[26]

Accommodation was an issue that needed to be resolved urgently. The space occupied by the Office before the transfer to the Air Ministry was about 34,000 square feet, made up of 26,000 square feet in the building on Exhibition Road and about 8000 in other premises, these being the BRO's home at 62 Camden Square, the premises at 15 Cromwell Road used for storing instruments, and a property at 47 Victoria Street used for an investigation of atmospheric pollution.[27]

Shaw recommended to a Cabinet committee on 17 July 1919 that the Office's administrative, operational and research work should all be carried out in one building

[25] A copy of this agreement was reproduced in the minutes of the Meteorological Committee's meeting on 30 July 1919. When the BRO was incorporated into the Office, responsibility for publishing *British Rainfall* devolved to the Office. Salter had become an assistant to Symons in 1897 and Mill's right-hand man by 1919.

[26] The *Monthly Circular of the Meteorological Office* was first issued on 20 June 1916. It contained official notices, short obituaries, announcements of appointments, news of publications and other information of interest to staff members and observers.

[27] A Committee for the Investigation of Atmospheric Pollution was appointed in 1912, with a brief to study all aspects of air pollution in the cities and industrial areas of the UK, as well as to propose remedial measures and recommend administrative actions. The work of this Advisory Committee came to be regarded as part of the duty of the Meteorological Office.

near Whitehall, with an exception made for the BRO, which would continue to be at 62 Camden Square, where it had been since the 1860s. He reckoned that a building which contained 36,000 square feet would be required, with the extra space needed to accommodate the additional staff who had joined the Office after the transfer to the Air Ministry. The figure of 36,000 was increased at the meeting to 39,000. As no building which provided so much accommodation would be available in the vicinity of Whitehall in the immediate future, the Air Ministry decided, in consultation with the Office, that 9000 square feet would be made available in an Air Ministry building on Kingsway (Canada House), specifically for the Office's administrative headquarters, printing office and Forecast Division.[28] In the event, the Marine Division moved to Kingsway, too. The Library and Statistics Division remained in the building on Exhibition Road, and the Instruments Division returned to Exhibition Road from 15 Cromwell Road. The top floor at Exhibition Road reverted to Science Museum use, and temporary provision was made in the building for chemical work on atmospheric pollution.

The constitution and functioning of the Meteorological Committee initially remained unchanged, and meetings were held on 25 June and 30 July. Then, by instruction of the Air Council, a meeting of the Committee was summoned for 15 October 1919. It was held at the Air Ministry, and those present were, in addition to Shaw:

• Major-General J E B Seely, Parliamentary Under Secretary of State for Air
• Major-General Sir Frederick Sykes, Controller-General of Civil Aviation
• Mr Henry McAnally, Secretary of the Air Ministry
• Messrs J S Ross and F G L Bertram of the Air Ministry
• Rear-Admiral Frederick Learmonth, Hydrographer to the Navy
• Dr Arthur Schuster, representing the Royal Society
• Sir Thomas Middleton, representing the Board of Agriculture and Fisheries
• Lieutenant Hugh Grant, Superintendent of the Admiralty's Meteorological Service

Only one member of the Committee did not attend. Colonel Lyons sent apologies for absence on account of illness.[29]

A new constitution for the Committee was agreed on at the meeting. Shaw would serve as chairman, and the Committee would include representatives from the War Office, Admiralty, Board of Agriculture and Board of Trade, plus two representatives from the Royal Society and two from the Air Ministry. Enquiries would be made

---

[28] The Forecast Division moved to Kingsway on 18 November 1919 and the Accounts Section and Director's Secretary moved two days later. The lithographic press was installed at Kingsway on 21 January 1920, and the *Daily Weather Report* was published and issued from there as from that date. The Marine Division transferred to Kingsway on 2 March 1920, and the Instruments Division moved back from 15 Cromwell Road to the building on Exhibition Road on 5 March.

[29] Lyons remained a member of the Meteorological Committee for many years. He was Director of the Science Museum from 1920 to 1933.

regarding representation of the Treasury and the Colonial Office.[30] The President of the Committee would be the Controller-General of Civil Aviation, to whom the Committee were already responsible.

Shaw and Lyons had both been involved in the process of forming a Department of Civil Aviation. Both had contributed to the documentation prepared by the Civil Aerial Transport Committee for a meeting of the War Cabinet on 29 November 1918. Shaw had submitted a long memorandum entitled *Meteorological Services in Aid of Aerial Transport* in response to one from Lyons entitled *Memorandum on Research in Regard to Meteorology*. The two men had been essentially in agreement, and both had pressed for services for civil aviation to be provided by the Office under its existing constitution. These services would include research and operational arrangements, and Shaw had suggested that meteorological advisers should be available at all aerodromes. The advisers would be trained by the Office, and Shaw had further suggested that the Imperial College of Science and Technology and/or the Universities or Technical Institutes of Edinburgh, Glasgow, Dublin, Liverpool, Cardiff, Southampton and Plymouth should be asked to set up a teaching centre for Aeronautical Engineering, Aeronautical Theory and the structure of the atmosphere.

However, the arguments of Lyons and Shaw were undermined to some extent by those put forward by G I Taylor and Lord Montagu of Beaulieu, who were both experts on aviation. In the document they wrote for the Civil Aerial Transport Committee, they set out succinctly and logically the meteorological requirements of civil aviation, concluding that the current climatological and forecasting work of the Office was adequate for aviation purposes but the present system for supplying what they called "knowledge of the momentary weather" was not. A dedicated Central Office would be required, one which provided a continuous weather service.

An announcement was made in the House of Commons on 12 February 1919 that the Government had decided to set up a Department of Civil Aviation under the control of the Air Council, and it was also announced that day that the appointment of Sir Frederick Sykes as Controller-General of Civil Aviation had been approved by King George V. An Air Navigation Bill was read a third time in the House of Commons on 27 February 1919 and passed by the Lords later that day. The resulting Air Navigation Act, which came into force on 1 April 1919, formally established the Department of Civil Aviation and legally sanctioned and regulated civil aviation.

The functions of Sir Frederick were clarified in the answer to a question asked in the House of Commons on 21 February. Among other things, Cecil Malone asked whether the Controller-General would be concerned with lighter-than-air as well as

---

[30] The Colonial Office subsequently named a representative, Mr J E W Flood, and so, too, did the Royal Society of Edinburgh. Dr E M Wedderburn became a member of the Committee at the beginning of December 1919 to represent not only that Society but also the Edinburgh branch of the Meteorological Office. Mr George Barstow, who had served as the Treasury's representative for almost twenty years, was replaced by Mr Ross of the Air Ministry's Finance Department.

heavier-than-air craft. Major-General Seely, the Under-Secretary of State for Air, gave the answer as "most certainly" and went on to say that the Controller-General would be charged with "the mapping out of air routes, with securing landing grounds, the supervision of Rules and Regulations for air travel; with the arrangements being made to link up this country, not only with the Continent, but with our Overseas Dominions, and so on". These many possibilities which involved heavier-than, as well as lighter-than, air machines, would all come under the purview of the Controller-General.

Miscellaneous Developments

In the war and the months that followed, much changed in the Meteorological Office. The conflict and its aftermath naturally determined the Office's priorities, but the Office's work was not entirely war-related. Some services for agriculture continued, for example, and so did the atmospheric pollution investigation. Routine climatological work continued, and the observatories continued to function. The demands of the war led to an increase in the number of professional and clerical staff employed by the Office, but one notable scientist left the Office during the war.

L F Richardson, the Eskdalemuir Superintendent, objected on conscientious grounds to the involvement of the Office in war. He was a Quaker and tried a number of times to persuade the Office to release him to join a Red Cross unit or the Ambulance Corps. Eventually, he gave notice that he would terminate his appointment in the Office on 15 August 1916 in order to join the Friends' Ambulance Unit in Flanders. His resignation was accepted, and the Gassiot Committee appointed in his place Dr A Crichton Mitchell, who had worked in India as Director of Public Instruction in Travancore until 1913 and since been engaged on research work in the Physical Laboratory of the University of Edinburgh. Richardson's appointment ended, in fact, on 15 May, and he then joined the Friends' Ambulance Unit.

Whilst at Eskdalemuir, Richardson developed ideas on weather forecasting by means of mathematical methods. In May 1916, he asked Shaw to communicate to the Royal Society the first draft of a book called *Weather Prediction by Arithmetical Finite Differences*. Generously, the Royal Society voted £100 towards the cost of publishing it, but Richardson first wished to add a practical example that would show how the method he proposed would work in practice. He took with him to France a set of surface and upper-air observations made at 7.00 a.m. GMT on 20 May 1910 and used them to test his method. He included the results in a revised version of his book, but the manuscript was unfortunately lost during the Battle of Champagne in April 1917 when sent to the rear for safe keeping. Great was the relief when it was rediscovered under a heap of coal some months later. The manuscript was delivered to Cambridge University Press in the autumn of 1918, and the book was published in 1922. Called *Weather Prediction by Numerical Process*, it is a 'classic' of the meteorological literature.

Figure 8.5. Lewis Fry Richardson. © Crown Copyright 2010, the Met Office.

After the war, Richardson was re-employed by the Office but not reinstated automatically, having earlier resigned. He wrote to Shaw from France on 28 October 1918, saying that he wished to continue his work on numerical weather prediction and be loaned "a small workshop, £200 for special apparatus and materials and the part-time services of a mechanic and an observer, and the use of a large field, and of ordinary pilot-balloon apparatus" to develop new methods of observing wind and temperature in the upper air. He also required a mathematician to assist him, as well as two clerks and several orderlies. Quite a request! Shaw replied on 5 November, a characteristically long letter, which Richardson must have considered encouraging. He promised to draw Richardson's request to the attention of the Meteorological Committee and said that he did not know how things would work out but thought it would be a pity to lose the advantage of his insight into meteorological problems.

The outcome was that Richardson joined W H Dines at Benson in March 1919 and worked on ways of improving his method of forecasting atmospheric changes by mathematical means. For this, he needed more observational data from aloft than were then available, so he carried out experimental work to test novel methods of measuring solar radiation, atmospheric turbulence and vertical distributions of temperature and wind. The transfer of the Office to the Air Ministry unsettled him, though. He was a pacifist. He felt unable to work for an institution that was associated with the armed services, so he resigned once again and left the Office at the end of August 1920, never to return. By 1922, when his *magnum opus* was published, he was Head of the Physics Department at Westminster Training College.

Sadly, at least five of the Office's staff who had joined up during the war died on active service, and many more were injured. There were also deaths from natural causes during and soon after the war. Two were particularly sad, for both men could have retired during the war but chose instead to continue serving the Office. They never enjoyed retirement. Both were in the Marine Division. William Allingham died on 24 January 1919, on his way to work. He was 68 and had served the Office since 1875. Captain Hepworth died on 25 February 1919 at his residence in Ealing. He was 69 and had served the Office since 1899.

Former Director Robert Scott died, too, on 18 June 1916, aged 83. And three who had worked under FitzRoy died during or soon after the war. One was Richard Curtis, who had joined the Meteorological Department of the Board of Trade in November 1861 and remained with the Office until his retirement in March 1912. He died on 21 May 1919, in his 73rd year. Another was James Staughton Harding, who had joined FitzRoy in 1855 and retired in 1906. He died on 11 January 1915, in his 76th year. The third was Frederic Gaster, who had served the Office from 1859 to 1909. He died on 10 September 1915, aged 73. Only one person who had worked under FitzRoy still survived: Richard Strachan.[31]

Yet another link with the early history of the Office was broken on 24 October 1916, when James Joseph Hicks passed away, eleven days short of his 81st birthday. He had supplied instruments to the Office since 1869. The original version of a most useful instrument which had only recently been invented was manufactured by J J Hicks & Co in 1915. This was the pilot-balloon slide rule, invented by F J W Whipple, who had become Superintendent of Instruments in 1912. By means of the slide rule, the work of reducing pilot-balloon observations was greatly simplified. A rather less happy reference to instruments can be found in *Meteorological Office Circular* No. 6, published on 21 November 1916. This contains a report that most of the meteorological instruments at Plumstead had been "broken up or stolen by roughs". The observer, Mr J G Waller, had also met with an accident which had prevented him from ascending to the sunshine recorder. Accordingly, all observations had been discontinued.

Confusion for observers, and annoyance, came just before the war with the introduction of absolute units. On the recommendation of the IMO's Commission for Scientific Aeronautics, the Meteorological Committee decided to record rainfall measurements in millimetres and barometric pressure readings in millibars (instead of inches in both cases). The Office used these units in the *British Meteorological and Magnetic Year Book* from January 1912 and the *Daily Weather Report* from 1 May 1914. And the BRO adopted millimetres in 1915. Instruments were recalibrated, with the dials of aneroid barometers labelled in centibars, not millibars, and the glasses for measuring rainfall labelled in millimetres. Curiously, though, the Office asked observers to report pressure readings in millibars, not centibars. No wonder there was confusion! In the Year Book, air temperatures were tabulated in Absolute degrees from 1 January 1912,

---

[31] Strachan died in 1924, at the age of 90. Gaster had been his assistant back in 1859.

this being considered helpful when barometer corrections were being applied, as the equation which governed these corrections required absolute units.

A development in 1916 which caused further confusion among observers was the introduction of Summer Time, which had first been discussed by Parliament in 1909 (see Chapter 7). The Summer Time Act, which received its Royal Assent on 17 May 1916, specified that, for general purposes, the time in Great Britain would be, from 2.00 a.m. Greenwich Mean Time (GMT) on 21 May 1916 to 2.00 a.m. GMT on 1 October 1916, one hour in advance of GMT. In Ireland, Dublin Mean Time would be replaced by GMT, the difference being twenty-five minutes.

The position of the Office over the introduction of Summer Time was set out in a very detailed circular issued by the Office on 18 May 1916. This gave notice that:

- In accordance with the Act, GMT would be used as theretofore in all Meteorological Office forms, records and books of instructions issued, for describing the times of the observations and other operations at the observatories and stations in connection with the Office.
- Until further orders, the hours of observation for the telegraphic reports of the Daily Weather Service would continue to be 01, 07, 13, 18 and 21 GMT (i.e., 2 a.m., 8 a.m., 2 p.m., 7 p.m. and 10 p.m. Summer Time).

This seemed clear, and observers who possessed two clocks were advised for their personal convenience to keep one of them set to Summer Time, the other to GMT. There was, however, confusion.

Many who observed for both the Office and the BRO were not sure exactly when their observations should be made, which led, in turn, to confusion among those in the Office and the BRO who processed the observations regarding exactly when certain observers had in fact made them! There were further problems, which were that the 07:00 GMT observations which were included in weather reports were made too late for the regular Post Office hours of delivery and postage, and the 18:00 GMT observations could not be communicated, except through a very few telegraph offices, because the offices had by then closed for the day. Moreover, it proved impossible to maintain the evening reports from the health resorts at the usual hour. They had to be made an hour earlier. But the concept of Summer Time was here to stay, and the confusion among observers naturally diminished over the years.

A more agreeable development was announced in a letter from Sir Thomas Heath of the Treasury. Dated 12 December 1918, it informed Shaw that His Majesty's Government had approved an extension of annual leave for the year 1919 in "appreciation of the public spirit with which civil servants, both permanent and temporary, have responded to the demands made by the existence of a state of war, in many cases sacrificing a considerable amount of their normal holiday". The Meteorological Committee agreed that extra leave should be granted to staff of the Office on the same terms as for permanent civil servants and approved an order which placed on record their appreciation of the high standard which had been maintained and the manner in which difficulties had been overcome.

We may wonder how the Office would, or could, have met the demands of war had its scientific state remained as lacking in vitality as in 1900. How fortunate it was that a man of vision had taken charge in 1900. In Shaw's time at the helm, the Office had changed out of all recognition. His policy of bringing in science graduates had been amply justified by the level of success obtained when this policy had been tested under conditions of active service. But how much of Shaw's time had been taken up with the composition of letters and memoranda since 1900? Most of his missives had been long and detailed, though always clear and direct. How much of his time could have been spent more productively? And how much of the time of those who replied to his letters could have been better spent? However, we should speculate no more. His time at the helm would surely soon end. He was 65 years of age when the Office became part of the Air Ministry.

9

The Inter-War Period

Two exploits in 1919 made headlines around the world. In June, John Alcock and Arthur Whitten Brown won the prize of £10,000 offered by the *Daily Mail* for being the first to fly non-stop in an aeroplane from any point in North America to any point in Great Britain or Ireland in under seventy-two hours. The following month, there was another triumph of British aviation. His Majesty's Airship R.34 made the first round-trip crossing of the Atlantic by air.

The Meteorological Office was one of the organizations which provided technical support for these ventures. For the flight of Alcock and Brown, meteorologists and instruments were sent to Newfoundland, the Azores and Lisbon, along with equipment for observing the upper air. For the flights of the airship, two RAF meteorological officers equipped with kites were assigned to the battle cruisers *Tiger* and *Renown* stationed north and south of the usual tracks of Atlantic steamships, providing observations which proved useful not only for providing information about flying conditions at heights up to about 600 metres but also for the production of synoptic weather charts of the North Atlantic. Moreover, the airship carried a meteorologist, Guy Harris, who had, in the spring of 1919, investigated the upper air over the North Atlantic with kites he flew from the cargo steamer *Montcalm* on voyages between London and New Brunswick.

While intrepid aviators were widening their horizons, civil aviation of a routine nature was developing. Passenger and mail services between London and Paris started in January 1919, and the Office responded by setting up local centres to provide meteorological reports, weather forecasts and climatological information. The intention was that twenty of these 'distributive stations' would be set up at appropriate aerodromes, each staffed by a professional meteorologist and trained technical assistants whose duties would be to make necessary local observations and collect simple meteorological reports from other places in the area. They were also to receive by wireless telegraphy or ordinary telegram the necessary data for the preparation of synoptic charts and to advise the aviation services in the area. Because time would

223

be needed to train staff for all of the planned stations, the decision was made that priority would be given to the provision of reports and forecasts for the routes between London, Paris and Brussels.

The first distributive station opened at Hounslow Aerodrome on 1 December 1919, staffed by one Professional Assistant and one Technical Assistant, and additional stations opened at Cranwell in Lincolnshire and Lympne in Kent on 14 January 1920, Pulham in Norfolk on 19 January, Calshot in Hampshire on 2 February, Felixstowe in Suffolk on 3 February, Biggin Hill, also in Kent, on 16 February, and Didsbury, near Manchester, on 24 February. The air terminal for London and its meteorological station moved from Hounslow to Croydon on 28 March 1920.[1]

The flights across the Atlantic were not the only notable feats of aviation in the months after the war ended. Aviation was opening up the world and meteorological support was essential. The Office recognized the need for greater knowledge of the atmosphere along aeroplane and airship routes and duly carried out investigations which greatly advanced aviation meteorology and climatology.

Meteorological Services Are Resumed

After the Armistice, a restoration of normal peacetime meteorological activities did not occur immediately. Colonel Lyons, then Acting Director of the Office, received a letter dated 19 December 1918 from Dr T Hesselberg, Director of Det Norske Meteorologiske Institut, Kristiania. Deutsche Seewarte in Hamburg wanted to know whether meteorological telegrams from English stations could be sent from Kristiania to Hamburg. The Germans had assured Hesselberg the telegrams would be used only for the agricultural weather service. Lyons refused the request, saying in his reply, dated 20 January 1919, that he had been instructed no communications with Germany were yet permitted in the UK, except on official matters. However, he added, telegrams would be sent freely as soon as this policy had been withdrawn.

Agricultural forecasting recommenced in Britain in April 1919, and harvest forecasts were reintroduced in the summer of that year. Provision of weather forecasts to the public began on 15 November 1918, only four days after the Armistice was signed, and weather charts began to appear again in newspapers in early 1919, first in the *Morning Post* on 1 January, later in *The Times* and the *Daily Telegraph*. The *Daily Weather Report* was redesigned and enlarged in 1919 to comprise, as from the issue for 1 April, a four-page British Section, a four-page International Section and a two-page Upper Air Supplement. The Office's 'Press Table' for journalists and other members of the public was partially restored in 1919 and then expanded after

---

[1] The estimates for the financial year 1920–1921, discussed at the Meteorological Committee's meeting on 24 March 1920, show provision for twenty distributive stations, located at Croydon, Lympne, Felixstowe, Grain, Andover, Calshot, Howden, Cranwell, Didsbury, Cattewater, Baldonnel, Orkney, Malta, Shoeburyness, Larkhill, Glasgow, Liverpool, Edinburgh, Biggin Hill and Guernsey or Jersey.

the move to Kingsway. A new series of Meteorological Office papers called *Professional Notes* began in 1918, covering scientific and technical aspects of meteorology, and publication of *Geophysical Memoirs* recommenced the same year. The first of the *Professional Notes* was by David Brunt, on 'The Inter-Relation of Wind Direction and Cloud Amount at Richmond (Kew Observatory)', and the first *Geophysical Memoir* published since 1914 was by Sir Napier Shaw, on 'The Travel of Circular Depressions and Tornadoes and the Relation of Pressure to Wind for Circular Isobars'.

The minutes of the Meteorological Committee's meetings show that the dominant issue in the first six months after the war ended was the future of the Office. Would control pass to DSIR or would it pass to the Air Ministry? Once that decision had been made, the Committee's attention naturally turned to arrangements that needed to be made in consequence of the Office's transfer to the Air Ministry, but a number of other matters needed to be addressed, too, particularly the reintroduction of services suspended or curtailed during the war, staff appointments, the incorporation of the BRO, the work of the Advisory Committee for Atmospheric Pollution, and services for civil and military aviation at home and abroad. The latter included provision of observers and weather forecasts for the Expeditionary Forces to North Russia and the regular supply of data to the Meteorological Sections of the Royal Engineers at Cologne, Wimereux, Murmansk and Archangel.

In the midst of the uncertainty over the future of the Office, a letter was received from Sir Arthur Robinson, Secretary of the Air Ministry. It was dated 24 March 1919 and addressed to the Secretary of the Office. It read as follows:

I am commanded by the Air Council to state that, in connection with the new Department of Civil Aviation, it will be necessary to reorganize the existing Meteorological Branch at the Air Ministry. The Air Council is desirous of securing for this Branch a person possessing the necessary expert knowledge and organizing capacity. The Controller-General of Civil Aviation has recently had an interview in France with Colonel Gold who has expressed his willingness to accept the position if it can be arranged. He would come on condition that he had the option of returning to the Meteorological Office if either (a) he found the work uncongenial or (b) he was found unsuitable in the organizing capacity for which his services are required during the first six months.

Robinson went on to say that he understood Gold would be demobilized shortly as his services would no longer be required by the War Office. He was commanded, therefore, to apply formally to the Office for Gold to be allowed to work for the Department of Civil Aviation on the stated conditions.

Though Lyons was still Acting Director, it was Shaw who replied. He informed Robinson that Gold had been retained on the staff of the Office as Superintendent during the period of his employment as Commandant of Meteor "with the understanding that upon demobilization he would be available to return and take up duty at the Office on the terms applicable to members of the Civil Service who joined the Army with the sanction of the Head of Department in which they were employed". He went on to say that the Committee recognized Gold's experience might be of assistance to the country through the Air Ministry at this juncture and stated that the Committee wished to assist in "the coordination of the various services for their common purposes". However, he said, the Committee could not consent to placing Gold's services at the disposal of the Air Ministry without knowing what the nature of the relationship between the Office and the Meteorological Branch of the Air Ministry would be and whether the proposed arrangement had the approval of H.M. Government. Shaw did not appear at all enthusiastic, and it seems the Controller-General did not pursue the matter. Nothing came of Robinson's enquiry. Gold returned to the Office.

A result of the Office being transferred to the Air Ministry was the establishment of a hierarchy of professional posts in the Office, viz. Director, Assistant Director, Superintendent, Assistant Superintendent, Senior Professional Assistant and Junior Professional Assistant. Shaw remained in post as Director, and two Assistant Director appointments took effect on 1 October 1919. Lempfert became Assistant Director for Administration and Contributive Stations, with overall responsibility for the observatories and other stations which contributed observations to the Office. Gold became Assistant Director in overall charge of forecasting, with two superintendents answerable to him, these being J S Dines as Superintendent of the Forecast Division at Kingsway and A H R Goldie as Superintendent of the Distributive Stations. Commander Louis Alfred Brooke-Smith RNR became Marine Superintendent on 3 November 1919. The other superintendents at the end of March 1920 were F J W Whipple (Statistical Division), C Salter (BRO), R Corless (Instruments), A Watt (Edinburgh Office), A Crichton-Mitchell (Eskdalemuir Observatory), L H G Dines (Valentia Observatory) and D Brunt (Army Services). In addition, two long-serving scientists held the rank of Assistant Director, viz. C Chree (Kew Observatory) and W H Dines (Benson Observatory).

A notice to meteorological officers of the RAF who were about to be demobilized was approved by the Committee in late 1919. This specified that the normal qualification for appointment as a Senior or Junior Professional Assistant in the Office was a degree with First-Class Honours in Mathematics or Physics. However, the course for naval officers at Dartmouth or the course for military officers at Sandhurst or Woolwich would be accepted as an alternative. Those with the required qualifications would be allowed to continue in or rejoin the Office on the following terms:

- Those under 25 years of age would be appointed Junior Professional Assistants with seniority to count from the date of their first commission as a meteorological officer in the RAF, RFC or Royal Naval Air Service.
- Those 25 years of age or over would be appointed Senior Professional Assistants, with seniority to count from the first day of April next following their 25th birthday if their commission dated back before that, or from the date of their first commission as a meteorological officer if already 25 years old at the time of commission.

The notice went on to say that appointments as Assistant Superintendent to take charge of certain stations were open to professional assistants appointed under these regulations.

A large increase in the Office's staff numbers between March 1913 and March 1920 reflected the wide recognition of meteorology's great importance during the war. The Forecast Division expanded more than any other section. Its strength at the end of March 1913 was a superintendent, two assistants, five clerks, a probationer and a boy clerk. The team of forecasters at Kingsway at the end of March 1920 comprised an assistant director, two superintendents, eighteen professional assistants, six other assistants, five clerks, a technical assistant, a clerk computer, two probationers, four boy clerks and an office keeper, and other members of staff were involved in forecasting at the distributive stations. The Forecast Division was renamed 'Forecast Service' in 1919, when it was split into two sub-sections, with one at Kingsway, the other concerned directly with aviation services and the training of personnel for distributive stations.

Not only the number of staff engaged in weather forecasting increased greatly between 1913 and 1920. So, too, did the quantity and quality of observations. However, there was little corresponding improvement in forecast accuracy. The basic reason was that forecasters still used the model put forward by Abercromby in the 1880s (see Chapter 5). The model which Shaw and Lempfert had published in 1906 in their paper on *The Life History of Surface Air Currents* (see Chapter 6) had been ignored. Its practical and theoretical importance had not been appreciated. David Brunt considered this paper "Shaw's greatest written contribution to meteorology" and went on to say that "it could have made meteorological history had it only been followed up".[2]

But it was not followed up. Instead, a group of meteorologists in Norway published, in 1919, a model of extratropical weather systems which would eventually be adopted by forecasters the world over. The Bergen group, led by Vilhelm Bjerknes, the foremost hydrodynamicist of the day, recognized that such systems, like all natural systems, grow to maturity and subsequently decay and also contain sloping transition zones which separate 'air masses' of different temperature and humidity. Introducing nomenclature evocative of the Great War, they called these zones 'fronts'; and they showed that precipitation tends to be concentrated in them. According to the Bergen group, a continuous wavy boundary encircles the hemisphere in middle latitudes, this boundary separating the cold air of high latitudes from the warmer air of lower latitudes. They called this boundary the 'polar front' and explained that 'families of depressions' in different stages of development progress along it.[3]

[2] See Brunt's paper titled 'A hundred years of meteorology' in *The Advancement of Science*, No. 30, September 1951, 11 pp.

[3] Vilhelm Bjerknes began to study extratropical weather systems in the latter years of the nineteenth century, using both synoptic and theoretical methods, and he published a visionary paper in 1904 titled 'Das Problem der Wettervorhersage, betrachtet vom Standpunkte der Mechanik und der Physik', published in *Meteorologische Zeitschrift*,

Bjerknes had long known Shaw, and each had admired the other's work. Indeed, Shaw had written to Bjerknes in September 1913 to tell him about Richardson's ideas on numerical weather prediction (NWP). Given their mutual interest in the structure of extratropical weather systems, and the similarity of the flow model of Shaw and Lempfert to the frontal model of the Bergen group, it was probably not surprising that Shaw received an invitation from Bjerknes in May 1920 to go to Bergen to inspect the new methods of forecasting which had been developed under his direction. The invitation was accepted, and a delegation consisting of Shaw, Goldie, Douglas, Richardson and the Air Ministry's Controller of Communications (L F Blandy) arrived in Bergen on 19 July.[4] Richardson was specially invited by Bjerknes because of their common interest in NWP.

The annual report of the Office for the year ending 31 March 1921 declared the results of the visit "of great importance" and noted that the proceedings of the conference had been set out in a paper of great value to the Office's forecasting staff.[5] The next step, one might think, would have been for the forecasters to adopt the frontal model of the Bergen group, but this did not happen. Douglas explained why in a paper he published in 1952:[6]

Fronts were not published in the *Daily Weather Report* till 1933, and there was good reason for this caution, since before that time there was no generally high standard of frontal analysis. Some years earlier a number of countries started publishing fronts, and there was vast diversity between them. A great deal of this was due to sheer bad analysis, which indicated a marked degree of backwardness in the standard of synoptic meteorology. J. Bjerknes informed me that some of the published monstrosities which were supposed to be fronts made him quite ashamed, as he felt some responsibility owing to his international propaganda in favour of marking fronts on charts.

Douglas went on to say that fronts defied precise definition. They varied greatly in structure and were often difficult to identify with certainty. Forecasters therefore erred on the side of caution, using frontal analysis internally but not marking fronts on the weather charts that were issued to the public.

Weather charts had long been published in the *Daily Weather Report* and in newspapers (except during the war years) but they were also, from April 1920, displayed in a front window on the ground floor of Empire House, Kingsway. Ten feet high and six

---

Vol. 21, pp. 1–7. Long before the Great War broke out, he had become the leading dynamical meteorologist of his day, but his synoptic studies did not bear fruit fully before 1917, when he returned to his native Norway from Germany. There, in Bergen, he formed a team of meteorologists which moved to the town's Geophysical Institute on 1 July 1918 and became known as the Bergen School of Meteorology. The classic paper that appeared in 1919 was by his son, Jacob Bjerknes, then aged 21, 'On the structure of moving cyclones', published in *Geofysiske Publikasjoner* (Vol. 1, No. 2, 8 pp.). The paper was also published in the *Monthly Weather Review* (1919, Vol. 47, pp. 95–99). The term *front* was not used in the paper. It was introduced later on in 1919.

[4]   They travelled free of charge, thanks to the generosity of the Bergen Steamship Company.

[5]   There are copies of this paper in the National Meteorological Archive at Exeter and in the Napier Shaw Collection held by the University of Cambridge.

[6]   See Douglas, C K M, 1952. 'The evolution of 20th-century forecasting in the British Isles', *Quarterly Journal of the Royal Meteorological Society*, Vol. 78, pp. 1–21.

feet wide, they were exhibited daily and showed weather reports from all the principal meteorological stations in the British Isles and a few on the Continent. A note about these charts in the May 1920 issue of the *Meteorological Magazine* reported that they had attracted a great deal of interest. However, the note went on to say, "Some of the eager enthusiasts do not at first realize that the map represents the reported weather at the time specified, and not the meteorologists' estimate of the weather to be". Such an estimate was, the note said, "indicated by the exhibition of the latest forecasts".

Scottish Meteorological Society Difficulties

The scheme of cooperation that had been agreed on in April 1913 between the Office and the Scottish Meteorological Society (see Chapter 7) worked smoothly through the war, but the Society could not continue the arrangement thereafter without a new financial agreement with the Office. A meeting of the advisory committee which supervised the Office's Scottish Branch was held in the Society's rooms on 16 April 1920 to discuss the matter. Shaw chaired it, and the others who attended were Mr A Watt, Dr E M Wedderburn, Professor R A Sampson, Sir Robert Greig (an authority on agriculture), Dr C G Knott (a mathematician and seismologist), Sir Leslie MacKenzie (representing the Scottish Board of Health) and Mr A A Norris (representing the Fishery Board for Scotland). Dr Wedderburn was now a member of the Meteorological Committee, representing the Royal Society of Edinburgh and the Scottish Branch.

After reviewing the agreement of 1913 and discussing the difficulties now faced by the Society, the advisory committee drew up proposals to be laid before the Meteorological Committee. These proposals were approved by the Committee on 28 April 1920 and were as follows:

- The duties of the Edinburgh Branch as endorsed by the Meteorological Committee on 30 April 1913 would be taken over by the Office, officially as from 1 April 1920.
- The work was to form part of the duty of the Office, and the staff who had been employed by the Branch would become staff of the Office. Those employed would be Andrew Watt in the rank of Assistant Superintendent with one year's seniority, Miss Madge Crawford and Miss Alison Murray as clerical assistants, and Captain J Crichton as a Senior Professional Assistant.
- The Meteorological Committee would maintain an office in Scotland for the discharge of the duties hitherto assigned to the Branch and such other duties as may be approved by the Committee.
- The Scottish Meteorological Society would hand over its collection of books, maps and other material to the Meteorological Office in trust for the maintenance of a meteorological library at the Office in Scotland with freedom of access for the Society's members and the general public under suitable conditions.
- There would be an advisory committee, which would comprise the Director of the Meteorological Office (as Chairman) and members nominated by the Board of Agriculture for

Scotland, the Fisheries Board for Scotland, the Scottish Board of Health, the Royal Society of Edinburgh, the Royal Society of London and the Scottish Meteorological Society.
• The advisory committee would have the power to submit to the Meteorological Committee any proposals for the development of meteorological science.

It was agreed that Dr A Crichton Mitchell, the Superintendent of Eskdalemuir Observatory, would serve as Superintendent of the Office in Scotland for two years from 1 April 1920. And it was further agreed at the Meteorological Committee's meeting on 23 June 1920 that there would be two representatives of Scottish universities on the advisory committee.

A primary purpose of the Scottish Meteorological Society had been to place on a permanent and satisfactory footing a meteorological organization for Scotland. As this had now been achieved, the Council of the Society decided that the Society's residual functions would best be served by amalgamation with the Royal Meteorological Society. This was agreed at the Annual Business Meeting of the Scottish Meteorological Society on 17 December 1920 and approved at the Annual General Meeting of the Royal Meteorological Society on 19 January 1921. At the same time, 124 Members of the Scottish Meteorological Society became Fellows of the Royal Meteorological Society.

When incorporation took place, the Royal Meteorological Society created the office of Vice-President for Scotland, a position on the Society's Council which exists to this day. In addition, to commemorate the amalgamation and perpetuate the memory of Alexander Buchan, the two societies agreed that some of the funds of the Scottish Meteorological Society that had been transferred to the Royal Meteorological Society would be devoted to a prize which would be awarded biennially for important original contributions to meteorology published in the *Quarterly Journal of the Royal Meteorological Society*. It was first awarded in 1925 and has been won by a number of members of staff of the Meteorological Office. The first recipient was W H Dines. Unfortunately, though, he was not able to receive the prize in person, owing to ill health. His son J S Dines received it on his behalf.[7]

New Directions for Sir Napier Shaw

Soon after the war ended, Shaw made a move to revive the International Meteorological Organization (IMO), whose activities had been suspended since 1914. He had been the serving President of the International Meteorological Committee (IMC) when war broke out and was therefore the appropriate person to take the initiative. He wrote to the available members of the old Committee on 3 June 1919, inviting them to an informal meeting in London, specifically to consider the Meteorological Codex,

[7] W H Dines died at Benson on 24 December 1927, aged 72.

which represented the international agreements concluded before the war. He pointed out that this Codex was no longer sufficient and would have to be revised to meet new requirements, particularly aviation. The meeting took place at the Office in South Kensington from 3 to 9 July 1919, attended by the directors of the meteorological services of France, Holland, Italy, Denmark, Norway, Sweden and India, as well as the Superintendent of the Admiralty's Meteorological Service and a representative of the U.S. Weather Bureau. The Secretary for the meeting was Ernest Gold. The delegates agreed that the IMO should resume where it had broken off in 1914 and again play a leading role in international meteorology.

The principal outcome of the meeting was the Extraordinary Conference of Directors of Meteorological Institutions which was held at Paris from 30 September to 6 October 1919. In all, twenty-one countries were represented, but the United States and Japan were unfortunately not, as their representatives were unable to attend.[8] Shaw was elected President of the Conference, and the delegates decided to re-establish an international meteorological organization very much on the lines of that which existed before the war. Thus it would comprise Conferences of Directors, the IMC and a number of Commissions. Shaw was re-elected President of the IMC and elected President of the Commission for the Réseau Mondial.[9] Gold was elected President of the Commission for Weather Telegraphy.

Before the Paris meeting, a Conference of Meteorologists of the British Dominions was held in London, convened by the Meteorological Committee and chaired by Shaw. It took place from 23 to 27 September and was attended by representatives of the meteorological services of New Zealand, Australia, Egypt, South Africa, Canada, Ceylon and India, together with the Superintendent of the Meteorological Service of the Admiralty (H D Grant), the Office's Superintendent of Statistics (F J W Whipple), the Superintendent of Army Meteorological Services (D Brunt), Captain Stiles of the Kite-Balloon Testing Station at Kingston, R B Mackenzie of the Air Ministry's Meteorological Service, and Lord Montagu of Beaulieu. It was also attended by C E P Brooks, who was in charge of records from colonial stations at the Office. The subjects discussed included arrangements for the exchange of surface and upper-air meteorological information by wireless telegraphy, instruments and equipment for the investigation of the upper air to meet the needs of aerial navigation, meteorological needs for the main air routes of the world, the selection of stations of the Réseau Mondial for the general study of the climatology of the globe, and the study of tropical hurricanes.

Shaw was 65 years of age when he revived the IMO. His capacity for work was extraordinary for a man of his age. His energy seemed boundless. He showed no

---

[8] The American representatives failed to obtain the sanction of Congress in time for them to attend.

[9] Réseau Mondial, literally 'world network', was the name given to the international system of cooperation whereby meteorological observations from around the world were exchanged and published.

sign of leaving the meteorological stage, but he would have to step down soon as Director of the Office. He had reached the official retirement age. In his time at the helm, he had changed the Office out of all recognition. He had turned it from an organization that was scientifically moribund into one that led the meteorological world in prestige and resources. Beginning with Lempfert in 1902, he had succeeded in attracting well-qualified scientists, and the demands of war had enabled him to recruit a greater number of capable science graduates than would probably have been the case otherwise. He had broadened contacts with academia at home and abroad, and the Office had, under his leadership, embraced new technology (wireless) and new applications (notably, aviation and air pollution). He was highly respected in all parts of the meteorological world. This notwithstanding, his touch at home during and immediately after the war had not always been sure. He was a scientist and believed the Office should essentially be a scientific institution. He was one of the great line of Victorian natural philosophers. He never came to terms fully with the transfer of the Office to the Air Ministry.

The courteous and considerate gentleman who was regarded with respect and affection at home and abroad retired from the Office in 1920. He stepped down as Director on 6 September, but 'retirement' is hardly an appropriate term to describe the next few years of his life.[10] He transferred to the Imperial College of Science and Technology to pursue his long-standing interest in meteorological education. As from 1 September 1920, he became part-time Professor of Meteorology in the Department of Aeronautics, where he joined Sir Richard Glazebrook, who had become Head of the Department and Professor of Aviation the previous year. When Shaw joined the Department, he became the first holder of a chair in meteorology in the UK.

The Department of Aeronautics had quarters in a building on Exhibition Road close to the South Kensington section of the Office, but they were rather cramped. To overcome this difficulty, the Office provided Shaw with accommodation and facilities in the building on Exhibition Road that had become the home of the Office in 1910. He was indeed fortunate, for he had available to him a work room, a lecture room and ready access to the Office's Library. He could also use the roof for instruction in observation and was able to take advantage of the instruments maintained in the building by the Office's Instruments Division.

At the request of the College's authorities, moreover, members of the Office's staff were seconded to the College to act as instructors or lecturers. A letter from Henry McAnally of the Air Ministry to the Secretary of the Treasury dated 9 July 1920 shows that the Ministry supported close cooperation between the College and the Office and offered to assist "in every possible way the advancement of a science so closely connected with the interests of aviation". He went on to propose that the salaries of

---

[10] Shaw's five-year period of service came to an end officially on 31 March 1920, but he agreed to remain in office until the process of appointing his successor was completed.

seconded members of staff be paid by the College. Months passed and nothing was heard from the Treasury. Indeed, Shaw's work at the College had begun by the time a reply did arrive, in a letter dated 11 November 1920, but there was no problem. The Treasury raised no objections to anything in McAnally's letter and agreed that staff who were seconded could count their service with the College as service with the Office for purposes of increments and superannuation.

The Meteorological Committee learned at their meeting on 26 January 1921 that Miss Elaine Austin had been seconded to the College for one year as from 11 November 1920 and that David Brunt would deliver a course of lectures in the Department. Miss Austin would serve as lecturer, demonstrator and Shaw's personal assistant. Brunt would be a visiting lecturer, and so, too, would C T R Wilson, who had worked with Shaw in the Cavendish Laboratory in the 1890s (see Chapter 6). He was now Reader in Electrical Meteorology at Cambridge.

Shaw retired in the autumn of 1924, whereupon Miss Austin returned to the Office.[11] However, his collaboration with her continued and helped bring to fruition the four-volume *Manual of Meteorology* which he had begun in 1918–1919. His link with the Office did not end when he left the Imperial College. He remained Chairman of the Advisory Committee on Atmospheric Pollution until 1925, having chaired it since 1912, when it was formed as the Committee for the Investigation of Atmospheric Pollution. It was reconstituted and renamed in 1917 and, at the same time, placed under the aegis of the Office, where it remained until April 1927, when transferred to the DSIR.

A New Director

The new Director of the Office was Dr George Clarke Simpson, a physicist with a strong interest in atmospheric electricity. Like his predecessor, he was not a stranger to the Office when he became Director. He had carried out experimental work for the Office in 1905 (see Chapter 6), and the Office had supplied him with meteorological instruments and other equipment for Scott's *Terra Nova* expedition to the Antarctic (also noted in Chapter 6). He became Director on 6 September 1920.

Born at Derby on 2 September 1878, he had attended Derby Diocesan School and then, in 1894, entered his father's retail business. Three years later, he had entered the Victoria University at Manchester and gained First-Class Honours in Physics in 1900. He had remained in the university two more years, studying the resistance of bismuth to an alternating current in a strong magnetic field, before proceeding, with a scholarship, to the Geophysikalisches Institut of Göttingen University. After a year there, tackling the problem of the earth's permanent negative charge, he had spent a

---

[11] Shaw's successor as Professor of Meteorology at the Imperial College was Sir Gilbert Walker, who had previously, since 1904, directed the India Meteorological Department.

Figure 9.1. George Clarke Simpson. © Crown Copyright 2010, the Met Office.

year in Lapland, studying atmospheric electricity. He had been appointed lecturer in meteorology at the Victoria University in 1905 and proceeded from there, the following year, to the India Meteorological Department (IMD), where, with the rather grand title of 'Imperial Meteorologist', he was one of three graduate assistants, the others being John Patterson and James Hermann Field.[12] He arrived at Simla in November 1906 and almost immediately began to study the electricity of rain, a research activity which developed into a lifelong interest in the electricity of thunderstorms.

In March 1912, whilst in the Antarctic, he was recalled to India to deputize for the Director of the IMD, Gilbert Walker, whose health had broken down.[13] He arrived at Simla in August 1912 and thereafter remained with the IMD until 1920, apart from three spells of absence. During the first of these, from March to May 1916, he served in Mesopotamia with the British Expeditionary Force, advising on meteorological problems. In the second, from 1917 to 1919, he was seconded to the Indian Munitions Board. In the third, from February to June 1920, he served in Egypt and the Sudan as a member of the Egyptian Government's Nile Projects Commission.

By 1920, his list of publications was impressive. He had published numerous papers on atmospheric electricity and also, in 1919, published the first two (of three) volumes

---

[12] Patterson was Director of the Canadian Meteorological Service from 1929 to 1946. Field succeeded Walker as Director of the India Meteorological Department in 1924 but sadly died only four years later.

[13] Whilst in the Antarctic, where he was considered a genial and friendly character, Simpson was given the nickname 'Sunny Jim', after the trademark character on the 'Force' breakfast cereal packet.

of a major work called *British Antarctic Expedition 1910–1913: Meteorology*. He was now held in high regard the world over and had, in fact, been awarded the degree of Doctor of Science *honoris causa* by the University of Sydney in 1914 and elected a Fellow of the Royal Society in 1915.

For those who hoped and expected that the dedication to science shown by Shaw would continue, this augured well. The Office would surely enhance its scientific reputation under Simpson. In the event, he did indeed publish papers in learned journals throughout the eighteen years he was Director of the Office, on a range of topics which included the monsoon of southern Asia, weather forecasting, past climates, thunderstorms, solar and terrestrial radiation, atmospheric electricity, and the meteorology of polar regions. He also published Meteorological Office *Professional Note* No. 44 in 1926 on 'The Velocity Equivalents of the Beaufort Scale', following up work he had carried out in 1906 (see Chapter 6).

Alas, the Office's emphasis on research during Simpson's time as Director was less than in Shaw's, and it is often asserted that he did not encourage his staff to undertake research. However, the previous two decades had been a time of unprecedented change, partly because of Shaw's scientific zeal and partly because of the war. It was probably inevitable that such a pace of change could not be maintained. But Shaw's zeal and the Great War were not the only determining factors. One of Simpson's priorities was to make the Office an integral part of the Air Ministry. Another was to create an institution that would serve the growing demands of aviation, and it is possible, too, that those who appointed Simpson had been irritated by Shaw's intransigence over the position of the Office as a scientific institution.

Some were surprised, and a number disappointed, that Ernest Gold did not become Director. As R C Sutcliffe and A C Best noted in their obituary of him in the *Biographical Memoirs of Fellows of the Royal Society* (1977, Vol. 23, pp. 115–131), "his record, qualifications and prestige seemed impeccable". Moreover, they pointed out, Simpson was his senior by only three years, and Gold had always enjoyed a warm professional friendship with Shaw. Under Simpson, Gold continued as an Assistant Director but, as Sutcliffe and Best put it, "without the old warm friendship with his Director". And they went on to say that "the position was aggravated when Simpson, an impatient and decisive man, took to himself the direct administrative control of the various branches of the Office, leaving Lempfert and Gold rather out on a limb". Although Simpson's administrative arrangements were not to Gold's liking, they did, as Sutcliffe and Best observed, relieve him of detailed administration and "allowed him to devote a great deal of attention to his international work which, until Simpson retired in 1938, was undoubtedly his major contribution to the profession".

Simpson, too, played a full part in the work of the IMO. He was a member of the IMC and its Executive Council for many years, and he also served as President of the Réseau Mondial from 1921 to 1946 and President of the Commission for Bibliography from 1933 to 1946. When he retired as Director of the Office he was elected an Honorary Member of the IMC, becoming only the third person to receive this honour, the others being Sir Napier Shaw of the UK and Dr Ewoud van Everdingen of the Netherlands. He retained this honorary membership until 1951.

Under New Management

The first meeting of the Meteorological Committee after Simpson became Director took place on 27 October 1920, with a contentious matter on the agenda, the question of Shaw's retirement grant. The essence of the problem was that the sum of £2000 offered to Shaw represented an annuity of £225 per annum, whereas Scott had retired on an annual pension of £400 per annum back in 1900. Shaw was understandably unhappy about this, and the Committee agreed with him. The matter had been unresolved for several months. At their meeting on 23 June 1920, for example, the Committee had agreed that "the Air Council should be moved to approach the Treasury whereby Parliament should be asked to vote a more adequate sum". However, the Treasury had declined to increase the grant, whereupon the Committee had decided, at their meeting on 28 July 1920 (whilst Shaw was away in Norway), that the Treasury letter should not be regarded as final. But the Treasury again declined, and the Committee again refused to let the matter drop, agreeing at their meeting on 27 October that the Controller-General of Civil Aviation, the Committee's President, should approach the Secretary of State for Air. Whether or not he did so, we do not know, but the grant was never again mentioned in the minutes of the Committee's meetings.

The papers for the meeting on 27 October also included a long memorandum from Shaw entitled 'Office Staff'. Dated 2 September 1920, it was one of the last documents he wrote before he stepped down as Director. It focused on deficiencies in the organization of the Office but went further than merely stating them. It also contained suggestions and recommendations. Perhaps he thought he was being helpful. Perhaps he thought Simpson would benefit from his insight and experience. Who knows? The memorandum gives the impression that he was trying to do more than bring a problem to Simpson's notice. He appeared to be trying to tell his successor how to organize the Office!

Shaw stated that the most obvious deficiency in the organization of the Office was "the want of administrative assistance for the Director". The administration had, he said, increased during the war and the administrative staff diminished, as a result of which the administrative work alone was now far more than could be "got through in the time available by the Director and the assistance at his disposal". He therefore suggested that an assistant director "with adequate meteorological experience" should be made responsible for administration and personnel and that this person should have two deputy assistant directors, each with the grade of superintendent. He went on to suggest that the allocation of an assistant director to take charge of administration and personnel full-time implied the appointment of a new assistant director who would be responsible for observatories, anemograph stations, climatological stations, rainfall stations, colonial stations and special expeditions. A solution, he proposed, was to regard W H Dines as Honorary Assistant Director and use the assistant directorship thus released "for the purpose of easing the burden of administration".

Later in the memorandum, Shaw pointed out that a gifted mathematician had been appointed Librarian at South Kensington. He was Dr Harold Jeffreys, a Senior Professional Assistant in the Office's Secretarial Division since 1918.[14] His duty was, Shaw said, to maintain an up-to-date catalogue and carry out "such other ordinary duties attaching to the post of librarian as any librarian undertakes". He was also expected to keep the Director and the scientific staff informed of the subject matter of material added to the library and to advise on material which needed to be acquired. Because so much that was published in meteorology contained what he called "intricate mathematical treatment", Shaw considered that "the opportunity of securing the services of a mathematician of distinction for the position ought to be recognized by assigning an assistant superintendentship to it and in the course of time a superintendentship". The Office now had, he pointed out, specialized libraries at Kew, Eskdalemuir and elsewhere, in addition to the library at South Kensington, which had been expanding since the earliest days of the Office and now contained a wide range of material.

Jeffreys recalled his time in the Office when interviewed in September 1984.[15] Asked how he came to be the Office's Librarian, he said this:

It began with Shaw wanting somebody that he could consult on questions of hydrodynamics. He did not know anything much about hydrodynamics himself. I suppose he was a pal of Newall's [the astronomer H F Newall] and he asked Newall if he could produce somebody that could help him with the questions of hydrodynamics, and Newall suggested me. I went to the Meteorological Office as his personal assistant, as I suppose I was called. I stayed with him till he retired and then he thought the new Director would not want a personal assistant, and he tried to find a job for me and the only thing he could think of was to make me Librarian. That was all right as far as it went but it turned out that the job was mostly answering silly questions about what was the weather at such and such a place at such and such a time. Nobody told me, but it turned out that this was actually the most important part of the job because there was liable to be a court case about somebody's car having skidded and what was the state of the road. But I did get sick of it.

Jeffreys also recalled in the interview that he had worked in the same room as G I Taylor when he joined the Office, their room next door to Shaw's.

---

[14] Jeffreys published a number of papers on dynamical meteorology, fluid dynamics and physical oceanography in the period 1917–1936, almost all of them in leading journals. Otherwise, his large output of books and papers reflected his main interest, solid-earth geophysics. He joined the Meteorological Office from Cambridge University in 1918 (after gaining a Doctor of Science degree from the University of Durham the previous year), and he returned to Cambridge in 1922, where he remained until he retired in 1958. He was knighted in 1953.

[15] Since the early 1980s, the Royal Meteorological Society has interviewed distinguished meteorologists and oceanographers. The recordings and transcripts are held in the National Meteorological Library at Exeter.

Shaw's memorandum is informative today as it provides insight into organizational shortcomings in the Office when Shaw stepped down as Director. However, the response of the Committee to the memorandum may well have been a collective sigh. Shaw was yesterday's man. If there was any discussion of the memorandum, it appears to have been brief. The minutes merely state that the Director "did not feel in a position to act upon Sir Napier Shaw's suggestions", and it was agreed that the matter should be "left over". The memorandum was not mentioned again in the minutes of the Committee's meetings, nor the suggestions contained in it. However, a restructuring of the administration did in fact take place a year later.

As from 1 January 1922, assistant directors were given new responsibilities. Lempfert was given administrative charge of personnel, preparation of annual parliamentary estimates, publications, library, stationery, establishment of new stations, and personnel arrangements at stations. Gold was placed in charge of meteorological observations, instruments, equipment and stores, works and buildings, collection and distribution of information, and quality-control inspections of stations. Below Lempfert and Gold, there were superintendents who dealt with matters appertaining to their division but were required to refer to their respective assistant director any matter that concerned new policy or new expenditure. Assistant directors could, if necessary, pass questions to Simpson for decisions.

The new structure seems to have added to Simpson's administrative workload, rather than reduce it. As Gold put it in his obituary of Simpson, published in 1965 in the *Biographical Memoirs of the Royal Society* (Vol. 11, pp. 157–175): "This [horizontal structure] cast directly on himself some burdens which the vertical division cushioned, and for the first few of his eighteen years as Director he was wholly occupied with the problems of administration; he had no spare time for research". This is reflected in Simpson's publications. Until 1928, when he published a series of research papers in the *Memoirs of the Royal Meteorological Society* on terrestrial radiation, his papers were almost all expository.

Gold went on to say that the process of making the Office an integral part of the Air Ministry took much time, thought and discussion. The Ministry was complex and developing, responsible for military, naval and civil aviation, including airships. The Office was much more broadly based than the Ministry. It catered not only for aviation but also for seafarers, farmers and the general public. It had long possessed a good deal of freedom to pursue interests and investigations of its own choice whilst, at the same time, providing forecasts and other services to users. However, the old leisurely approach to meteorology had been swept away, partly as a result of the war and partly by the transfer of the Office to the Air Ministry. Shaw's focus on scientific research had suddenly come to seem passé. There was now much greater emphasis on meteorological applications and operations than on research. The central importance of meteorology in both wartime and peacetime had been recognized, and the Office had become an integral part of a government department whose specific responsibility was aviation. It could no longer be considered primarily a scientific institution, and Shaw continued to grumble about this for years afterwards.

Research was not discouraged, though. On the contrary, the Committee agreed at their meeting on 27 October 1920 that the estimates should contain a separate item called 'Research'. And throughout Simpson's time as Director, staff were encouraged, and sometimes required, to write up their work, whether the work be operational or investigative. Many were the books and papers published by staff of the Office during Simpson's time. Furthermore, the Monday Discussions continued. Simpson reported at the meeting on 27 October that he had asked Shaw to continue the series of these fortnightly meetings for the discussion of important contributions to meteorological literature, particularly those by colonial and foreign meteorologists. Shaw had agreed and already arranged a series of ten discussions in the period from November 1920 to March 1921.

The restriction of an annual grant which altered little, if at all, from year to year had also been swept away by the transfer to the Air Ministry. The Office now enjoyed a new financial situation. Money was more readily forthcoming than hitherto. Irresistible demands could now be met without the need to pare other budgets. However, the influence of Simpson's upbringing showed. He had learned from his parents the importance of prudence when running a business. That was, of course, commendable, but, in the economic crisis of the early 1930s, he had no choice.

He succeeded in the challenging task of fitting the Office into the Air Ministry. The Secretary of State for Air for most of the 1920s, Sir Samuel Hoare, was certainly impressed by him, saying in his book *Empire of the Air* (Collins, 1957):

He was one of those pleasant and agreeable people who could explain to me in simple words that I could understand what he was attempting to do. Thanks to him, I was able to follow with growing interest his pioneering activities at a cardinal moment in the development of meteorology.

The minutes of the Committee's meeting on 27 October 1920 show that Simpson was not afraid to make changes. He reported at that meeting correspondence with the Radio and Research Board, a body that had been formed earlier in the year by DSIR to direct any research of a fundamental nature that was required and any study which had a civilian as well as a military interest. The Board had suggested that the Office should extend its radio investigation at South Farnborough in accordance with a scheme put forward by Robert Watson Watt, this being that the thunderstorm enquiry which had begun during the war (see Chapter 8) could reach a satisfactory stage only if pursued as part of an investigation into the origin, nature and travel of 'atmospherics' (electrical disturbances in the atmosphere typically caused by lightning).

The Committee expressed great interest in the proposed research and recognized that it could ultimately have important meteorological significance. However, Simpson showed that he was willing to discontinue work which had been carried on for some time should there be good reason to do so. The Committee accepted his advice that the problem of detecting atmospherics required specialized knowledge of the technique of radio research which was outside the sphere of meteorology and could not properly be supervised by the Office. Three possible alternatives to Watt's scheme were discussed by the Committee:

- Continue the radio work carried out at South Farnborough and extend it to include general research in atmospherics.
- Restrict the work at South Farnborough to the development of coils for thunderstorm location and leave the investigation of atmospherics and other radio problems to the new Board.
- Discontinue all radio research by the Office and leave such work, including the location of thunderstorms, to be carried out by the Board, provided that the Office could utilize and adapt the results obtained.

The Committee rejected the first two of these alternatives and adopted the third, agreeing that the Board could use the radio station at South Farnborough and that Watt could be seconded to the Board if that were desired. The work was transferred to DSIR at the beginning of 1921 and Watt seconded to the Board at the same time. This secondment continued until 1925, when he and his assistant, F J Herd, who had also been seconded, were transferred to DSIR on a permanent basis.

Yet another development reported at the Committee's meeting on 27 October 1920 was the establishment of a geophysical observatory at Lerwick, Shetland. The need for such a station in Shetland had been suggested before the war, but there had been no progress until May 1919, when the Office had received a request from the Norwegian Government for cooperation in a special study of meteorological, magnetic and auroral conditions of high latitudes in connexion with an expedition by the explorer Roald Amundsen to the ice field on the Arctic Ocean. The Committee had suggested that cooperation with the Norwegians might take the form of a British geophysical expedition to Jan Mayen, including meteorologists. However, the Royal Society had rejected the idea, saying that the information before the Committee was not sufficient to convince them that an application to the Government for the necessary funds could be justified. The most practical method of meeting the wishes of the Norwegians, the Royal Society said, would be to establish a station in Shetland equipped for making meteorological observations and for investigating the aurora, terrestrial magnetism, earth currents and atmospheric electricity, and they further decided that such a station would "also be of value if permanently established".

Another reason for establishing an observatory in Shetland was that the Office, the body mainly responsible for making magnetic measurements in the British Isles, needed to find another station to replace Kew Observatory, where the electrification of the nearby railway had seriously affected the instruments that were used for measuring atmospheric electricity and terrestrial magnetism (see Chapter 6). The only observatory where hourly values of the magnetic elements were obtained and tabulated was the one at Eskdalemuir. Measurements of magnetic forces were made at Valentia only once a week. The need for additional geophysical observatories was now urgent, especially one well to the north of Eskdalemuir, for it had been recognized that variations of the components of geomagnetic forces increased greatly from south to north across the British Isles. The assumption that observations from one station were representative of forces over the whole of Britain was no longer tenable.

Dr Crichton Mitchell, the Office's Edinburgh Superintendent, visited Shetland to search for a suitable site for the proposed observatory and reported that a wireless station near Lerwick appeared suitable, not least because it had ample, though somewhat basic, residential accommodation. His recommendation was approved, and the station subsequently transferred from the Post Office to the Air Ministry on condition that the wireless apparatus was maintained and kept available for the Post Office to use in the event of an emergency. The new observatory was opened on 7 June 1921, staffed by a Senior Professional Assistant (Jock Crichton), two Technical Assistants (both recruited locally) and a wireless telegraphy operator. The observatory was fully equipped with meteorological instruments from the outset and routine synoptic weather observations were made, with the existing wireless apparatus used for transmitting them.

The observatory soon became an important centre for geophysical research. Besides meteorological observations, the routine included observations of the aurora, terrestrial magnetism, earth currents and atmospheric electricity. In addition, measurements of the luminosity of the night sky were made from 1921 to 1928, along with studies of the aurora by means of spectroscopic photography.[16] Radio direction-finding equipment installed by DSIR was used for recording atmospherics from 1924 to 1929, with Lerwick Observatory one of three base stations, the others being at Aboukir in Egypt and the Radio Research Station at Ditton Park near Slough in Buckinghamshire. Furthermore, G M B Dobson of the University of Oxford, who had worked at Upavon and South Farnborough before and during the war (see Chapters 7 and 8), twice set up instruments at Lerwick Observatory to measure atmospheric ozone. In 1926 and 1927, he employed a Féry spectrograph he himself had built, and from 1940 to 1943 he used a photo-electric spectrophotometer of his own design.

The minutes of the Committee's meeting on 24 November 1920 note that Simpson had visited Lerwick but had not been able to visit Edinburgh and Eskdalemuir, as storms had delayed him in Shetland. Travel plans when visiting Lerwick Observatory have always needed to be flexible. Weather systems often bring strong winds and thick cloud to the Shetland Islands. Delays of several days because of stormy seas or weather-related airport closures can occur even today for those who visit the Observatory.

[16] These investigations were initiated by Lord Rayleigh, who visited the observatory soon after it opened.

The minutes of the meeting on 24 November note, too, that authority for the public-ation of *Professional Note* No. 17 had been given by the Committee. The title of this paper, by E V Newnham, was self-explanatory: 'Report on the Thunderstorm Which Caused Disastrous Floods at Louth on 29th May 1920'. The storm had devastated parts of Louth, a town in Lincolnshire, and shocked the nation. Twenty-three people had died. Newspapers had devoted much space to the disaster, and questions had been asked in the House of Commons. The flooding caused by the storm had occurred unexpectedly and with great suddenness. The inevitable questions had been asked. Had the storm been forecast? Why had the flooding occurred so suddenly? Why had it been so severe? There had been storms of similar intensity in the area on previous occasions without devastating consequences.

A question was asked in the House of Commons on 16 June 1920 by Nicholas Doyle. He asked the President of the Board of Trade if he would arrange for a scientific investigation into the flood event at Louth. The Minister of Health replied, saying that he understood the Office had "already undertaken an inquiry into the meteorological aspects of this occurrence". It had indeed. Newnham had carried out a thorough analysis of the meteorological conditions which had occurred on the day of the storm and previous two days and had also investigated rainfall amounts, finding that the fall had exceeded 120 mm in under three hours slightly to the west of Louth. The reason for the disaster had been that trees and other debris had created dams which had suddenly given way, releasing large volumes of debris-laden water all of a sudden. Newnham did not comment on the accuracy of the weather forecast but did not need to. The people of Louth agreed that thunderstorms had been forecast for the afternoon in question, and the matter of whether a storm of such intensity had been forecast was not an issue. The state of weather forecasting in 1920 was such that predictions of the intensities and exact locations of storms were not possible, and this was widely appreciated. The forecasters were thus exonerated.

Steady Progress

At the meeting of the Committee on 26 January 1921, the Chairman (Sir Frederick Sykes, Controller-General of Civil Aviation) reported that Simpson's probationary period of twelve months was proving inconvenient for the Government of India because his post with the IMD was being held open. Accordingly, Sykes considered it desirable to come to an early decision with regard to a permanent post being offered. The Committee agreed and resolved to recommend to the Air Council that Simpson be confirmed in his appointment. This was duly approved, and Simpson thereafter served as Director for another seventeen years. His time at the helm was generally a period of steady progress for the Office, driven partly by developments in aviation and partly by other factors. Shadows were cast, though, by economic clouds, these being darkest during the Great Depression of the early 1930s. And by 1938, when

Simpson stepped down as Director, the clouds of war had been gathering for some years. Indeed, military preparedness remained a priority for the British government throughout the 1920s and 1930s, with significant implications for the Office. For example, knowledge and understanding of air motions in the atmospheric boundary layer were fundamental in respect of chemical warfare. Studies of dispersion needed to be made, making use of the new science of turbulence, which had been pioneered by G I Taylor and L F Richardson.

The minutes of the Committee's meeting on 26 January 1921 state that the Treasury had sanctioned the seconding of Meteorological Office staff to the War Office Experimental Station at Porton Down in Wiltshire and that N K Johnson, Senior Professional Assistant, had already taken up duty there. The station had opened in 1916, its remit being to conduct research and development in respect of chlorine, phosgene and other poison gases used by the military. Three years later, the War Office had set up a committee chaired by Sir Arthur Holland to consider the future of chemical warfare and defence, and the Cabinet had agreed in 1920 to the committee's recommendation that work on chemical defence should continue at Porton Down, with the assistance of civilian scientific staff supplied by the Meteorological Office. By the late 1920s, the number of the Office's staff seconded to the station had risen to four, including Johnson, who was promoted to the grade of Assistant Superintendent in 1928.

An important development in the early 1920s was the commencement of radio broadcasting to the general public by the British Broadcasting Company Limited (BBC).[17] Daily broadcasts from the London station 2LO began on 14 November 1922 and included that same day the first weather forecast for the public broadcast on radio, read by an announcer from a script prepared by the Meteorological Office. The broadcasting of forecasts for the public on radio became a daily service on 26 March 1923, with the forecasts considered important by farmers and coastal fishermen in particular.

Transmission of gale warnings to ships on the North Atlantic began in 1911, using Morse Code, but ceased in 1914 and did not recommence immediately after the war ended. Not until 1 June 1921 were weather bulletins for shipping again broadcast regularly. Then, the transmission of them twice a day from the wireless telegraphy station at Poldhu in Cornwall began. The bulletins comprised forecasts in plain language for the UK's western approaches, along with observations in code from Blacksod Point in County Mayo, Stornoway, Holyhead, Scilly and Dungeness.[18] Wireless reports from ships recommenced soon after the war and proved increasingly

---

[17] The commercial British Broadcasting Company Limited was dissolved on 31 December 1926 and its assets transferred to the non-commercial and Crown Charted British Broadcasting Corporation.

[18] These were not secret codes. They were groups of letters and numbers agreed by the IMO for simplifying and standardizing the reporting and exchanging of weather and upper-air variables, particularly temperature, pressure, humidity, visibility, present weather, past weather, cloud types and amounts, and wind speeds and directions. Such a coding system is still used by meteorologists the world over and has, indeed, evolved in recent decades to facilitate the exchange of data by computers (by use of binary data formats, for example).

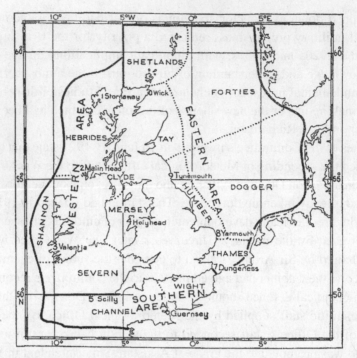

Figure 9.2. Shipping forecast sea areas and coastal stations, 1924. © Crown Copyright 2010, the Met Office.

valuable to weather forecasters. The annual report of the Meteorological Committee for the year ending 31 March 1922 notes that 1472 reports had been received during the year in question from vessels between the 100 fathoms line west of the British Isles and longitude 54°11'W. Of these, 274 had been received within an hour of observation, 307 between one and two hours of observation and 288 between two and four hours of observation.

The broadcasting of weather bulletins for shipping in the form we know today as the 'shipping forecast' began on 1 January 1924. Called 'Weather Shipping', bulletins were broadcast twice daily at 09:00 and 20:00 GMT from the Air Ministry station GFA in London on a wavelength of 4100 metres, using continuous-wave transmission, and could be received 2000 miles away. As many of the smaller ocean-going vessels and some coasters could only receive spark transmissions, however, the broadcasting of bulletins by both spark and continuous wave was introduced by the end of 1924. A further development came in October 1925, when the broadcasting of bulletins from the BBC station 5XX at Daventry began, twice daily on a wavelength of 1600 metres. The bulletins included a summary of the weather situation over the British Isles, actual observations in code at ten British stations (Malin Head, Valentia, Holyhead, Scilly, Guernsey, Dungeness, Yarmouth, Tynemouth, Wick and Stornoway), forecasts of wind and visibility for the twelve hours following the times of the observations for the sea areas around the British Isles, and the 'outlook', which contained a general statement of the weather expected after twelve hours.

The Daventry transmitter was also used from 1 July 1925 for broadcasting forecasts specially for farmers.

Another development for seafarers was the introduction of a new Meteorological Office magazine, *The Marine Observer*, first published in January 1924. The editor was the Marine Superintendent, Brooke-Smith, whose suggestion it was that the magazine was needed, to replace the *Monthly Meteorological Charts* which had been published for many years. The backs of these charts had long been used as the principal means of communication between seafarers afloat and the Marine Division ashore, but this had become unsatisfactory, as the faces of the charts had become so full of information that the usefulness of the charts had become impaired. Moreover, circulation of the charts was limited to the 500 selected ships of the UK's Corps of Voluntary Marine Observers, while the information and instructions were useful to all seafarers.

The principal aims and objectives of *The Marine Observer* were to provide information useful to navigation concerning winds, weather, climate, currents, derelicts and ice and to stimulate interest in observation and the practice of meteorology at sea. Another aim was to provide the Office with a means of acknowledging the voluntary labours of observers and of making some return for their efforts. FitzRoy would surely have approved. He had always been keen to show his gratitude to seafarers.[19]

The Meteorological Committee considered on 28 April 1920 a memorandum by Brooke-Smith in which he reviewed the work of the Marine Division. This raised questions about the efficiency of various aspects of the Division's procedures and contained a number of suggestions. After ceasing during the war, meteorological work at sea had resumed, and the time was now ripe for taking stock of procedures, priorities and so forth. The Committee agreed it was desirable to form a Sub-Committee for Marine Meteorology and appointed as its members Lempfert, Lyons, Brooke-Smith, the Hydrographer of the Navy (Rear-Admiral F C Learmonth) and representatives of the Board of Trade and the Colonial Office.[20]

In his memorandum, Brooke-Smith reported that there were 9176 meteorological logs in the Marine Division, from all parts of the ocean, the earliest dating from 1837. However, he said, few of these logs had "been exhausted into data books", and the extraction of observations had, moreover, been indiscriminate, so that data were incomplete from many Marsden squares. He proposed a new system of extraction. From 1 April 1920, logs would be classified on receipt, and from those deemed 'excellent' or 'very good' values of barometric pressure, temperature, wind, weather

---

[19] *The Marine Observer* was published monthly until December 1932 and thereafter quarterly, in January, April, July and October each year. Publication ceased with the outbreak of the Second World War but resumed in July 1947 and then continued quarterly until July 2003, when it ceased permanently.

[20] A feature of the administration of the Meteorological Office after its transfer to the Air Ministry was a proliferation of committees and sub-committees.

and surface currents would be extracted completely and published yearly in data books for all parts of the ocean. At the same time, the extraction of data from old logs would continue. Brooke-Smith also proposed that steps be taken to induce ship owners to supply tested mercurial barometers as part of the standard equipment of new ships and, further, that the number of sets of tested meteorological instruments supplied to ships by the Office be brought back to pre-war levels as soon as possible. In order to optimize the work of the Marine Division, he asked that the staff be brought up to establishment and increased. The Sub-Committee agreed with all of Brooke-Smith's recommendations, and the Meteorological Committee approved them.

In part, the reason why meteorological observations of the highest quality were required from seafarers was the same as in FitzRoy's time, to improve knowledge of weather, prevailing winds and ocean currents and thereby increase safety, decrease the risk of damage, reduce voyage times and save expense. Also, however, reliable observations from ships at sea were essential for weather forecasting purposes, and not only for improving forecasts for seafarers. The rapid growth of aviation during and after the war had highlighted the vital need for forecasts specifically prepared for aviators. Forecasters possessed an insatiable desire for reliable observations (as they do to this day) and especially required observations from ships on the Atlantic, given that weather systems tend to approach the British Isles from the west.

Brooke-Smith showed in his memorandum that he was keen to improve the training of marine observers. To this end, he had secured the cooperation of the HMS *Conway*, HMS *Worcester* and Pangbourne College training establishments and also prepared a manual of instruction for midshipmen, apprentices and cadets of the mercantile marine. Titled *Weather Forecasting in the Eastern North Atlantic and Home Waters for Seamen*, this was published by the Office in 1921. He had also designed and published for training purposes a 'Cadets' Meteorological Log' and begun to prepare a new edition of the *Marine Observer's Handbook*, which he said was very much needed. Cloud identification plates had been distributed to all observing ships.

To facilitate the extraction of data from logs, the use of a Hollerith Electric Sorting and Tabulating System using 45-column punched cards was trialled. The idea for using it had come from C S Durst, a Senior Professional Assistant. The trial proved a great success, for it improved and increased data extraction and relieved the technical assistants of much monotonous clerical work which, at the same time, helped bring about an increase in the general output of the Division, both with regard to the information for the practical use of navigators and airmen and for scientific use by meteorologists. From 1 May 1921, the Hollerith system was employed fully, greatly improving the accuracy and speed of data reduction. Enquiries about the use of the system were received from a number of meteorological services abroad, and the system was quickly adopted by the Dutch.

The Sub-Committee for Marine Meteorology met on 13 May and 17 June and reported to the Meteorological Committee on 23 June 1920. The above proposals

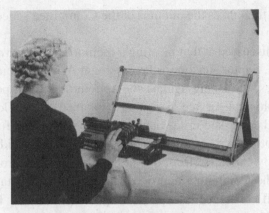

Figure 9.3. Punching Hollerith cards. © Crown Copyright 2010, the Met Office.

and recommendations were all approved, and so, too, was the proposal that Port Meteorological Officers be appointed, their broad function being to act as agents for the Meteorological Office and foster an interest in meteorology among marine observers. The first Port Meteorological Office was opened at Liverpool on 18 April 1921, with Lieutenant-Commander G F H Lloyd in charge, graded Senior Professional Assistant.

Meanwhile, civil aviation was developing rapidly. By 1923, the meteorological station at Croydon Aerodrome had grown into a large establishment where eleven assistants were employed, and the service had become practically continuous day and night. Every hour, meteorological observations were despatched to, and collected from, the aerodromes along each of the aerial lines radiating from that station. The information thus obtained was communicated to pilots by reports exhibited on the aerodromes and by special 'ground signals' which were laid out in such a way that they could be read from the air. In addition, aircraft carried radio telephones, with which pilots made frequent calls for meteorological information during flights. The provision of accurate and up-to-date replies to these calls necessitated considerable organization and close cooperation between the meteorological and radio staff on aerodromes.

Overseas, too, meteorological services for aviation developed. In 1920, for example, the Air Council informed the Meteorological Committee that civilian meteorologists were required for the maintenance of an establishment at Malta and that civilian meteorologists would soon be required farther afield for the air routes which converged at Cairo. The need for meteorologists at Malta was considered urgent by both the Admiralty and the Air Ministry, and a meteorological office was duly established there in 1922, with Dr W A Harwood as Superintendent. He had carried out investigations of the upper air with kites and balloons in the opening decade of the century (see Chapter 7) and subsequently worked in India for the IMD. His appointment caused unrest in the Office, though, as the minutes of the Committee's meeting on 25 January 1922 show.

The Committee discussed at that meeting a memorial signed by forty-eight Assistant Superintendents, Senior and Junior Professional Assistants out of a total of fifty-eight in these grades, protesting against the appointment of a person from outside the British meteorological service. Simpson explained to the Committee that each Superintendent had been asked if he wished to go to Malta and, furthermore, that the possibility of promoting one of the junior grades had been considered. He had taken this action before any candidate from outside the British meteorological service had been considered. The Committee approved the action he had taken and instructed him to meet the staff concerned to explain the position to them. They also approved the appointment of Harwood as Acting Superintendent at Malta. Another base establishment was formed in late 1926, 'Meteorological Office Middle East', with James Durward the Superintendent. This had headquarters at Heliopolis in Egypt and outstations at Aboukir in Egypt, Amman in Jordan, Abu Sueir in Egypt and Ramleh, today in Israel.

As the years went by, aircraft flew higher, faster and more efficiently, and routes to all parts of the world developed. Correspondingly, meteorological services evolved and expanded to cater for both civil aviation and the RAF. An Aviation Services Division was formed in the Office in 1925, with Frank Entwistle as Superintendent. Regular air routes were supplied with data and forecasts, and requests for data concerning meteorological conditions which affected flying were received from various parts of the world. Sir Alan Cobham, for example, consulted the Office on several occasions in August 1926 during his flight from England to Australia. Meteorological support was also provided for air displays at Hendon, Biggin Hill and elsewhere and for the King's Cup, Schneider Trophy and other great air races of the 1920s and 1930s.

Another flight in the 1920s which received special meteorological reports and forecasts from the Office was that undertaken by Sir Samuel Hoare and his wife from London to India and then back to Cairo. Hoare, the Secretary of State for Air, said in his book *Empire of the Air* that the flight was arranged so that he departed in the Christmas parliamentary recess of 1926 and returned home in time for the Air Estimates in March! When he asked the Office about weather prospects, his "friend Dr Simpson" gave him "depressing diagrams and statistics" which showed that a "very high percentage of wind and fog" was to be expected at the end of December, "and a very low percentage of fine weather". Hoare devoted several chapters of his book to the flight to India and back to Cairo and gave another account of the adventure at a special meeting of the Royal Meteorological Society on 28 April 1927.[21]

---

[21] 'Flight to India of the Secretary of State for Air', *Quarterly Journal of the Royal Meteorological Society*, 1927, Vol. 53, pp. 233–240.

In his talk, Hoare said that he and his wife were "very nervous about one thing and one thing alone". They feared that when they arrived at Croydon Aerodrome before sunrise on 26 December they would find a thick London fog and "sit there for hour after hour" and then have to go "ingloriously" back to the house they had just left to the amusement of their friends who had all gone to see them off. In the event, the morning of departure was cold, clear and windy, but Simpson's advice proved correct overall. Indeed, Hoare said, they experienced during most of the flight even worse weather than might have been expected: gales, squalls, head winds, fog, heavy rain and sand storms. He wished to make clear that he attached great importance to the development of meteorological science. The flight, he said, was not inherently dangerous and had been made punctually in the face of the worst possible weather. The assistance of meteorologists was crucial for the opening up and development of commercial air routes from the UK to all parts of the world.

The Rise and Fall of Airships

Hoare said in his book that he wondered when he took office in 1922 whether it was worth considering airships at all. Service opinion in the RAF was, he said, on the whole opposed to them. As he put it, "airships had proved very vulnerable when used in the war for offensive purposes, and there seemed little use for them in the strategy and tactics of attack that formed the basic principle of Air Force thought". The Air Estimates, he said, had been "cut to the bone" and there was a very real danger of the funds available for the RAF being further reduced "by any new expenditure on ambitious and visionary plans that did not strengthen our striking power". However, he said, "there was a devoted company of enthusiasts" who firmly believed in airships. Their view was that such craft could be invaluable for naval reconnaissance and long-distance air transport.

One of these enthusiasts was an engineer and former naval officer, Dennistoun Burney, who became Member of Parliament for Uxbridge in 1922. He had the previous year proposed to Vickers Limited that commercial airships be constructed for establishing airline services between Britain and distant parts of the globe. Hoare commented in his book that "the possibilities appeared at the time very attractive". No aeroplane had yet succeeded in crossing the Atlantic from east to west, whereas an airship had (the R.34), and the existing heavier-than-air machines were low-powered, noisy and uncomfortable. Moreover, flying boats had almost ceased to exist and there was then no plan for an Empire air service of any kind.

Despite the optimism from some quarters, the future for airships looked bleak in the early 1920s. The Treasury's view was that the financial state of the country was so parlous that airships had to be scrapped if they could not be disposed of otherwise. The view of the Commercial Airships Committee, which had been set up by the Air Council in May 1920 to consider the disposal of the rigid airships in the

Air Ministry's possession and their possible utilization for commercial transport, was that the question must be laid before the Cabinet for decision. They pointed out that a decision to scrap airships would be unpopular with the public, given that some £40,000,000 had been spent on the formation of an airship service. The Air Council felt unable to take the responsibility of ordering the scrapping of airships, which they believed would ring the knell of British civil airship enterprise.

The Cabinet agreed at their meeting on 29 July 1920 that "the Air Ministry should make further endeavours to enter into arrangements with commercial firms for the disposal to them of the airships by sale or gift, on the basis that the Government would undertake no financial liability, and that, failing such arrangements, the airships,should be kept in store or scrapped". The Air Ministry issued an order on 20 September 1920 that all work on airships was to cease, whereupon there followed a progressive contraction of airship operations. This was not, though, the end of the airship story. An Air Conference was held in London on 12, 13 and 14 October 1920, attended by representatives of twenty countries under the auspices of the Air Ministry, and the whole of the final day was devoted to airships. The view of the conference was that the aeroplane and the airship were not rivals but allies. The unanimous opinion was that airships had a future for long-distance travel.

In contrast to the Government's lack of enthusiasm, commercial interest in airship development grew, diminished only temporarily by the disaster of 24 August 1921, when airship R.38 broke in two over the River Humber near Hull and crashed in flames, killing forty-four of the forty-nine on board, including technical experts, key RAF personnel and American naval officers and ratings. The loss of many experts was keenly felt three years later, after a change of government brought a more positive attitude to airship development. The new administration of Ramsay MacDonald decided in May 1924 to launch the Imperial Airship Scheme, under which an airship route from Britain to India would be established, and the Air Ministry was authorized to proceed with airship research and development. MacDonald's government agreed that two craft would be constructed, one under the direction of the Air Ministry at Cardington, its primary function naval reconnaissance, the other at Howden in Yorkshire by the Airship Guarantee Company, a new venture formed for the purpose by Vickers, Armstrong and Burney. The loss of the experts in the R.38 disaster was a serious blow because, as Hoare said in *Empire of the Air*, "there were really not enough skilled men to divide between two widely separated efforts of construction". The airships were bigger than any previously constructed (both more than 230 metres long) and presented technical difficulties which had not been resolved completely by the time they were launched.

As Hoare stressed in his book, an accurate knowledge of the weather was exceptionally important for aircraft that were lighter than air. An airship's cruising speed might be doubled, he said, if use could be made of favourable winds, and navigators might be able to change course to avoid weather disturbances such as tropical cyclones. We may add that airships flew at a height of approximately 1500 feet, which is within the so-called boundary layer of the atmosphere, where irregular and unexpected air motions commonly occur. Moreover, airships could be difficult to control in windy weather, especially when leaving and regaining their mooring masts, and there was a danger of them being torn from their moorings during vigorous squalls.

Encouraged by Hoare, the Office formed an Airship Division at Cardington (on 1 January 1925) to work closely with the Air Ministry's Directorate of Airship Development to investigate meteorological conditions on the proposed airship routes between England, Egypt and India. Its Superintendent was Maurice Alfred Giblett, a mathematician who had served as a meteorologist with the British Expeditionary Force in northern Russia in 1919 and later that year joined the Forecast Division of the Office as a Professional Assistant.

In the words of Lempfert, in his review of the work of the Airship Division up to 1930, Giblett "threw himself into his task with characteristic energy".[22] A comprehensive programme of work was soon under way. A retrospective series of daily synoptic weather charts for the twelve months which began on 1 April 1924 was compiled to study day-to-day variations of weather over routes from Britain to North America, the Middle East, South Africa, India, Australia and New Zealand. Measurements of upper winds and upper-air temperatures at points along the proposed airship routes were collected and analysed, and the occurrence of thunderstorms along these routes was investigated. Weather records at meteorological stations on airship routes were analysed, and case studies of regional and local weather phenomena were made. Diurnal variations of wind and thermal stability conditions in the lower atmosphere at Cardington and other places on airship routes were also studied. Much of the work was written up. Many 'Airship Meteorological Reports' were produced, and climatological charts for airship routes were drawn.

Giblett's work took him abroad a number of times. He was away from November 1925 to January 1926, for example, on a tour of duty in Malta, Egypt, Italy and France to investigate sources of meteorological information which might be drawn upon and which, when airship operations began, could be used for meteorological reporting and forecasting. Besides carrying out his duties as technical adviser, he took advantage of every opportunity which presented itself for obtaining meteorological data and discussed with local authorities the meteorological ground arrangements that would be required for the operation of airship services.

From October 1925 to March 1926, Jacob Bjerknes was attached to the Office to promote the use of the Bergen School's theory of cyclones. Giblett had long been impressed by the work of the Norwegians on the structure of depressions and counted himself a close friend of Jacob. Indeed, the two men collaborated in 1924 whilst

---

[22] 'The scientific work of the Meteorological Office, Cardington', *Quarterly Journal of the Royal Meteorological Society*, 1931, Vol. 57, pp. 119–131.

both were visiting the U.S. Weather Bureau, the fruit of this collaboration being a joint paper titled 'An Analysis of a Retrograde Depression in the Eastern United States of America', published in the *Monthly Weather Review* in November 1924 (Vol. 52, pp. 521–527). On the return voyage from the United States, they collaborated again, this time demonstrating the practicability of weather forecasting at sea. They arranged for the ship's radio operator to obtain weather reports from other ships, along with meteorological data from Europe and North America, using the data to draw synoptic weather charts on board. Whilst Giblett was at Cardington, he and his staff kept detailed records of all fronts which passed over and thus became experts in the behaviour of fronts, which Lempfert called 'frontology'. Much attention was therefore paid to the dangers to airships posed by weather disturbances associated with fronts.

Giblett was involved in a further innovative use of radio on 20 March 1929, when he and Cardington colleagues collaborated with Robert Watson Watt during a lecture the latter was delivering in London to the Royal Meteorological Society. In the course of his lecture, he demonstrated the first transmission of a weather chart by wireless telegraphy, this being the chart for 18:00 GMT that day. The chart was drawn at Cardington and reproduced in the lecture room 90 km away. A development of this work took place during the night of 23–24 April 1930, when the General Post Office connected Cardington to the Rugby Wireless Telegraphy Station by land line. A weather chart of the North Atlantic on the scale of 1:2,000,000, covering the area from Greenland to the Azores and from Denmark to the Great Lakes, was transmitted by land line from Cardington to Rugby and then broadcast by Rugby's high-power transmitter, from where it was re-received at Cardington.

By 1929, the two great airships were nearing completion, R.100 at Howden, R.101 at Cardington. The former made her maiden flight on 16 December 1929 and followed this with a successful flight from Cardington to Canada and back between 29 July and 16 August 1930, with Giblett on board. The story of R.101 was far from happy. There were technical problems, including some which endangered lift and stability, and serious doubts about her airworthiness remained when she set off from Cardington for Karachi at 6.24 p.m. on 4 October 1930 with Lord Thomson (the Secretary of State for Air), Sir Sefton Brancker (Director of Civil Aviation) and Maurice Giblett aboard. The weather forecast was poor, and strong winds, low cloud and heavy rain were indeed soon encountered. Disaster struck just after 2.00 a.m. near Beauvais in northern France. The airship hit the ground and burst into flames. Only six of the fifty-four on board survived. Thomson, Brancker and Giblett perished. The life of a very promising meteorologist had been cut short. He was only 36 years of age.

The R.101 disaster brought to an end British airship development. Despite its promise, R.100 was first grounded, then abandoned and later, in November 1931, sold for scrap. However, the Office's Airship Division was not disbanded immediately. Its first task was to prepare data for the official inquiry into the disaster, at which

Simpson gave evidence. Thereafter, the Division's staff completed the investigational and scientific work they had been conducting before being required to devote their whole attention to the operation of the two great airships. Then, in late 1931, the Airship Division was disbanded, but balloon activities did not cease at Cardington. Kite balloons were flown, and No. 1 Balloon Training Unit was formed at Cardington on 1 January 1937, following the decision of the Air Staff to proceed with plans to deploy 450 barrage balloons in defence of London.[23]

Meteorology from the Air

After the Great War, meteorological observations were made increasingly from aircraft. They were invaluable because they supplemented the upper-air measurements obtained by means of meteorographs attached to balloons. A disadvantage of the balloon method was that the observations were useful to forecasters only if the instruments attached to the balloons were recovered quickly enough. Forecasters needed observations made as recently as possible, and so, too, did aviators. No longer was the making of upper-air measurements the leisurely pursuit it had been before the war, when the data obtained from aloft were used solely for research purposes. There was now a need for current observations.

In the early 1920s, the RAF and the Royal Aircraft Establishment flew meteorological sorties whenever possible, but these flights did not meet the need for a regular series of observations. To this end, a special RAF Meteorological Flight of two aeroplanes equipped with the necessary instrumentation was established at Eastchurch in Kent on 1 November 1924 and transferred to Duxford in Cambridgeshire in January 1925.[24] Ascents were made daily (weather permitting) and commonly reached heights of 15,000 feet. These THUM (Temperature and Humidity) ascents proved invaluable for weather forecasting and were also useful at Shoeburyness for calculating trajectories of artillery shells, work which had hitherto relied on kite-balloon observations.

The Office has continued to use aircraft for observational purposes to the present day, particularly for research purposes and weather reconnaissance. The use of aircraft is expensive, though, so routine methods of obtaining upper-air data that are cheaper and more flexible than those which use aircraft are desirable. In this respect, the invention and development of the radiosonde in the late 1920s and early 1930s came as an important advance for practical weather forecasting. The apparatus consisted of a balloon that lifted specially designed instruments which transmitted values of temperature, pressure and humidity to ground stations by means of radio-telegraphy.

---

[23] There is still today a Meteorological Office research group at Cardington, studying boundary-layer meteorology and surface processes.

[24] The Meteorological Flight was further transferred to Mildenhall in Suffolk in 1936, when another Flight was formed at Aldergrove.

Balloons were tracked by means of theodolites and the directions and speeds of upper winds thus obtained.[25]

An advantage of the radiosonde was that it could gather data from greater heights than aircraft could reach. However, the Office was slow to make use of these probes. Indeed, in December 1935, when making a case to the Air Ministry for a second Meteorological Flight in either Scotland or Northern Ireland, Simpson declared there was no means other than aeroplane flight for obtaining full information about the upper air in time for it to be useful for forecasting. An additional Meteorological Flight was, in fact, formed at Aldergrove in Northern Ireland in 1936, but the Office, in collaboration with the NPL, commenced development work on radiosondes the same year. Despite considerable progress, though, and the announcement in 1939 that radiosondes would soon be launched routinely at Lerwick Observatory, the Office still did not have an operational radiosonde when the Second World War broke out.

The Automobile Association (AA) was formed in 1905 to serve the needs of motorists. After the Great War, it pioneered the use of aircraft for road traffic control purposes and in 1928 established an Aviation Department in response to the increasing popularity of private and club flying. One of the department's purposes was to inspect, assess, describe and list suitable landing grounds, many of which were by the sea or in places of beauty popular with tourists. Another was to provide information about current flying conditions at these landing grounds and elsewhere. Accurate and frequent information about the weather was considered essential, both to the owner pilot and the crews of the traffic control aircraft. To this end, the AA's 'Airmet' service was instituted on 18 July 1931 in association with the Office to provide weather reports and forecasts. It became operational two months later.

Information about the *raison d'être* of this service was provided in the Office's annual report for the year ending 31 March 1932, in which it was pointed out that the supply of meteorological information to aerodromes on RAF and civil aviation routes from Great Britain to the continent had increased considerably over the years, but the needs of owner pilots had not been met satisfactorily. The chief difficulty was that radio reports of weather forecasts and current weather were issued in Morse code, and many private owners who had wireless fitted to their aeroplanes were unable to receive and interpret Morse messages. To overcome this difficulty, the AA had approached the Air Ministry with a scheme for cooperation to supply information for the owner pilot by radio-telephony. As a result of cooperation between the AA, the General Post Office and the Meteorological Office, a scheme had been worked out and the first

---

[25] It is not clear who actually invented the radiosonde. A Russian, Pavel Molchanov, first reported his design of a radio-meteorograph in 1927 at a meeting of the IMO's Commission for the Exploration of the Upper Air held at Leipzig. He described the device in a paper published in the proceedings of that meeting and also in a paper published in 1928 in the journal *Beiträge zur Physik der freie Atmosphäre* (Vol. 14, pp. 39–48). However, he did not fly his device until 30 January 1930. Some claim that an American, William Blair, first flew a primitive form of radiosonde in 1924, but his work was not published until 1931. Others claim the Frenchman Robert Bureau first flew a radiosonde on 7 January 1929.

Figure 9.4. Making an Airmet broadcast in the 1930s. © Crown Copyright 2010, the Met Office.

steps to apply it taken during the year. The Post Office had issued a licence to the AA's radio station at Heston Airport, allowing the station's staff to issue by wireless telephony such weather messages as were provided by the Meteorological Office.

The staff in charge of the station intercepted the most relevant of the coded forecasts and from them produced messages in plain language which provided the information that was specifically required by owner pilots. These messages were then issued by radio-telephony at thirty minutes past each hour from 9.30 a.m. to 6.30 p.m., except 2.30 p.m. Similarly, the meteorological reports which were broadcast in code from Croydon Airport were intercepted and re-issued in plain language. The messages broadcast from Heston could be received more than a hundred miles away by the crews of aircraft fitted with the necessary radio facilities and could sometimes be received by means of light portable radio equipment. Moreover, messages could be received by all and sundry, not just aviators. Many of the general public who had been accustomed to telephoning the Air Ministry for weather information ceased to do so and instead relied on Airmet.[26]

The radio station at Heston was intended to be the first of a chain of similar AA stations across Britain that would allow weather reports to be received at any aerodrome in the British Isles and also by aircraft in flight in any part of the country. Such a chain did not, in fact, materialize, but Airmet was extended to most of the British Isles a few years later. The Office took over the service from the AA on 10 July 1935 and thenceforth broadcast weather messages for pilots from a special studio built in the Office's Forecast and Aviation Division in London, using a transmitter at the Borough Hill Radio Station in Northamptonshire.

The wavelength of the transmission remained the same as at Heston (1186 metres), but the power of the transmitter was considerably greater than at Heston. Thus the transmissions could be heard on a good selective radio receiver over most of the

[26] For detailed information about Airmet, see 'The AA wireless station at Heston', published in the magazine *Flight* on 15 July 1932, pp. 673–674.

British Isles. Reports from a number of meteorological stations on the continent were added on 2 September 1935, at the request of pilots who wished to fly across the English Channel, and the whole service was reorganized on 15 February 1936 to provide full information for both RAF and civil aviation. Weather reports, weather forecasts, air navigational warnings and weather warnings were then broadcast daily, including Saturdays and Sundays, according to a regular schedule which commenced soon after 7.00 a.m. and continued until 8.00 p.m. when British Summer Time was not in force, 9.00 p.m. when it was.

Meteorological Education and Training

The Office's annual report for the year ending 31 March 1936 stated that the year in question had been memorable because it had seen the introduction of revised scales of pay on the lines recommended by the 'Committee on the Staffs of Government Scientific Establishments' chaired by Harold Carpenter. New scales had been introduced as from 1 April 1935, along with the various grades of Technical Officer and Assistant which Carpenter had proposed. The new employment arrangements stipulated that recruitment for the Technical Officer class would normally be from university graduates who held Honours degrees in Mathematics or Physics, while the level of education required for the Assistant class would be that of the Intermediate Bachelor of Science degree. The Carpenter Committee had, in fact, reported as long ago as 1930, but the Office had not been able to implement its recommendations any earlier because of financial difficulties arising from the Great Depression of the early 1930s.

One of the main recommendations of the Carpenter Committee was that rates of pay should not be improved in any establishment until that establishment had scrutinized its working arrangements. The purpose of the scrutiny was to bring to light any work currently performed by Technical Officers which could be carried out satisfactorily by staff of lower grades. In this respect, the Office's forecasting work was considered carefully and a modification of Carpenter's recommendations found necessary. Under the system which had been in operation for fifteen years, forecasting was entirely in the hands of professional staff who had been recruited from universities, the underlying reason being that weather forecasting was a branch of meteorological science which demanded the high degree of scientific training provided by an honours degree.

At the same time, though, the Meteorological Committee recognized that experience played an important part in the more routine aspects of forecasting. They agreed that Assistants who had acquired such experience in the course of long service to the Office should continue to be utilized to the full. Accordingly, Assistants who had gained this experience would be employed on forecasting duties provided steps were taken to give them adequate training in the scientific principles on which the practice of forecasting was based. Assistants who had been trained in this way could be assigned to forecasting posts under the supervision of higher grades, while Assistants who had previously been trained would be allowed to carry out forecasting without such supervision.

Further consideration of Carpenter's recommendations led the Meteorological Committee to the conclusion that all new entrants to the Office needed to be given appropriate training, and the Committee also concluded that existing members of staff needed to be trained when undertaking new responsibilities on transfer from one division of the Office to another. To this end, in the autumn of 1935, a Technical Officer who had been relieved of his ordinary duties to become an instructor gave a course on the theory of forecasting to six Assistants who had all in the course of long service gained a great deal of practical experience in the Forecast Division. The course, given in the Office at South Kensington, lasted four months, after which the Assistants were given posts which required them to prepare forecasts as part of their regular duty. But whilst staff were being trained, they were not available to carry out their normal work. As a result, shortages of staff occurred in some divisions of the Office. The reorganization of the Office which followed the Carpenter Committee, though necessary, did therefore create some temporary inconvenience.

The arrangement with the Imperial College of Science and Technology whereby the Office provided accommodation, facilities and lecturers did not continue beyond 1924, when Sir Napier Shaw retired as professor. By then, the Office could no longer spare the space or staff, and after Shaw's retirement very few members of the Office's staff pursued courses at the College anyway. On 1 October 1934, the Superintendent of the Office's Army Services Division, David Brunt, became Professor of Meteorology. He believed the department should be a training ground for the graduates in physics and mathematics who were recruited by the Office and, as part of his endeavours to bring this about, pressed for the appointment of additional staff. He was soon to be joined by another member of the Office's staff, Percival Albert Sheppard, who was appointed Reader in May 1939.[27]

Plans for the development of the department had to be suspended when the Second World War broke out. Brunt and Sheppard were released immediately by Imperial College and seconded to the Air Ministry to open a Meteorological Office Training School, at Berkeley Square House, London. Their responsibility was to train all forecasters for the Office, service and civilian. For the war that had seemed inevitable for some time, the Office had made provision for the recruitment or secondment of a large number of scientists to it. A Meteorological Section of the RAF Volunteer Reserve had been formed in May 1939, and the recruitment of officers and airmen had begun. Airmen were posted in small groups to RAF stations, where the staff of local meteorological offices provided their training. Officers were trained by Brunt

---

[27] Sheppard joined the Office in 1929 and was employed first at Kew Observatory as a Junior Professional Assistant. He later worked at Porton Down for some years, one of his colleagues there being Oliver Graham Sutton, a future Director of the Meteorological Office.

and Sheppard. Their work began on 15 September 1939, which is an important date historically, for it marks the official opening of the Office's Training School.

Before the autumn of 1935, there was no systematic centralized training of staff in the Office. Any training that was given was ad hoc in nature. Recruits were posted to the places where they were required to serve and considered to be under training until declared competent in their respective grades by their officers in charge. They read recommended textbooks but otherwise learned by watching their colleagues at work. Then, in February 1936, at Croydon Airport, the Office set up a Training and Special Investigations Section of the Overseas Division, with S P Peters in charge, assisted by two trained forecasters, D F Bowering and E S Tunstall.[28] The *primum mobile* of this development was that the rate of recruitment to the Office had increased suddenly in the latter part of 1935. A number of Technical Officers had been appointed in a short space of time. They needed to be trained.

The principal reason for the increase in strength was that the Office had been warned in 1935 that forecasts would be required for an experimental transatlantic flying-boat service which was scheduled to start in the spring of 1937. It would be experimental because the idea of regular commercial flights across the North Atlantic Ocean was at best bold – some would say foolhardy. Pioneering flights across the Atlantic had excited the public, but there had also been tragic failures. Many aircraft and crew had disappeared when attempting to cross the ocean, and the weather had been blamed for these failures. Little was known about flying conditions over the Atlantic, which meant that the old pattern of staff training would no longer suffice. There was no experience on which to draw. Research and investigation needed to be carried out. However, existing staff of the Office were fully committed. None could be spared. Forecasters could not be redeployed to undertake the work.

To address the matter, the Meteorological Committee took the advice of Simpson.[29] He was given permission to recruit mathematics and physics graduates direct from the universities to be trained as forecasters and carry out investigations of Atlantic weather. Under Peters, they were introduced to theoretical meteorology through a course of lectures supplemented by private study and, in parallel, taught to make weather observations, carry out pilot-balloon ascents and plot weather charts.[30] One of the team, David Arthur Davies, then studied the weather over the Atlantic at first hand by means of visual, instrumental and pilot-balloon observations.[31] In all, he

---

[28] Peters, a physicist, joined the Office in 1923 after working with C J P Cave privately. His first posting was to RAF Cranwell in Lincolnshire, as a Junior Professional Assistant, and he moved to Cardington in 1925 on promotion to Senior Professional Assistant. After the R.101 disaster, he served at RAF Worthy Down in Hampshire for a number of years and became an instructor in the summer of 1935.

[29] He was now Sir George Simpson, having been knighted in June 1935.

[30] Recollections of the first such training course have been provided by Patrick Meade in 'Transatlantic civil aviation – the first training course for scientists in the Meteorological Office', published in 1986 in the *Meteorological Magazine* (Vol. 115, pp. 193–199).

[31] Davies was from 1955 to 1979 Secretary-General of the World (formerly International) Meteorological Organization. He was knighted in June 1980, thus becoming Sir Arthur Davies.

made sixteen crossings of the Atlantic on a cargo ship, the *Manchester Port*, the first in November 1936 and the last in October 1937.

Others of the team contributed to an intensive examination of the weather that might be encountered over the Atlantic Ocean.[32] They prepared a daily sequence of Atlantic weather charts for a complete year, using a post facto dataset, and from these, along with a set of working charts for a period of ten years obtained from the Office's Forecast and Aviation Services Division, compiled climatological charts from which optimum routes were established and flight times estimated. Peters also trained a member of the Iraqi Meteorological Service, two graduates engaged by the Colonial Office for service in Singapore and the Sudan, and a number of graduate recruits to the Meteorological Office, some of whom were to serve at new RAF stations, others to undertake gunnery and sound-ranging duties with the Royal Artillery.

In the latter part of 1936, C J Boyden was posted to Croydon to become responsible for training staff at all levels. He was an experienced forecaster who had worked for a year in the Office's London headquarters on Kingsway whilst Jacob Bjerknes was visiting the Office. Bjerknes arrived on 14 December 1935 and stayed for five months, paying particular attention to frontal analysis of northern hemisphere charts, focusing especially on the development and progress of fronts over the Atlantic Ocean. His work was therefore complementary to that of the group at Croydon. At that time, as Boyden said in an article entitled 'Meteorological Office training scheme: the first ten years', published in 1986 in the *Meteorological Magazine* (Vol. 115, pp. 190–192), British forecasters still varied greatly in their enthusiasm for fronts. Even in 1935, the superintendent of the Forecast and Aviation Services Division, Richard Corless, had sounded unenthusiastic about them when speaking to a Conference of Empire Meteorologists. However, fronts had begun to appear on charts in the *Daily Weather Report* in 1933, and British forecasters had started applying frontal analysis routinely by the time the Croydon group started work.

In his article, Boyden recalled how he came to be posted to Croydon, saying that he found a note waiting for him one evening in 1936 when he reported for night duty. It was from Lempfert, and he asked to see Boyden in the morning. Boyden said that he was astonished to be told he was to open a training school. He went on to say that the next few days were hectic, and he realized that the best he could hope for was to prepare lectures for a week or two and trust he could keep ahead of the class for the six months the course was to last. Working from a two-page syllabus by Peters, he said, he set about his task, basing his initial notes largely on the four volumes of Shaw's *Manual of Meteorology* and another book now considered a 'classic', Brunt's *Physical and Dynamical Meteorology*, published in 1934. These

---

[32] An account of the work carried out by Davies and other members of the team was published in 1938 by Frank Entwistle, in a paper titled 'Atlantic flight and its bearing on meteorology', in the *Quarterly Journal of the Royal Meteorological Society* (Vol. 64, pp. 355–389). When the work was carried out, Entwistle was Superintendent of the Overseas Division.

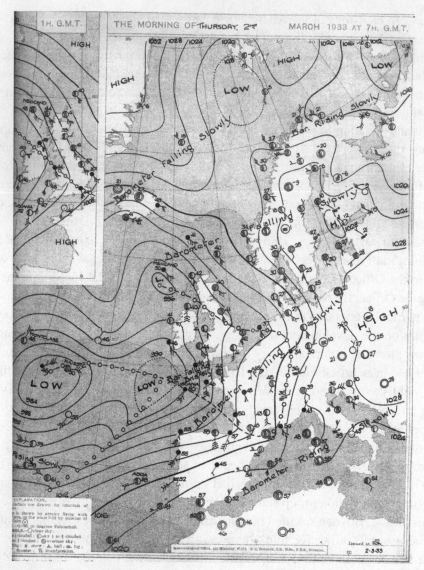

Figure 9.5. Chart in the *Daily Weather Report* showing fronts, 2 March 1933, 07:00 GMT. Fronts were first shown on charts the previous day. Warm fronts are shown as dashed lines, cold fronts as lines of open circles, occluded fronts as alternate dashes and open circles. © Crown Copyright 2010, the Met Office.

were supplemented later, Boyden said, by *Some Problems of Modern Meteorology*, a compilation of important papers published in the *Quarterly Journal of the Royal Meteorological Society*, and a short book on weather forecasting (*Prévision du Temps par l'Analyse des Cartes Météorologiques*) by Jacques Van Mieghem, published in 1936.[33]

---

[33] Professor Van Mieghem, a Belgian meteorologist, founded the World Meteorological Organization's Education and Training Programme in 1965.

In due course, Boyden faced eight students, all veterans, all strangers to him, including one who had joined the Office before he had been born (1908)! At the beginning of the course, he referred to a matter which had caused annoyance in the Office for years. He explained it thus in his article:

In relation to this initial class, it is important to realize that Meteorological Office staff were divided between those who were recruited from the universities and the ordinary assistants, most of whom joined straight from school. Officially, the graduates were the forecasters and the rest were not. Regardless of what happened at outstations, the distinction rankled in the minds of many non-graduates, who, with their years of experience, knew how important they were in the functioning of outstations.

Boyden appealed to the students to maintain an open and friendly relationship with each other, saying that the success of the course depended on it. His words were heeded. The response of the students was, Boyden said, "magnificent". As he pointed out, they were aware their careers depended on their success in the course.

Croydon Airport did not remain the home of the training scheme for long. Aircraft engines were tested for an hour or more almost every day in a nearby hangar, so that lecturing and study were impossible. Boyden and his colleagues and students moved in February 1937 to rooms above a Lyons tea-shop close to South Kensington station. Meanwhile, some of the newly recruited graduates who had formed the Training and Special Investigations Section of the Overseas Division had completed their studies and proceeded to outstations. By and by, in the summer of 1937, Peters and three of the graduates transferred to Foynes in western Ireland and another, Patrick Meade, to the flying-boat base at Hythe near Southampton, these being the places from which Imperial Airways and Pan American Airways intended to operate transatlantic flights. Boyden and his assistants remained in London, continuing to train staff of the Office who were graduates and those who were not.

By the summer of 1937, the Irish Free State had its own meteorological service, established on 1 April that year and directed by Austen Nagle, who had previously been a Technical Officer in the Naval Division of the Meteorological Office. According to the minutes of the Meteorological Committee's meeting on 7 July 1937, however, Valentia Observatory and the transatlantic base at Foynes continued to be run by the Office, on an agency basis, with the salaries of staff recovered from the Irish Free State. Technically, therefore, Peters and his team were seconded to the Irish Meteorological Service.

Further Notable Developments

The story of the years Simpson served as Director included a number of other notable developments, one of them the second International Polar Year (IPY), a research programme which involved scientists of forty-four nations. It took place exactly fifty years after the first IPY with a rationale not fundamentally different from that for the first. In the 1932–1933 IPY, the emphasis was again, as in 1882–1883 (see Chapter 5), on meteorological, geomagnetic and auroral observations in high latitudes. According to V Laursen, in his article about the second IPY published in the July 1982 issue of the *WMO Bulletin* (Vol. 31, pp. 214–222), the IPY was proposed by Dr J Georgi, a German meteorologist, at a meeting in Hamburg in November 1927. The idea was referred first to the President of the IMC, then to Simpson, who was not only Director of the Meteorological Office but also (from 1921–1946) President of the IMO's Commission for Réseau Mondial and Polar Meteorology. An ad hoc sub-commission was formed, with Simpson one of its members, and they prepared a formal proposal which was adopted by the Organization's Conference of Directors at a meeting held at Copenhagen in September 1929. Thereafter, several bodies cooperated with the IMO to plan the IPY, most notably the International Union of Geodesy and Geophysics.

The IPY went ahead despite the worldwide financial crisis of the early 1930s, which at times threatened the viability of the whole undertaking. It began on 1 August 1932 and lasted for thirteen months, with considerable British involvement, supported by a government grant of £10,000 spread over three years. Simpson and the Norwegian oceanographer H U Sverdrup took on overall responsibility for meteorological instruments, and Professor E V Appleton of the University of London led an expedition to Tromsö in Norway to investigate the layers in the upper atmosphere which are important in respect of the transmission of radio waves. In addition, special auroral observations were made in Scotland, and a Senior Professional Assistant of the Office, James Martin Stagg, led a six-man expedition to Fort Rae on the Great Slave Lake in Canada to make surface and upper-air meteorological observations, study the aurora and measure terrestrial magnetism and atmospheric electricity. Stagg was then working in the Office's directorate in London but had been in charge at Lerwick Observatory in 1929 and 1930 and, before that, had worked at Kew and Eskdalemuir Observatories.[34] The question of re-opening the Ben Nevis Observatory for the IPY was considered but rejected. The expense could not be justified.

Another notable development in the 1930s was the launch of the world's first regular high-definition public television service, by the BBC, on 2 November 1936, a service that was used for weather broadcasting straightaway. In the words of the annual report of the Office for the year ending 31 March 1937:

An experiment was made between 2 and 28 November of issuing weather charts by the television process and accompanying them by a spoken description of the chart and a short forecast. This experiment was made at the Alexandra Park Television Station of the British Broadcasting Corporation, and as the radius of the broadcast is limited the forecast referred to London only. The experiments indicated that there were possibilities in this method which might have to be reviewed later when television becomes one of the amenities in the lives of the majority of the inhabitants of the country.

[34] Another member of the IPY expedition to Fort Rae was P A Sheppard.

The first weather chart was shown on 3 November, with an anonymous hand drawing isobars onto the map whilst a disembodied voice read the forecast. Prescient indeed was the assertion that television would become an amenity for the British public. Weather presentations on television are taken for granted nowadays.

A further notable development was the establishment of a meteorological tele-printer network in 1937. The need for this had become clear by 1935, when it had been realized that the distribution of weather reports by means of radio alone would eventually prove inadequate in peacetime and certainly be impracticable in wartime. The frequency and number of reports by wireless telephony from the Borough Hill Radio Station had increased as the RAF had expanded and night-flying developed. Saturation would be reached before the expansion of the RAF had been completed. Moreover, radio broadcasts would have to cease in wartime. Following discussions between the Chief of the Air Staff, the Directorate of Signals, the General Post Office and the Meteorological Office, Treasury approval for the creation of a teleprinter net-work was granted in August 1937. The network would link the Meteorological Office to twelve RAF stations which provided weather reports for the broadcasts made from Borough Hill. By early October, all were linked except Abbotsinch, which was con-nected soon afterwards. There followed a gradual expansion of the network, limited by the availability of land-lines and the supply of teleprinters. The number of stations on the network increased to twenty by October 1938 and forty-eight by September 1939, when war broke out.

The clouds of war had been gathering for some years, and plans and preparations for such an eventuality had been made carefully and methodically. By September 1939, the Office was much better prepared for the demands of war than it had been for the conflict which began in 1914. Much of the credit for this was due to Simpson, who had responded so ably and professionally to the challenges placed before him by the Air Ministry and others. By the time war broke out, however, he had retired, and a new Director was in post.

10

The Clouds of War

Sir George Simpson retired on 2 September 1938, his sixtieth birthday. Who would succeed him? This matter had been addressed as early as 21 February 1936, in an internal minute from J B Abraham, an Assistant Secretary in the Air Ministry, to J A Webster, a Principal Assistant Secretary.[1]

Abraham had ruled out Lempfert and Whipple on grounds of age and ruled out Gold, too, but not on grounds of age, even though he would have been 57 when Simpson retired.[2] Rather, there were misgivings about him. Abraham doubted he had "the administrative ability requisite in the Director of a big department" and commented that his temperament was "rather difficult". Abraham also ruled out Corless, Entwistle, Goldie, J S Dines and E G Bilham, the five in the Principal Technical Officer grade.[3] His conclusion was that none had shown the all-round scientific and administrative ability a Director needed. Simpson's successor would need to be someone not currently a member of the Office's staff. In Abraham's view, the field of selection was "rather restricted". Brunt might be a possibility but appeared to "lack the personality and all-round ability which was desirable". Perhaps the most suitable successor was Watson Watt. He was, Abraham said, "a scientist of great ability and exceptional energy".

Webster passed on Abraham's minute to the Air Ministry's Permanent Secretary (Sir Christopher Bullock), suggesting that Sir Christopher "may like to talk it over with Simpson some time in the near future". Simpson had been giving thought to the possibility of delegating "some of the more purely scientific functions of the Office, such as seismology and terrestrial magnetism, to some other scientific body". If something of the sort was desirable, it would naturally affect the qualifications

---

[1] There is a copy of this minute in the National Archives (AIR 2/10737, File 17722).
[2] Lempfert was born on 7 October 1875, Whipple on 17 March 1876.
[3] Bilham was head of the Office's British Climatology Division in 1936.

264

Figure 10.1. Nelson King Johnson. © Crown Copyright 2010, the Met Office.

required of the next Director. And also, Webster said, the enormous growth in the Office's forecasting work and the close connections which the Office had to maintain with the RAF, the Royal Navy and civil aviation made it desirable that the person chosen "should have strong administrative as well as scientific qualifications".

A note from Abraham to Sir Christopher dated 1 April 1936 shows that Dobson was also considered. In the event, though, Nelson King Johnson became Director on 3 September 1938. He was no stranger to the Office, having served at Shoeburyness from 1919 to 1921 and then, on secondment to the War Office, taken charge of the meteorological section of the Experimental Station at Porton Down, where he had investigated the physics of the atmosphere very close to the ground. He had left the Meteorological Office in 1928, when appointed Director of Experiments at Porton Down, and later moved to London, to become Chief Superintendent of the War Office's Chemical Defence Research Department. Born at Canterbury on 11 March 1892, he had been educated at the Simon Langton School in that city and proceeded to the Royal College of Science in South Kensington, where he had gained an honours degree in physics in 1913 and thereafter remained for a year as an assistant demonstrator in spectroscopy. He had then taken up astronomy under Sir Norman Lockyer at Sidmouth Observatory in Devon but given up this career in 1915 to join the Royal Flying Corps. It is said that his experiences as a pilot influenced his decision to join the Meteorological Office.

The Office's annual report for the year ending 31 March 1939 included only the briefest of references to Simpson's departure, stating that the retirements of Simpson, Lempfert, Whipple and Brooke-Smith had taken place within a period of a few months.[4] There was no other reference to Simpson. The report went on to note that Corless and Goldie had been promoted to fill the two vacancies in the Assistant Director grade, and Captain C E N Frankcom had been appointed Marine Superintendent.[5] It also stated that Miss Dorothy Chambers had received an MBE in the 1939 New Year's Honours in recognition of the valuable services she had rendered to the Office.[6]

She had become Simpson's personal secretary in 1922 and remained so until he retired. She had also served as Secretary to the Meteorological Committee since 1936 and as Secretary to the Office's Polar Year Committee in the early 1930s. When interviewed by Jim Burton in September 1983, she recalled going with Simpson to a meeting of the IMO in Warsaw in 1935 and remembered in particular the Nazi influence being obtrusive on the German delegation. She recalled, too, that Sir Arthur Schuster had been a major influence on Simpson. He had been a member of the Meteorological Committee for many years, from 1901 to 1932. A further recollection of Miss Chambers was that when Johnson became Director of the Office, administrative arrangements were changed. His requirement for secretarial assistance was purely for a shorthand typist, as a result of which she was transferred to the Office's Personnel Section, where she remained until her retirement in January 1954.[7]

The Pre-War Years

The meeting of the Committee on 29 November 1938 was the first after Johnson took office. At it, the promotions of Corless and Goldie were approved, and Johnson reported that he had taken the opportunity of the retirements of Lempfert and Whipple to reorganize the duties of his three Assistant Directors. The work of the Office could now, he advised, be grouped into synoptic meteorology (home and overseas), British and overseas climatology, ancillary work (i.e., the supply of personnel, instruments, buildings, etc.), and observatories and research. Furthermore, to strengthen the Synoptic Division, which had become too large for one Principal Technical Officer, he proposed to split it and form a new and separate Division concerned solely with RAF requirements.

---

[4] Brooke-Smith retired on 3 December 1938, Lempfert on 31 December 1938 and Whipple on 31 March 1939. Another who retired soon afterwards was J S Dines, on 30 June 1939.

[5] Frankcom took up his appointment on 16 January 1939. A Master Mariner and Captain in the Merchant Navy, he served in the Royal Navy during the war as Commodore of North Sea Convoys. After the war, he was known as Commander Frankcom, in recognition of his rank in the Royal Naval Reserve.

[6] As mentioned in Chapter 8, Miss Chambers had joined the Office in 1913, aged 20.

[7] Miss Chambers' brother-in-law was Austen Nagle, Director of the Irish meteorological service.

As from 1 January 1939, the Office was restructured thus, with the various divisions each given a descriptor:

- Under the Assistant Director for Climatology and Ancillary Duties (Corless):

    M.O.1 – Marine Division
    M.O.3 – Climatology Division
    M.O.4 – Army and Instruments Division
    M.O.10 – Administrative Division (including the Training School)

- Under the Assistant Director for Synoptic Meteorology (Gold):

    M.O.2 – Headquarters Forecasting and Civil Aviation Division
    M.O.5 – Overseas Division
    M.O.6 – Royal Air Force Division

- Under the Assistant Director for Observatories and Research (Goldie):

    Edinburgh Office (Aberdeen, Eskdalemuir and Lerwick Observatories)
    Kew Observatory

The descriptors M.O.7, 8 and 9 were not used. The designation of the Marine Division as M.O.1 recognized the original maritime purpose of the Office.

The Committee discussed on 29 November 1938 the research to be undertaken by the Office. Johnson stated that he had "already taken steps to draw up a list of the problems requiring solution" and given some thought to priorities. He was able to assure the chairman that "civil aviation problems would not have to take second place". Sydney Chapman expressed the view that the trend of Johnson's proposals was excellent but thought the Office needed to receive advice from scientists on its research programme. Doubts were expressed by Sir D'Arcy Thompson (Professor of Natural History at St Andrew's University). He was not convinced a government department could undertake research successfully. Nevertheless, Chapman's view prevailed, and Johnson was asked to put forward a proposal for the formation of a Meteorological Research Committee which would act as an advisory body to the Office.

The Committee also considered a proposal made by the Royal Meteorological Society that university institutes for meteorological research be established. Johnson reported that the Office had sent to the Society a reply to the effect that the Air Ministry would support an appeal to the Government for funds for establishing such institutes, with pure research the preserve of the institutes and applied research carried out in the Office. Besides undertaking research, he said, "such institutes would provide training grounds for future members of the Office staff". Chapman pointed out, however, that a professorship of meteorology had existed in the University of London for many years and, moreover, that a scheme had already been laid before the University's authorities for the establishment of a meteorological institute at South Kensington. He foresaw a possible financial difficulty, in that some might argue that money required for an institution which would benefit the Air Ministry should not come from the general education fund of the nation. The Committee resolved that Johnson should discuss the matter with Sir Henry Tizard, the Rector of Imperial College, and "urge the interest of the Air Ministry".

Little came of approaches to Tizard at this time. Plans had to be shelved when war broke out. However, the seeds had been sown. The idea of a meteorological institute at Imperial College gained support from both the College and University of London authorities, and Brunt's sub-department became independent of the College's physics department in 1939. Furthermore, as we saw in Chapter 9, the training of graduates for the Meteorological Office was entrusted to Brunt and Sheppard immediately when the war began. Meanwhile, questions about the Office's role in meteorological research had been asked in the House of Commons.

On 10 February 1937, whilst Simpson was still Director, Frank Markham had asked the Under-Secretary of State for Air whether the increase of staff which had occurred in recent years would permit the Office to devote more time to research and less to routine work. The Under-Secretary, Sir Philip Sassoon, had replied "I hope that may be so". Markham had then asked the Under-Secretary whether he considered that the science of weather forecasting had been sufficiently developed, or was sufficiently accurate, to provide adequate information to defence forces relying on a foreknowledge of weather conditions for their successful operation. Sassoon replied that the science of weather forecasting already afforded information of much value to the defence forces, and the possibilities of its further development were kept under constant study and review. The riposte of Markham was that this answer was vague and uninformative, but Sassoon was not minded to say more.

Markham was on his feet again on 24 February 1937, asking the Under-Secretary of State for Air a series of questions about the Office, particularly about weather forecasting and research. In one question, he asked whether any staff of the Office were engaged in full-time research in weather forecasting, and whether any arrangements were being made to provide more intensive researches into this subject. Sir Victor Warrender, Financial Secretary to the War Office, told him that it was not practicable "in the present conditions of pressure arising from the urgent needs of expansion" for the time of any members of the Office's staff to be devoted wholly to research in weather forecasting. A considerable amount of research on this and other meteorological subjects was, however, carried out constantly, and as soon as the pressure abated and fully trained meteorologists were available, he hoped it would be possible to provide for more intensive research into weather forecasting. Markham persisted. There seemed to be an impression, he said, that no research had taken place since before the Great War and that the present situation was not due to the defence plan. Warrender's rejoinder was that research was "going on all the time", but it was limited because of the pressure which existed on the present staff.

Not satisfied, Markham asked the Under-Secretary of State for Air on 3 March 1937 whether any special funds or facilities were given to superintendents of the Office to carry out investigations or research work and whether the results of such investigations were published by the Office. Warrender replied at length, saying:

No question arises of making special payments to members of the staff of the Meteorological Office, who are all whole-time civil servants, in respect of investigations or research work carried out by them. Provision is, however, made in Air Estimates to provide payments to investigators outside the Office at universities and elsewhere, and to the National Physical Laboratory for special investigations. As regards facilities for research at the Meteorological Office, the normal work of a large proportion of the staff is in the nature of research, especially at the observatories, and the sum total of research work carried out by the staff is very considerable, as will be seen from the list of papers published by them which is quoted in the annual report of the Director of the Meteorological Office. The Office itself publishes two series of monographs: *Geophysical Memoirs* and *Professional Notes*. During the ten years 1927 to 1936, 35 papers have been published in the former and 29 in the latter, and many of these deal with outstanding research work in meteorological science.

The widely held perception (then and now) that Simpson discouraged research is clearly not correct.

Markham asked yet more questions about the Office on 22 June 1938. In one of them, he asked the Secretary of State for Air whether he was satisfied the present staff of the Office was adequate for its duties, and what numbers were engaged in weather forecasting. The Under-Secretary of State for Air, Harold Balfour, replied, informing Markham that, as the result of the expansion of the RAF and of civil aviation, a very large increase in the staff of the Office had been called for and was being met as rapidly as suitable personnel could be recruited and trained. There were at present eighty-seven persons engaged in making day-to-day weather forecasts and thirteen under training. Those numbers would be increased during the next twelve months. Markham persisted with a supplementary question. Did the proposed expansion give sufficient time for the senior staff to devote adequate time to their research duties? Balfour said that these duties formed a heavy strain on the senior members of the Office, but he had been told the matter could be dealt with adequately without further expansion. Still Markham persisted, asking whether it was not a fact that the senior staff had been for years swamped with routine work and that the present expansion did not provide enough staff to tackle research problems. Balfour responded by saying that the present expansion aimed at training other experts for the work in order to reduce the routine work that needed to be carried out by senior staff.

In another question on 22 June 1938, Markham turned his attention to long-range forecasting, asking Balfour whether he was aware that weather forecasts for a fortnight or longer were being made both in Britain and in Germany and had "attained a high percentage of accuracy". Would he consider taking steps to ensure that the forecasts

Figure 10.2. Campbell-Stokes sunshine recorder. The glass sphere focuses the sun's rays onto a graduated card, thus burning a track on it. Photograph by Malcolm Walker.

of the Office were supplemented by comparable long-distance forecasts? Balfour did not agree that the accuracy was high. He said he was aware of the long-range weather forecasts that were being attempted by various methods in many different countries, but none of these methods had so far "attained the accuracy which would justify the issue of such forecasts in this country".

Why Markham was so interested in meteorology and seemingly so antagonistic towards the Office is not known. He continued to ask questions in the House of Commons. On 1 March 1939, he asked the Secretary of State for Air whether he was aware that in some holiday resorts in southern and eastern England sunshine records were manipulated. Balfour replied that isolated instances of the manipulation of sunshine records had been discovered from time to time, but it was almost always easy to detect and such data were withheld from the Press and from official publications and records. No data were published by the Office without critical examination. Markham was not satisfied, asking whether it was not a fact that misleading information had indeed been given to the public. Balfour denied this, whereupon Markham, true to form, persisted. Again, though, Balfour supported the Office, saying that the instruments and methods of recording at stations, as well as the results, were examined by officials of the Office.

Meanwhile, the staff of the Office and the Government had more pressing matters on their minds. The clouds of war were gathering, and the first priority was to make sure the Office was fully prepared. A large number of additional meteorological stations had been established at aerodromes, and new staff had been recruited and trained to man them. The international crisis of September 1938 had served to focus attention on the importance of providing a reserve of trained personnel to cope with additional demands that may arise.

There had also been a number of technical developments in the latter years of the 1930s. In particular, trials of radiosonde apparatus had been carried out, and attempts to dissipate fog by artificial means for the benefit of aviation had been made. Considerable progress on the design of a cloud searchlight had also been made, to help observers ascertain the heights of cloud bases during hours of darkness, and Gold, in collaboration with the Instruments Division, had designed a visibility meter, an important instrument for observers at airfields. Furthermore, investigations had begun in 1938 to find out why certain day-to-day weather forecasts lacked accuracy. According to the 1939 annual report, it was hoped to discover the factors which affected certain meteorological situations which were imperfectly understood and thus reduce the number of occasions on which satisfactory forecasts could not be made. Weather forecasts were clearly not as accurate as Frank Markham thought!

War Breaks Out

International tension mounted in the 1930s until, on 1 September 1939, Germany invaded Poland. Two days later, France and the UK declared war on Germany. Other nations became involved, and the war became a global conflict. The meeting of the Meteorological Committee on 4 July 1939 was the last for seven years, and no annual reports were issued during the war years. The wartime activities of the Office were controlled directly by Johnson and his senior staff, working closely with Government ministers and officials and leaders of the armed forces. To some extent, the pre-war activities of the Office continued. To a much greater extent, though, the Office's work was driven by wartime demands. On 1 September 1939, full security restrictions on the use of meteorological information were imposed. All broadcasts of synoptic weather information *en clair* were forbidden, and limits on the issue of weather news to the general public were introduced.

Numerous articles about meteorology during the Second World War have been published in meteorological periodicals. They contain experiences and recollections of meteorologists and also provide accounts, case studies and explanations of particular meteorological activities, problems, decisions and operations. Conference proceedings, too, contain personal recollections and historical accounts, notably *Meteorology and World War II*, published by the Royal Meteorological Society (Second Edition, 2004).[8]

---

[8] This volume is a compilation of the papers presented at two conferences, one in October 1986, the other in October 1988. All of the papers were tape-recorded by the Imperial War Museum, and the tapes now form part of the National Archives.

The principal source of information about the wartime work of the Office is the official report entitled *The Second World War, 1939–1945, Meteorology*, published by the Air Ministry in 1954 (AP1134, 570 pages plus twenty-seven appendices). It is not, though, a complete record of British wartime meteorology, as some work during the war remained secret for decades. Indeed, the report (hereafter called AP1134) was itself classified 'Confidential' until November 1975, when downgraded to 'Unclassified'. Even then, some wartime meteorological work remained unknown to the public, especially that which proved so crucial at Bletchley Park for breaking the codes used by the Germans. The articles in periodicals provide a welcome supplement to the information contained in AP1134. In addition, many documents and records relating to the Office's wartime work are preserved in the National Archives at Kew and the National Meteorological Library and Archive at Exeter.

For the Army and the RAF, the Office was the body solely responsible for meteorological advice and information. For the Royal Navy, the Office's responsibility was limited to the supply of basic reports and technical deductions from them to the Naval Meteorological Service, which in turn was responsible for advising the Admiralty and for supplying forecasts to His Majesty's ships and naval shore establishments. An exception was made for Combined Headquarters, for which the Office was responsible for providing technical staff and technical advice. To meet the needs of the Army and the RAF, a number of meteorological offices had already been established by the time war broke out. Others were subsequently established at most operational airfields, as well as at specialized Army Units and the headquarters of Commands, Groups and Corps. In Commands overseas, the Chief Meteorological Officer was effectively the meteorological theatre commander, dependent for guidance and supplies on the Office's headquarters in London, but with full technical responsibility for meteorological advice supplied to the armed forces in his theatre.

Preparations for a wartime dispersal of the Office to the provinces were made in 1938, around the time of the Munich crisis, and these plans were amended in early 1939. The original plan was that the Office's administrative headquarters would be moved to Southport in Lancashire. Then, in February 1939, the evacuation destination was changed to Tetbury in Gloucestershire. In the event, the administrative headquarters remained in London throughout the war, but most divisions of the Office were, in fact, moved away from the capital.

The planned destination of the Forecasting Division was Dunstable, Bedfordshire, only thirty-five miles from London, five miles from the RAF's Communications Centre at Leighton Buzzard and twelve miles from the decryption centre at Bletchley Park. Moreover, Dunstable was conveniently placed in relation to Post Office land-line trunk routes, and the site chosen for the Division's new premises was on high ground, so that wireless reception would be good. Plans to construct a combined forecasting and communications centre at Dunstable were prepared in the summer of 1939. However, the international situation deteriorated so rapidly that the new buildings were not ready for occupation when war broke out. As a temporary measure, the Forecasting Division was evacuated to Birmingham.[9] The move took place on 27 August 1939 at only three days' notice and was accomplished without disturbing the flow of current synoptic information to outstations. When the time came for the move to Dunstable, in February 1940, the weather could hardly have been worse, further complicating an already complex operation. Service had to be maintained for twenty-four hours a day without interruption, which was achieved by staff moving in stages, the last contingent travelling by car on roads that were deep in thawing snow.

The Marine and Climatology Divisions and the Instruments and Army Division moved in November 1939 to Wycliffe College at Stonehouse, Gloucestershire.[10] Their accommodation in South Kensington was required for other purposes. A stock of essential instruments and equipment was transferred from South Kensington to a store in Cheltenham in the last few days of August 1939, and historic and otherwise important books and records from the Library, together with logbooks and Hollerith cards from the Marine Division and further instruments and stores from the Instruments Division, were transferred to Stonehouse in November 1939. When the Marine Superintendent went on active service, in November 1940, his deputy, Jack Hennessy, took charge of M.O.1.[11]

The Training School remained at Berkeley Square House under Brunt and Sheppard until June 1940 and then, in the autumn of that year, transferred to Barnwood, on the outskirts of Gloucester, with Boyden in charge. Brunt returned to Imperial College and Sheppard remained with the Office, first to take charge of meteorological services concerned with civil defence and later to organize and administer the Office's programme of upper-air observations in the European Theatre and elsewhere. From 1940 onwards, according to Hannah Gay, in *The History of Imperial College London, 1907–2007* (Imperial College Press, 2007, p. 247), Brunt "appears to have carried out a number of secret missions, mostly across the Atlantic". Imperial College records show that his salary was paid by the Government for almost the entire duration of the war.

A number of emergency arrangements came into force when war broke out. Goldie took charge of the Marine, Climatology and Instruments and Army Divisions, as well as the sections of the Administrative Division not concerned with Personnel.

---

[9] The code name for the evacuation headquarters of the Central Forecasting Office (CFO) and Communications Centre at Birmingham was 'ETA' (the Greek letter η), and this name continued to be used for the Dunstable headquarters until 1947, when dropped in favour of CFO. No definition of ETA is given in AP1134, but official files in the National Archives suggest that 'Evacuation Temporary Accommodation' seems likely. Alternatively, the E of ETA may have stood for 'Emergency'.

[10] Staff and students of the College relocated to Lampeter in Wales for the duration of the war.

[11] Commander Hennessy joined the Meteorological Office in 1919 after serving in the Merchant Navy and, later, the Royal Navy. He made a number of valuable contributions to the work of the Marine Division (see the obituary of him in *Weather*, 1954, Vol. 9, p. 185). In particular, he was an expert in the use and care of the Hollerith system of punched cards, with which he had been associated since the system's introduction to the Marine Division in 1921. He made much use of the cards when producing the marine climatological atlases which were prepared during the Second World War. He was a Commander by virtue of holding that rank in the Royal Naval Reserve.

Thus Corless was relieved of all aspects of the work he had hitherto directed, except Personnel. This decision showed foresight, for it recognized that the Office was likely to expand during the war, which it certainly did. The number of staff increased from 763 in March 1939 to 6760 in August 1945. The Office and the Air Ministry had anticipated that recruitment and training would be a time-consuming task. It was indeed, but it did not become a full-time job, and Corless resumed full control of the Administrative Division in November 1939, when Goldie and the Marine, Climatology and Instruments and Army Divisions moved to Stonehouse.

An additional responsibility was taken on by Goldie in 1941 when the Air Ministry set up the Meteorological Research Committee (MRC) that had been proposed before the war. He became responsible for the administration of this committee. A number of problems which came to the fore in 1940 and 1941 led several prominent scientists to offer to serve on the committee. One problem was that condensation trails (contrails) were made by aircraft at great heights and thus revealed the whereabouts of the aircraft to the enemy. The exact conditions for the occurrence of the trails needed to be investigated. In addition, there was a revival of interest in the problem of fog dispersal by artificial means, and the naval authorities pressed for an inquiry into the possibility of forecasting the weather for periods of a week to a month ahead.

The MRC was appointed by the Secretary of State for Air on 7 November 1941, with the following members: S Chapman (Chairman), D Brunt, G M B Dobson and G I Taylor, along with the Director of the Meteorological Office, the Director of the Naval Meteorological Service, the Director of Scientific Research in the Ministry of Aircraft Production, a representative of the Air Staff and a representative of the Director-General of Civil Aviation. The terms of reference of the committee were to advise the Secretary of State for Air regarding the general lines along which meteorological research should be developed, as well as to advise and assist in the carrying out of investigations and research within the Office and to receive reports on meteorological investigations carried out in the Office or on behalf of the Air Ministry and make recommendations for further action.

The MRC held its first meeting on 10 December 1941 and thereafter met another thirty-three times during the war. Sub-committees met on numerous occasions to deal with special questions. One of the MRC's first initiatives was to establish, through the Air Ministry, a special Meteorological Research Flight to conduct atmospheric research from aircraft. It was formed at Boscombe Down in Wiltshire in August 1942, and one of its important early tasks was to investigate the atmospheric conditions which favoured the formation of contrails.[12] The MRC was not disbanded when the war ended.[13]

---

[12] For further information, see 'A short history of the Meteorological Research Flight, 1942–92', by W T Roach, published in 1992 in the *Meteorological Magazine* (Vol. 121, pp. 245–256).

[13] See 'The history of the Meteorological Research Committee', by F J Scrase, published in 1962 in the *Meteorological Magazine* (Vol. 91, pp. 310–314).

Simpson offered his services immediately when war broke out, an offer readily accepted by Johnson. He was made Superintendent of Kew Observatory and also placed in charge of the work of the Edinburgh Office and the observatories at Aberdeen, Lerwick and Eskdalemuir, thus releasing Goldie to concentrate on his new responsibilities. At Kew, Simpson replaced Stagg, who had not long been in post as Superintendent, having succeeded Whipple on 1 April 1939. After the International Polar Year, Stagg had served under Goldie in the Edinburgh Office, and immediately prior to taking the post at Kew he had completed a tour of forecasting duty with the RAF in Iraq. Shortly before the war began, he was given various jobs concerned with putting the Office on a war footing, and from these jobs his chief task became the management of weather services for the Army, with much of his time spent in ensuring the RAF and the Army worked effectively together.

Whilst at Kew, Simpson was able to find time to engage in science and invention. He completed investigations of thunderstorm clouds he had begun some years earlier and, in addition, working with the physicist E G Dymond, remodelled the unsatisfactory British radiosonde and arranged not only for its manufacture but also for its distribution to six launching stations in the British Isles and some overseas. This device proved extremely important during the war as a means of obtaining upper-air information. Simpson also modified apparatus which was used for measuring the electrification of rain and in 1948 published his work in a memoir called 'Atmospheric electricity during disturbed weather' (*Terrestrial Magnetism and Atmospheric Electricity*, Vol. 53, pp. 27–33).

Another emergency arrangement that was implemented when war broke out was the posting of a liaison officer for duty with the Office Nationale Météorologique in Paris, a posting reciprocated later in September 1939 when an officer from Paris took up a similar position in London. A further arrangement was that a senior officer from the headquarters staff of the Office (Entwistle) was commissioned in the RAF in October 1939 with the rank of Group Captain to serve as Meteorological Officer-in-Chief in the Field, with responsibility for meeting all requirements of the Army and the RAF in the Field.[14] Another senior officer (Charles Ernest Britton) was commissioned as a Wing Commander for liaison duties in London in respect of all RAF Volunteer Reserve (RAFVR) matters.

The work of the IMO and its Secretariat continued during the war, though on a much reduced scale. Some commissions continued to be active, particularly those concerned with scientific matters, such as the Commission for Solar Radiation and the Commission for Terrestrial Magnetism and Atmospheric Electricity. In addition, a small amount of progress was achieved by the Commission for the Réseau

---

[14] Entwistle joined the Office in May 1915 (see Chapter 8). He was Superintendent of the Aviation Services Division from 1926 to 1935 and then in charge of the Overseas Division until September 1939 (see Chapter 9). He left the Office in February 1946, when appointed chief of the Meteorological Section of the Provisional International Civil Aviation Organization.

Mondial and the Commissions for Climatology, Maritime Meteorology and Agricultural Meteorology, mainly in the form of preparation of statistical tables.

Work also continued on the drafting of a convention which was intended to secure official status for the IMO as an inter-governmental body. The need for such a convention had been agreed on at the meeting of the Conference of Directors held at Warsaw in 1935, with the initial drafting work entrusted to the President of the IMC, Theodor Hesselberg, and the Director of the French Meteorological Service, Philippe Wehrlé. Their draft had been revised at the meeting of the IMC held at Berlin in June 1939 and an advanced version known as the Berlin Draft produced. A commission was formed to refine this Draft and produce a definitive version for consideration at the meeting of the Conference of Directors which was scheduled to take place at Washington, DC, in 1941. That meeting did not take place. The war intervened.

The IMO had no secretariat before 1928. Then, a small administrative office with no policy-making functions was established, its principal task being to reduce the managerial burdens borne by the President and Secretary of the IMC. Pending a decision on the permanent location of the secretariat, accommodation was provided in the headquarters of the Royal Netherlands Meteorological Institute at De Bilt. A resolution was passed at the meeting of the Conference of Directors at Copenhagen in 1929 that the secretariat would eventually be located in Switzerland, but no date for implementation of this decision was set. Eventually, the secretariat took the matter into their own hands. They departed hastily from De Bilt soon after the war began and removed to Lausanne, Switzerland.

Military Operations during the War

A problem in the early stages of the Great War had been that commanders of land and air forces had been casual about the value of meteorological advice. This was no longer the case. In the words of Stagg, in 'The Meteorological Office and the Second World War' (*Meteorological Magazine*, 1955, Vol. 84, pp. 178–183):

By the outbreak of the Second World War, meteorological advice for military operations on land and sea and in the air had become so much appreciated that the meteorological officer was brought on to the commander's staff at the planning stage, and he could influence decisions on even the largest scale of operations. History will also record that it was more than a coincidence that two of the major advances in weather forecasting were related to those two wars: the development of the idea of fronts and the introduction of three-dimensional analysis as a regular and indispensable procedure in the preparation of forecasts.

So well did the Air Ministry, the Office and the armed forces prepare for war that many decisions made before September 1939 stood the test of the war basically unchanged.

Employees of the Office were all civilians before the war, apart from a small number who were commissioned in the Meteorological Branch of the RAFVR during the summer of 1939. The general policy until 1942 was that reservists would serve in uniform in the actual theatres of operations, and other staff of the Office would be civilians. By 1942, many meteorological offices overseas were staffed almost wholly by personnel in uniform, and from 1 November 1942 forecasters overseas, except those in South Africa, Canada and the United States, became RAF officers holding CC commissions.[15] The reason for this change in policy was that the practicalities of security, travel, accommodation and messing had made the employment of civilians overseas a source of difficulty and sometimes of danger to the staff themselves. The wearing of uniform by meteorological officers was extended to certain Home Commands on 1 April 1943. In the words of AP1134, this was "mainly because the closest possible liaison was needed between meteorological officers charged with briefing duties and the aircrews and other service personnel whom they were briefing, especially for operational flights". Most meteorologists were in uniform by April 1943, but headquarters staff of the Office remained civilians throughout the war.

Women were involved in a wide range of meteorological duties from August 1941 onwards, some as civilians, others as officers and airwomen of the Women's Auxiliary Air Force. In Stagg's words, "they were debarred only from service overseas and from forecasting duties at operational stations, though even there a relaxation was made in the latter half of the war when they took equal part with their male colleagues in the briefing of aircrews at operational training airfields".[16]

The huge expansion of the Office's staff during the war created problems and difficulties. The growth of the Meteorological Branch of the RAFVR was particularly rapid in the opening days and weeks of the war. The additional staff all needed to be trained. Officers were trained by Brunt and Sheppard at Berkeley Square House. So, too, were civilian forecasters and assistants and a number of airmen. Other recruits were posted to RAF stations all over the UK and received their training in local meteorological offices. The expansion brought with it difficulties created by a considerable dilution of experience in working with a science that was still largely empirical. These difficulties, as Stagg put it, "were certainly not diminished by the range of operational activities on which the Office was called upon to advise or by the necessity for deploying the available staff in small and widely dispersed units". It was also not possible in 1939, Stagg said, to anticipate the radical changes in forecasting practice that were to result from experiments then in progress at Larkhill on the measurement of upper winds by means of the radio-triangulation method.[17]

---

[15] CC (Civil Component) was a Reserve class for civilians employed by the RAF who would be commissioned on mobilization.

[16] A personal account of life as a meteorologist in the Women's Auxiliary Air Force, by Beryl Spear, can be found in *Meteorology and World War II* (edited by Brian D Giles, published by the Royal Meteorological Society in 2004).

[17] The basis of this method was that the movements of pilot balloons which carried small radio transmitters were tracked by two radio direction-finding stations at the ends of a base line. The great benefit of the method was that information about the upper atmosphere became available when cloud and visibility conditions were such

Most commands, sections, units and other military groups required meteorological advice and information. Moreover, civil aviation continued to a limited extent during the war, with meteorological support required for internal flights, routes to neutral countries and routes in many other parts of the world. At home, a number of Government departments required meteorological advice and information, including the Ministries of Fuel and Power, War Transport, Home Security and Agriculture and Fisheries. The needs of the RAF particularly stretched the capabilities of forecasters, with the operational requirements of Bomber and Coastal Command especially important. Bombers had to take off safely and land without difficulty eight or ten hours later, and their crews needed to fly to and from target areas without getting lost or straying over heavily defended areas. They also needed to be able to find and identify targets so as to attack them with precision. Coastal Command aircraft maintained anti-submarine patrols and shadowed and protected convoys of merchant ships. These flights sometimes lasted more than twelve hours and, until the later stages of the war, relied on dead reckoning. Thus it was essential, and could literally be vital, that weather forecasts were reliable, especially forecasts of upper winds.

Radiosondes proved invaluable during the war. In all parts of the world, a great improvement in upper-wind forecasting occurred, brought about to no small extent by the availability of reliable upper-air data obtained by means of radiosondes.[18] Better methods of measuring winds and temperatures aloft, and the development of weather reconnaissance over the North Sea and the Atlantic, made it possible to construct routine upper-air charts for the British Isles and Atlantic with much greater precision than had previously been possible. However, much remained to be discovered about the structure of upper winds. Forecasting errors with unfortunate consequences sometimes occurred, the most notorious example being the drift and scatter of an Allied bomber stream during a raid on Berlin during the night of 24–25 March 1944, caused by strong winds which had not been forecast. Seventy-two bombers were lost, fifty of them destroyed by anti-aircraft fire, the others shot down by fighter aircraft.

The cyclone model of the Bergen School which the Office used for weather forecasting placed emphasis on air masses and fronts but did not include information about upper winds, though Bergen meteorologists had, in fact, suggested the existence of strong winds in the upper atmosphere as early as 1933.[19] And more than a decade before that, C K M Douglas had found from observations of cloud movements that winds of more than 100 miles per hour were common in the upper

---

that visual methods were useless for tracking balloons over any distance. The experiments at Larkhill were so successful that similar stations were set up elsewhere in the British Isles.

[18] For details of radiosonde development during the war and other methods used to obtain upper-air data, see AP1134, pages 439–445.

[19] See *Physikalische Hydrodynamik, mit Anwendung auf die dynamische Meteorologie*, by V and J Bjerknes, H Solberg and T Bergeron (Berlin: Springer, 1933, 797 pages).

atmosphere.[20] The fact that upper winds could be so strong therefore did not come as a surprise to meteorologists, but the narrowness of the bands of strong winds did. So, too, did the association of these bands with the fronts of weather systems. The winds came to be known as 'jet streams' after the war.

To advance knowledge and understanding of the upper atmosphere and thereby improve forecasts of upper winds, a special Upper-Air Unit was formed at Dunstable on 1 January 1943 under Sverre Petterssen, a Norwegian meteorologist of distinction who had been closely associated with the Bergen School for many years. He had become Professor of Meteorology at the Massachusetts Institute of Technology in 1939 and been invited, in August 1941, by the Commander in Chief of the Norwegian Air Forces, to help Norway's government in exile in England by serving as a weather forecaster. He was transferred to the Office at Dunstable on 1 January 1942 to develop techniques of upper-air analysis and forecasting and remained with the Office until 1945, playing an important role as a weather forecaster, not least as a member of the team which helped produce the celebrated forecast for D-Day, 6 June 1944.[21] This forecast proved to be one of the most crucial ever made.

For the Allied invasion of Europe, code-named Operation Overlord, a meteorological support group advised the Supreme Commander, Dwight Eisenhower. The fundamental tasks of the meteorologists were to help him and his planning staff:

- Obtain meteorological information about the likelihood of specified weather conditions in locations and at the times laid down in the plans
- Interpret this meteorological information, particularly in light of changes which were made in the provisional operational plans from time to time

The Chief Meteorological Adviser to Eisenhower was James Stagg of the Meteorological Office, and his deputy was Donald Yates of the U.S. Army Air Forces (USAAF). They were based in the Supreme Headquarters of the Allied Expeditionary Force and supported by teams of meteorologists in the Central Forecasting Office at Dunstable, the Admiralty in London, and Widewing, the forecasting centre of the USAAF at Teddington, Middlesex.[22]

Johnson's choice of Stagg may seem curious. Stagg was neither a forecaster nor a climatologist, though he had gained some experience of forecasting in Iraq shortly

[20] See 'Observations of upper cloud drift as an aid to research and to weather forecasting', by C K M Douglas (*Quarterly Journal of the Royal Meteorological Society*, 1922, Vol. 48, pp. 342–356).

[21] For information about the life and career of Petterssen (1898–1974), see 'Sverre Petterssen, weather forecaster', in the *Bulletin of the American Meteorological Society*, 1979, Vol. 60, pp. 182–195. See also Petterssen's autobiography, first published in English in 2001 (*Weathering the Storm: Sverre Petterssen, the D-Day Forecast, and the Rise of Modern Meteorology*, edited by James Rodger Fleming and published by the American Meteorological Society, 329 pp.). It was originally published in Norwegian in 1974.

[22] The team at Dunstable was led by Sverre Petterssen and Charles (C K M) Douglas. The briefing team in the Admiralty consisted of Instructor Commanders Geoffrey Wolfe and John Thorp of the Royal Navy and Instructor Lieutenant Larry Hogben of the Royal New Zealand Navy. The Widewing team was led by Lieutenant Colonels Ben Holzman and Irving Krick of the USAAF.

Figure 10.3. Group Captain James Stagg. © Crown Copyright 2010, the Met Office.

before the war. He was a geophysicist. However, he was known to be an excellent administrator and manager, the primary attributes required for the job in question. His was not a straightforward task of using his own judgement. In the words of R C Sutcliffe, he had "the trying duty of welding into a mutually acceptable coherent statement the different opinions of both British and American forecasters, men of the highest scientific standing in the forecasting art, working in separate centres and connected by telephone".[23] As Sutcliffe said, Stagg's task might have been no great burden had the weather been average for early June, but it was exceptionally unsettled, which "called for all the judgement that he could bring to bear".

Never before D-Day had so much depended on a weather forecast, and the stormy weather of preceding days made the task all the more difficult. The decision to postpone the invasion from 5 June to the following day was made on meteorological

---

[23] These words have been taken from Sutcliffe's obituary of James Martin Stagg published in 1976 in the *Quarterly Journal of the Royal Meteorological Society* (Vol. 102, pp. 273–274). Reginald Cockcroft Sutcliffe (1904–1991) joined the Office in 1927 as a Professional Assistant. He had previously gained a degree in Mathematics with First Class Honours at the University of Leeds and a PhD in Statistics at the University College of Wales, Bangor. He was much involved in aviation meteorology in the late 1920s and in the 1930s, both as a forecaster and an instructor, and also began to conceive and develop fundamental ideas on dynamical meteorology during the later years of the 1930s.

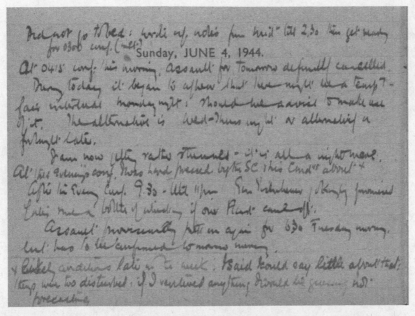

Figure 10.4. Stagg's diary for Sunday 4 June 1944, in which he said that he did not go to bed on the Saturday night but wrote up notes for a conference at 3 a.m. He also noted that "it began to appear that there might be a temporary fair interlude Monday night" and that the assault was "provisionally put in again for 6.30 Tuesday morning". © Crown Copyright 2010, the Met Office.

advice during the early hours of the fourth, though at that time the weather prospects for the sixth looked far from promising. By the evening of the fourth, however, forecasters were able to hold out hope of a lull in the stormy conditions on the sixth and thus help Eisenhower decide to launch the invasion that day. The meteorological advice proved correct and the invasion was successful.

There were, though, tensions and rifts between the Dunstable, Admiralty and Widewing teams in the days leading up to D-Day, and arguments have continued ever since over whose forecast for the day itself was right and why.[24]

The reality is that the state of weather forecasting capability in 1944 was such that trying to forecast the weather several days ahead was an almost impossible

---

[24] Many works have focused on the meteorology of D-Day. See chapter 14 of AP1134 and Stagg's own book, *Forecast for Overlord* (London: Ian Allan, 1971, 128 pp.). See also Petterssen's autobiography (*Op. cit.*, footnote 21). Other notable publications include *With wind and sword – the story of meteorology and D-Day, 6 June 1944* by S Cornford (Meteorological Office, 1994, 132 pp.); *Some meteorological aspects of the D-Day invasion of Europe, 6 June 1944* (Proceedings of a Symposium held on 19 May 1984 at Fort Ord, California), edited by R H Shaw and W Innes, published by the American Meteorological Society (170 pp.); 'Forecasting for the D-Day landings' by C K M Douglas (*Meteorological Magazine*, 1952, Vol. 81, pp. 161–171); and 'Meteorological services leading to D-Day' by R J Ogden (Royal Meteorological Society *Occasional Papers on Meteorological History*, No. 3, 2001, 24 pp.).

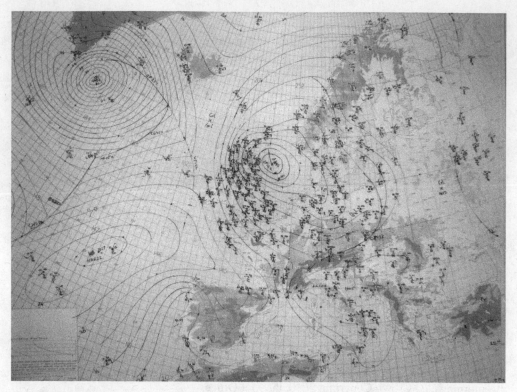

Figure 10.5. The weather map for 13:00 GMT on D-Day, 6 June 1944. Note that observations were available from Ireland and other neutral countries and from the Atlantic, Germany, Norway and other parts of occupied Europe. © Crown Copyright 2010, the Met Office.

task. Even forecasts beyond twenty-four hours in unsettled weather were considered impracticable by most forecasters, including Douglas.[25]

The forecast for D-Day was an extraordinary achievement which should be lauded. Those who argue otherwise should bear in mind the state of meteorological knowledge and understanding in 1944, along with the lack of observational data compared with today and the difficult conditions under which forecasters worked in wartime. Eisenhower was certainly grateful. He sent a personal letter of thanks to the forecasters who had taken part in the telephone conferences, and Stagg was made an Officer of the Legion of Merit by the President of the United States.

Routine services provided by the Office for the RAF and the Army included the supply of weather forecasts and meteorological information to bomber and fighter

---

[25] Long-range forecasts (for up to twelve days ahead) had been prepared on a trial basis at Dunstable from October 1942 to March 1943. The opinion formed by Douglas and other senior forecasters was that the forecasts had not been sufficiently accurate for them to be used for any practical purposes. Forecasts based on persistence of the existing type of weather had given better results. See *Meteorological Research Papers* 86 and 88, by J Wadsworth and C K M Douglas, respectively, published in April 1943 by the Meteorological Research Committee (copies in the National Meteorological Library, Exeter).

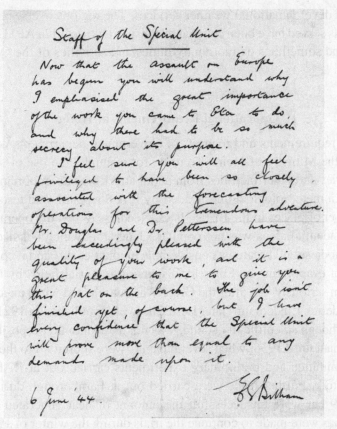

Staff of the Special Unit

Now that the assault on Europe
has begun you will understand why
I emphasised the great importance
of the work you came to Oka to do,
and why there had to be so much
secrecy about its purpose.

I feel sure you will all feel
privileged to have been so closely
associated with the forecasting
operations for this tremendous adventure.
Mr. Douglas and Dr. Petterssen have
been exceedingly pleased with the
quality of your work, and it is a
great pleasure to me to give you
this pat on the back. The job isn't
finished yet, of course, but I have
every confidence that the Special Unit
will prove more than equal to any
demands made upon it.

6 June '44.

E G Bilham

Figure 10.6. Appreciative letter from E G Bilham, 6 June 1944, to staff of the Women's Auxiliary Air Force who provided duplicate charts for Douglas and Petterssen. Bilham was Assistant Director (Forecasting). © Crown Copyright 2010, the Met Office.

crews and other forces in the field. They also included the training of airborne troops and the provision of advice for parachute and glider operations. The great many special services provided by the Office included, in January and February 1945, the establishment and operation of a weather forecasting and meteorological observing system for the meeting of Churchill, Roosevelt and Stalin at Yalta in February of that year.[26]

Staff of the Office were posted to RAF units and theatres of war in many parts of the world to observe, analyse and forecast the weather and to advise commanders and other military personnel. They often worked side by side with staff of other meteorological services and in some places helped train meteorologists of other

---

[26] For the story of the setting up of a meteorological station at Yalta and the provision of a weather forecast for Churchill's flight from Malta to Yalta, see 'Transport Command operations and the Yalta Conference', by Arthur Davies, in *Meteorology and World War II* (edited by Brian D Giles, published by the Royal Meteorological Society in 2004). This was the same Arthur Davies mentioned in Chapter 9.

countries and develop national weather services. The regions overseas where British meteorologists served have been covered in considerable detail in AP1134, along with the diverse and sometimes extraordinary duties and activities of the Office's staff in these regions.

Operational Requirements at Home

The wartime requirements and requests of Government departments were many and various. For the Ministry of Home Security, for example, the use of smoke screens to hide factories and other facilities from aerial attack was an important issue. It was examined by a meteorological unit headed by P A Sheppard, and tests were carried out at a number of places in the British Isles. Needless to say, the operation of smoke screens in residential areas was not popular. Generators which belched forth smoke and half-burnt diesel oil created a considerable amount of dirt and inconvenience.[27]

The idea of evaporating and thereby dispersing fog at airfields by means of heat was first taken seriously just after the Great War but rejected when an analysis by the physicist Frederick Lindemann (later Lord Cherwell), published in 1921, showed that the method, though scientifically sound, was not an economic proposition. The idea then lay dormant until 1936, when several papers were submitted to the Aeronautical Research Committee and preliminary experiments carried out at RAF Martlesham Heath in Suffolk. Further trials were carried out at Farnborough during the winter of 1938–1939 but were unsuccessful; the amount of heat generated was far from sufficient. Plans were made to continue the trials during the winter of 1939–1940, but nothing came of them. The war intervened. The idea then lay dormant again until the autumn of 1941, when the War Cabinet discussed the problem of fog at airfields and set in train an investigation into possible dispersal methods. However, the cost of fuel was soon considered prohibitive, so the idea was dropped. Then, in 1942, concern grew over the number of bombers which had crashed whilst trying to land in fog when returning from missions. Lord Cherwell, now the Government's chief scientific adviser, revived the idea of fog dispersal and the Prime Minister, Winston Churchill, agreed that the matter needed to be explored again. He ordered, on 26 September 1942, that experiments be carried out by the Petroleum Warfare Department (PWD) of the Ministry of Fuel and Power.

The Office became involved at an early stage. A meteorologist was posted to the PWD to advise and assist on the meteorological aspects of fog dispersal. Trials were carried out near Staines, and the NPL (National Physical Laboratory) conducted experiments in a wind tunnel. Soon, the prototypes of operational systems were

---

[27] For information about wartime smoke-screen work, see AP1134, pages 215–219. For an account of his own involvement in smoke-screen activities in 1940 see 'Smoke screens – the early years', by R J Ogden, published in *Weather* in 1986 (Vol. 41, pp. 283–287).

installed at airfields. A coke-burning system at RAF Lakenheath in Suffolk proved a failure, but a petrol-burning device at RAF Graveley in Cambridgeshire was successful (though expensive). By December 1943, seven airfields possessed FIDO (as the device had come to be known), and by May 1945 fifteen airfields possessed it. But what did FIDO stand for? There is still uncertainty over this. The official version given to the Editor of *The Oxford English Dictionary* in 1958 was that the acronym stood for 'Fog Intensive Dispersal Operation'. Another suggestion was 'Fog, Intensive Dispersal of'! Whatever the truth of the matter, FIDO served its purpose well during the war but turned out to be expensive. Diversion to an airfield that was clear of fog was a much cheaper option.[28] The meteorological section of the PWD was disbanded in October 1945.

Concern during the winter of 1940–1941 that the use of poison gas against large towns in the UK was possible led the Office to develop, in conjunction with the Ministry of Home Security, a system of warnings of weather suitable for this form of attack. Brief indications of the weather to be expected during hours of darkness in the various civil defence regions were issued from 15 January 1941 onwards. Initially, they were issued to the Ministry of Home Security two hours before blackout, but later on, from October 1941, they were issued at 16:30 hours. The warnings were distributed via the Ministry's communication network, including, at the end of the message, a plain-language statement of the weather expected.

A further concern was addressed in the summer of 1941. This was the possibility that the enemy might try to set fire to standing crops and stack-yards by means of incendiary weapons. Again the Office became involved, again in cooperation with the Ministry of Home Security. A system of warnings was developed, based on the state of ripeness of the crops in relation to the weather expected. The issuing of warnings began on 21 July 1941 and continued through the harvest seasons of 1941 and 1942. However, they were discontinued in 1943 because their usefulness was no longer considered sufficient to justify the work involved in their preparation and distribution.

In the early days of the war, individual farmers sometimes requested weather forecasts, but the government felt unable to grant these requests, fearing the practice would spread rapidly and thereby compromise the security arrangements of the Air Ministry. By 1942, though, an additional six million acres of land were under cultivation, and the safe harvesting of grain crops was vitally important. This being so, the Ministry of Agriculture and Fisheries suggested to the Air Ministry in July 1942 that some way of issuing meteorological advice without violating security requirements needed to be devised. A code system was introduced to indicate broadly the expected suitability of conditions the following day, as well as an indication of the further outlook.

---

[28] For the story of fog dispersal work, see AP1134, pages 213–215. See also the two-part article entitled 'Fog dispersal at airfields', by R J Ogden, which includes his own experiences of FIDO. It was published in *Weather* in 1988 (Vol. 43, pp. 20–25 and 34–38).

For example, the code word 'dog' meant no rain before sunset next day with reasonably dry air and some sunshine, and 'buy' meant that the weather would probably continue fair or become fair or good for some time. The issuing of harvest forecasts in this form began in England and Wales on 10 August 1942 and continued until the last day of the following month. Then, in 1943 and 1944, forecasts were issued from 1 June to 30 September each year. Forecasts for Scotland began on 16 August 1943 and continued during the harvest season of 1944.

The need for continued enforcement of meteorological security at home was questioned in Parliament and the Press in March 1945, "in view of the changed war situation and the handicap imposed on farmers and fruit growers" (as it was put in AP1134). The point was taken. The BBC started to issue frost warnings on 1 April 1945, this being the first time since September 1939 that Meteorological Office weather forecasts which referred to the UK had been broadcast.

Weather forecasts were especially important during the war for the managers of gas and electricity supply companies, who needed to gauge probable loads on their supplies and make appropriate arrangements. And electricity companies additionally required warnings of thunderstorms, which could affect or even disrupt power supplies. Though weather forecasts were not available in the Press or broadcast by the BBC during the war, the *Daily Weather Report* was still published by the Office. However, the information published in it was out of date by the time gas and electricity companies received copies. Agreement was reached in May 1940 that the Central Electricity Board could receive a duplicated weather chart of the British Isles and a general forecast. This was collected daily by messenger, but even this involved some delay, a difficulty that was overcome in December 1942, when a new procedure was introduced. The daily forecast was telephoned each morning to the National Control Centre of the Board in a special confidential code.

This arrangement continued until January 1945, when representatives of the Board, the Office and the Ministry of Fuel and Power agreed on a procedure that would help the Board deal with the morning peak of electricity consumption. A forecast would be telephoned each evening at 8 p.m. to the Board's National Control Engineer, giving expected temperature, weather, wind direction and force for the following morning at 8 a.m. in the London area and inland in Great Britain south of the River Tay. On the basis of the forecast and knowledge of the capacity available, the Engineer arranged with the Ministry of Fuel and Power for the Press and the BBC to issue a Fuel Warning when he considered consumption of electricity likely to reach a dangerously high level. No reference to the weather was included in these warnings. Security considerations ruled out anything that might give the enemy information about current weather.

Arrangements similar to those made with the Central Electricity Board were made with gas supply companies. An additional factor which particularly affected gas supplies was a ban on the use of central heating in certain classes of premises in the interests of fuel economy. In the setting up of this ban, which came into force in the autumn of 1942, the assistance and advice of the Office was central. Implementation arrangements were reviewed in February 1943 to consider how to take into account short-term temperature variations, which were especially a problem in spring and early autumn. Warm spells could occur in March and April and cold spells in May and September. The Office supplied climatological information which helped the Ministry of Fuel and Power decide annual dates when the use of central heating would be banned, and the Office also supplied weather forecasts which allowed the ban to be implemented flexibly. Permission to use central heating during cold spells could, for example, be granted if the weather forecast indicated that such a course of action was called for.

The Ministry of Transport informed the Air Ministry in 1938 that they could dispense with weather forecasts in wartime! However, the winter of 1939–1940 turned out to be severe in many parts of the British Isles. Transport services were often disrupted. As early as November 1939, the Southern Railway sought the help of the Office, and it soon became clear that other railway companies, as well as bus companies and local authorities, needed meteorological advice. At the end of January 1940, the Transport Priority Committee of the Ministry of War Transport asked for help. The weather events of the last week of that month were exceptional, with gales and deep snow in the south-east of England and, at the same time, thick glazed frost from Hampshire to North Wales and in the south-west of England. Traffic was paralysed. The outcome was that arrangements were made for the general weather inference to be telephoned from the Office to the Ministry of War Transport each morning at about 9.30 a.m. and for weather reports and district weather forecasts to be sent to the Ministry and to the Railway Executive Committee.

Another severe winter in 1940–1941 stretched to the full the resources of local authorities, which were the bodies principally responsible for maintaining safe conditions on roads. Again a code system was devised, this one consisting of:

'Steel', meaning snow expected, becoming icy with traffic.
'Copper', meaning night frost following thaw or rain, producing patches of ice.
'Gold', meaning thaw expected, probably only temporary.

From 18 December 1941, snow warnings in code were telephoned to Divisional Road Engineers, Road Liaison Officers and County and County Borough Surveyors, a total of 260 individuals. The warnings were broadcast on the meteorological teleprinter network and then passed to authorized recipients from the most conveniently situated meteorological station in each of the areas affected. As the war progressed, the system of warnings was modified from time to time. In February 1942, for example, it was decided that the headquarters of the London Civil Defence Region would distribute warnings on their telephone network to most of the recipients in the London area. This saved the Office much time that would otherwise have been spent in making individual telephone calls.

Internal Reorganization

As the war progressed, changes in the administration and organization of the Office became necessary. The RAF's demands for meteorological services increased greatly, and the Office's synoptic work became ever more important and crucial. In consequence, Gold was upgraded to the status of Deputy Director in May 1940. A special section for Investigations was created in July 1940 to respond to the many inquiries from the Army and the RAF for technical information or advice, and special provision for meteorological work for the Ministry of Home Security was made. Two new posts of Assistant Director were established in November 1941, and a new section was formed on 1 March 1942, to be responsible for all questions of security and the codes and ciphers used in the Office.

Another reorganization of the Office took effect on 1 August 1942, after which:

- The divisions in London were those concerned with personnel, training, general nontechnical services, and the meteorological requirements of the Army and military and civil aviation, together with two divisions which were directly under Johnson and Gold, viz. Investigations, and Security, codes and ciphers.
- The divisions at Dunstable were those concerned with weather forecasting and analysis and the collection and distribution of synoptic information.
- The divisions at Stonehouse were the Marine Division, the Climatology and Library Division and the Instruments and Equipment Division.

A further partial reorganization occurred in November 1943, when a new branch was established at Dunstable with responsibility for obtaining upper-air observations and for developing upper-air observational techniques and new methods of analysis.

The headquarters staff of the Office numbered 160 in September 1939. By 1945, it numbered 490, of whom 120 were in London, 190 at Dunstable and 180 at Stonehouse.

The work at Stonehouse was in part a continuation of that carried out before the war, but it was also, to a much greater degree, a response to wartime demands. The Climatology and Library Division continued to collect and tabulate British climatological data and to give advice to specialist bodies. However, publication of *British Rainfall* and the *Monthly Weather Report* was delayed until the war was over, and, to save paper and manpower, publication of the *Meteorological Magazine* ceased after the fall of France in May 1940.[29] The resources of the Library were much in demand for investigative and other purposes. Loans increased from 3515 in the year ending 31 March 1940 to more than 37,000 in each of the last three years of the war.

---

[29] The last printed issue of the *Meteorological Magazine* was that for June 1940. However, a typescript substitute for this publication was produced throughout the war, complete with diagrams and photographs. Copies can be viewed in the National Meteorological Library at Exeter. Publication of the *Meteorological Magazine* on a monthly basis commenced with the issue for January 1947. Publication of *Marine Observer* was also suspended during the war. The issue for October 1939 was the last until July 1947.

An important part of the work of the Climatology and Library Division was the production of a series of handbooks on the weather over the oceans. Several of the handbooks were completed before the war, and others were completed during the war, with priority given to handbooks for the main theatres of war.

The main work of the Marine Division during the war was the preparation of climatological, ocean current and sea-ice atlases, which were needed by the Admiralty for operational purposes. At the same time, efforts were made to maintain the Voluntary Observing Fleet of the Merchant Navy and recruit ships to replace those lost in the war. In 1939, the number of 'Regular Observing Ships' was 360. These were ships of the British merchant fleet equipped with certified instruments loaned by the Office, with officers of these ships making observations which were broadcast to meteorological services for weather forecasting purposes and also to other ships. In addition, there were 600 'Supplementary Ships', which recorded observations but transmitted them in an abridged form.

Though the Office could not obtain data regularly from merchant vessels during the war, Port Meteorological Officers and Merchant Navy agents maintained contact with ships of the Voluntary Observing Fleet. They continued to visit ships whenever possible and endeavoured to recruit ships. They also replenished the meteorological equipment aboard the ships, the aim being to restart the service provided by the Fleet as soon as possible after the war ended. Despite heavy losses of British ships during the war and enemy raids on ports, the number of Regular Observing Ships in May 1945 was 272 and the number of Supplementary Ships 208.

The Instruments Division supplied not only the Office but also the Naval Meteorological Service and (in the later stages of the war) the Dominions and Allied Services. They also developed new equipment and instruments. Total annual expenditure on meteorological equipment increased from an average of £7800 in the financial years 1929–1933 to £37,000 in the year 1939–1940 and £225,000 in the year 1944–1945. Only then did it fall back a little, to £212,000 in 1945–1946. Items which were supplied in great numbers during the war included radiosondes (30,000), pilot and sounding balloons (2,000,000), thermometers (30,000) and aneroid barometers (18,000). Difficulties were sometimes experienced: factories of suppliers were bombed, consignments of stores were destroyed whilst awaiting shipment at docks and supplies were lost as a result of enemy action at sea.

When the war in Europe ended, in May 1945, plans were made for the staff and students of Wycliffe College to return to their premises at Stonehouse from their wartime quarters at Lampeter in Wales. However, the Office's divisions at Stonehouse could not return to their premises in South Kensington and central London. They had grown so much during the war that a new home had to be found for them. A building adjoining His Majesty's Stationery Office at Harrow was chosen, and they moved into it in the last week of August 1945.

Wartime Experiences and Reminiscences

Glimpses of life at Stonehouse by several who were stationed there have been included in an article by R P W Lewis on 'The Meteorological Office at Stonehouse 1939–45', published in 1986 in the *Meteorological Magazine* (Vol. 115, pp. 18–31). In Lewis's words: "the abiding impression produced either by reading reminiscences of the staff who worked at Stonehouse or by talking to them is that their wartime 'exile' was, on the whole, a happy one". They worked hard but lived a full and varied life, making their own entertainment and raising funds from sporting events, concerts and other activities. The senior staff contributed greatly to the welfare and camaraderie, none more so than Goldie and his wife.

Elaine Austin, Sir Napier Shaw's faithful assistant, recalled that, when air-raids began, she and colleagues took it in turns to carry out fire-watching duties at night. When not on duty, she returned to her lodgings, where her hosts provided bed, breakfast and evening meal, plus dinner on Sundays, all paid for by the Government. Like her colleagues, she took her main meal in the Office canteen. She further recalled that a unit of the Home Guard was formed when invasion was expected and she was given training in how to throw a grenade![30] Her work was concerned mainly with the production of the handbooks on the weather over the oceans, and her duties took her to London at least once a fortnight to visit the Investigations Branch and the Naval Meteorological Service. She mentioned that Commander Hennessy organized the move from London to Stonehouse and also found billets for all members of staff who did not wish to find their own.

Oliver Ashford served at Stonehouse from June 1943 to August 1945, in charge of the section of the Instruments Division responsible for the design and development of new and improved instruments. He had been Superintendent at Lerwick Observatory from 1937 to 1943. In his reminiscences of life at Stonehouse, he remembered his 'office' being an enormous science laboratory, big enough for at least ten normal offices. Attempts were made by other branches to acquire some of the space, but he won the day by arguing that the whole laboratory was needed for displaying a vast collection of meteorological instruments. In fact, many of these instruments had no current interest, and several were subsequently put on display in the Office's museum! Ashford also mentioned that he had to wait for furnished accommodation to be allocated and in the meantime lived in lodgings whilst his family remained in Scotland. Such were the deprivations of wartime.

Life in the Office's headquarters at Victory House in central London during the war was quite a contrast to the relative tranquillity of Gloucestershire. This is illustrated by the reminiscences of Mary Buchanan, published in *Weather* in May 1994 (Vol. 49, pp. 181–185). She joined the Office in March 1940, as an Assistant, to work for Ernest

---

[30] Miss Austin could not join the Home Guard, as it was an all-male group.

Gold. Somewhat unpromisingly, she was told the girl she was replacing had left with a nervous breakdown brought on by Gold shouting at her. To Mary's surprise, though, she found on her first encounter with him that he was "a little old man" who peered at her "over half-moon glasses perched half-way down a long thin nose" (see Figure 7.2). She went on to say that he "carried the largest share of all decisions and had a character no-one could ignore", adding that he could be kind, bad-tempered, petty, infuriating or childish and that some were afraid of him, but, so far as she knew, no one disliked him.

Mary said that Gold's work load was enormous. He personally wrote all Synoptic, Special Synoptic and Top Secret Instructions and had a succession of visitors, notably the Director, the Assistant Directors, staff about to go abroad, staff returning from duties abroad and, indeed, many others visiting Victory House, including Americans and Russians. He arrived at work around 9.30 each morning and never left before 8 p.m. As regards his sartorial arrangements whilst commuting, Mary had this to say:

He was allowed a petrol ration so he used to drive to work from Hampstead Garden Suburb in an ancient open Alvis which had a petrol tin on the running board and the bonnet done up with a leather strap. He wore a balaclava helmet with a bowler hat on top, wristlets, mittens and gloves. One morning, on hearing that the anemometer on the roof was sticking, he climbed the mast wearing all his driving outfit!

Mary mentioned that Gold loved to invent code words, among them 'baratic' and 'okta', these being abbreviations, respectively, for 'barometric pressure distribution over the Atlantic' and 'eighths of sky obscured by cloud'. On more than one occasion, however, he used real words and thereby contravened the regulations of Air Ministry Intelligence. One morning, Mary said, the telephone rang and she was asked if "peas 1800, lentils 0200" meant anything to her. When she replied that this was a fog warning, the officer making the call was furious, informing her that no one could use code words which had not been issued officially. She was given to understand the warning had been broadcast by mistake on the operational teleprinter network, and Groups had been telephoning to ask what it meant. The outcome was that Gold was given a list of words which were permitted. She was much surprised in 1951 when her husband, then based at Aldergrove in Northern Ireland, talked about 'doing a Bismuth', a code word she herself had chosen during the war for a Meteorological Reconnaissance Flight which took place over the Atlantic north and north-west of Ireland. Despite the occasional lapses when he used real words, Gold was, Mary said, very security conscious, and with good reason. He handled many documents marked 'Top Secret' and personally made sure they were in due course completely destroyed.

Gold and Johnson were the only people allowed to have their own shorthand typists. Everyone else had to send for someone from the typing pool when they wanted to dictate. This could have its funny side. Mary remembered Stagg finding one of his papers bore the title "Meteorology for parrot shooting" (instead of parachuting).

Perhaps it was not surprising that most of the senior staff preferred to write, rather than dictate!

Mary recalled that she did not have much to do with the Director, except when Gold was away. She commented that Johnson was a more straightforward man to deal with and always kind and considerate. She said he virtually lived in Victory House, sleeping on a bed in his office and going home for three days every second weekend. His wife was evacuated to Minehead in Somerset during the war while he remained in London. Mary spent many a night in Victory House, too, sleeping in a basement shelter. Being a qualified First Aider, she joined the Victory House Air Raid Precautions Group. Gold was a stubborn man who refused to go to a shelter during daytime air-raids, many of which involved flying bombs. When he was particularly busy before the Normandy landings, he asked Mary if she would mind staying in her office to help him get through the work that needed to be dealt with. She agreed and sometimes had to shelter under the knee-hole of his desk with her tin hat on! Despite a number of air-raids, the only first aid she ever had to administer was to dress a boil on an arm of the Director of Accounts.

Recollections of working at Dunstable during the war have been provided by Brian Audric, who considered the town "quite a pleasant place to be".[31] As at Stonehouse and Victory House, a Meteorological Office unit of the Home Guard was formed after the fall of France in 1940, and most of the male staff joined it. The Office emerged from the war unscathed, so the alternative CFO which had been built at Monks Risborough in Buckinghamshire was never required. The only significant damage to the CFO's buildings at Dunstable during the war was caused by staff carelessness. It occurred, Audric reported, when a discarded match set fire to a basket of paper which in turn set light to the ceiling above. The local fire brigade attended, and repairs to the building were needed. On another occasion, an assistant fell asleep in a leatherette-covered armchair whilst smoking. This time, though, local resources proved sufficient. The chair was taken outside and doused.

When Audric arrived at the CFO (on 1 April 1940), he was surprised to find that no hourly observations of the weather were made. He thought there may have been a barometer and a barograph, but there were no instruments outside. His recollection was that the only forecaster who showed any interest in the local weather was C K M Douglas. Quite late on in the war, Audric said, a meteorological enclosure with a rain gauge and Stevenson screen was created, and a pressure-tube anemometer was set up at a suitable distance from the buildings. Audric remained at Dunstable until the autumn of 1947 and then left the Office.[32] His work at Dunstable from 1940

---

[31] 'The Meteorological Office Dunstable and the IDA Unit in World War II', by Brian Audric, published in 2000 by the Royal Meteorological Society (*Occasional Papers on Meteorological History*, No. 2, 17 pp.). See also 'The Meteorological Office in Dunstable during World War II', by Omer Roucoux (*Weather*, 2001, Vol. 56, pp. 28–31).

[32] Audric was employed for the remaining 33 years of his career in the laboratories of the National Chemical Laboratory and the National Physical Laboratory at Teddington, Middlesex.

to late 1943 was crucial to the war effort and remained top secret for several decades after the war.[33]

Accounts of wartime meteorological activities at outstations at home and abroad can be found in a number of publications, but articles which contain reminiscences of meteorologists in the field in war zones are few in number. One such was written by Gren Neilson, who told in an article published in *Weather* in 1992 (Vol. 47, pp. 430–435) of his experiences when in action in North Africa and (from November 1943) Italy. His work consisted mainly of making pilot-balloon ascents to provide information about the upper air for the artillery and was generally routine in nature. There was, however, a break from routine on 15 December 1943, when two senior members of the Office visited his unit, Group Captain Stagg and Wing Commander Meade.[34] Stagg had by then been designated Chief Meteorological Officer for Operation Overlord and was making familiarization visits to units in the field. Neilson commented in his article that AP1134 contains no reference to the presence of a meteorological unit at Anzio and wondered whether any meteorological personnel other than himself and a fellow aircraftman were on the beach head.

A graphic account of wartime meteorological work has been supplied by Ken Anderson.[35] It is based on his personal diary and tells the story of his experiences in central and western Belgium and the Nord-Pas-de-Calais in May 1940 whilst close to the front line and during the retreat to Dunkirk. In the most perilous circumstances, Anderson and other members of a special Weather Unit made meteorological observations, including pilot-balloon ascents, and succeeded in passing them by telephone to the headquarters of the 2nd Survey Regiment, Royal Artillery. For bravery, defiance, resourcefulness and disregard for personal safety, this story of meteorological activities in wartime has no equal.

The wartime experiences of Morris Albert Oliver and John Oswald Weston were also remarkable. Oliver was a member of the crew of a bomber which had to be abandoned over Belgium on 3 June 1940 when an engine caught fire. He was captured and taken to Stalag Luft III. Weston joined the Meteorological Branch of the RAFVR in August 1939 and led an adventurous life as a meteorologist (mainly in the Sudan and Egypt) before being captured at Sternes on Crete on 29 May 1941. From a Prisoner of War camp on that island, he was moved to Stalag Luft VIII B in October 1941 and Stalag Luft III in May 1942.

---

[33] Wartime recollections of the Office at Dunstable have also been provided by a Norwegian, Per Sundt, a clerk with an insurance company who escaped to Shetland from Bergen in August 1941. He was trained as a meteorologist in the Office's Training School at Gloucester and posted from there to Dunstable in May 1942. See 'A Norwegian at ETA, 1942–1945', by J C P Sundt and B J Booth, published in *Weather* in 2010 (Vol. 65, pp. 160–165).

[34] Meade referred to this visit in 'Operation TORCH and follow-up operations: meteorological aspects', published in the conference compilation *Meteorology and World War II* (*op. cit.*, this chapter, p. 8).

[35] 'Weather services at war', by K D Anderson, published in 2009 by the Royal Meteorological Society (*Occasional Papers on Meteorological History*, No. 7, 17 pp.). There are copies of this in the National Meteorological Library, and the paper is available online on the Royal Meteorological Society's website.

Oliver and Weston began a meteorological collaboration soon after the latter's arrival at Stalag Luft III. On Weston's initiative, they set up a weather station and taught fellow prisoners meteorology. Permission was granted for the station to be set up, but no instruments were forthcoming. However, a couple of fellow prisoners constructed for them a thermometer screen to Meteorological Office specification, and another made a rain gauge out of Red Cross food tins. Others made a nephoscope, an aneroid barometer and an anemometer, though it hardly needs to be said that none of these instruments met the Office's rigorous calibration standards![36] Bribery of guards helped Weston and Oliver obtain thermometers, and they acquired through the British Red Cross Brunt's *Physical and Dynamical Meteorology* and two of the Office's publications, *A Short Course in Elementary Meteorology* by W H Pick and *Meteorology for Aviators* by R C Sutcliffe.[37] Oliver used these as textbooks for the course on general meteorology which he taught to some fifty fellow prisoners, and Weston used the *Admiralty Weather Manual* when teaching synoptic meteorology. Moreover, the two of them corresponded with J S Dines about meteorological matters. Weston and Oliver were transferred to Stalag Luft VI at the end of June 1943 and continued their meteorological activities there, still using equipment made at Stalag Luft III.[38]

Forecasts and Observations

A fundamental requirement of weather forecasting is a reliable knowledge of the current state of the atmosphere. A wartime priority was, therefore, to obtain observations from far and wide by whatever means possible. The importance of doing so was paramount. Weather forecasting was a vital part of the war effort at home and abroad.

Shortly before the war, there were forty-nine principal observing stations in the UK and seven in Ireland, these being stations which made weather observations according to criteria agreed by the IMO. The observations were supplied on a regular basis to the CFO at Dunstable for use in the construction of the charts from which forecasts were produced. The number of these stations increased steadily as new operational airfields were opened until, by the end of the war, there were eighty-nine such stations in the UK and eight in Ireland. In addition, there were dozens of auxiliary stations in the UK, these being rainfall and observing stations maintained by private individuals, municipal authorities and others who made monthly returns of data to

[36] A nephoscope (from the Greek *nephos*, meaning cloud) is an instrument used for ascertaining the directions and speeds of cloud movements.

[37] William Henry Pick served in the Meteorological Section of the Royal Engineers in the Great War and joined the civilian staff of the Meteorological Office in 1920. He was posted to RAF Cranwell and published *A Short Course in Elementary Meteorology* the following year. This book was a basic textbook for meteorology students for many years and was updated and reprinted many times.

[38] For details of the work of Oliver and Weston, see 'Meteorology behind the wire' by J O Weston, published in the *Quarterly Journal of the Royal Meteorological Society* in 1945 (Vol. 71, pp. 424–426) and 'Meteorological research!' by M A Oliver, published in *Weather* in 1947 (Vol. 2, pp. 226–230).

the Climatological Branch of the Office. Though the observations from these stations were not generally of sufficient quality to be used by forecasters, they could prove useful for chart-drawing purposes and for providing assistance in meeting aviation requirements.

Meteorological reports from land stations outside the British Isles were much less plentiful during the war than before. They were classified in both Allied and Axis countries, and severe security restrictions were placed on the broadcasting of them *en clair* by radio. As soon as the war began, the flow of observations from the parts of Europe controlled by the enemy dwindled almost to nothing, and the availability of observations from other parts of the world decreased progressively as countries were occupied by forces of the Axis powers. Thus the flow of observations from France, Norway, Denmark and the Low Countries was reduced to a trickle in 1940 as these countries succumbed to the Germans. However, surface and upper-air observations from Spain, Portugal, the Azores, Iceland, Canada and the United States continued to reach Dunstable throughout the war. They were mostly transmitted by cable and distributed within Britain by means of the teleprinter network. Observations continued to be received from the Republic of Ireland, too. In the early part of the war, normal Post Office lines were used. Then, in 1941, a teleprinter link was established between Dunstable and the offices of the Irish Meteorological Service at Collinstown near Dublin.[39]

There was concern for a while in 1939 and 1940 about the possibility of British weather observations leaking to Germany through Ireland, as documents in the British National Archives at Kew show (BJ5/83). Concern was expressed by Gold in a letter to the Dominions Office dated 14 November 1939 and by Johnson in a memorandum to the Director of Intelligence dated 20 November 1939. However, an analysis of tapped telephone calls allayed fears. An investigation by the Postal and Telegraph Censorship authorities confirmed that ordinary telephone calls between Ireland and the UK contained "a relatively high proportion of references to weather conditions", and calls which contained such references were analysed to ascertain whether the references occurred regularly on any particular call numbers. But it was decided in the end that fears appeared to be groundless.

Meteorological charts plotted at Dunstable during the war show that observations from Germany, the Atlantic Ocean and countries occupied by Axis forces did indeed reach forecasters in Britain, despite the imposition of security restrictions by the enemy. The full story of how these observations came to be available has yet to be told, but much of it is now known. Clandestine meteorological services operated in Poland and Belgium, for example, and observations were received from agents in other

---

[39] Arrangements for obtaining observations from neutral and allied countries varied from country to country. For details, see AP1134, pp. 459–469. For information about the supply of meteorological observations from Ireland during the war, see 'Ireland, the Irish Meteorological Service and The Emergency', by D De Cogan and J A Kington, published in *Weather* in 2001 (Vol. 56, pp. 387–397).

occupied countries.[40] More importantly, though, observations from Germany became available through the skilful work of the meteorological section of the Government Code and Cypher School at Bletchley Park. In the words of R P W Lewis, "British Intelligence was able to supply the Central Forecasting Office at Dunstable with information of such good quality, so fast, that surface and upper-air charts could be plotted, covering the whole of Axis-occupied Europe".[41]

The crucial work of a special unit called IDA, which carried out meteorological decoding work at Dunstable, has been described by Brian Audric, a member of the unit from April 1940 to late 1943.[42] As he showed, the story of how the decoding was carried out was a remarkable one, involving cryptographers and meteorologists at Bletchley Park working closely with senior staff of the Office at Dunstable.

Given that weather systems tend to approach the British Isles from the west, the imposition of radio silence on the merchant ships of the British and their allies during the war created a serious problem for forecasters. They were deprived of weather reports from the Atlantic and other oceans, especially the eastern Atlantic, a crucial area for those who were forecasting for the British Isles. A French weather ship, *Carimaré*, was withdrawn when the war began. Since the spring of 1938, this ship had been stationed 800 miles west of the Azores, providing surface and upper-air observations routinely from an area not much visited by merchant ships.[43]

Questions remain to this day over sources of data used at Dunstable for plotting Atlantic weather charts. On some of these charts, analyses of isobars are surprisingly detailed, but no observations are shown. On other charts which show considerable isobaric detail, a few observations are shown, but not enough to justify the detail. Meteorologists at Dunstable needed to ensure that the enemy did not begin to suspect their codes had been broken, so it was necessary to be cautious over the plotting of observations on charts, which was facilitated by the use of transparent overlays to divorce observations from analyses. It is known that observations came from German meteorological reconnaissance aircraft and from American weather ships, and it is possible that some came from German submarines, too, but there appear to have been other sources of data not yet identified.

A lack of weather observations from the Atlantic was anticipated before the war began, and thought was given by the Office to possible ways and means of obtaining

---

[40] For information about clandestine meteorological activities in occupied countries, see two articles published in *Weather*: 'Polish meteorology in the Second World War', by Z Bartkowski (1992, Vol. 47, pp. 123–126), and 'Meteorology in Belgium, 1940–44', by R J Ogden (1993, Vol. 48, pp. 265–267).

[41] See 'The use by the Meteorological Office of deciphered German meteorological data during the Second World War', by R P W Lewis, published in 1985 in the *Meteorological Magazine* (Vol. 114, pp. 113–118).

[42] See *The Meteorological Office Dunstable and the IDA Unit in World War II* by Brian Audric (*op. cit.*, footnote 31). See also '[George McVittie's] War work, 1940–1945', by E Knighting, published in 1990 in *Vistas in Astronomy* (Vol. 33, pp. 59–62). This article mentions that the Bletchley group started to break the Japanese meteorological cipher in 1943.

[43] See 'Ship weather stations solve secrets of ocean blanks', an anonymous report published in November 1938 in *Popular Mechanics* (Vol. 70, No. 5, pp. 669–670).

the much-needed information. The use of meteorological reconnaissance flights was considered, but an acute shortage of aircraft and aircrew made such an eventuality impracticable. Attempts to persuade the Admiralty to permit the transmission of weather observations from a certain number of ships failed, as a consequence of which a joint Anglo-French meeting was held at the Office Nationale Météorologique on 4 October 1939 to address the matter. The outcome was a recommendation that weather reports from the Atlantic area might be provided by fast armed trawlers or submarines.

The Marine Division of the Office acted quickly, submitting proposals within a few days, the essence of them being that two large trawlers be bought or chartered, one of them to be on patrol in a selected area, the other on stand-by, both of them offensively armed. The observation area would be bounded by latitudes 45°N and 55°N and longitudes 15°W and 35°W, with the trawler sending reports by wireless telegraphy at agreed intervals and patrolling at full speed in this area in order to minimize the risk of interference by enemy vessels. Each trawler would spend nineteen days at sea and nine days in port, the nineteen days made up of five days on passage to and from the observing area and fourteen on patrol within the designated area.

Nelson Johnson put these proposals to the Admiralty on 14 October 1939, along with a French proposal to use submarines. However, the Admiralty did not consider the proposals justifiable and expressed themselves in favour of using aircraft, whereupon Johnson pointed out to the Director of the Naval Meteorological Service (Captain Garbett) that although simple meteorological observations were available from operational aircraft they came from areas relatively close to land. Regular reports from a weather ship were essential.

The matter rested there until 4 March 1940, when the Air Council suggested to the Admiralty that two oceanographic research vessels then in the Port of London, *Discovery II* and *William Scoresby*, be used as weather ships. The Admiralty replied that the ships were not available for the proposed duties but did concede that weather ships were desirable and promised to review the situation later. The Air Council pressed the matter of weather ships with the Admiralty on 12 July 1940, and planning began on 22 July at a meeting attended by Johnson and Garbett. A number of decisions were made at this meeting, the main ones being the speed, range and armament of the ships; the personnel on board; the area to be patrolled; the meteorological equipment needed; the timing of observations; and radio transmission arrangements. It was also agreed that the ships would have a dual role as weather ships and U-boat hunters. However, many of the decisions were flatly rejected by the Admiralty, and permission for the ships to be heavily armed and sail under the White Ensign was refused. Moreover, in a complete volte-face, the Admiralty insisted that the weather ships should be merchant vessels chartered on a contract basis.

After further discussions between the Air Council, the Office, the Admiralty and the Naval Meteorological Service, it was agreed, during August 1940, that the weather ships would be defensively armed, not offensively. Two vessels were chartered by the Admiralty at the beginning of September 1940 to serve as weather reporting ships, the SS *Arakaka* and the SS *Toronto City*. Each was armed with a four-inch gun in the stern, two Hotchkiss machine guns, one anti-aircraft kite and four anti-aircraft rockets. Gold agreed on the telephone with Garbett that the code name for these ships would be 'Panther', the name selected by Mary Buchanan.

The *Arakaka* set out on her first voyage on 16 September 1940, and the *Toronto City* on hers on 3 October 1940. The former completed five voyages, the latter six, supplying by radio detailed weather reports which were distributed only to the principal meteorological offices which received decrypted German reports. In addition, detailed scientific reports were submitted after each voyage by staff of the Office who were on board, covering the exposure and efficiency of meteorological instruments, upper-air observations, the measurement of sea temperature, wind determination from state of sea, and the measurement of sea and swell. The authors of these reports were Sidney Portass (attached to the *Arakaka*) and Stanley Proud (attached to the *Toronto City*). Both ships were sunk by German submarines, the *Arakaka* by U-77 on 22 June 1941 at 47°00'N 41°40'W, and the *Toronto City* by U-108 on 1 July 1941 at 46°56'N 30°26'W. Sadly, both men lost their lives.[44]

Meanwhile, on 10 February 1940, the Americans, not yet involved in the war, had established two ocean weather stations, one about 600 miles northeast of Bermuda, the other about 800 miles southwest of the Azores. The primary function of these stations was initially to provide weather reports and beacons for trans-Atlantic civil aviation, but a third was added later that year 500 miles north-east of Newfoundland to act as a guard and weather ship for the benefit of aircraft flying direct to the UK. In July 1942, to serve the growing needs of military aviation flying between Labrador, Newfoundland and the UK via Greenland and Iceland, two more stations were established, one between Labrador and Greenland, the other between Iceland and southern Greenland. Three more stations were established in mid-Atlantic by the Americans during 1944, the last becoming operational just before D-Day.

At the same time, the Royal Navy deployed two frigates to act as temporary weather ships between 20°W and 25°W in latitudes 42°N to 59°N, but when these were withdrawn at the end of June 1944 they were replaced by a single ship operating between 20°W and 25°W in latitudes 49°N to 53°N. By the time the war in Europe ended, in May 1945, sixteen ocean weather stations had been established by the Americans on the Atlantic north of latitude 15 N, their *raison d'être* being to supply weather observations and provide air-sea rescue facilities. Thus the observational coverage of the North Atlantic Ocean improved considerably by the end of the war.

---

[44] For an outline of the story of the UK's meteorological reporting ships and a review of the observational work and scientific investigations carried out aboard the *Arakaka* and the *Toronto City*, see AP1134, pp. 434–439, and Appendix 20. See also 'The ill-fated first UK weather ships', by R J Ogden, published in *Weather* in 1995 (Vol. 50, pp. 381–384). The reports submitted by Portass and Proud are preserved in the National Archives.

Observations from the American vessels were available to forecasters at Dunstable both before and after the United States entered the war.

Though weather observations from the Azores, Greenland, Iceland and a few ships on the Atlantic reached Dunstable during the early part of the war, there were too few of them to provide the observational coverage of the eastern North Atlantic desired by forecasters. The Office pressed in 1939 for the establishment of special reconnaissance flights to supplement the Meteorological Flights which had been based at Mildenhall and Aldergrove since 1936. However, nothing came of the Office's representations because of a shortage of aircraft and aircrew. Then, in the summer of 1940, RAF Bomber Command voiced great concern about the accuracy of weather forecasts for flights over the North Sea and northern Germany. Furthermore, forecasts of Atlantic weather were required by the RAF and the Royal Navy.

The Office repeatedly emphasized the need for observations over the Atlantic, North Sea and Germany, and their case was strengthened when Intelligence revealed that the Germans were making meteorological reconnaissance flights over the seas surrounding the British Isles. Eventually, action was taken. The Expansion and Re-equipment Policy Committee of Coastal Command's Liaison Section decided on 4 October 1940 that three meteorological flights would be formed, to make observations over the sea. Two flights would make observations over the North Atlantic, the third over the North Sea. The bases of the flights would be at Bircham Newton (Norfolk), St Eval (Cornwall) and Aldergrove (Northern Ireland). Blenheim aircraft would be used.

In November 1940, Johnson asked a Czech meteorologist, Eric Kraus, to establish and organize the flights. Kraus had been a graduate student of Jacob Bjerknes at Bergen before the war and joined the Meteorological Branch of the RAFVR in the summer of 1940. In his own words:[45]

I was asked to recommend an operational procedure and a data reporting code for Gold's approval. I suggested regular flight tracks over the Atlantic in a north-westerly and south-westerly direction. The North Sea was added to this on Gold's insistence. My suggested plan of operations was to fly out at a low level to a distance to be determined by the range of the aircraft, then make a spiral ascent to the 500-millibar level [about 5500 metres above sea level] and then return on a glide path to save fuel and stretch out the aircraft endurance. All this was rather readily accepted, but the proposed code turned out to be much more controversial. In his time, Gold had devised innumerable codes and standards for the international meteorological community. He considered himself – rightly – an expert in this field. He was also a passionate adherent of the imperial as distinct from the metric system. We fought heatedly over trivia, until I felt that it was unbecoming for a young man to be so dogmatic with this old, experienced gentleman. I then proceeded to accept everything he said, but that too was the wrong policy. Gold loved to argue.

---

[45] 'How the meteorological reconnaissance flights began', by E B Kraus, published in 1985 in the *Meteorological Magazine* (Vol. 114, pp. 24–30).

Kraus's first posting was to Aldergrove, where, on his arrival in January 1941, he found there were the expected Blenheims, and ground crews, too, but no aircrews. To while away time, he flew convoy missions with a local Anson squadron and borrowed some of their pilots during their spare time to calibrate meteorological instruments on the Blenheims. He also trained as an air navigator, after which, at the end of February 1941, he was posted to Bircham Newton, carrying out from there, on 14 March, the first operational meteorological reconnaissance flight. At the end of the following month, he transferred to St Eval and flew meteorological sorties over the Bay of Biscay, on one occasion diving down with the Blenheim's machine gun blazing to attack a U-boat which turned out to be a whale! He continued to play a key part in the establishment and expansion of the reconnaissance flight network and was directed, in November 1941, to set up a flight based in Iceland (a country occupied by British forces since April 1940). Until the autumn of 1941, he was able to act on his own initiative to establish and organize flights. Then, he was placed formally under the command of a Coastal Command Wing Commander.

As the war progressed, the effort devoted to meteorological reconnaissance increased. By the latter part of 1944, there were nine flights, with one south-westwards out of Iceland (Reykjavik), one westwards out of Gibraltar and the rest over the waters around the British Isles from bases at St Eval in Cornwall, Brawdy in Pembrokeshire, Tiree in the Hebrides, Wick on the north coast of Scotland, and Docking in Norfolk, as well as Aldergrove and Bircham Newton. Each flight had a code name (Nocturnal, Epicure, Sharon, Allah, Rhombus, Mercer, Bismuth, Recipe and Magnum), all selected by Mary Buchanan, and the flights were made on fixed courses which could, nevertheless, be altered to suit operational requirements.

For the most part, flights were made at an altitude of about 1800 feet, and observations were made every fifty nautical miles (i.e., roughly every twenty minutes). The observations included temperature, humidity, cloud type and amount, height of cloud base, turbulence, icing and visibility, while wind direction and speed were computed by the aircraft's navigator. Ascents to the 500-millibar level were also made (to obtain an upper-air temperature sounding), and descents to near sea level were also made (to check surface barometric pressure and reset the altimeter). Observations were initially made by navigators, because trained meteorologists were not available, but in September 1942 the Meteorological Air Observer Section of the General Duties Branch of the RAFVR was formed. Staff of the Office were recruited and, from the spring of 1943, trained in the technique of making observations from aircraft. The first officers were posted to meteorological squadrons and flights in June 1943, followed by non-commissioned officers soon afterwards.[46]

---

[46] For details of British meteorological reconnaissance flights, see *Even the Birds Were Walking: the Story of Wartime Meteorological Reconnaissance*, by J A Kington and P G Rackliff, published in 2000 by Tempus Publishing Ltd (224 pp.). See also AP1134 (pp. 446–450), 'Meteorological reconnaissance flights', by R J Ogden, published in the *Meteorological Magazine* in 1985 (Vol. 114, pp. 108–113), and 'Wartime long-range meteorological flights',

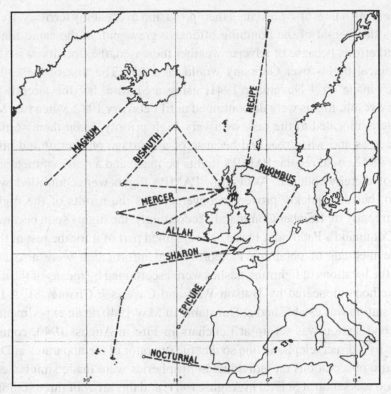

Figure 10.7. Meteorological reconnaissance flights, September 1944. © Crown Copyright 2010, the Met Office.

Other meteorological flights were made during the war, notably the ascents to a height of about 30,000 feet to obtain upper-air temperature and humidity data known as THUM flights. They continued to be made at Mildenhall and Aldergrove during the early part of the war, but the Mildenhall flights were transferred to Bircham Newton in the autumn of 1941 and the Aldergrove flights to Ballyhalbert (County Down) in December 1944. THUM ascents at Gibraltar began in May 1942 and in the Azores in January 1944. Pressure and temperature ascents known as PRATA flights, to heights of about 12,000 metres, were also made, at Wick, Bircham Newton, Aldergrove and Manston, using Spitfires. Besides supplementing radiosonde observations, these PRATA flights provided high-altitude information about ice formation on aircraft and the formation of condensation trails.

The need for meteorological reconnaissance over enemy territory was recognized as early as 1938. However, the shortage of aircraft and trained aircrews early on in the war which prevented the provision of reconnaissance flights over the waters

by J B Yates, published in *Weather* in 1986 (Vol. 41, pp. 61–63). For details of German reconnaissance flights, see *Wekusta: Luftwaffe Meteorological Reconnaissance Units and Operations 1938–1945*, by J A Kington and F Selinger, published in 2006 by Flight Recorder Publications Ltd (256 pp.).

around the British Isles also prevented their provision over enemy territory. Eventually, though, as the weight of the bombing offensive grew and, at the same time, losses and wasted efforts because of adverse weather increased, the decision was taken that meteorological flights over Germany would be made. The first of these 'PAMPA' flights was made on 7 November 1941, using a Spitfire, but the aircraft failed to return. As a result, flights were discontinued until February 1942, when two Mosquito aircraft were allocated to the task with very high priority. Even then, sorties could not be made as and when required because of a shortage of aircraft and pilots. Not until May 1942 could regular PAMPA flights be made and a range of meteorological information thereby obtained. At first, the PAMPA flights were controlled by Coastal Command, but this quickly proved inconvenient, as the results of the flights were required urgently by Bomber Command. Accordingly, the flights soon became part of Bomber Command's Pathfinder Force and remained part of it for the rest of the war.[47]

Remote methods of obtaining meteorological information were also used. For example, the locations of lightning flashes were ascertained by means of the direction-finding methods pioneered by Watson Watt and Cave (see Chapter 8). A lightning detection station was established at Dunstable in May 1940 on an experimental basis, and a second station was set up at Leuchars in Fife in August 1940, connected to Dunstable by a direct telephone line so that the operator of the apparatus at Dunstable could ensure observations on individual atmospherics were made simultaneously. A third station was set up at St Eval in August 1941, and this array of three stations there-after remained operational throughout the rest of the war. The information provided by the apparatus proved extremely valuable to Bomber Command, especially for the location of fronts over the oceans and enemy territory.

Another remote method of obtaining meteorological information resulted from radar operators noticing that noise in returned echoes was associated with weather elements such as rain, sleet and snow. Radar was pioneered by Robert Watson Watt in the 1930s and employed as a military and meteorological tool during the war under conditions of great secrecy. A shower was tracked by a 10-cm radar at a range of seven kilometres off the English coast on 20 February 1941, and meteorologists quickly came to recognize on radar displays the differences between showers, thunderstorms and continuous precipitation. They also learned to identify fronts and tropical storms. Trials which began at Larkhill in 1943 showed that the directions and speeds of upper winds could be measured by tracking weather balloons which had metal-coated radar reflectors hanging from them, and the routine use of radar for this purpose began at Shoeburyness in Essex and Fazakerley near Liverpool in 1944. By the end of the

[47] For further information about PAMPA flights, see AP1134, pp. 452–454, 'Farewell to PAMPA', by R B Birchmore and R H M Vere, published in *Weather* in 1946 (Vol. 1, pp. 249–253), and 'The PAMPA flights', by R J Ogden, published in the *Meteorological Magazine* in 1992 (Vol. 121, pp. 257–259).

war, the Office had established a radar research station at East Hill, near Dunstable, equipped with a 10-cm radar.[48]

A further important meteorological advance of the war was a product of necessity. The Allies became so concerned about losses of life during landing operations on Pacific beaches that in 1942 they commissioned two oceanographers, the Norwegian Harald Ulrik Sverdrup and one of his students, the American Walter Munk, to devise a means of anticipating the dimensions of waves on beaches. By the following year, the two men had derived wave-forecasting relationships of operational value which allowed wave heights to be predicted from a knowledge of wind speed, fetch (the distance the wind has blown over the water) and duration (the length of time the wind has been blowing). Hitherto, the Beaufort scale of wind force had been used to make forecasts of waves on beaches, but with no great success. When deriving their formulae, Sverdrup and Munk developed a theoretical approach advanced in 1925 by Harold Jeffreys, who had worked for the Office from 1918 to 1922 (see Chapter 9).[49]

Accurate forecasts of sea conditions on the beaches of France prior to and during the landings on D-Day were vital to the success of the assault by the Allies. The wave-forecasting technique of Sverdrup and Munk was used to calculate sea, swell and surf conditions for the D-Day and subsequent beachhead operations. As noted earlier in this chapter, the decision to proceed with the Normandy landings during so unsettled a spell of weather was a bold one. The decision to apply a wave-forecasting model that was so new and relatively untried as that of Sverdrup and Munk was equally bold. It is testimony to the work of Stagg's team and of Sverdrup and Munk that wind, weather and wave conditions on the English Channel and beaches of Normandy were forecast so well.[50]

Peace Returns

At last, the war ended. The armed forces of Nazi Germany surrendered on 8 May 1945, and hostilities ceased in the Far East on 15 August 1945. Soon after the war in Europe ended, the redoubtable Frank Markham was on his feet again in the House of

[48] See 'A short history of radar meteorology', by R R Rogers and P L Smith, pp. 57–98 in *Historical Essays on Meteorology 1919–1995*, edited by J R Fleming and published by the American Meteorological Society in 1996 (617 pp.). See also AP1134, p. 444.

[49] Jeffreys published his work in two papers 'On the formation of water waves by wind' in the *Proceedings of the Royal Society (A)*, one in 1925 (Vol. 107, pp. 189–206), the other in 1926 (Vol. 110, pp. 241–247). The work of Sverdrup and Munk was published in 1947 in *Wind, Sea and Swell: Theory of Relations for Forecasting*, U.S. Navy Publication No. 601 (44 pp.). For an overview of the wave-forecasting work of Jeffreys, Sverdrup and Munk and others, see *The Ocean-Atmosphere System*, by A H Perry and J M Walker, published in 1977 by Longman (160 pp.).

[50] A wave-forecasting technique advanced by Commander C T Suthons of the Royal Navy was also tested. However, the technique of Sverdrup and Munk proved more accurate. See the paper titled 'Sea, swell and surf forecasting for Operation Overlord', by C C Bates, in *Some Meteorological Aspects of the D-Day Invasion of Europe, 6 June 1944 (op. cit.*, footnote 24, pp. 30–38).

Commons, this time, on 14 June, asking the Secretary of State for Air if meteorological statistics for the years 1939 to 1944 might now be published in summarized form. The Secretary of State, Harold Macmillan, informed the House that the *Monthly Weather Report* for each of the years in question would shortly be available to the public on application to His Majesty's Stationery Office. In addition, he said, copies were being supplied to the Royal Meteorological Society for distribution to Fellows, and the regular supply to the Society was being resumed.

During the war, Markham had continued to take a close interest in matters meteorological. In the House of Commons on 29 February 1944, for example, he had repeated an assertion he had made before the war: that there had been a shortage of trained meteorologists in the UK in those days and still was. He had compared the current situation in Britain with that in the United States, where, he said, between 3000 and 4000 young men were being trained as meteorologists. Furthermore, their training was more thorough than that provided in Britain. In America, he said, there was a staff of trained meteorologists in every city, their task being "not only to observe and register the public air but to see that the public air was kept clean, both by industrial and by private users". Britain, he went on, was losing "one of the greatest assets God has given us, unpolluted sunshine" because no Minister had "taken the responsibility of keeping our air clean".

At this point, Markham was ruled out of order, because he had strayed from the subject of the debate, the Air Estimates. However, he refocused his comments, saying that he thought the Secretary of State for Air (then Sir Archibald Sinclair) "should improve the publications of the meteorological side of his Ministry". In particular, he said, it was beyond question that many observers in coastal resorts had "deliberately faked" their sunshine figures "to produce higher totals and thus attract more trippers to their towns". He suggested that "the Minister, who was not only the Air Minister but the Minister for meteorology, should see that such records and observations were the truth, the whole truth and nothing but the truth".

In reply, the Under-Secretary of State for Air, Harold Balfour, informed Markham that the Air Ministry was indeed "the guardian of the public air" and, more importantly at that time, "endeavouring to defend [the British people] from the onslaughts of the enemy and endeavouring to command the air over the enemy's soil". He could not accept Markham's criticisms. He had been fortunate enough, he said, on more than one occasion, "to have the opportunity of going down to Bomber Command, or of seeing the flight plan made up for a flight from Newfoundland to Scotland". If he (Markham) had "seen the confidence that the captains of the aircraft placed in the meteorological forecasts given them, on which they based the whole of their flight plans" he did not think Markham "would have cast that aspersion of inefficiency on that service which had given such vital help in [Britain's] bomber offensive". Balfour rebuked Markham for his comments about the publication of meteorological statistics during the war. Had he "gone into the matter further", Balfour said, "he would have found the continued suppression of some reports was not because the information might be of direct operational information to the enemy but because [the British authorities] broadcast that information to His Majesty's ships and the Fighting Forces overseas, and if it were published it would help the enemy break down our ciphers".

Markham lost his seat at the General Election held in July 1945. Nevertheless, matters meteorological were raised in the House by other Members, one subject recurring time and time again, that of the delayed demobilization of the men and women who had been serving in the Meteorological Branch of the RAFVR. There was particular dissatisfaction over the length of time taken to release schoolteachers. Also, as Thomas Brooks put it on 21 November 1945, there was a suspicion that some meteorological offices at home and abroad were "more liberally manned than they need be". The first annual report of the Office published after the war, covering the period August 1945 to 31 March 1947, made no reference to delayed demobilization. It merely stated that service personnel had been released in accordance with the age and length of service scheme.

The annual report stated that the transition from a wartime to a peacetime organization had been difficult for the Office for a variety of reasons, the most serious being staffing problems. Owing to the demands of the fighting forces, the Office had expanded greatly during the war, and by August 1945 some 90% of the 6760 staff had been in uniform. Since then, military requirements had fallen progressively, largely because British and foreign airfields had been closing. However, there had been, at the same time, a rapid increase in the demands of other interests, particularly civil aviation. And a new military obligation had been recognized: the UK and its allies were now engaged in a 'Cold War'. As the report went on to say, the pre-war strength of the Office would have been inadequate to deal efficiently with all its post-war responsibilities. Accordingly, it had been found necessary to recruit new staff to replace temporary war entrants. The latter had been leaving in accordance with the age and service release scheme but, the report said, it was gratifying to record that a number of them had returned to the service of the Office after their release leave.

At the end of the war, staff of the Office were serving in many places around the world, including places in territory which was not British or normally under British influence. Many commitments in such places continued after the cessation of hostilities. Moreover, in countries liberated from enemy occupation, the Office helped re-establish meteorological services by attaching liaison officers and supplying equipment. The annual report for the period August 1945 to March 1947 lists the many countries in question and provides information about the considerable amount of assistance given.

As in the conflict of 1914 to 1918, many of the Office's staff received awards or decorations for their contributions to the war effort, and many were mentioned in dispatches. Chief among the honours was one announced on 2 June 1943. The Office's Director was to be made a Knight Commander of the Order of the Bath and thus become Sir Nelson Johnson KCB. The recognition was well deserved. Sir Nelson's devotion to duty had been exceptional. His wartime workload had been enormous, and he had virtually lived in his office in London throughout the war. He must have hoped that he could relax a little when the war ended. Alas, it was not to be.

11

Aftermath of War to Forecasting by Numbers

Sir Napier Shaw died on 23 March 1945, aged 91. In his lifetime, meteorology had been transformed, partly through his scientific insight, inspiration and leadership, but also through technological developments such as wireless telegraphy, aircraft, the radiosonde and radar. He had given the UK a leading position in international meteorology and laid the foundations which had been built upon in peacetime and wartime to form the institution the Meteorological Office had become by 1945, a scientific and technological body of national and international importance.[1]

Further Reorganization of the Meteorological Office

Hardly had the war ended when there came yet another reorganization of the Office, this one the outcome of an inquiry into the constitution of the Scientific Civil Service. The inquiry had been carried out during the war by a committee chaired by Sir Alan Barlow, a Treasury under-secretary, and its recommendations had been presented to Parliament in September 1945 in a White Paper (Cmd. 6679). The Office was reorganized on the lines the Barlow Committee recommended. Out went the grades that had been introduced in the 1930s in response to the report of the Carpenter Committee (see Chapter 9). In came the grades of Scientific Officer, Experimental Officer and Scientific Assistant, and with them came revised scales of pay.

The main objects of Barlow's recommendations were to create a unified Scientific Civil Service with common classes, standards, pay scales and conditions of service. There was also a recommendation that staff of the various departments of the civil service would be interchangeable. Scientific Officers would normally be honours

---

[1] A number of those who reviewed Shaw's life and work drew attention to the Monday evening discussions he had introduced to the Office in 1905, modelled on seminars held in the Cavendish Laboratory. See, for example, a note of appreciation by Gold published in the 1945 volume of the *Quarterly Journal of the Royal Meteorological Society* (Vol. 71, pp. 192–193). These discussions had not taken place during the war but were revived in February 1947.

graduates who would concentrate on purely scientific work and provide scientific leadership. Experimental Officers would be responsible for routine work and possess the practical capabilities and experience that would relieve Scientific Officers of work which did not call for a high degree of theoretical knowledge or scientific initiative. Scientific Assistants would be required to help in laboratories and with field experiments and serve as weather observers. Common standards would be achieved by centralized recruitment of staff through the Civil Service Commission.

Barlow's recommendations were broadly welcomed. However, the inability of the Office to take advantage of one of them was regretted by some, this being the recommendation that there should be greater interchange between government staff and universities. In the words of the editorial in the December 1945 issue of *The Observatory*, a British periodical devoted to astronomy (Vol. 66, pp. 165–166):

The staff of the Meteorological Office will not share in these academic interchanges owing, as the White Paper puts it, to a "lack of opportunity for interchange of employment in their special field". It is unfortunately true that meteorology is a much neglected study in British universities. Yet in the hands of men such as Brunt and Dobson – to mention two British names only – meteorology has proved to be a subject eminently suitable for university study. It is difficult to see how theoretical meteorology can reach in this country the development it has attained in Scandinavia, Germany or the United States if the practical meteorologist cannot spend some years on a university staff and if there are few academic meteorologists to address practical problems in government service.

The new grades and revised scales of pay were introduced in the Office on 1 January 1946, with recruitment through the Air Ministry at the same time replaced by central recruitment through the Civil Service Commission. By the end of the period covered by the first annual report published after the war (August 1945 to March 1947), 434 members of the established staff had been assimilated, 104 in the Scientific Officer grade and 330 in the Experimental Officer grade. In addition, eighty-two of the ninety-three staff with temporary posts who had been selected for nomination by the Civil Service Commission had been offered established appointments. Assignments to the Office were also made by the Civil Service Commissioners through the Reconstruction Competition, an evaluation of those who had joined the Scientific Civil Service on a temporary basis during the war to determine what permanent employment, if any, they could be offered. By the end of March 1947, fifteen in the Scientific Officer grade and forty-seven in the Experimental Officer grade had been appointed.

All new staff received special training in meteorology. By the end of the war, the Office's training facility was in London, partly at Kilburn and partly in Cornwall House on the South Bank. The Training School had endured a nomadic existence during the war, with moves from Gloucester to various places in London at different times, including the upper floors above a furniture shop in Oxford Street, and courses had also been run in Edinburgh. The nomadic existence continued after the war, with a second period in premises on Oxford Street from July 1946 to April 1947, followed

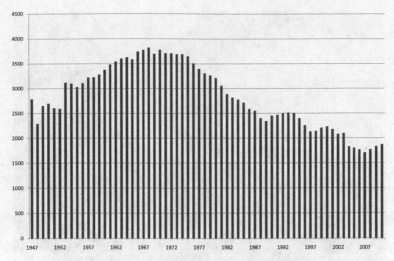

Figure 11.1. Meteorological Office staff numbers 1947–2010. The figure in August 1945 was 6760.

by a move to Alexandra House in Kingsway and later, in August 1951, a move to Stanmore.[2]

After the war, the Training School provided courses not only for new entrants but also for specialists. Courses for forecasters lasted six months, of which fifteen weeks were at the School and the remainder of the time at outstations (where trainees benefited from the supervision of experienced forecasters). Courses for scientific assistants lasted three months, with six weeks at the School and the remainder of the time at an outstation. Courses were also provided for officers of the Merchant Navy and for personnel of the Meteorological Air Observer Section of the RAF's General Duties Branch. Courses for specialists included radiosonde and radar courses, the former at Downham Market in Norfolk, the latter at War Office training schools.

Revisions of the Office's internal organization were made necessary partly by the transition from wartime to peacetime operations and partly by implementation of the White Paper's recommendations. In March 1947, responsibilities and duties were as follows:

- *The Director* was assisted by a Deputy Director and five Assistant Directors, with each of the five in charge of a number of branches. Directly responsible to the Director were the head of the Edinburgh Branch Office and the Assistant Directors in charge of personnel and research.

- *The Deputy Director* was in charge of the sections of the Office immediately concerned with synoptic meteorology and the supply of forecasts to the services, civil aviation, government

---

[2] For information about Meteorological Office training facilities in the 1940s, see two articles by R J Ogden: 'Meteorological Office training of assistant staff': 1939–51', published in 1986 in the *Meteorological Magazine* (Vol. 115, pp. 200–213), and 'Training at the Meteorological Office, 1936–71', published in 1992 in *Weather* (Vol. 47, pp. 349–353).

Figure 11.2. Weather briefing at Dunstable for senior scientists, *c.* 1946. © Crown Copyright 2010, the Met Office.

departments and the public. Under his control were the Assistant Directors in charge of forecasting, services and civil aviation.

- *The Assistant Director (Forecasting)* was in charge of the CFO at Dunstable and responsible for the collection and distribution of synoptic data, the issue of basic forecasts, and investigations into the technical problems of weather forecasting. The CFO also supplied the weather forecasts broadcast regularly by the BBC and maintained a small section in London which provided information to the press and the general public.
- *The Assistant Director (Services)* was responsible for meeting the requirements of the RAF, the Army and the Ministry of Supply, both at home and overseas.
- *The Assistant Director (Civil Aviation)* was responsible for the requirements of civil aviation, both at home and abroad.
- *The Assistant Director (Personnel)* was responsible for personnel and establishment matters and for works, services, finance and other routine administration.
- *The Assistant Director (Research)* was responsible for coordinating all research, including all matters concerned with the Meteorological Research Committee (MRC). He also controlled the four observatories and was, as a temporary measure, in charge of the Marine, Climatology and Instruments Branches. The observatories were at Kew, Eskdalemuir, Lerwick and Aberdeen.[3]

---

[3] Supervision of Aberdeen Observatory by the Meteorological Office ceased in 1947 when George Aubourne Clarke retired. He had been appointed Meteorological Observer in 1903, following meteorological training at Kew Observatory. He died on 13 February 1949, aged 69. He was a very fine photographer and his entire collection of cloud photographs was bequeathed to the Royal Meteorological Society. These photographs are now in the National Meteorological Archive at Exeter, along with an equally fine set of meteorological photographs by C J P Cave, who died in 1950.

By March 1948, the administrative structure had altered again, to reflect increased responsibilities of the Office. The Director was now assisted by a Principal Deputy Director (Dr Stagg) and, below him, three Deputy Directors and nine Assistant Directors.

- Under the Deputy Director (Forecasting), there were five branches: synoptic observations and communications; upper-air observations; operational forecasting; short-range forecasting research; long-range forecasting research.
- Under the Deputy Director (Services), there were ten branches: RAF overseas; RAF at home; Flying Training Command, Army, and Ministry of Supply; civil aviation at home; civil aviation overseas; British climatology; world climatology; agricultural meteorology; marine meteorology; Library.
- Under the Deputy Director (Research), there were four branches: radiosonde and aircraft instruments; general instruments; special investigations; and a branch which covered general research, the MRC, the Edinburgh Branch Office, and the observatories at Kew, Lerwick and Eskdalemuir.

In all, there were twenty-two branches, the remaining three being the Training School, the Personnel Branch and a branch concerned with the provisioning, accounting, testing and calibration of instruments. All three were supervised by the Assistant Director for Personnel and Supply, who was responsible directly to Dr Stagg.

Revival of the International Meteorological Organization

After the war, the immediate priority for the IMO was to restore international meteorological cooperation and refocus attention on matters that had been in abeyance since 1939. These were the main items on the agenda for the extraordinary meeting of the Conference of Directors which was held in London from 25 February to 2 March 1946. Soon, however, IMO began to respond to the meteorological and technological developments that had taken place since 1939. The importance of meteorology had been demonstrated over and over again during the war, and the scientific and technological developments of the war years had set in train a meteorological revolution that would soon take place. The density of the global network of meteorological stations had increased during the war, and developments in radio, radar and radiosondes had made possible the establishment of aerological stations capable of making meteorological observations up to heights of twenty kilometres or more in the atmosphere. The coming expansion of civil aviation, with ever-longer flights over continents and oceans, was clear to all, and the need for reconnaissance flights and ocean weather ships was also recognized.

During the next phase of its development, IMO moved ahead under the leadership of Sir Nelson Johnson, who was elected President of the IMC and President of IMO's Executive Council at the London meeting in February 1946. He was not the only member of the Office's staff to become an officer of the IMO. Gold was re-elected President of the Commission for Synoptic Weather Information, Frankcom was elected President of the Commission for Maritime Meteorology and numerous other members of staff were elected members of commissions.

The changed meteorological world after the war was evident in the agenda for the Conference of Directors meeting which took place at Washington, DC, in September 1947. The 400 resolutions reflected the wide range of topics considered at the meetings of the technical and regional commissions in Toronto the previous month. These topics included codes, units, diagrams, symbols, instruments, methods of observation, climatological statistics, station networks, telecommunications, the safety of air navigation, publications, education, professional training, meteorological research, legal issues and administrative matters. Moreover, relationships between IMO and the International Civil Aviation Organization, the International Telecommunication Union and the International Ice Patrol had been discussed, too. The issue of greatest importance to IMO after the war was, however, its status and structure.

Endeavours to resolve the matter began as early as February 1946, when the Conference of Directors asked the IMC to prepare a new constitution which would make IMO an inter-governmental body. At the same time, the Committee was asked to review the composition and functions of the technical commissions and consider how IMO might become associated with the newly formed United Nations (UN) without compromising IMO's independence and character. The Berlin Draft as revised during the war (see Chapter 10) was used as a basis for the new Convention.

The task of drafting the Convention was carried out at the meeting of the IMC held at Paris in July 1946, and the so-called Paris Draft 1946 was adopted, this being a draft World Meteorological Convention with two annexes, one containing draft General Regulations, the other draft Technical Regulations. These three drafts were then submitted to national meteorological services and their governments for perusal. Comments were considered at the Washington meeting of the Conference of Directors, along with four new drafts of the Convention, submitted by Canada, France, the UK and the United States. Eventually, after many disagreements had been settled and concerns over the independence and worldwide character of the new body had been allayed, the World Meteorological Convention was approved unanimously and a resolution adopted that the new organization should seek affiliation with the UN.

The World Meteorological Convention was signed on 11 October 1947 and came into force on 23 March 1950.[4] In the interim, the work of the IMC continued, in particular by ensuring that the many resolutions and recommendations of the Washington meeting were implemented. The Committee also secured recognition for itself

---

[4] By strange coincidence, both Sir Napier Shaw and Sir Nelson Johnson died on 23 March, Shaw in 1945, Johnson in 1954.

by the UN as the preparatory body for the new organization and accordingly took part in relevant UN activities. The last meeting of IMO's Conference of Directors took place in Paris in March 1951, and IMO formally became the World Meteorological Organization (WMO) on the 17th of that month. For the opening Congress of WMO in 1951, Sir Nelson was elected President, a well-deserved recognition for his skilled, careful, patient and unselfish contribution to both the revival of IMO after the war and the creation of WMO. He was also elected a member of WMO's Executive Committee, the body that controlled the finances of WMO, directed the work of the Secretariat and gave provisional approval to recommendations made by WMO's technical commissions.[5]

Research Developments

Research was a subject close to Sir Nelson's heart, and he contributed an article on this subject to the first issue of the Royal Meteorological Society's new magazine, *Weather*, launched in May 1946. In it, he said that private research was undoubtedly important. As he put it: "Many of the most valuable contributions to science have been made by individual investigators working by the light of their intuitive genius, and it must never be forgotten that the quality of the results obtained in any kind of research must depend in the last resort upon the intellectual ability of the investigators". However, he said, there were certain types of investigation which could not be carried out effectively without a certain amount of organization. Endeavours to solve some problems required, for example, the use of aircraft or extensive workshop facilities.

He devoted much of the article to the work of the MRC, which he had been keen to set up. His pleasure when Air Ministry approval for the formation of this Committee reached him was recalled by P A Sheppard:[6]

I shall always remember the morning in 1941 when Air Ministry approval for the formation of the Committee reached Sir Nelson Johnson. He called me into his room at Victory House to share his good news and tell me some of his hopes for research in the Office. Johnson was not given to emotional displays but that morning he was like the proverbial dog with two tails. It was not of course just the formation of the Committee itself that occasioned the happiness but the implication it carried that meteorological research was at last to be a recognized function of the Office, no longer to be left to the insufficient efforts of enthusiastic members of its staff at their postprandial dining tables.

Sir Nelson stated in his article that the policy adopted by the Committee was to concentrate upon meteorological problems that were more or less closely concerned with

---

[5] For a comprehensive review of the work of the IMO, see 'The history of the International Meteorological Organization, 1872–1951', by H G Cannegieter, published in 1963 in *Annalen der Meteorologie*, Neue Folge Nr.1, 280 pp.

[6] 'Notes by the chairman of the Meteorological Research Committee', by P A Sheppard, published in 1962 in the *Meteorological Magazine* (Vol. 91, pp. 335–337).

practical questions. Investigations of a long-term nature or purely scientific character would continue to be organized by the Royal Society, through its Gassiot Committee. In 1946, five members of the MRC were members of the Gassiot Committee: Sir Nelson himself, Brunt, Chapman, Dobson and Taylor.[7] Other members of the Gassiot Committee included Gold and Simpson. The importance of the Office's research work had been recognized in official quarters, Sir Nelson said. The Air Ministry had "recently obtained from the Treasury a grant of £20,000, spread over a period of five years, to assist in the financing of meteorological investigations carried out under the direction of the Gassiot Committee".[8] Financial aid had also been given by the Royal Society itself.

An exploration of the lower stratosphere was one of the research projects directed by the MRC, for which it was necessary to develop an instrument to measure the small amounts of water vapour present at those heights and temperatures. To carry out this work, organized research was necessary, making use of the Meteorological Research Flight (MRF), which possessed large aircraft equipped as flying laboratories, as well as small aircraft which were capable of climbing into the stratosphere. The problems that were being investigated by the MRC included fog prediction, the dynamics and physical structure of depressions and the internal structure of the towering clouds which produced large hailstones. The formation of such hailstones implied the existence of strong upcurrents within the clouds, and lightning was associated with the clouds, too, as well as severe turbulence within and below the clouds. Aircraft safety was therefore an important consideration underlying this work.

Within a few years of the war ending, a research ethos had developed, as the Office's annual reports show. The report for the year ending 31 March 1949, for example, mentions collaboration with the Gassiot Committee, the NPL, the University of Oxford's Clarendon Laboratory and research establishments of the Ministry of Supply, including an atmospheric pollution investigation controlled by DSIR's Fuel Research Station. The Office's research programme involved the Gassiot Committee, the MRF and the Office's three observatories, and the programme was wide-ranging, embracing studies of ice accretion on aircraft, microclimatology near the ground in relation to agriculture, and the temperature and humidity structure of the upper troposphere and lower stratosphere. Instruments were being developed for a variety of purposes, including, in connection with the proposed Severn Bridge, one for ascertaining the strength of the vertical component of the wind. Research in physical meteorology included measurements of the sizes of raindrops and, in collaboration with British European Airways, studies of bumpiness experienced by aircraft flying in clear (cloud-free) air at heights of 20,000 feet or more.

---

[7] Taylor was knighted in 1944, thus becoming Sir Geoffrey Taylor.
[8] After the Second World War, the Air Ministry was very much research-minded.

Much effort was devoted to weather forecasting research, with particular emphasis on the physical processes that were fundamental to the forecasting problem and the application of new knowledge to forecasting procedure and technique. The work of the new Forecasting Research Division was at an early stage when the annual report for 1948–1949 was issued, so, as the report stated, the stage of important results had barely been reached. However, the need to focus on the structure of frontal regions had been recognized. In the words of the report, "nature is much more complex than the classical frontal models suggest, and it is necessary to establish the facts more thoroughly". Attempts to forecast the weather four days ahead began on an experimental basis in October 1948 and more systematically in February 1949. The report for 1948–1949 stated that early efforts were "not discouraging". The report for the year ending 31 March 1952 stated that such forecasts could be made "with a measure of success but not with uniform reliability" and added, somewhat doubtfully, that it remained to be decided whether the potential value of the forecasts was sufficient to justify the expense of employing extra staff on this work at that time.

Developments in Weather Forecasting

In his Presidential Address to the American Meteorological Society in December 1946, H G Houghton stated, with reference to weather forecasting in the United States, that there had been no significant improvement in the accuracy of short-range forecasting in the preceding thirty to forty years.[9] The number of elements forecast, the frequency of forecasts and the types of forecast had all increased greatly, but, he said, there had been no corresponding increase in accuracy, and he went on to say that there appeared to be no immediate prospect of an objective method of forecasting based entirely on sound physical principles. All this was true of the UK, too, but it is nevertheless clear from the opening words of a lecture given by Gold in April 1947 that he did not consider the accuracy of weather forecasts merited as much criticism as some asserted:[10]

In view of recent criticism of forecasts, I make the following offer: On any occasion (a) when rain is forecast in the BBC 5.55 p.m. forecast for S.E. England and there is no rain in London before 6 p.m. next day, or (b) when no rain is forecast and there is rain before 6 p.m. next day, I will pay a guinea to any person who on his part undertakes to pay me half a guinea on those occasions when rain is forecast and there is rain or when no rain is forecast and there is no rain. Contract to be for one year and accounts settled monthly.

---

9 Houghton's words are quoted on page 16 of a major review by C K M Douglas entitled 'The evolution of 20th-century forecasting in the British Isles', published in 1952 in the *Quarterly Journal of the Royal Meteorological Society* (Vol. 78, pp. 1–21). 'Short-range' meant a day or two ahead.

10 'Weather forecasts', the Symons Memorial Lecture delivered before the Royal Meteorological Society on 16 April 1947, by E Gold, published in 1947 in the *Quarterly Journal of the Royal Meteorological Society* (Vol. 73, pp. 151–185).

In his lecture, Gold referred to crude checks which had been made of the forecasts for south-east England broadcast by the BBC at 5.55 p.m. during the previous twelve months and reported that forecasts of precipitation could be considered correct four times out of five.

One of the Office's forecasters, George Cowling, recalled that forecasting in 1946 was "labour-intensive, difficult and time-consuming and results could be disappointing".[11] Communications were slow, facsimile was limited, and observations were far from plentiful. By mid-1946, he said, "ship reports were averaging about a dozen for each of the four main synoptic hours" and these reports were "mainly restricted to a very few shipping lanes, while transmission errors sometimes resulted in false positions or false pressures which could be very misleading in the absence of cross-checks". After the war, he added, the number of routine meteorological reconnaissance flights had been reduced to two, with both flights based in Northern Ireland. Radiosondes were launched four times a day from Larkhill, Lerwick, Downham Market and Fazakerley and twice a day from Stornoway, Penzance, Belfast and Leuchars. Soundings were also made at Valentia in south-west Ireland and many other places outside the British Isles. Locations of thunderstorms were ascertained by means of the 'sferics' techniques which had been developed during the war, while weather radar information was available from the East Hill research station near Dunstable.[12]

Cowling mentioned that R C Sutcliffe had found time during the war to develop techniques for weather forecasting based on fundamental principles of dynamical meteorology. Thus he had built upon mathematical foundations he had laid in the late 1930s. His work was published in 1947 in a seminal paper which shed new light on processes which governed the development of extratropical weather systems, a paper which would be cited time and time again over the years as dynamical meteorology progressed.[13] Cowling commented that he and his colleagues on the forecast bench at the Air Headquarters of the British Air Force of Occupation at Bad Eilsen in northern Germany in 1946 were careful to give due consideration to Sutcliffe's ideas. He was the Chief Meteorological Officer at Bad Eilsen and keenly interested in forecasting, which Cowling said was "an interest not necessarily shared by all meteorologists"!

Towards the end of his lecture, Gold asked "What of the future?" By 1947, it had been appreciated by most leading meteorologists that the future of weather forecasting lay to a great extent in better understanding of the physical and dynamical processes at work in the atmosphere, and Gold was well aware of this, though he did not say so in so

[11] 'A forecaster recalls – 1946', by George Cowling, published in *Weather* in 1996 (Vol. 51, pp. 151–156).

[12] Thunderstorm location techniques were developed by Cave and Watson Watt during and after the Great War (see Chapters 8 and 9). For information about such techniques after the Second World War, see 'Sferics', by C V Ockenden, published in 1947 in the *Meteorological Magazine* (Vol. 76, pp. 78–84).

[13] 'A contribution to the problem of development', by R C Sutcliffe, published in 1947 in the *Quarterly Journal of the Royal Meteorological Society* (Vol. 73, pp. 370–383).

many words. Rather, he put forward a suggestion he attributed to C S Durst, wondering whether "a meteorological electronic brain could be constructed to deal in an orderly manner with the immense amount of information available daily and provide much more accurate prebaratics".[14] A pre-requisite for the construction of such a brain would be what he called "a new set of 'Richardson Rules' of computation". Thus he showed that Richardson's *Weather Prediction by Numerical Process*, published in 1922, had not been forgotten (see Chapter 8). He was hinting to his audience that the day when Richardson's dream might become a reality was approaching. Perhaps he was being prescient. More likely, as the second most senior person in the Office, he had heard about a development in the United States.

The likelihood of Gold having this development in mind is suggested by the penultimate paragraph of the published version of his lecture:

It might eventually be possible to put at the forecaster's disposal the means to make his forecast come right – to adjust the atmosphere by means of an appropriate release of atomic energy in the right place to counteract any tendency on its part to stray from the prebaratic. The energy in atomic bombs is comparable with the energy of depressions – one thousand of the modest size already made would provide as much energy as that of a depression 1,000 miles across and 10 millibars deep – but it is clear that the selection of the right place for release would call for all the skill of the synoptic meteorologist.

Gold was close to retirement, and he did indeed retire on 31 October 1947, aged 66, but his mind was clearly still as active as ever.

A key figure in the development in question was the Swedish meteorologist Carl-Gustaf Rossby, who, like Sutcliffe, rejected the view of most weather forecasters that dynamical meteorology had no relevance to them. Rossby had studied mathematical physics at the University of Stockholm and then spent two years in the Bergen Meteorological Institute, where he had become thoroughly acquainted with the polar front theory of mid-latitude depressions. He was not only a competent mathematician and physicist; he was also thoroughly grounded in synoptic meteorology. He had been largely responsible for the establishment of meteorology programmes at the Massachusetts Institute of Technology and, later, the University of Chicago, and during the Second World War he had been the leading figure in the meteorological training programme of the U.S. Army Air Forces. In around 1940, he and his collaborators had begun to make hand-calculated numerical predictions of the weather using a very simple mathematical model of the atmosphere.

An important meeting took place in Washington, DC, on 9 January 1946. Convened by Francis Reichelderfer, the U.S. Weather Bureau's Chief, it was supposed to be secret, but a detailed account of it appeared in the *New York Times* two days later![15]

[14] 'Prebaratic' was a term introduced by Gold in 1942 as an abbreviation for a chart showing the forecast distribution of isobars and fronts over the North Atlantic and western Europe.

[15] For details of the early days of numerical weather prediction in the United States, see *Weather by the Numbers*, by Kristine Harper, published in 2008 (The MIT Press, 308 pp.). See also 'A history of numerical weather prediction

There were a dozen meteorologists at the meeting, some of them military men, and there were two guests: John von Neumann, a mathematician from the Institute for Advanced Study in Princeton, and Vladimir Zworykin of RCA, who had invented the scanning television camera. They had come to explain their startling proposal, that the electronic digital computer planned by Neumann might be used to forecast and ultimately *control* the weather. Perhaps Gold had this mind when he speculated on the use of atomic bombs to "adjust the atmosphere".

From 1943 to 1945, Neumann had worked on the Manhattan (atomic bomb) Project at Los Alamos in New Mexico, where he had become interested in numerical methods of solution and the use of calculating aids. In 1944, he had become aware of the first general-purpose electronic computer, the Electronic Numerical Integrator and Computer (ENIAC), and become involved in the planning of a more powerful computer, the Electronic Discrete Variable Arithmetic Computer (EDVAC). His objective had been to build a powerful computer to advance the mathematical sciences, but he was not a meteorologist. Only a few weeks before the meeting on 9 January 1946 did he come to consider meteorology a principal area for application of the proposed computer. By May 1946, however, he had decided that the weather forecasting problem certainly was such a principal area. For the most part, the meteorological community did not show much interest in the idea of controlling the weather, but many meteorologists were excited by the possibility of forecasting by numerical means.

Rossby used his connections and influence in the spring of 1946 to encourage academic institutions and agencies to progress the idea of numerical weather prediction (NWP). Thus the Meteorology Project was born, supported by funding from the U.S. Navy's Office of Research and Inventions. Soon, with meteorologist Hans Panofsky, Neumann developed electronic methods of weather-map analysis, but the accuracy of the machine they used was not sufficient. An attempt was made to produce forecasts using real data by numerical integration, but this attempt was also abandoned. For a couple of years, the Meteorology Project did not seem to be making much progress. Then, in July 1948, Jule Charney became the Project's chief meteorologist and began to use a model he had designed. He was a mathematician who had been greatly influenced by Rossby at the University of Chicago and continued to be influenced by him whilst working with Neumann on the Meteorology Project.

The first results of NWP were made by hand computation, using Charney's model. They were disappointing, but he persevered and by early 1949 believed his method sufficiently accurate to justify its use in day-to-day forecasting. The first forecast made by means of ENIAC was produced in April 1950. Rossby thought Charney's method "extraordinarily promising", and Jerome Namias, another leading meteorologist, was

in the United States', by P D Thompson, published in 1983 in the *Bulletin of the American Meteorological Society* (Vol. 64, pp. 755–769); 'The birth of numerical weather prediction', by A Wiin-Nielsen, published in 1991 in *Tellus* (Vol. 43AB, pp. 36–52); and 'Weather forecasting from woolly art to solid science', by P Lynch, pp. 106–119 in *Meteorology at the millennium* (edited by R P Pearce), published by Academic Press (*International Geophysics Series*, Vol. 60, 255 pp.).

keen to introduce the method in the U.S. Weather Bureau. However, the method did not at that stage become a part of standard forecasting procedure, but it did bring home to many meteorologists and others that successful numerical forecasting would not be long coming.

At last, the work of Richardson was no longer merely of academic interest for want of computing power, and Charney himself wrote in 1950 that Richardson's 1922 book on weather prediction by numerical process had become of the greatest importance. For years, practising forecasters had considered this book nothing more than an interesting intellectual curiosity. It is today considered a classic work.[16]

In the UK, developments in the United States did not pass unnoticed. The possibilities of using electronic computing machines in meteorology was discussed formally on 25 May 1948 at a meeting held jointly by the Office and the Imperial College Department of Meteorology, the meeting attended by R C Sutcliffe, C S Durst, G C McVittie and E T Eady.[17] The following year, a Meteorological Office Forecast Research Division was set up at Dunstable, headed by Sutcliffe under the overall direction of Goldie, who had become the Office's Deputy Director (Research) in 1948. Others in the Division then or soon after included E Knighting, A G Forsdyke, H H Lamb and a gifted young mathematician, F H Bushby.

There was an important development for British NWP in October 1951: Bushby attended a course at Cambridge on the use of the University's EDSAC computer, EDSAC standing for Electronic Delay Storage Automatic Calculator. By the end of 1951, he and an assistant, Mavis Hinds, were using LEO 1, a copy of EDSAC, which had been built by Messrs J Lyons, the caterers, at their Cadby Hall headquarters in Hammersmith, London.[18] LEO 1 was the world's first business computer, LEO standing for Lyons Electronic Office.

Independent tests of Charney's method were carried out at Stockholm in 1950 under Rossby's supervision. Tests were carried out by two members of the Office, Bushby and another gifted mathematician, John Sawyer. They were not as enthusiastic as Rossby and Namias, reporting that "significant success had been achieved but the success was definitely less than that achieved by conventional forecasting methods on the same charts". Charney's model was initially too demanding for British computers, so the Office used a model based on Sutcliffe's development equation.

An international argument raged in the early 1950s over the merits of the two approaches.[19] R S Scorer of the Imperial College Department of Meteorology attacked the approach of Charney in 1950 in a provocative letter to the *Journal of Meteorology*.

---

[16] Richardson lived to see his dream of forecasting the weather by means of mathematical methods start to become a reality. He died on 30 September 1953.

[17] For a detailed account of the early days of numerical weather prediction in the UK, see 'Early operational Numerical Weather Prediction outside the USA: an historical introduction. Part III: endurance and mathematics – British NWP, 1948–1965', by Anders Persson, published in 2005 in *Meteorological Applications* (Vol. 12, pp. 381–413).

[18] See 'Computer story', by M Hinds, published in 1981 in the *Meteorological Magazine* (Vol. 110, pp. 69–81).

[19] For details, see Persson's paper (*op. cit*, footnote 17).

Figure 11.3. C K M Douglas, the greatest British synoptic weather forecaster of all time, early 1950s. © Crown Copyright 2010, the Met Office.

He certainly needled Charney and Rossby, as Charney's indignant and scornful reply showed. The argument continued into 1951 at a meeting on dynamical methods in synoptic meteorology held at the Royal Meteorological Society on 17 January 1951, when Scorer called Charney's approach "utterly useless". In return, in a letter to Rossby, Charney called Scorer "a fool".

Scorer was certainly not the only British meteorologist critical of the work in the United States. Indeed, the greatest British synoptic forecaster of all time, C K M Douglas, considered in 1952 that, "on the whole, the prospects of computing the future weather, with or without electronic machines, looked remote".[20]

But he conceded that the introduction of quantitative elements did represent a real advance, because it led to a more precise understanding of atmospheric processes. The annual report of the Office for the year ending 31 March 1952 put it thus:

Mathematical research led to a system of equations which might lead to a mathematical forecast, and some progress was made in planning to use an electronic calculating machine. The main difficulty remains, however, that the mathematics are only approximate and there is still no evidence that the computer will do as well as the conventional forecaster.

[20] *Op. cit.*, footnote 9, this chapter.

Figure 11.4. C K M Douglas inspecting work in the Central Forecasting Office at Dunstable, late 1940s. © Crown Copyright 2010, the Met Office.

The report for the year ending 31 March 1953 was a little more optimistic:

An electronic computing machine was used to carry out a number of calculations which may lead to a system of mathematical forecasting based on the dynamics of the atmosphere. Results so far are of great theoretical interest and not without promise of practical value.

The following year, from 21 September to 2 October 1954, a large military manœuvre took place in Sweden. During this, the first real-time operational numerical weather predictions were made, using the Swedish BESK computer (BESK meaning, in English, Binary Electronic Sequence Calculator).[21] Comparisons with independent purely subjective forecasts showed that the numerical predictions were clearly the more accurate. The age of operational NWP had dawned, but not yet for the UK.

Floating Meteorological Observatories

Observations from ships became available soon after the war in the Far East ended. Radio transmission of weather reports from merchantmen was resumed on 1 November 1945, and the number of messages received each month from vessels on the North Atlantic increased rapidly thereafter. The number received at Dunstable in the first month was 601, rising to 1910 in July 1946 and 2374 in January 1947. Mariners made observations at the main synoptic hours (00:00, 06:00, 12:00 and 18:00 GMT)

[21] For the story of NWP in Sweden, see 'Early operational Numerical Weather Prediction outside the USA: An historical introduction. Part I: Internationalism and engineering NWP in Sweden, 1952–69', by Anders Persson, published in 2005 in *Meteorological Applications* (Vol. 12, pp. 135–159).

Figure 11.5. The teleprinter room at Dunstable. © Crown Copyright 2010, the Met Office.

and transmitted them to the appropriate shore station. From ships east of 40°W, they were transmitted to Portishead, Wick, Malin Head, Humber or Valentia, whence they were relayed to the Central Telegraph Office of the General Post Office and forwarded by teleprinters to Dunstable. Observations made west of 40°W were transmitted to Halifax in Nova Scotia or Gander in Newfoundland and reached Dunstable through a radio link between Montreal and Prestwick.

Reports from ships were essential to forecasters but subject to the disadvantage of tending to be concentrated along the main shipping routes. They needed to be supplemented by observations from under-reported areas of the ocean. The time had come to reconsider an idea first suggested in the nineteenth century, floating meteorological observatories (see Chapter 4). Representatives of the Office proposed at a meeting of the Meteorological Division of the Provisional International Civil Aviation Organization (PICAO) held at Dublin in March 1946 that thirteen ocean weather stations be established on the North Atlantic, and this was agreed at a PICAO conference held in London in September 1946, attended by delegates from Belgium, Canada, Denmark, France, Iceland, Ireland, The Netherlands, Norway, Portugal, Spain, Sweden, the UK and the United States.[22]

The UK undertook to finance and operate two of the stations, with one of them at 60°00'N 20°00'W (about 250 miles south of Iceland) and the other at 53°50'N 18°40'W (about 350 miles west of Ireland). Britain also agreed to assist Norway and

[22] The Provisional International Civil Aviation Organization ceased to be provisional in April 1947 and thereafter became the International Civil Aviation Organization (ICAO).

Sweden maintain a third station at 66°00'N 2°00'E (about 300 miles west of Norway). Each station would make full meteorological reports eight times daily, while upper-air data would be obtained by means of radiosondes four times a day. In addition, the ships on the stations would be provided with navigational aids to assist aircraft in flight, and they would be equipped to carry out rescue duties should the need arise, such as a ship in distress or an aircraft having to make a forced landing on the sea. Meteorological data would be shared with all other nations, including the Soviet Union, despite the Cold War.

Four corvettes were converted and duly took up their duties at the two British stations. They were designated 'Ocean Weather Ships' and bore the names *Weather Observer*, *Weather Recorder*, *Weather Watcher* and *Weather Explorer*. The first to take up position was *Weather Observer*, which was renamed by the Secretary of State for Air, Philip Noel-Baker, at a ceremony in the London Docks on 31 July 1947. It had formerly been HMS *Marguerite*, built in 1941. The first observations from *Weather Observer* appeared in the International Section of the *Daily Weather Report* on 8 August 1947, and she started reporting from 53°50'N 18°40'W (called Station J) two days later.[23] Commissioning of the remaining three ships followed soon afterwards: *Weather Recorder* on 4 October 1947, *Weather Watcher* on 27 November 1947 and *Weather Explorer* on 4 February 1948.

The weather ships were operated and administered by the Office, in consultation as necessary with the Admiralty, the Ministry of Civil Aviation and the Ministry of Transport. They were based at Greenock and spent approximately 27 days at sea (21 of them on station), followed by fifteen in port. The intention was that two ships would always be on their station and two in port, but this was sometimes not achieved. There was, for instance, a break of about ten days in manning the station at 60°00'N 20°00'W (called Station I) in the summer of 1948, when, to quote from the annual report of the Office for the year ending 31 March 1949, diversions were made "to cooperate in aircraft-safety arrangements for the outward and return flights to the USA via Iceland of a flight of RAF Vampire aircraft". And on 11 January 1948, *Weather Recorder* rescued the crew of a Norwegian ship off the Scottish coast. Those who manned the weather ships were all civilians, including the officers. The complement numbered fifty-two, made up of fourteen on deck, eleven in the engine-room, ten in the radio section, ten in the catering section and seven meteorological staff.

In an article in the July 1948 issue of *Weather*, Commander Frankcom, the Office's Marine Superintendent, reported that the winter of 1947–1948 had been stormy.[24] Nevertheless, there had been no instance of a weather ship failing to maintain her station or carry out essential duties, even though the ships and their crews were new

[23] For an introduction to *Weather Observer*, see the unsigned report entitled 'Ocean Weather Ships', published in the September 1947 issue of *Weather* (Vol. 2, pp. 263–266). The other British station was known as Station I.
[24] 'The operation of ocean weather ships', by C E N Frankcom, published in *Weather* in 1948 (Vol. 3, pp. 212–215).

to the task in hand. This was remarkable, because operational conditions in heavy weather could be very difficult, as J E Brown, the Senior Meteorological Officer of *Weather Recorder*, emphasized in another article published in the July 1948 issue of *Weather*.[25] The launching of radiosondes in winds of Force 9 or more was particularly problematic. With the ship heading into wind, he explained, the downdraught abaft the balloon shelter was sufficient to prevent the balloon rising immediately. On occasions, he said, it fell almost to sea level fifty to a hundred yards astern of the vessel, immersing and smashing the radiosonde transmitter.

By March 1949, surface and upper-air observations were being received from eleven weather stations on the North Atlantic, including the one at 66°00'N 2°00'E (called Station M) which had become fully operational in September 1948. It was operated by the Norwegian Meteorological Institute with financial support from Sweden and the UK. Unfortunately, the number of stations on the North Atlantic soon had to be reduced, for reasons of economy. The number decreased to ten in 1949 and nine in 1954, with four stations operated by the United States (one jointly with Canada) and five by European nations, including Stations I, J and M. The four British weather ships were replaced by frigates during the period 1958 to 1961. Renamed *Weather Adviser*, *Weather Monitor*, *Weather Reporter* and *Weather Surveyor*, they maintained Stations I and J for many years.

From February 1953 onwards, in collaboration with the UK's National Institute of Oceanography, instrumental records of sea state on the open ocean were obtained by means of a shipborne wave recorder attached to the hull of *Weather Explorer*.[26] The records were obtained initially at Stations I and J but soon obtained additionally from Stations A and K, the former at 62°00'N 33°00'W (maintained jointly by France, Norway, the UK and the Netherlands), the latter at 45°00'N 16°00'W (maintained jointly by France, the UK and The Netherlands). *Weather Explorer* was on station for about two-thirds of the year, her time distributed evenly between the four stations. The wave recorder that was fitted to *Weather Explorer* was transferred in May 1958 to this ship's replacement, *Weather Reporter*.

The station operated jointly by the USA and Canada (Station B) was in a place of particularly inclement weather (56°30'N 51°00'W), in the Davis Strait between Labrador and southern Greenland, but the meteorological observations and measurements of sea-surface temperature made there proved invaluable in the 1960s to those in the Office and others who were studying the factors that needed to be considered when forecasting the weather a month or more ahead. Stations maintained solely by

[25] 'The sea-going meteorologist's aspect of work in ocean weather ships', by J E Brown, published in *Weather* in 1948 (Vol. 3, pp. 216–218).

[26] The National Institute of Oceanography was founded in 1949 and renamed Institute of Oceanographic Sciences in 1973. It is now part of the National Oceanography Centre at Southampton. For information about the wave recorder, see *Of seas and ships and scientists: the remarkable story of the UK's National Institute of Oceanography*, edited by A S Laughton *et al* and published in 2010 by The Lutterworth Press (350 pp.). The first recorder was attached to the Royal Research Ship *Discovery II* in 1952.

the United States at 52°45'N 35°30'W (Station C) and 45°00'N 45°00'W (Station D) were important for the same reason.

Meteorological Services for the Public

As soon as the war in Europe ended, the BBC resumed radio broadcasts of weather forecasts for the general public, transmitting them immediately before scheduled news bulletins. Warnings of gales, snow and frost were broadcast as soon as they were received from Dunstable, and forecasts for the East Anglian Herring Fishing Fleet and the Scottish Herring Fleet were broadcast when necessary in due season. The Office also issued information routinely to the press, including district weather forecasts, London observations, and morning and evening reports from health resorts. From 1 February 1947, specially prepared weather maps were made available to the press every evening. These showed the actual synoptic situation at 12:00 GMT and the prebaratic chart for noon the following day.

The 'Airmet' service (see Chapter 9) recommenced on 7 January 1946, with broadcasts from 07:00 to 18:00 GMT in winter and 06:00 to 21:00 GMT in summer. Each hour, actual weather reports from forty airfields in the UK were broadcast, using a frequency of 244.7 kHz. Unfortunately, the revival of the service was short-lived. The broadcasts ceased on 14 March 1950 because no provision had been made for Airmet in the 1948 Copenhagen Plan, an international agreement which allocated channels for radio broadcasting. The station at Kalundborg in Denmark had been assigned an almost identical frequency, 244.9 kHz, and it used a powerful transmitter.

Protests followed. Airmet had been useful not only to aviators but also to farmers, seafarers, schoolteachers, astronomers, housewives and many others. The press, the Office and the Royal Meteorological Society received a spate of letters, and questions were asked in the House of Commons. The President of the Society, Sir Robert Watson-Watt, spoke at length and with great feeling about the demise of Airmet in the Presidential Address he delivered to the Society on 16 May 1951.[27] A petition organized by the Society attracted 21,535 signatures and was presented to Parliament on 26 July 1951.[28] The authorities were unmoved. The service was not restored.

Some people were, however, resourceful. They found a substitute for Airmet. They made use of the weather bulletins for shipping that were issued by the Office and broadcast by the BBC. From these, they were able to obtain a general picture of the weather over and near the British Isles, particularly after April 1956, when improved bulletins were introduced. Though the improved bulletins still contained less detail than the former Airmet broadcasts, the general weather synopsis and the

---

[27] 'Meteorology and aviation', by Sir Robert Watson-Watt, published in 1951 in the *Quarterly Journal of the Royal Meteorological Society* (Vol. 77, pp. 552–568).

[28] See 'Airmet petition', a report published in the December 1950 issue of *Weather* (Vol. 5, pp. 403–405). See also 'Presentation of the Airmet petition', a report in the August 1951 issue of *Weather* (Vol. 6, p. 227).

meteorological reports from coastal stations enabled anyone with a basic knowledge of meteorology to construct a map of the existing distribution of isobars and fronts over the British Isles and eastern parts of the North Atlantic, together with an indication of the weather situation's probable development.[29]

Meanwhile, television (TV) broadcasting had recommenced after the war. On 29 July 1949, weather charts appeared again on TV, the first of them at the end of that evening's programmes, broadcast from a BBC studio in London. Two captioned charts were displayed, one showing isobars 'today', the other the distribution of isobars forecast for 'tomorrow'. The BBC's duty announcer, out of vision, read a forecast prepared and supplied by the Office.

A TV transmitter at Sutton Coldfield near Birmingham came into operation on 17 December 1949, followed by transmitters at Holme Moss near Huddersfield on 12 October 1951, Kirk o'Shotts near Glasgow on 14 March 1952, Wenvoe near Cardiff on 15 August 1952, and other places soon afterwards. Thus the daily broadcasts of weather charts and forecasts reached most people in the UK fairly quickly, but a number of remote and thinly populated areas of the nation remained without TV for some years. The transmitting station in Shetland, for example, opened on 15 April 1964, with the start of transmissions of BBC TV six days later.

The style of presenting weather charts that was introduced in 1949 was not considered 'good television'. In November 1953, the BBC accepted the opinion of the Office that live presentations would be preferable and pressed for the appointment of someone from the Office who could become a TV personality.[30] However, to quote Britain's first TV weatherman, George Cowling, "the idea of a civil servant becoming a personality was anathema to both the Office and the Air Ministry and they wisely argued that the job could not possibly be done seven days a week by one person".[31] Senior Meteorological Officers around the country were asked to nominate possible TV weathermen, for which a qualification was that spectacles must not be worn, as they would reflect into the camera's lens. Two of the Office's staff were selected, George Cowling and Tom Clifton. They made successful trial presentations on closed-circuit TV just before Christmas 1953 before an audience of BBC and Meteorological Office directors, after which, on 4 January 1954, they were introduced to reporters from national and regional newspapers at a press conference held in Broadcasting House. Cowling made the first live presentation a week later, at 7.55 p.m. on 11 January.

[29] See 'Using the shipping forecasts' by K Blowers, published in 1957 in *Weather* (Vol. 12, pp. 87–91). See also six articles in *Weather* by C E Wallington which explained in detail how to use the information broadcast in shipping bulletins to draw weather charts. The articles were published in the August, September and October 1964 and January, February and March 1967 issues of the magazine and reprinted in a booklet entitled *Your own weather map* (Royal Meteorological Society, 1967, 36 pp.). An expanded and updated version of this booklet, by M W Stubbs, was published by the Society in 1983 (66 pp.).

[30] See 'TV weatherman', by George Cowling, published in 1957 in *Weather* (Vol. 12, pp. 22–26).

[31] 'First weatherman on BBC TV', by George Cowling, published in 1994 in *Weather* (Vol. 49, pp. 347–349).

Camera trials showed that the working charts of the Office were unsuitable for television because of the shine from their paper and their weak colouring and outlines. Accordingly, special charts were printed on thin card with a matt surface, with the sea shown in pale grey and land in dark blue. Isobars and fronts were drawn on the charts with quick-drying black ink, using a thick, felt-nibbed fountain pen.[32] During the actual broadcast, the forecaster used sticks of compressed charcoal for demonstration purposes. Television was, of course, in black and white in those days. Colour TV came later.

A critical review of the TV presentations appeared in the March 1957 issue of *Weather* (Vol. 12, pp. 70–71). The unnamed author considered them "sound but a bit unimaginative" and went on to say that he thought it particularly welcome that the impersonality of the announcer of the radio was not being followed. He was also glad the forecasts were not being described as 'official', a word which, in his opinion, gave a spurious air of accuracy to many people. He suggested that "the cult of personality could be encouraged still further", adding:

Would it be so expensive in these days to have a permanent TV link with the Central Forecasting Office, so that viewers can see the actual forecasters, who would be able to add the personal touch so much better than the spokesman now seen?

As for the charts themselves, the author's view was that these were "merely simplified versions of the forecasters' own synoptic chart", showing isobars and fronts. He suggested that the charts should "have on them a direct representation of the weather" which the forecaster could show by using stipple rollers to shade areas of frontal rain and showers. He asked if some use could be made of still photographs or film to show, say, fair-weather cumulus when this was likely to occur. Perhaps "an inch or two of space could be found occasionally in the *Radio Times* to illustrate a few cloud forms or to explain the meanings of some of the terms used".

Besides radio and TV broadcasts and the weather forecasts published in newspapers, the Office provided meteorological services for the RAF, civil aviation, the Army, the Royal Navy, the Merchant Navy, fishing fleets, British Railways, gas undertakings, a number of Commonwealth countries and various other bodies.[33] Services listed in the annual report of the Office for the year ending 31 March 1952 included:

- For the British Electricity Authority, daily forecasts of meteorological conditions that would affect the loading of the electricity grid, as well as special week-end temperature forecasts and a bi-weekly broad outlook for four days ahead.

---

[32] For details of how and when charts were prepared for use on TV, see the two articles by Cowling cited in footnotes 30 and 31.

[33] During the Festival of Britain, the Office maintained a weather forecasting unit in the Festival of Britain Dome of Discovery pavilion on London's South Bank and published, twice daily from 4 May to 30 September 1951, a special *Souvenir Weather Report and Forecast*. More than 150,000 copies were sold to visitors, the price per copy being threepence.

- For agriculture and horticulture, frost warnings for dissemination by County Branch Officers of the National Farmers' Union to fruit growers and market gardeners, as well as warnings of snowfalls or drifts expected on high ground in northern England, for dissemination by these Officers to sheep farmers.
- For the Docks and Inland Waterways Executive, warnings of persistent frosts that were likely to produce ice on canals in the Wolverhampton area.
- For road engineers in government departments, local authorities and motoring associations, warnings of the onset of meteorological conditions likely to cause difficult road conditions.
- For the Ministry of Fuel and Power, warnings of sharp falls in barometric pressure that might affect colliery operations.
- For ceremonial occasions, forecasts as required, such as, in February 1952, the special forecasts supplied to the Earl Marshall's office and the Household Cavalry in connection with the funeral of King George VI.

The Office also provided meteorological services in a number of foreign countries. In the British Zone of Germany, for example, the Office continued to be responsible in the early 1950s for the provision of meteorological services to meet the needs of British Forces and the British High Commission. Plans to transfer control of meteorological activities in Allied sectors of Germany to the authorities of the Federal Republic of Germany were, however, well in hand by the early 1950s. A law that brought about the fusion of the meteorological services of the three western zones of Germany was passed by the Federal Parliament in November 1952, and the central administrative unit of the new Federal German Meteorological Service, Deutscher Wetterdienst, was established in February 1953, at Offenbach am Main, near Frankfurt.

A meteorological office was maintained in Austria, too, at Schwechat (near Vienna), with a British meteorologist in charge of Austrian forecasters and assistants. Moreover, meteorological services in many parts of the Middle East were supported or controlled by the British for quite some time after the war. In early 1953, for instance, technical control of meteorological services in Libya continued to be exercised by the Director of the Office, with staff of the Office continuing to fill posts for which suitably trained staff of local origin were not available. From some countries, though, British staff were being withdrawn. From Sudan, for example, one of the two forecasters who were provided by the Office to meet RAF requirements at Khartoum was withdrawn during 1952, and in Eritrea technical control over meteorological services was exercised by the Director of the Office until September 1952, when the territory was federated with Ethiopia and British staff were withdrawn.

Severe Weather Events

Three severe weather events in the early 1950s received much coverage in the press and on radio and TV. The first of them was an intense and prolonged rainstorm over the hills of Exmoor in south-west England, where, on 15 August 1952, some

230 millimetres of rain fell in under twenty-four hours on already sodden ground. A number of villages were flooded, with the place worst affected being Lynmouth, on the coast five to ten miles north of the area of heaviest rain. It was devastated. Torrents of water poured off the moors, carrying boulders, trees, carcasses of animals and dozens of vehicles. Bridges and buildings were demolished, and thirty-four people died.[34] As with the disaster at Louth in 1920 (see Chapter 9), questions were asked. Was the storm forecast? If not, could it have been?[35] Heavy and prolonged rain was indeed forecast. However, as in 1920, the state of weather forecasting in 1952 was still such that predictions of the intensities and precise locations of storms were not possible.

An intriguing speculation has persisted for many years, that rainmaking experiments over southern England contributed to the meteorological conditions which caused the storm over Exmoor. Rainmaking was certainly topical at the time. In fact, the whole of the July 1952 issue of *Weather* was devoted to the subject, which served to fuel the speculation. It is recorded in the annual report of the Office for the year ending 31 March 1953 that more than thirty cloud exploration flights had been carried out by aircraft of the MRF during the year in question. However, there is no mention of rainmaking in that annual report. The UK's practical involvement in this activity began in the spring of 1954, when the MRC concluded that the growing interest in, and importance of, the subject was such that a trial should be carried out.[36] It should be stressed that there is no evidence that the storm which brought devastation to Lynmouth in 1952 was anything but a natural occurrence.

The second severe weather event occurred in December 1952. A smoke-laden fog shrouded London for five days and brought with it premature death to thousands and inconvenience to millions. Road, rail and air transport were brought almost to a standstill. This infamous fog has come to be known as 'The Great Smog', the term 'smog' being a portmanteau word meaning 'fog intensified by smoke'.

The weather was cold in early December 1952, as it had been for some weeks. To keep warm, the people of London had burnt large quantities of coal in their grates, causing smoke to pour from the chimneys of their houses. Particles and gases had been emitted from factory chimneys, too, and winds from the east had brought pollution from industrial areas on the Continent. Early on 5 December, an anticyclone brought to the London area atmospheric conditions that were ideal for the formation of 'radiation fog', and a smoky layer of it 100–200 metres deep formed.[37] Elevated spots such as Hampstead Heath were above the fog. From there, the hills of Surrey and Kent could

---

[34] See *The Lynmouth Flood Disaster*, by E R Delderfield (Newton Abbot: Raleigh Press, 1953, 164 pp.).

[35] For a detailed discussion of the meteorological conditions, see 'The 1952 Lynmouth floods revisited', by J B McGinnigle, published in 2002 in *Weather* (Vol. 57, pp. 235–242).

[36] For details of this trial, see 'The Meteorological Office experiments on artificial rainfall', by B C V Oddie, published in *Weather* in 1956 (Vol. 11, pp. 65–71).

[37] At night, when skies are generally clear of cloud, the ground cools. This occurs because more heat is radiated from the ground than is received from above, there being no sunshine at night. If there is no wind, dew forms

be seen. In some parts of London, the fog was so thick that people could not see their own feet! Not until 9 December did the fog clear.

The death toll of about 4000 was not disputed by the medical and other authorities, but exactly how many people died as a direct result of the fog will never be known. Some suffered already from chronic respiratory or cardiovascular complaints. Without the fog, they might not have died when they did. The number of deaths in Greater London in the week ending 6 December 1952 was 2062, which was close to normal for the time of year. The following week, the number was 4703. Mortality from bronchitis and pneumonia increased more than sevenfold as a result of the fog.[38]

The Office predicted the anticyclonic conditions and the formation of fog. What they could not forecast was the extent to which the smoke from the chimneys of London would aggravate the fog problem. Legislation followed the Great Smog of 1952 in the form of the City of London (Various Powers) Act of 1954 and the Clean Air Acts of 1956 and 1968. These Acts banned emissions of black smoke and decreed that residents of urban areas and operators of factories must convert to smokeless fuels. As residents and operators were necessarily given time to convert, though, fogs continued to be smoky for some time after the Act of 1956 was passed, but nothing on the scale of the 1952 event has ever occurred again. In the early 1960s, winter sunshine totals were 30% lower in the smokier districts of London than in rural areas around the capital. Today, there is little difference.[39]

The third severe weather event occurred at the end of January 1953, in the form of one of the worst storms to visit the UK in the twentieth century. Indeed, three disasters resulted from this one storm. First, on 31 January, a car ferry, the *Princess Victoria*, on passage from Stranraer in Scotland to Larne in Northern Ireland, sank with the loss of 133 lives. Only forty-one of the passengers and crew survived. Later that day, winds of almost hurricane-force blew down more trees in Scotland than were normally felled in a year. And then, from Yorkshire to the Thames Estuary, coastal defences were pounded by the sea and gave way under the onslaught. Sea level was raised more than two metres, and huge waves were created. Almost 100,000 hectares of eastern England were flooded, and 307 people died. In The Netherlands, fifty dykes burst and 1800 people drowned. Along the coast of eastern England, there have been many failures of coastal defences because of raised sea level and accompanying large waves. However, no previous event compared with that of 31 January and 1 February 1953. It was the greatest on record for the North Sea *as a whole*.[40]

(i.e., condensation in the air which is in contact with the ground). If there is a very light wind, a shallow layer of condensation known as radiation fog forms. It is typically 100–200 metres deep.

[38] For personal experiences of the 1952 Great Smog, see 'London Particulars and all that', by R J Ogden, published in 2000 in *Weather* (Vol. 55, pp. 241–247).

[39] This is true of visible pollution but not of micro-particles and gases, particularly ozone and nitrogen oxides. Levels of these are considerably greater in London today than in the surrounding countryside.

[40] For details of the events of 31 January and 1 February, see *The Great Tide*, by H Grieve, published in 1959 by Essex County Council (883 pp.). See also the June 2005 issue of the *Philosophical Transactions of the Royal Society*

The risk of flooding was predicted by the Office and the Dutch storm-tide warning service, in that forecasts of exceptionally high water levels were issued several hours before they occurred.[41] Nevertheless, the Waverley Committee that was appointed by the British government to inquire into the disaster recommended that a 'Flood Warning Organization' be set up. This advice was accepted and the name 'Storm Tide Warning Service' adopted later. Its Dutch counterpart had been formed soon after the event of January 1916, when vast areas of the Netherlands were inundated. The Storm Tide Warning Service has been operated by the Office from the outset and is today called the 'Storm Tide Forecasting Service'.

A New Director

Sir Nelson Johnson retired on 28 August 1953. He had served the Office well, but the heavy workload he had borne during his fifteen years in charge had taken its toll. By the time he retired he was not a well man, suffering from Parkinson's disease. His retirement was short. He took his own life on 23 March 1954.

The choice of a new Director was made in advance of Sir Nelson's retirement by a sub-committee of the Meteorological Committee made up of Professor G M B Dobson (the Committee's Vice-Chairman), Sir David Brunt (Secretary and Vice-President of the Royal Society), Sir Ronald Ivelaw-Chapman (Deputy Chief of the Air Staff) and Mr I V H Campbell (the Assistant Under-Secretary of State for Personnel in the Air Ministry).[42] There had been three candidates: Instructor Captain P Bracelin, Dr J M Stagg and Professor O G Sutton.

Bracelin was Director of the Naval Weather Service and sat on the Meteorological Committee and the MRC. Stagg had enjoyed a long and distinguished career in the Office, and Sutton was also well acquainted with the Office. He sat on the Meteorological Committee, the MRC and the Gassiot Committee and had worked for the Office in the early part of his career. He had joined the Office in 1928 as a Junior Professional Assistant and been seconded in 1929 to the War Office Experimental Station at Porton, where he had become head of the Meteorological Section in 1938, Superintendent of Research from 1942 to 1943 and Superintendent of Tank Armament Research from 1943 to 1945. He had then been Chief Superintendent of the Radar Research and Development Establishment at Malvern from 1945 to 1947 before moving to the Military College of Science at Shrivenham, near Swindon, where he had become the

---

(Vol. 353A), in particular 'The Big Flood: North Sea storm surge', by A McRobie, T Spencer and H Gerritsen (pp. 1263–1270) and 'The east coast Big Flood, 31 January – 1 February 1953: a summary of the human disaster', by P J Baxter (pp. 1293–1312).

[41] The Meteorological Office forecast the storm winds and raised sea level operationally by conventional methods. However, one of the earliest of the Office's computer forecasts showed a severe north-westerly gale across the British Isles and North Sea. See Hinds (*op. cit.*, footnote 18).

[42] Brunt was knighted in 1949 and retired from Imperial College in 1952. The chair in meteorology at the College passed to P A Sheppard and Brunt was then granted the title of Professor Emeritus.

Bashforth Professor of Mathematical Physics in 1947 and Dean of the College in 1952.

The sub-committee interviewed Stagg and Sutton on 18 March 1953 and submitted their report, dated 24 March, to George Ward MP, who was the Parliamentary Under-Secretary of State for Air and Chairman of the Meteorological Committee.[43] The sub-committee agreed that Bracelin "was an admirable administrator" but had produced "no evidence he had made any outstanding contribution to meteorology or advanced any original ideas". They therefore concluded that "no useful purpose would be served by interviewing him". They felt, moreover, that "it would be wrong in principle to go outside the academic field and the Scientific Civil Service, unless no really suitable candidate could be found from either of these two sources".

The view on Stagg was that he "lacked certain of the essential qualities required of the Director, particularly the breadth of outlook required to foresee future demands on the Meteorological Office and the ability to direct research". They felt that he was "unlikely to inspire his staff with the enthusiasm which was essential for an efficient and progressive organization" and considered that he was "far better and indeed well suited for the post of Principal Deputy to the Director". Sutton, on the other hand, possessed "the requisite breadth of outlook and the necessary vision and scientific ability to direct research into the proper channels, as well as considerable administrative experience". Furthermore, he had "wide contacts both in the scientific world and in the Civil Service", which the sub-committee considered "most desirable for a Director of the Meteorological Office". His appointment was recommended unanimously.

The announcement of Sutton's appointment was made by the Air Ministry on Sunday 12 April 1953, and the following comment appeared in *The Times* the next day:

The naming of the successor to Sir Nelson Johnson should arouse interest in a nation that talks continuously about its weather. It would be pleasant to think that fishermen from Finisterre to Forties were this morning pausing at their nets to murmur regrets at the departure of a trusted friend; that gardeners and gymkhana organizers and all those millions who know of the Meteorological Office only through wireless and newspaper weather forecasts were penning notes of appreciation. Actually this almost anonymous chief of the most impersonal part of the vast Air Ministry machine is little known to the general public he has served with distinction.

One of Sir Nelson Johnson's responsibilities, before he lays down his office, will be the forecasts of the Coronation weather. Nine Junes ago, when the Normandy landings were imminent, he carried an even heavier burden without faltering.[44]

---

[43] There is a copy of this report in the National Archives at Kew (AIR2/10737, Document 41A).

[44] As correctly forecast, the weather in London on Coronation Day, 2 June 1953, turned out to be chilly with a brisk northerly wind and sporadic outbreaks of rain.

Figure 11.6. Oliver Graham Sutton. © Crown Copyright 2010, the Met Office.

The editorial went on to note that Sutton would "inherit a distinguished tradition" and preside at the centenary ceremonies of the Office. The Office was "everybody's servant".

Oliver Graham Sutton was 50 years of age when appointed Director of the Office. Born on 4 February 1903 at Cwmcarn in Monmouthshire, he had been educated first at the local elementary school and then, from 1914 to 1920, at Pont-y-waun County School. He had proceeded from there to the University College of Wales at Aberystwyth, where he had graduated in 1923 with first-class honours in pure mathematics with subsidiary applied mathematics, physics and chemistry. After that, he had spent two years at Jesus College, Oxford, where he had gained the research degree of Bachelor of Science for his work on series of orthogonal functions. He had then taught mathematics in a school in Cardiff for a year before becoming, in 1926, an assistant lecturer in mathematics in the University College at Aberystwyth. His interest in meteorology had stemmed from a visit by Brunt to Aberystwyth, and he had joined the Office on 26 March 1928, his first posting being to Shoeburyness, where he had worked under Brunt's direction on meteorological corrections to gunnery procedures.[45]

---

[45] Brunt was also a graduate of Aberystwyth.

After his move to Porton (where he had served initially under Nelson Johnson), Sutton had established an international reputation for his work on atmospheric turbulence and smoke diffusion. His publications included *Atmospheric Turbulence*, published in 1949, and a popular book, *The Science of Flight*, published in 1948. The book which an obituarist called "his crowning achievement", *Micrometeorology. A Study of Physical Processes in the Lowest Layers of the Earth's Atmosphere*, was published in the year he became Director of the Office.[46] He was elected a Fellow of the Royal Society in 1949.

He took the helm of the Office on 1 September 1953 and did indeed preside at the centenary celebrations, but they took place in 1955. No one in the Office seemed to know that the year of foundation had been 1854. In the special issue of the *Meteorological Magazine* that was published in June 1955 (Vol. 84, pp. 161–198), there was a message from the Secretary of State for Air which asserted that 1955 was the centenary year. Furthermore, G A Bull stated incorrectly in the opening paragraph of his otherwise excellent 'Short history of the Meteorological Office' (pp. 163–167) that the Office had been founded early in 1855 and its title changed to 'Meteorological Office' in 1863. In fact, FitzRoy took up his duties on 1 August 1854 and the title was changed formally on 25 February 1867.

Sutton contributed a Foreword to the special issue, noting that, when FitzRoy started work, "it was difficult to disentangle fact from folklore in the study of weather, and meteorology hardly existed as a science". It was, he went on,

... a far cry from the nineteenth century, with official interest in meteorology represented by a handful of staff in a small London Office, to the modern weather service with its headquarters, outstations, observatories, weather ships and research flights. Aviation and telecommunications have greatly extended the range and power of operational meteorology, but at the cost of an increasing complexity of organization. The paucity of university support for meteorology has compelled the Meteorological Office to build up its own research division, to the great profit of all concerned. Our activities now touch almost every aspect of national life.

Already a Commander of the Order of the British Empire (CBE), appointed in June 1950, Sutton received a knighthood in the 1955 New Year's Honours List.

The Brabazon Report

Yet another review of the Office's work and organization commenced in 1955. This time, the committee of inquiry was chaired by the aviation pioneer Lord Brabazon of Tara, its remit being, as announced in the House of Commons on 25 April 1955 (by George Ward), "to review the organization of the Meteorological Office in relation

---

[46] Obituary of Sir Graham Sutton (1903–1977), by R C Sutcliffe, published in 1978 in the *Quarterly Journal of the Royal Meteorological Society* (Vol. 104, pp. 539–540).

to current and future requirements". The other members of the committee were Sir Charles Darwin, Major R H Thornton, Sir Folliott Sandford and Mr J R Simpson.[47]

As with previous inquiries into the work of the Office, Members of Parliament became impatient over the time the committee took to produce their report. On 23 February 1956, Geoffrey de Freitas asked the Secretary of State for Air when he expected to receive a report from the Brabazon Committee. In reply, the Secretary of State, Nigel Birch, said that a considerable amount of evidence had been submitted, and it was impossible to say when the committee would report. The following month, however, a first draft of the report appeared, followed in June by a second draft. Then, in August 1956, the final draft was submitted to the Air Ministry. It was a typewritten document which ran to forty-seven pages and contained 146 paragraphs and eleven appendices.[48]

The terms of reference of the committee were set out in Appendix A. Besides the remit announced on 25 April 1955, the committee were "to have regard to the international obligations of the UK as a member of the World Meteorological Organization, and to the requirements of all users – aviation and non-aviation – service and civilian – in peace and in relation to war". The committee had to assume that the Office would continue to form part of the Scientific Civil Service and that the structure, grading, standards and conditions of service would conform with those of the Scientific Civil Service as a whole. They were also to "have regard to statements relating to promotion and career prospects contained in the White Paper of 1945 (Cmd. 6679)". The committee were authorized to take evidence from Air Ministry and RAF witnesses, as well as from other government departments and from representative organizations such as the National Farmers' Union. With a view to ensuring harmonious industrial relations, the staff side of the Air Ministry Departmental Whitley Council had to be consulted, too. In total, as listed in Appendix B, fifty-four bodies were invited to submit evidence.

The committee held twenty-eight meetings, which included visits to the CFO at Dunstable, the branches of the Office at Harrow and the meteorological forecasting office at London's Heathrow Airport.[49] Individual members of the Committee also visited the forecasting offices at a number of RAF stations, as well as the Edinburgh Branch of the Office and the Imperial College Department of Meteorology. In addition, Sir Charles Darwin visited the U.S. Weather Bureau.

---

[47] Darwin was a physicist who had made valuable contributions to hydrodynamics. He had served as Director of the NPL during the war. His father, Sir George Darwin, had served on the Meteorological Committee in the early years of the century (see Chapters 7 and 8). Thornton was a member of the board of the British Overseas Airways Corporation and a member of the Air Registration Board. Sandford was Deputy Under-Secretary of State in the Air Ministry. Simpson was a senior official of the Treasury.

[48] There are copies of the Brabazon Report in the National Archives at Kew (AIR20/9417) and the National Meteorological Archive at Exeter.

[49] For information about the establishment and operation of the forecasting office at Heathrow Airport, see 'The meteorological forecasting office at Heathrow', by R J Ogden, published in 1998 by the Royal Meteorological Society (*Occasional papers on meteorological history*, No. 1, 43 pp.).

In their report, the committee said that the post-war period had seen no falling off in aviation requirements. Indeed, they reported, "the development, on the one hand, of civil aircraft operating over long distances with high premiums for regularity, safety and economic loading, and, on the other, of fighters operating at very great speeds and with limited endurance, had set the meteorologist a number of new problems". Moreover, involvement with the International Civil Aviation Organization, WMO and the North Atlantic Treaty Organization had added materially to the commitments of the Office, particularly those of senior staff. In parallel, the committee reported, there had been "a growing public interest in the application of forecasting and climatology to agricultural and other purposes". Furthermore, "the very successful television forecasts arranged by the BBC with the assistance of the Meteorological Office", along with the forecasts broadcast on the radio and published by the Press, telephone weather services, pamphlets such as *Your Weather Service* and *Weather and the Land*, instruction given in schools, and the activities of the Royal Meteorological Society had all "helped in building up an interest and market for meteorology" which had not existed a few years ago.[50]

Since the war, the committee pointed out, there had been a steady increase in expenditure on the national meteorological service. This had resulted in part from general rises in costs and salaries but also from an increase in services provided by the Office and the development and use of new types of equipment, such as the radiosonde, facsimile reproduction, the radar scanner and the electronic computer. The cost of the Office to the state was now of the order of 4 to 5 million pounds a year and tending to rise, though not all of this expenditure was reflected in Air Votes, which did not include the cost of staff belonging to the Department of the Permanent Under-Secretary who were engaged wholly or partly on establishment, finance and other work connected with meteorology, nor any of the cost of operating the RAF squadron at Aldergrove in Northern Ireland, which was largely engaged in making the meteorological flights over the North Atlantic. Furthermore, expenditure on accommodation occupied by the Office was a final charge to the Ministry of Works, and expenditure on stationery and printing a final charge to Her Majesty's Stationery Office.

The factors which had led to the appointment of the Brabazon Committee were stated in the committee's report, one of them being that the Office was "experiencing serious difficulty in recruiting and retaining staff in adequate numbers". Another was that Sutton wished to improve and extend the services available to non-aviation users through regional offices situated in the provinces. In addition, doubts had been expressed by the Chairman and some of the independent members of the Meteorological Committee about the value of that Committee as at present constituted.

---

[50] These pamphlets were prepared by the Meteorological Office in association with other government departments and published by Her Majesty's Stationery Office.

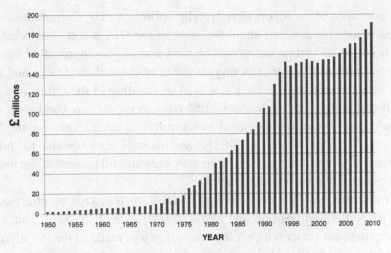

Figure 11.7. Meteorological Office turnover in millions of £ from the year ending 31 March 1950 to the year ending 31 March 2010.

Another important factor was that the Office's branches were somewhat scattered in and around London, with some at Victory House in Kingsway, others at Harrow, Dunstable and Stanmore. The Brabazon Committee felt that "the whole organization would unquestionably gain not only in efficiency but in morale if it could be assembled in one place", and "economies of staff should also be possible". To this end, the Committee were very glad to report that approval had been given to a plan for developing a new headquarters for the Office in the new town at Bracknell in Berkshire.[51]

In their report, the Brabazon Committee considered all aspects of the Office's work, but especially forecasting techniques, research, services to various user groups, administrative structure, finance, and recruitment and retention of staff. As regards forecasting, the Committee considered that "the acid test for a meteorological organization costing some £4M or £5M a year must clearly be the measure of success achieved in improving the technique of forecasting and applying the data obtained for the benefit of the various users".

They found that forecasts had been "about 80% right" in the 1930s and were still only about 80% right. They were optimistic over the potential of NWP for improving the quality of weather forecasting and glad to learn in this respect that approval had recently been given to the purchase of an electronic computer for use in the Office. The Committee noted that the Office was experimenting with four-day forecasts but found "no evidence that research into the problems of long-range forecasting was about to enter a decisive or significant phase".

---

[51] Bracknell was first mentioned as a possible site for the Office's headquarters in November 1954, and the final decision was taken in July 1955. The move was part of the government's strategy to move departments out of central London.

The conclusions of the report were broadly optimistic. The Committee did not find any serious cause for concern about the state of the Office and concluded that "the present shortages of certain categories of staff and the deployment of headquarters branches were to some extent reacting on the efficiency of the Office and its ability to accept new commitments". They felt that the standing of the Office as a scientific institution and in the eyes of the public had risen in recent years and were confident there would soon be an improvement in recruitment, particularly for the Scientific Officer class. They also hoped the Director and his staff would be able to "bring home to the universities the challenge which meteorology should present to the imagination of the first-class scientist or mathematician".

The Committee had "no major points to raise on the scale of effort devoted to research" or to the composition and terms of reference of the MRC, but suggested that the arrangement under which a separate grant was made to the Royal Society for fundamental research should be discontinued at the end of the current five-year period. After that, they recommended, all grants for research should be channelled through the MRC. They further recommended that the Monday discussions should continue and that Scientific Officers be encouraged to attend symposia arranged by the Royal Meteorological Society and the Imperial College Department of Meteorology.

A number of staffing and organizational recommendations were made, among them the possibility of graduates being recruited into the Experimental Officer class. So far as higher posts were concerned, a recommendation was that there should be two people graded Chief Scientific Officer, one responsible for forecasting and services, the other for research. In addition, there should be an Assistant Secretary responsible for administrative and general duties. They believed Sutton was under-graded and ought to be placed on a par with the Directors of the NPL and the Royal Aircraft Establishment.

The Brabazon Committee concluded that the Meteorological Committee as then constituted did not perform any useful function. Their view was that it should be replaced by a small committee of not more than five members, of whom one should be the Chairman of the MRC and another a scientist. The others, they suggested, should preferably be laymen who did not represent any particular bodies. The purpose of the committee should be to keep under review the progress and policy of the Office and the scale of expenditure on the meteorological service. It should also ensure that adequate contact was maintained between the Office and those who used its services. They recommended that each member of the committee should receive an honorarium.

Many months passed before the government reacted officially to the report of the Brabazon Committee. Then, in the House of Commons on 28 June 1957, the Secretary of State for Air, George Ward, said that he was now in a position to inform the House of the Committee's conclusions and the decisions the government had taken as a result. Essentially, the government had accepted all of the Committee's recommendations.

He reported that the Committee had seen "no reason to question the wisdom of the decision taken in 1919 and reaffirmed in 1945 to entrust responsibility for the State Meteorological Service to the Air Ministry as the department with the major user interest". The close association of the Office with aviation since 1919 had been "to the mutual advantage of both parties". The Committee judged, Ward said, that the standing of the Office as a scientific institution was high. Moreover, the work being carried out on the development of numerical methods of forecasting had been welcomed by the Committee. So, too, he said, had the decision to install an electronic computer in the CFO and the plan for establishing a combined headquarters at Bracknell. In addition, the recommendation of the Committee that the higher posts be reorganized had been accepted and the status and salary of the Director raised, with Sir Graham's title altered to 'Director-General' with effect from 1 July 1957.

Ward informed the House of Commons that a recommendation of the Committee was still under discussion, this being that grants for research which emanated from the Air Ministry should normally be channelled through the MRC. However, the government had accepted the Committee's advice that a system of research grants for post-graduate study in meteorology was desirable. Indeed, the Department of Scientific and Industrial Research had been approached and found willing to consider awards under the scheme it administered. Ward hoped the number of candidates for research grants in meteorology would increase.

The government agreed with the Brabazon Committee that the Meteorological Committee as constituted in 1919 was too large and had ceased to perform any very useful function. As originally constituted, it had included representatives of the various government departments with a user interest in meteorology, as well as the universities, the Royal Society and the Royal Society of Edinburgh. In view of the close relations which existed at working level between the Office and the other bodies concerned, Ward said, the government had accepted the recommendation that the existing committee be replaced by an advisory committee of not more than five members. Their number would include the Chairman of the MRC ex-officio (then Sir Charles Normand), and Lord Hurcomb would chair the new Committee. He was a former chairman of the British Transport Commission.[52] The names of the other members of the new Advisory Committee were announced by the Air Ministry in August 1957. They were Sir Austin Anderson (Chairman of the Orient Steam Navigation Company), Sir David Brunt (Professor Emeritus, University of London) and Colonel N V Stopford-Sackville (Chairman of the Northamptonshire Agricultural Executive Committee).

---

[52] Sir Charles Normand (1889–1982) was a distinguished meteorologist who had spent the whole of the period from 1913 to 1944 in India, apart from a few years during the Great War when he had served in Mesopotamia. He had been the head of the India Meteorological Department from 1927 to 1944, his title being Director-General of Observatories. He had been knighted in 1945, the year he returned to the UK, and had served as President of the Royal Meteorological Society from 1951 to 1953.

Progress at Home and Abroad

Whilst its future was being determined, the Office continued to develop and expand at home and play a full part in international meteorology. The Director-General was the Permanent Representative of the UK with WMO, and various members of staff served on WMO panels and commissions. In 1957, for example, the Head of the MRF, R J Murgatroyd, joined a panel of experts to study meteorological developments required by the expected world-wide introduction of commercial jet aircraft, and Sutcliffe became President of the Commission for Aerology. In addition, Stagg attended a meeting of the Meteorological Committee of the Standing Group of the North Atlantic Treaty Organization held in Washington, DC, in May 1957, and Sutton attended the General Assembly of the International Union of Geodesy and Geophysics at Toronto in September 1957. At home, members of the Office served on national committees and supported learned societies, especially the Royal Meteorological Society. Sutcliffe, for example, was President of the Society from 1955 to 1957, and so was Stagg from 1959 to 1961.[53] Sawyer was Editor of the Society's *Quarterly Journal* from 1958 to 1961 and served as President of the Society from 1963 to 1965.

As regards scientific meteorology, the principal undertaking of the 1950s was an ambitious research enterprise which brought about international cooperation in a world-wide programme of observation that embraced nearly all branches of geophysics. This was the International Geophysical Year (IGY), which formally extended from 1 July 1957 to 31 December 1958, though some parts of the observational programme commenced well before July 1957 and some continued through 1959. Meteorologists and other scientists from more than sixty countries undertook a cooperative investigation that involved observations and studies of the sun and the earth's interior, crust, oceans and atmosphere, and scientists from more than 100 countries took part in the meteorological programme.

The IGY grew out of a proposal made socially at a meeting of earth scientists held in Washington, DC, in April 1950 that a third International Polar Year should take place not fifty years after the last one but twenty-five years (i.e., in 1957–1958). The International Council of Scientific Unions (ICSU) adopted the idea in 1951 and appointed a committee in 1952 to organize the venture. This committee decided that the scope of the project should be widened to embrace the whole earth.[54] In the event, a number of international scientific bodies joined with ICSU to conduct the enterprise, with WMO bearing responsibility for the meteorological programme and the collection, reproduction and subsequent distribution of the data. The cost of the venture was shared by the various participating bodies and nations. The choice of

---

[53] Stagg retired from the Meteorological Office in September 1960.

[54] For a detailed assessment of the objectives and achievements of the Polar Year programmes and an outline of the evolution of the IGY concept, see 'From Polar Years to IGY', by N C Gerson, published in 1958 in *Advances in Geophysics* (Vol. 5, pp. 1–52).

1957–1958 proved an excellent one, for this period coincided with a sunspot maximum of record intensity. A partial repetition of the IGY programme took place in 1964–1965, in the so-called International Quiet Sun Years (IQSY), which coincided with the next sunspot minimum.[55]

During the IGY, the normal observational programmes of the Office continued. These included surface and upper-air observations; measurements of radiation, ozone and the earth's magnetism; and studies of atmospheric chemistry. There were also six special periods called 'World Meteorological Intervals', each lasting ten days, with one period in each quarter year. During these periods, the Office made supplementary upper-air soundings at the equinoxes and solstices and, at the same time, made special efforts to attain greater heights. Other activities undertaken by the Office during the IGY included the measurement of ozone at Habbaniya in Iraq and the recording of solar and terrestrial radiation at Aden, Malta and Stanley (Falkland Islands) and on the British weather ships on the North Atlantic.

Stagg and Sutton served on the British National Committee for the IGY, as national correspondents for geomagnetism and meteorology, respectively. With H W L Absalom, Stagg also served on the Antarctic Sub-Committee, helping to organize a venture which formed a considerable contribution of the UK to the IGY, this being the Royal Society's Antarctic Expedition, which was based at Halley Bay (75°31'S 26°36'W) from 1957 to 1959. Several members of the Office took part in this expedition, studying meteorology, geomagnetism, glaciology and seismology.

A Royal Society soirée was held on 28 June 1957 to mark the inauguration of the IGY, and the Office demonstrated at it equipment used for locating thunderstorms by means of radio direction-finding. In addition to that technique, radar was used increasingly in the 1950s to study thunderstorms, notably by a team from the Imperial College Department of Meteorology led by a former member of the Office, F H Ludlam, with the support of staff of the Office's radar station at East Hill.

In the late 1950s, Ludlam studied thunderstorms intensively, using not only radar but also a network of voluntary observers across southern England who recorded, whilst storms were in progress, hail size, lightning frequency, and rain and hail duration. He also used sferics observations supplied by the Office. Out of his work came a revolutionary three-dimensional model of a severe storm that occurred over southern England in 1959.[56] This model showed the flow pattern of updraughts and downdraughts in the storm and the areas affected by rain and hail, and it inspired further radar-based studies of weather systems. Meanwhile, at Malvern in Worcestershire from 1958 to 1960, J R Probert-Jones of the Meteorological Office used a Doppler

[55] For an introduction to the work and achievements of the IGY and IQSY, see the second edition of *The Planet Earth*, edited by D R Bates and published in 1964 by Pergamon Press (370 pp.).

[56] See 'The role of radar in rainstorm forecasting', by F H Ludlam, published in 1960 in *Nubila* (Vol. 3, pp. 53–75). See also 'Air flow in convective storms', by K A Browning and F H Ludlam, published in 1962 in the *Quarterly Journal of the Royal Meteorological Society* (Vol. 88, pp. 117–135).

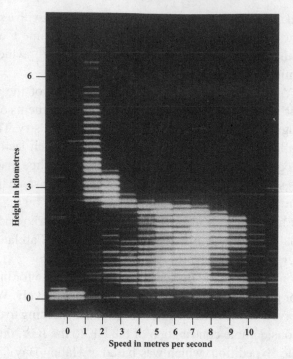

Figure 11.8. Doppler radar matrix display showing a vertical profile of echoes from precipitation over Worcestershire whilst continuous rain was falling. The display shows raindrops falling with a speed of between 5 and 8 metres per second below a height of about 2.5 kilometres and snowflakes falling with a speed of between 1 and 2 metres per second above a narrow band which contained a mixture of rain drops and melting snowflakes. © Crown Copyright 2010, the Met Office.

radar to measure the vertical velocities of raindrops and hail in showers and thereby deduce patterns of air motion in clouds.[57]

Other special investigations carried out by the Office in the latter part of the 1950s were focused mainly on problems connected with the operation of military and civil aircraft at increasing altitudes. In the words of the annual report for the year ending 31 March 1958, "enquiries from Service Departments, the Ministry of Transport and Civil Aviation, and operating companies concerned wind speed and direction and the density and temperature of the atmosphere, at heights up to 50,000 feet, in many parts of the world". Another problem was put thus in the same annual report:

Some time was spent on questions connected with the operation of Britannia aircraft, particularly in assessing the likelihood, in different seasons of the year and on different routes, of an encounter with the special cloud conditions believed responsible for the malfunctioning of the

[57] For an overview of this work, see 'A history of radar meteorology in the United Kingdom', by J R Probert-Jones, pp. 54–60 in *Radar in Meteorology* (edited by D Atlas), published in 1990 by the American Meteorological Society (806 pp.).

engine of this aircraft. Recent investigations within the Office into the structure and physics of cloud were of great value in making this assessment.

Problems of atmospheric diffusion and pollution were also tackled in the late 1950s, notably an investigation into the incidence of sulphur dioxide pollution near a generating station of the Central Electricity Authority. And a very serious occurrence was the serious fire that occurred in the nuclear reactor at Windscale in Cumbria on 10 and 11 October 1957. As a consequence of the release of radioactive material, a study was made of the Atomic Energy Authority's requirements for meteorological observations and forecasts.[58]

More mundane but no less important were the services provided by the Office for a wide range of users. Weather forecasting was, of course, the main preoccupation, with services offered routinely to the general public and to mariners, farmers and the operators of aircraft. Services specially for the RAF included forecasts for V-bombers carrying live nuclear weapons; and other special forecasts in the late 1950s included forecasts for pigeon races which started from places outside the British Isles, seasonal forecasts for horticultural concerns in relation to the raising of food crops, and forecasts of the likelihood of icing on the conductor rails of electrified railways in the south-east of England.

Warnings of weather hazards such as fog, snow and thunderstorms were issued principally through daily newspapers and the BBC's radio and TV services, while visual warnings of gales over coastal waters were still provided by means of cones hoisted at appropriate places along coasts, as in the 1860s. It was normal daily practice for the CFO to prepare weather forecasts for two to three days ahead. Outlooks for a day or so further ahead were issued to the press and the BBC whenever the degree of confidence in them was sufficiently high.

An innovation by the BBC in October 1957 was a five-minute talk at 1.00 p.m. on Thursdays by one of the TV weather presenters immediately after a new weekly programme for farmers. These talks proved to be popular with many viewers other than those in the agricultural world for whom they were intended. By 1957, forecasts and other meteorological information were supplied not only to the BBC but also to commercial TV companies. In addition, weather information was made available through an automatic telephone weather service which had begun in London on 5 March 1956. In the year ending 31 March 1958, this service was available in eleven major towns and cities and the total number of calls made was nearly five million. In that same year, an itinerant meteorological exhibition was displayed in museums and

[58] The report of the official inquiry into the Windscale event was published in 1957 as Parliamentary Paper Cmnd.302, entitled *Accident at Windscale No.1 Pile on 10th October 1957* (HMSO, 25 pp.). The inquiry was chaired by Sir William Penney. For meteorological information about the event, see two papers published in 1959 in the *Quarterly Journal of the Royal Meteorological Society* (Vol. 85): 'Deposition of iodine-131 in Northern England in October 1957', by A C Chamberlain (pp. 350–361), and 'The travel and diffusion of the radioactive material emitted during the Windscale accident', by J Crabtree (pp. 362–370).

Figure 11.9. London Weather Centre window display of current charts and other meteorological information, Kingsway, 1959. © Crown Copyright 2010, the Met Office.

art galleries in Birmingham, Coventry, Bristol, Cardiff, Cheltenham, Worcester and Northampton, where it was inspected by large numbers of visitors, including school parties. By the middle of 1960, weather centres had been opened in three cities. The first, on Kingsway in London, opened on 31 August 1959, the second in Glasgow on 14 December 1959 and the third in Manchester on 10 June 1960.

According to the Office's annual report for the year ending 31 March 1958, much of the work of the assistant directorate of Instrument Development lay in "the steady but unspectacular modification of existing instruments, to improve accuracy and reliability and to reduce the cost of purchase and maintenance". An example given in the annual report was the production of specifications for a precision aneroid barometer which it was admitted might cost as much initially as a mercurial barometer but was less liable to damage in transit and cheaper to repair if damaged. Radiosonde design was also being reviewed critically, with a view to equipment superior to that currently in use being developed for mass production at economic rates. This was a matter of major importance for the Office, as the number of radiosonde ascents made each year from weather ships and from stations at home and overseas was more than 12,000.

An event that was to prove enormously important for meteorology was the launch of the first earth-orbiting artificial satellite, *Sputnik 1*, by the Soviet Union on 4 October 1957. This was a Soviet contribution to the IGY, and the satellite provided information about temperature, density and other physical conditions at great heights

Figure 11.10. Making a weather observation on the Air Ministry roof, Kingsway, London, 1959. © Crown Copyright 2010, the Met Office.

Figure 11.11. Checking a rain-gauge on the Air Ministry roof, Kingsway, London 1959. Weather radar in the background. © Crown Copyright 2010, the Met Office.

Figure 11.12. Launch of radiosonde at Lerwick Observatory, 1963. Photograph by Malcolm Walker.

in the atmosphere.[59] There followed *Sputnik 2* on 3 November 1957 and *Sputnik 3* on 27 April 1958, the latter carrying a large array of instruments for geophysical research. The first American artificial satellite, *Explorer 1*, was launched on 31 January 1958, followed by *Explorer 2* (which failed to enter orbit) on 5 March 1958 and *Explorer 3* on 26 March 1958. These satellites, too, were part of the IGY programme and also helped advance knowledge of the high atmosphere.

The first successful weather satellite was *TIROS 1*, launched by the U.S. National Aeronautics and Space Administration on 1 April 1960.[60] It was designed to test experimental techniques for obtaining television footage of cloud patterns, doing so from a polar orbit which ranged in altitude from 700.9 to 753.62 km. For the first time, meteorologists could easily relate cloud distributions to patterns of isobars and fronts on weather charts, which they were especially keen to do for depressions and tropical cyclones. Hitherto, information about the whereabouts and extent of areas of stormy weather over the oceans far from land had been available only if seafarers provided it, and seafarers tended to avoid severe storms if at all possible. The images from this satellite revealed a large degree of organization in the cloud systems of

---

[59] The orbit of *Sputnik 1* was elliptical, ranging from 230 km altitude at perigee to 950 km at apogee.
[60] The acronym TIROS was derived from Television Infra-Red Observation Satellite.

extratropical depressions, which had been suggested by meteorologists but never before confirmed.[61]

The images from *TIROS 1* were not, in fact, the first pictures of clouds obtained from high altitudes. Such pictures had been taken from vertical-sounding rockets as early as 1949. However, pictures that embraced the full extent of weather systems had never before been obtained.[62] Meteorologists were quick to realize the potential of weather satellites.[63] Not only could information about cloud distributions be obtained; so also could temperatures of the upper surfaces of clouds, ground temperatures where there was no cloud, movements of clouds, concentrations of water vapour, and perhaps precipitation amounts. Some of these possibilities were well ahead of their time, though *TIROS 2*, launched on 11 November 1960, carried a five-channel medium-resolution scanning radiometer which was designed to measure radiation from the earth and its atmosphere.

The Office's First Electronic Computer

The mathematical models that formed the basis of NWP were refined progressively during the 1950s in a number of countries, including the UK. Moreover, the first attempts were made to produce a realistic numerical simulation of the atmospheric circulation over the whole earth. In 1956, indeed, the Royal Meteorological Society awarded its newly instituted Napier Shaw Memorial Prize to Norman Phillips of the Institute for Advanced Study at Princeton (United States) for his paper on 'The General Circulation of the Atmosphere: A Numerical Experiment', published in 1956 in the Society's *Quarterly Journal* (Vol. 82, pp. 123–164).[64]

Throughout the 1950s, Sawyer, Bushby, Knighting and others of the Office continued the work begun in 1951 to develop numerical models for predicting the weather, but they did not have their own computer. As noted earlier in this chapter, they first used LEO 1. Then, from early 1954, they had available to them a machine significantly faster than LEO, the Ferranti Mark 1 computer at the University of Manchester, which they used for several years, typically through two nights each alternate week. In December 1954, the MRC's Synoptic and Dynamical Research Sub-Committee recommended that the Office should have its own electronic computer for the purpose of carrying out research in numerical forecasting, and a year later the government granted the Office funds for the purchase of a Ferranti Mercury machine. Nicknamed 'Meteor', this machine would be the Office's first electronic computer. Its arrival

---

[61] See 'Cloud pictures from satellite TIROS 1', by S Fritz and H Wexler, published in 1960 in the *Monthly Weather Review* (Vol. 88, pp. 79–87).

[62] See 'Satellite meteorology – fancy and fact', by W K Widger, published in 1961 in *Weather* (Vol. 16, pp. 47–55).

[63] See 'Meteorological satellites', by R Frith, published in 1961 in *Weather* (Vol. 16, pp. 364–370).

[64] The term 'general circulation of the atmosphere', or 'global circulation', refers to the complete description of atmospheric flow patterns and energetics over the whole earth.

was eagerly awaited and originally expected in 1957. However, delays occurred and parts of the computer did not begin to arrive at Dunstable until the autumn of 1958. Installation quickly followed and Meteor was ready for use in January 1959. Now, for the first time, the Office's scientists could make forecasts regularly by numerical methods, though still only on an experimental basis.

Meanwhile, staff of the Office had attended programming courses and written programs, not just for numerical forecasting and objective analysis but also for extracting automatically from teleprinter tapes surface meteorological reports from ships and some upper-air information. However, a cautious attitude had developed in the Office over the use of computers in forecasting. In his article on 'Forecasting and public services', for example, in the centenary issue of the *Meteorological Magazine* (June 1955, Vol. 84, pp. 192–196), S P Peters, the Office's Deputy Director (Forecasting), had written thus:

As regards the effect on forecasting of the use of electronic computers, it is too early to express any definite opinion, since the employment of them in numerical forecasting is at present only in the research stage. There are, however, some grounds for supposing that, so far as obtaining forecast charts for 24 hours ahead is concerned, the electronic computer will prove to be a valuable aid, and its adoption in forecasting a very significant milestone in the history of synoptic meteorology.

In his article on 'Scientific Research and Development' in the same centenary issue (pp. 183–187), the Office's Deputy Director (Research), Sutcliffe, had written approvingly of the developments in dynamical meteorology that were aiding forecasting, including those he himself had pioneered, but he had made no mention whatsoever of computers.

Sutcliffe was ambivalent about NWP.[65] On the one hand, he acknowledged in the Office's annual report for the year ending 31 March 1959 that computer-based methods were "yielding important results for the future of weather forecasting" and admitted in an article on 'The Future of Weather Forecasting' published in *Weather* in 1959 (Vol. 14, pp. 163–169) that the electronic computer was a tool that could help forecasters. On the other hand, he wrote in an article that was published in the *Yearbook of the Physical Society of London* in 1960 that he was "rather pleased to be able to say" that NWP was "not yet a great success story".[66] In his view, there was "no danger of the art of forecasting being entirely superseded". In an article entitled 'The basis of present-day weather forecasting', published in *Weather* in 1961 (Vol. 16, pp. 349–363), Sutcliffe's colleague M K Miles was rather dismissive, saying that methods had been developed in Britain for obtaining a chart of forecast isobars using an electronic computer. That was all he said about the use of computers.

[65] Sutcliffe was now a Fellow of the Royal Society, elected on 21 March 1957.
[66] See 'Weather forecasting as a problem in fluid dynamics', by R C Sutcliffe, published in the 1960 *Yearbook of the Physical Society of London* (pp. 9–15).

Figure 11.13. Operating the 'Meteor' computer at Dunstable. © Crown Copyright 2010, the Met Office.

Such a cautious attitude contrasted with the enthusiasm of the Brabazon Committee for the work being carried out in the Office on the development of numerical methods. It contrasted, too, with the positive response of the government to the report of the Brabazon Committee, but Sawyer, Bushby, Knighting and others were not deterred from continuing to refine their numerical methods.[67] Weather forecasting was fast being turned from an art into a science in the United States, Japan, Sweden and other countries, and the Office had no wish to be left behind.[68] As soon as Meteor was available for use, experiments were run on it, culminating in a full-scale real-time experiment in operational NWP in the winter and spring of 1960–1961. The aim was to produce for forecasters by 09:30 GMT a prebaratic for 06:00 GMT the next day. The aim was met on about 35% of occasions, and the forecast was ready by 09:45 GMT on a further 20% of occasions. Sometimes computer faults caused the delay or even abandonment of the forecast, and many of the faults were located quickly and corrected, but the conclusion was reached that a larger, faster and more reliable computer than Meteor was needed if numerical prediction was to be used operationally.

[67] Recognition for Sawyer came in March 1962 in the form of election to Fellowship of the Royal Society, for his distinguished contributions to dynamical meteorology, especially in the field of numerical forecasting. He also received, in 1962, the Buchan Prize of the Royal Meteorological Society for the period 1957–1961 for his outstanding contributions to the study of dynamical meteorology.

[68] For the story of NWP outside the United States, the UK and Sweden, see 'Early operational Numerical Weather Prediction outside the USA: An historical introduction. Part II: Twenty countries around the world', by Anders Persson, published in 2005 in *Meteorological Applications* (Vol. 12, pp. 269–289).

The Move to Bracknell

As with the Office's move to South Kensington half a century earlier, so it was with the move to Bracknell. A new building had to be designed and constructed. A progress report on the move resulted from a question asked in the House of Commons on 20 December 1957 by Arthur Skeffington. He asked the Secretary of State for Air whether land had been acquired for the building of the Office in Bracknell New Town. In addition, he wanted to know when work on the new building would commence and the staff be transferred. In reply, George Ward stated that approval had been given for the acquisition of the site at Bracknell, and he understood that the Minister of Works expected its purchase to be completed shortly. He believed that construction would start in the autumn of 1958 and the move be completed in 1961. The foundation stone was, in fact, laid on 28 October 1959, by Sir Cyril Hinshelwood, President of the Royal Society.

Sir Graham Sutton recorded in his Foreword to the annual report of the Office for the year ending 31 March 1958 that design work on the various units of the new headquarters building was nearing completion. He expected the first block to be ready early in the summer of 1960 and the whole group completed in 1961. The staff who were currently at Victory House, Harrow, Dunstable and Stanmore would all move to Bracknell, but not the staff in the London Airport Forecasting Office. In the event, the Training School remained at Stanmore until 1971. So great was the pressure of growth in other branches of the Office by 1961 that space could not be made available for the School in the new headquarters.

The buildings at Bracknell were planned by staff of the Office to satisfy both operational and research requirements.[69] In his Foreword to the annual report for the year 1961, Sutton reported that key service units such as the CFO and the Communications Centre were now in their new homes and occupied rooms specially designed for the purpose.[70] Furthermore, he added, the new laboratories, wind tunnels, freezing chambers, workshops and drawing offices represented a notable advance on the facilities previously available. The new premises should be adequate, he said, to meet foreseeable demands, and he also noted that the Office now had, for the first time in its history, an experimental ground in open country a few miles from Bracknell. In the annual report for 1962, he was able to say that the new premises at Bracknell were "now fully occupied and working smoothly". After many years of nomadic and divided life, the Office now had a unified headquarters again.

The move to Bracknell began in February 1961, with the Marine Division the first on the new site. Appropriately, the part of the building which housed this branch was named the FitzRoy Wing, thus commemorating the Office's first director. This

---

[69] The buildings themselves were designed by the Ministry of Works.
[70] This was the second annual report for a calendar year, rather than a financial year, the first being that for 1960.

Figure 11.14. The new headquarters of the Meteorological Office at Bracknell, 1962.
© Crown Copyright 2010, the Met Office.

wing also contained the offices of the Director-General and other directors, plus the branches concerned with aviation services, climatological services, climatological research, special investigations and services to the public. The main entrance hall to the headquarters building was in the FitzRoy Wing, with the National Meteorological Library adjacent to this hall.

Archives of the Office were preserved in a building on Eastern Road about 400 metres from the new headquarters building, in purpose-built accommodation which contained over two miles of shelving. This repository was completed in April 1962, and the existing separate archives at Harrow, Dunstable and Victory House were then brought together at Eastern Road. Later in 1962, following negotiations between the Office and the Public Record Office, the Lord Chancellor appointed the Office's repositories at Bracknell, Edinburgh and Belfast places of deposit under Section 4 of the Public Records Act 1958, with the archives at Edinburgh and Belfast holding records for Scotland and Northern Ireland, respectively. All other records were to be held at Bracknell.

The Dines Wing, on the western side of the main building, was named after W H Dines and housed the meteorological logbooks of voluntary observing ships and many millions of punched cards, each containing an observation from a ship or land-based

Figure 11.15. Sir Graham Sutton shows Her Majesty the Queen a component of a Skylark rocket at the Bracknell headquarters of the Meteorological Office, 25 June 1962. © Crown Copyright 2010, the Met Office.

meteorological station. This wing housed, too, the instrument design and development laboratories and the branches responsible for the observatories at Kew, Eskdalemuir and Lerwick. The third and tallest block was the Napier Shaw Building, which had nine storeys and housed the CFO on the top floor, the Communications Centre on the floor below and the electronic computer on the floor below that. Occupation of the new headquarters was completed on 30 September 1961, when the forecasting sections arrived from Dunstable. Communications facilities were duplicated for a time, by which means the old and new forecast rooms operated simultaneously during the morning of 30 September. Then, in the early afternoon, the service from Dunstable ceased. The move of the CFO thus took place without any break in service. The new buildings were formally handed over to the Air Ministry by the Ministry of Works on 1 November 1961.

Her Majesty Queen Elizabeth II and His Royal Highness the Duke of Edinburgh paid a visit to the new headquarters on 25 June 1962, during which Her Majesty was shown a meteorological logbook that had been maintained in 1890–1891 by her grandfather, King George V, when, as Prince George of Wales, he had commanded HMS *Thrush* on the North America and West Indies station. The Duke showed great interest in the work of the Office's newest branch, that devoted to study of the high atmosphere. Formed in 1960, this branch concentrated on levels above those reached by radiosonde balloons (about 35 km), making use of satellites and rockets. These rockets carried aloft payloads of meteorological instruments that were ejected and thereafter descended on parachutes which were tracked by radar to measure wind speeds and directions.

The annual report for 1960 noted that, in common with other government departments which employed scientific staff, the Office had continued to find it difficult to obtain sufficient new entrants with university degrees or, indeed, first-year university qualifications to fill all available posts. By contrast, the flow of school-leavers into the Scientific Assistant class had improved markedly, though the turnover had remained high. In the course of 1960, members of the executive and clerical classes who did not wish to transfer to Bracknell had been replaced gradually by those who had volunteered to do so.

In his Foreword to the annual report for 1961, the Director-General commented that the Office had for many years suffered from a lack of recruits to the Scientific Officer class, partly because of a widespread shortage of young mathematicians and physicists and partly because of the neglect of meteorology and other environmental sciences as fields of study in British universities.[71] However, he was happy to report that the situation appeared to be improving. In 1961, the Office had been able to make up most of the deficiency in the Scientific Officer class by recruitment, mainly through the Civil Service Commission. In spite of the best recruitment figures from school-leavers for many years, though, the high resignation rate and the ending of National Service had led to there being more vacancies in the Assistant class at the end of the year than at the beginning. In 1962, recruitment proceeded so well that the Office was almost at full strength by the end of the year, with the total number of staff in post 3601 at the end of the year.

With the move to Bracknell, and partly because of it, there came further progress and expansion. The age of electronic computers and weather satellites had well and truly arrived, and electronics had begun to move from the age of valves to an age of semiconductors. The day when the Office could begin to use NWP operationally could not be far in the future, but not without a more powerful computer.

---

[71] He could also have mentioned that many physics and mathematics graduates at the time were encouraged to seek employment in fields considered 'modern', such as nuclear physics or radio astronomy, rather than a science like meteorology, which was based on classical physics and thus considered relatively unexciting.

12

Global Meteorology

By the early 1960s, the Meteorological Office had become respected around the world for its scientific and technological capabilities and the progress it had made towards realizing Richardson's dream of forecasting the weather by mathematical methodology. It had also long been a leading member of the international meteorological community. And yet, an international issue remained unresolved. The universal use of the metric system in meteorology had been opposed by the Office's Director at Leipzig in 1872, and the matter had continued to simmer. A partial introduction of the system in the Office had come just before the Great War, when there had been a change from inches to millimetres for measuring rainfall and inches to millibars for recording barometric pressure (see Chapter 8), but the Office and the British public had continued to use the Fahrenheit scale of temperature.

Fahrenheit or Celsius

A move to clear up the matter came in 1953, when the Executive Committee of WMO decided that degrees Celsius should be used for coding temperatures in all upper-air reports, and another move came in 1955, when the Second WMO Congress passed a resolution that the metric system be adopted for all international exchanges of meteorological information. The Third WMO Congress, in 1959, resolved that meteorological services which had still not adopted the metric system fully should do so, at least in coded messages for international exchanges, at some time in the period 1959 to 1963.

The Office acted on the 1953 decision in 1956, thereafter using degrees Celsius in all communications to aircraft, but the 1955 resolution was disregarded. The 1959 resolution was, however, implemented. The Office adopted Celsius for almost all internal and international purposes, thus coming into line with the great majority of other countries and with most scientific bodies in the UK.[1] The change took effect on 1 January 1961, but still not for the public, for whom the Office continued to use degrees Fahrenheit.

The desirability of using Celsius in weather information for the public was, however, considered by the Office, but, before making any change, they decided to seek the views of industry, public utilities and the leading technical and scientific bodies, as well as the views of the principal makers of thermometers. They put to them a proposal that Celsius and Fahrenheit be used concurrently, except that they proposed that the term 'centigrade' be used, rather than 'Celsius', on the grounds that most people in Britain were not familiar with Celsius. When no objection was raised to the proposal, the Air Ministry authorized the use of both scales, and Sir Graham Sutton announced at a press conference on 5 December 1961 that centigrade would be introduced early in 1962. As a first step, both centigrade and Fahrenheit would be used, with Fahrenheit values given first. He hoped the order would be reversed in time and the use of Fahrenheit dropped eventually.

The use of centigrade in weather information for the public began on 15 January 1962, with Fahrenheit values given first. Nine months to the day later, the order was reversed, with centigrade given first. No decision was taken, though, on the omission of Fahrenheit values altogether. A possible connection of Common Market negotiations with the switch to metric units was raised in the House of Commons on 27 February 1963 by Frank Bowles. In reply, Hugh Fraser, the Secretary of State for Air, denied any such connection, which gained him a riposte from Mr Bowles that he did not always accept what government spokesmen said about the Common Market! Bowles went on to ask if there was any truth in the letter he had seen that morning, written by the Anti-centigrade Society, which claimed that Fahrenheit would be dropped altogether in May 1963? Could the Secretary of State assure him that Fahrenheit would never be dropped completely? Fraser's reply was that the use of Fahrenheit would cease only when acceptable to the public. In fact, it has never ceased, even though the UK has been metric officially since the 1970s. Weather presenters on radio and TV still occasionally give temperatures in degrees Fahrenheit.

Those who suspected that a switch to metric units had something to do with the Common Market were right. When the UK joined the European Economic Community (EEC) on 1 January 1973, the British government became obliged to adopt all EEC legislation and therefore to stop using all non-metric units, but this merely accelerated the metrication programme that had been underway in the UK since 24 May 1965,

---

[1] The Americans continued to use the Fahrenheit scale of temperature long after 1960, and it is said that the positions of their warships during the Vietnam War could be ascertained easily because the meteorological reports broadcast from these ships contained temperatures in degrees Fahrenheit.

when the British government announced that the country would be fully metric within ten years.[2]

Weather Forecasting Developments

The question from Mr Bowles was not the only weather-related one asked in the House of Commons on 27 February 1963. Another came from Gresham Cooke, who asked the Secretary of State for Air whether, "in view of the reliable forecasts put out by the United States Weather Bureau of British weather in January", he would seek guidance from American forecasters regarding the possibility of predicting any severe conditions next winter. The weather had been a matter of unusually great public interest for quite some time when Mr Cooke asked his question. The winter had been exceptionally cold since just before Christmas and overall turned out to be the coldest since 1740.

In his reply to Mr Cooke, Mr Fraser said that the U.S. Weather Bureau considered that its thirty-day predictions had shown "some modest success" but warned users that experience over a number of years had not shown that its 'outlooks' for the northern hemisphere as whole could be relied on as a guide to forthcoming weather over the British Isles, which lay in an area of particularly variable weather. Mr Cooke pointed out that the 30-day forecasts for December 1962, January 1963 and February 1963 which the Americans had issued for the British Isles had proved remarkably accurate. He believed the Office also produced thirty-day forecasts, but only for its own internal use. Would it not be "a good thing" for it to produce a similar 30-day forecast for the British Isles next October or November to provide a warning should another severe winter be threatened? Mr Fraser urged caution. Only thirty-two of 108 outlooks issued by the Americans had proved correct. Long-range forecasting was, as yet, far from perfect. He agreed that forecasts for the current winter had been extraordinarily accurate but thought this could have been a fluke.

Government pressure was, though, applied to the Office, and the annual report for 1963 contained a statement that a decision had been taken in the autumn to initiate a series of thirty-day 'weather prospects' which would be promulgated by regular publication of a monthly bulletin with a mid-month supplement.[3] This decision, it was stressed, "did not mean that a major breakthrough had been achieved in this formidable problem but that a practical system had been evolved which, judged by the experience of some five years, was capable of producing general statements about the weather of this country during the ensuing 30 days that were in tolerably good agreement with events on a majority of occasions".

[2] For an introduction to the use of the International System of Units (SI) in meteorology, see 'SI units in the Meteorological Office', by F E Lumb, published in the *Meteorological Magazine* in 1972 (Vol. 101, pp. 366–368).
[3] The decision to make thirty-day forecasts available to the public was announced by the Secretary of State for Air in the House of Commons on 13 November 1963.

To support this work, the Climatology Research Branch had been sub-divided. The purpose of one branch was to focus on synoptic climatology, this being the study of climate from the perspective of regional atmospheric circulation, principally by means of synoptic charts. This branch had been given responsibility for preparing the thirty-day prospects. A new branch was to focus on dynamic climatology, this being the study of climate from the perspective of atmospheric dynamics and thermodynamics, carried out principally through theoretical analysis of the general circulation of the atmosphere.

Publication of the new series of monthly bulletins, called *Monthly Weather Survey and Prospects*, began on 1 December 1963, and Sir Graham was able to say of the forecasts in the Office's annual report for 1964 that "a modest level of success" had been achieved, but he also pointed out that "long-range forecasting was the most difficult problem in meteorology for which, as yet, there was no secure physical basis".[4] The procedure used to produce the forecasts was basically twofold: to match periods at the time of year in question when weather types over the British Isles in previous years had resembled those of recent weeks, and to compare monthly-mean temperature distributions over the northern hemisphere with those of recent months. Charts of weather types extended back to 1873 and charts of temperature distributions back to 1881.[5]

Each forecast contained statements about the expected mean temperature and total rainfall for the whole month, together with additional information about the type of weather and incidence of, for example, snow, frost or thunderstorms. The accuracy of each forecast was assessed by a panel of Meteorological Office staff, and an analysis of the forecasts issued for the first 33 months was carried out. The conclusion was that · the long-range forecasts represented a solid achievement but were, as yet, far from reliable.[6]

Meanwhile, there had been important developments in respect of endeavours to produce reliable short-range (day-to-day) forecasts by means of numerical methods. A decision had been announced by the UK Atomic Energy Authority in August 1961 that a Ferranti Atlas electronic digital computer had been ordered for the National Institute for Research in Nuclear Science. It would be installed in the Institute's Rutherford High Energy Laboratory at Harwell in Oxfordshire, and a prototype was already being assembled at the University of Manchester. The machine at Harwell was expected to be ready for use early in 1964, and the Office was considered likely to be one of the computer's main users. Also, the Office had decided in April 1963 to place

---

[4] The forecasts were issued free to press and radio, while private subscribers were charged 1s. 2d. a copy or 13s. 6d. a year, inclusive of postage.

[5] See 'The problem of long-range weather forecasting', by J M Craddock, published in 1964 in *Weather* (Vol. 19, pp. 44–46).

[6] See 'The accuracy of long-range forecasts issued by the Meteorological Office', by M H Freeman, published in *Weather* in 1967 (Vol. 22, pp. 72–76).

an order for a KDF9 electronic computer made by English Electric Leo Computers Ltd. The need to replace Meteor with a larger and faster machine had become clear. Without it, the goal of using NWP operationally could not be achieved.

The KDF9 machine was nicknamed Comet and cost £400,000. The expectation was that it would be ready for use early in 1965, but, in the event, the machine was installed during the summer of 1965 and production of numerical forecasts began in September. Meteor was removed in October 1964 to permit the preparation of the computer laboratory for the installation of Comet, which left the Office without a computer at Bracknell for several months, during which time use was made of computer facilities at other institutions. Urgent work was carried out, mainly at night, on the Mercury computer in the Royal Aircraft Establishment at Farnborough. Work on the development of programs for use on Comet took place on the KDF9 machine in the NPL. Meteor was reassembled in the Chemical Defence Experimental Establishment at Porton, and arrangements were made for some of the computations formerly carried out on it at Bracknell to be performed there. As planned, considerable use was also made of the Atlas computers at Manchester and Harwell, particularly for research into the general circulation of the atmosphere in connection with long-range forecasting and for developing a weather prediction model designed by Fred Bushby and his colleague Margaret Timpson.

A New Director-General

Sir Graham Sutton reached the normal civil service retirement age of 60 in February 1963 but remained Director-General for another two and a half years. However, he did not then retire from public service. He had already taken up another appointment, as chairman of the Natural Environment Research Council (NERC), a new body that had been established by Royal Charter on 1 June 1965 under the Science and Technology Act 1965. This appointment had been announced in the House of Lords on 4 February 1965, when the Minister without Portfolio, Lord Champion, had informed their Lordships that Sir Graham had accepted the invitation of the Secretary of State for Education and Science to take on the chairmanship. For the time being, he said, the major part of Sir Graham's time would "continue to be devoted to his responsibilities as Director-General of the Meteorological Office".

Formation of the new Council was an outcome of an inquiry into the organization of civil science which had been carried out by a government-appointed committee chaired by Sir Burke Trend of the Treasury.[7] This committee had been set up in March 1962 and its report published in October 1963. The key recommendation was that DSIR, which had been set up in 1915 (see Chapter 8), should cease to exist and its

---

[7] Trend was Second Secretary of the Treasury when the inquiry began. He became the Cabinet Secretary at the beginning of 1963.

work be divided between two new research councils, one for science, the other for natural resources. The Conservative government accepted the report in November 1963 and the Labour opposition broadly concurred, but the government decided that formal implementation of the recommendations by means of an Act should not be rushed. A General Election had to be held no later than October 1964, which meant there was a possibility that the session of Parliament that began on 12 November 1963 would be cut short by a dissolution. In any case, some of the recommendations were expected to prove contentious and provoke lively debate in university, government and other science communities. Some recommendations did indeed turn out to be controversial, with the Institution of Professional Civil Servants particularly unhappy. They argued that DSIR should not be disbanded but, instead, expanded within a new Ministry of Science. Their arguments fell on deaf ears, though, and the decision to disband DSIR was upheld by the Labour government which came to power in October 1964.

The Royal Charter of NERC specified that the Council was empowered to "encourage and support research by any person or body in the earth sciences and ecology and in particular in geology, meteorology, seismology, geomagnetism, hydrology, oceanography, forestry, nature conservation, fisheries or marine and freshwater biology". Trend's Committee recommended that astronomy and space research be overseen by the Science Research Council and advised that the operational requirements of aviation were such that responsibility for the Office should remain with the Air Ministry (which became, on 1 April 1964, the Air Force Department of the Ministry of Defence). Nevertheless, Trend's Committee felt that close links should be established between the Office and NERC, and this advice was acted on by the government. In a Parliamentary Statement made in the House of Commons on 28 July 1964 by Quintin Hogg, Secretary of State for Education and Science, it was announced that NERC would provide policy direction and financial support for research in geomagnetism and seismology undertaken at the Office's observatories and would be "associated with" the Office's general programme of research.

Sutton stepped down as Director-General on 30 September 1965. He had achieved much during his twelve years in post. There had been major changes in the organization and scientific structure of the Office, and he had brought together at Bracknell the various branches of the Office which had been scattered in and around London. There had been much progress in meteorological technology, automation and the application of electronic computers, and he had kept the Office focused on the goal of being at the scientific forefront of meteorology.[8] After he resigned the

---

[8] When Sir Graham retired, he and Lady Sutton presented to the Office a silver rosebowl to be awarded annually for the most worthy contribution to the social life of the Office. He died on 26 May 1977. After the death of a former Director of the Office, Sir George Simpson, who died on 1 January 1965, his widow, Lady Dorothy Simpson, presented to the Office a silver cup in memory of her husband. Called the 'George Simpson Cup', this was to be

chairmanship of NERC, in 1968, he returned to Wales and lived in retirement at Swansea.[9]

The new Director-General, Basil John Mason, was only 42 years of age. For the past four years, he had been Professor of Cloud Physics in the Imperial College of Science and Technology. Before that, since 1948, he had been a lecturer in the Imperial College Department of Meteorology. Born in Norfolk on 18 August 1923, he had attended Fakenham Grammar School and proceeded from there in 1941 to the University College at Nottingham, where he had gained an external degree of the University of London in physics with First Class Honours. His studies had been interrupted, though, by war service, in the radar branch of the RAF at home and in the Far East. He had been called up early in 1944 and was unable to return to Nottingham before 1946 to complete his degree. After graduation, in 1947, he had remained at Nottingham for a year, as a Research Fellow, and gained a Master of Science degree for a dissertation on surface tension. Whilst at Imperial College, he had published a great many papers on theoretical and experimental aspects of condensation and precipitation and gained, in 1956, a Doctor of Science degree of the University of London. He had also published two books: *The Physics of Clouds* in 1957 (Oxford: Clarendon Press, 481 pages) and *Clouds, Rain and Rainmaking* in 1962 (Cambridge University Press, 145 pages). His election to Fellowship of the Royal Society had come in 1965, a few months before his move to the Meteorological Office.

His first day as Director-General was 1 October 1965, and he soon showed boldness. He decided the time had come for NWP to move from research mode to operational. He met opposition from senior directors of the Office, whose opinion was that a longer trial period was required, but he went against their advice and decided that numerical forecasts would be issued routinely twice a day from Monday 2 November 1965.

A press conference was arranged for that day, and all who attended not only saw the first operational weather forecast chart emerge from the computer but also were each given a personal copy of it. The forecast was accurate and the Office's first ever press conference a great success. Mason was delighted with the success but realistic about the capabilities of NWP, pointing out that weather forecasting remained as much an art as a science. Forecasters still had to interpret the output from the computer, using techniques which included intuition and personal experience. However, he said, forecasting was becoming a branch of applied mathematics, and the British people could look forward to a steady improvement in the accuracy of forecasts.[10]

awarded annually for the most meritorious achievement by a team or individual in any form of outdoor sport by staff of the Office. Another who died in 1965 was Sir David Brunt, on 5 February.

[9] Whilst in retirement, he wrote a book for the scientific layman, entitled *The Weather* (published in 1974 by Teach Yourself Books, 173 pp.).

[10] See 'Press Conference', an anonymous report published in the *Meteorological Magazine* in 1966 (Vol. 95, pp. 28–30).

Figure 12.1. Basil John Mason. © Crown Copyright 2010, the Met Office.

Figure 12.2. Dr Mason speaking at the Meteorological Office's first-ever press conference, 2 November 1965. © Crown Copyright 2010, the Met Office.

Figure 12.3. Computer print-out ('zebra chart'), 200-mb analysis for much of the northern hemisphere, 5 October 1966, 12:00 GMT. © Crown Copyright 2010, the Met Office.

To meteorologists, it came as no surprise that forecasting failures continued to occur from time to time. Others were expectant, though, and on 1 August 1966, in the House of Lords, Lord Erroll of Hale asked what steps were being taken by the government to correct the deterioration in the standard of the weather forecasts issued by the Office. In reply, Lord Shackleton said that he was not aware of any deterioration. In his opinion, the general standard of the daily weather forecasts issued by the Office had remained consistently good for many years. These forecasts were reckoned to be about 80% accurate. He expected improvements in the standard of the forecasts as experience was gained with the new computer (Comet) and as further sources of information, such as more cloud pictures from satellites, became available. Moreover, research was continuing to improve both short-range and long-range forecasts. Lord Erroll's rejoinder was that, as the accuracy was only 80%:

Would it not be possible for the Meteorological Office to issue an apology when they get their forecast completely wrong; and would it not be possible for them occasionally to admit that conditions are so variable that it is not possible to issue a reliable forecast at all, instead of the public being hoodwinked by the appearance of a dogmatic assertion about the forecast which is not likely to prove true in the event?

Lord Shackleton did not think it a practical solution to "apologise once every five forecasts". He considered Lord Erroll "rather naïve in thinking that forecasting can be an exact science".

The exchange in the House of Lords was reported in *The Times* the following day, in an article which included comments from Captain C C Jackson, secretary-general of the International Federation of Airline Pilots Associations. He did not consider that the forecasting service had improved since the war. He thought most pilots would say they were not getting a very good service. And then, all of a sudden, for no apparent reason, the article degenerated:

Farmers, working closely with their local outstations, are happier. But their needs tend to be simpler methods of forecasting – from the wind in the trees, from the colour of the sky, from the visibility, from the swarms of gnats. The traditional forecasts which come from local knowledge have their strong adherents. At Gleneagles, golfers have their own way of assessing the prospects. Looking from the hotel across the course to a certain spinney, they know that if they can see the spinney, it is going to rain. If they cannot see it, it is already raining.

Fortunately, this was only a temporary lapse into silly-season journalism. The author went on to draw attention to an important international venture, pointing out that plans for developing a World Weather Watch (WWW) were to be submitted to a WMO Congress in April 1967. This Watch, he reported, would involve all Member States of WMO and comprise a coordinated global observation system which would use conventional meteorological stations, artificial satellites, floating balloons, automatic weather stations, weather buoys and aircraft reports. There would also be a global telecommunications system to exchange information, as well as a data-processing system.

We consider the WWW and other important international developments later in this chapter. First, though, we note that many questions concerned with meteorology in general and weather forecasting in particular were asked in the House of Commons in the latter part of the 1960s by John Ellis, the Member for Bristol North-West. He had been a laboratory technician in the Office for fifteen years before his election to the Commons at the General Election held on 31 March 1966. He was as frequent an inquisitor of government ministers as Frank Markham had been in the late 1930s (see Chapter 10).

The reliability of weather forecasts was particularly called into question in September 1968, when heavy rain caused widespread flooding in Kent, Sussex, Surrey, Essex and East Anglia, with flooding especially serious at East Molesey in Surrey. Both John Ellis and Gresham Cooke wanted to know why there had been no advance warning of the deluge. According to *The Times* on 17 September, two days after the rainfall event, Mr Ellis had arranged to go to Bracknell to meet weather forecasters and senior officials. He was going "to take up the question why the official forecast had said 'cloudy, outbreaks of rain and thunder' and given no indication of the serious situation which was to develop". He wanted to ask the Office how forecasting had improved since the introduction of computers, including the new computer, which, he said, with ancillary equipment, had cost £475,000. Mr Cooke was disappointed that neither the short-range nor the long-range forecasts had provided any warning. He had, he told *The Times*, been the person who had, in 1963, persuaded the government to introduce "long-distance weather forecasts".

Dr Mason issued a statement in which he said that the computer at Bracknell was not being used to make routine mathematical forecasts of rainfall. It was, he said, very successful in forecasting many features of the weather, but there was nowhere a computer sufficiently powerful to make an accurate forecast of rainfall on a day-to-day basis. Research was nevertheless being carried out. Forecasts using the Atlas computer at Harwell, he went on, showed great promise in forecasting the amount of rain that would fall across the country for periods of twenty-four hours ahead, but the amount of computation was so great that it took eight hours on this computer to make a twenty-four–hour forecast. Making such forecasts operationally would not be possible without a computer at least twenty times faster than Atlas and eighty times faster than Comet.

A case was made soon afterwards by the Office to acquire another new computer. Mr Ellis asked in the House of Commons on 13 November 1968 if the Secretary of State for Defence would make a statement on the agreement in principle reached to install a new computer in the Office. In reply, Gerald Reynolds, the Minister of Defence for Administration, said that a more powerful computer was needed in the Office "in order to apply techniques" which were "under development for more accurate rainfall forecasting and forecasting for five to seven days ahead". He went on to say that practising forecasters considered that pressure charts produced by numerical means were consistently more reliable than those produced by traditional subjective methods.

With the arrival of the age of operational NWP, the demands of forecasters for reliable observations became ever more insistent. The numerical models that were being developed were becoming progressively more complex and required observations from ever more levels in the atmosphere. The model used operationally in 1965 used three levels. That being developed by Bushby and Timpson used ten. Thanks to modern technology, though, especially the growing availability of devices made from semiconductor materials, efforts to meet the insatiable demand of forecasters increased markedly. The quantity and quality of observations improved greatly in the 1960s.

Research Developments in the 1960s

The meteorological rockets that were mentioned in Chapter 11 produced data from ever greater altitudes. A height of more than sixty kilometres was reached by a rocket launched from the Outer Hebrides in early 1964, and a height of more than seventy kilometres was attained in early 1965. Trials of meteorological rockets took place in Australia, too, and in April 1963 a Skylark rocket sponsored by the Office reached an altitude of 230 kilometres.

Radiosonde development continued, too, with a completely new design of sonde undergoing flight trials by late 1965.[11] Moreover, Britain's first automatic weather station was tested at Bracknell in late 1963. In the words of a report published in *The Times* on 19 November 1963, this station "could soon be ready for action on some remote headland or island". The device had been "a long time coming compared with other major countries, because its work had been done cheaply and efficiently by the hundreds of part-time paid observers in lonely places". Its development had been hastened by lighthouse automation and "the need to fill other gaps in the observing network". The weather station, made to the designs of the Office, was probably, *The Times* said, "the most reliable in the world at its price", about £6000 for a complete station.

Images from satellites continued to improve during the 1960s. Soon, weather forecasters in Britain and elsewhere wondered how they had ever managed without these images, not least because the use of infra-red sensors allowed pictures of cloud systems to be obtained during hours of darkness. The ninth *TIROS* satellite was launched on 22 January 1965 and, like the seventh and eighth, remained operational for several years, not just the few hundred days of the earlier satellites. The tenth and last of the *TIROS* satellites was launched on 2 July 1965 and provided excellent satellite coverage during that year's Atlantic hurricane season.

The American *ESSA* satellite programme (*ESSA* standing for Environmental Science Services Administration) was an extension of, and complement to, the *TIROS* programme. The first of these satellites was launched on 3 February 1966 and functioned until deactivated on 12 June 1968, providing photographs of cloud cover specifically to support operational weather analysis and forecasting. Two more *ESSA* satellites were launched in 1966, another three in 1967, another two in 1968 and the

---

[11] An attempt in the 1950s to develop a 'radar sonde' (in collaboration with Mullard Ltd) proved a very costly failure for the Office. It is said that the files relating to this project were locked in a safe in Sutton's office away from prying eyes. The radar sonde was a development of the radiosonde system of measuring temperature, pressure and humidity by means of balloons released and tracked from ground stations. The idea of the radar sonde was that the entire process would be unified and made fully automatic by means of an interrogator and computing system on the ground and airborne transponders. The project failed because of poor performance and the advent of the transistor. An outcome of this débâcle was that the Office adopted a policy under Mason that expertise for any substantial project would be held in-house.

ninth in this series on 26 February 1969. Regular direct reception of pictures from these operational weather satellites began at Bracknell in February 1966 and at the Heathrow Airport meteorological office towards the end of that year.

The American *NIMBUS* satellite programme was more ambitious than the *TIROS* and *ESSA* programmes. The intention was that information would be obtained from the *NIMBUS* satellites not only about cloud cover but also about atmospheric temperature, atmospheric chemistry, sea-ice coverage, the earth's radiation budget and sea-surface temperature. The first was launched on 28 August 1964 and functioned until a solar panel on it failed twenty-six days later. Thereafter, however, the *NIMBUS* programme developed rather slowly. Nearly four years elapsed before the second satellite was ready for launch, and that failed two minutes after launch. Not until 13 April 1969 was the third *NIMBUS* satellite launched, but this one proved successful. It was the first satellite to make global measurements of temperatures at varying levels in the atmosphere day and night, and it remained operational until 1972.

The *ARIEL 2* and *ARIEL 3* research satellites were launched from sites in the United States. However, both carried British-built experiments, and *ARIEL 3* was, in fact, built in the UK. *ARIEL 2* carried experiments designed by the Office to make a global survey of the vertical distribution of atmospheric ozone. It was launched on 27 March 1964 and functioned satisfactorily until a tape recorder aboard it failed at the beginning of October of that year. *ARIEL 3* carried a Meteorological Office experiment to observe the distribution of molecular oxygen in the lower part of the atmospheric layer which extends from a height of about eighty-five kilometres to the atmosphere's outermost fringe (the layer known as the thermosphere). It was launched on 5 May 1967 and provided data until April 1969.

The statement in *The Times* on 14 August 1964 that there was "a glut of weather information available for the making up of those synoptic charts that puzzle us in the newspapers and on television screens" came as something of a surprise. A glut? Surely not. The demand of weather forecasters for observations remained insatiable. According to *The Times*, though, "with up to seventy airliners flying the Atlantic in a day and the regular passage of *TIROS* satellites", the flights made by RAF No. 202 Meteorological Reconnaissance Squadron were no longer needed. This squadron, based at Aldergrove in Northern Ireland, had been making regular flights over the North Atlantic since 1946 and had been the only such squadron still in service since 1950. It made its final flight on 31 July 1964. The real reason for disbanding it was cost saving.

As regards the code name 'Bismuth' for the squadron's operations, the article in *The Times* said that it was "one of the peculiarities of the services" that no one could be found who could explain why the name had been given. One cynic had suggested that "it might be due to the stomach upsets which were so often a part of these 'press-on-regardless' flights". The ignorance showed there had been a failure of corporate memory. As we saw in Chapter 10, Gold's secretary, Mary Buchanan, had chosen the name for the wartime meteorological flight which operated over the Atlantic north-west of Ireland, selecting it from a list of code words supplied by Air Ministry Intelligence.

By the middle of the 1960s, the aircraft of the Meteorological Research Flight (MRF) were showing their age and needed to be replaced if they were to continue supporting effectively the research programme of their parent body, the Office. The MRF, which was based at Farnborough, was equipped with three aircraft at the beginning of the 1960s: a Canberra for high-level work, a Hastings fitted as a flying laboratory and a Varsity, all acquired in the early 1950s. The Canberra was tragically lost in the North Sea in February 1962 when approaching the air station at Leuchars in eastern Scotland after a flight to the Arctic to measure distributions of water vapour and ozone in the lower stratosphere. It was, however, replaced by another Canberra, which made its first sortie on 7 April 1964.

The Hastings and Varsity were withdrawn from service in the late 1960s, and the Air Staff agreed in 1968 to the Hastings being replaced by a C-130 Hercules aircraft for the exclusive use of the MRF. The Varsity was replaced by a newer Varsity and the Canberra remained in service. Conversion of the Hercules took two and a half years, and the aircraft was not delivered until late 1973. This conversion involved the removal of the radar from the aircraft's nose and its relocation in a pod on the top of the aircraft, as well as the fitting of a nose boom seven metres long fitted with wind, turbulence and temperature probes. The case for acquiring the Hercules, nicknamed 'Snoopy' because of its long nose boom, rested largely on an invitation for the aircraft to participate in an international geophysical experiment in the 1970s (described later in this chapter), but another factor was that it could be redeployed in wartime as a transport plane.

During the 1960s, the MRF explored, *inter alia*, turbulence in cloud-free air at low levels over the Libyan Desert, cumulonimbus clouds in the Far East, sea-surface temperatures near Malta (using the Varsity's new radiation thermometer), wave clouds over the Pyrenees and, in April 1964, emissions from Surtsey, a volcanic island in Iceland (which started to erupt on 14 November 1963 and continued to erupt until June 1967).

The MRF also played an increasingly significant role in endeavours to obtain basic physical data for improving and validating the models used for NWP and climate modelling, and a number of external factors helped shape the MRF programme during the latter part of the 1960s. One of these was the pioneering work carried out by Keith Browning and his colleagues in the Office's Research Unit based in the Royal Radar Establishment at Malvern, Worcestershire. Their work consisted of radar-based investigations into the structures of weather systems and the physical processes occurring in the rain areas within the systems. Common areas of interest between the Malvern group and the MRF were the dynamics of fronts and clear-air turbulence, the latter being of interest not just to aircraft designers and operators, whose concern

Figure 12.4. The C-130 Hercules aircraft nicknamed 'Snoopy', showing its long nose boom and radar pod. © Crown Copyright 2010, the Met Office.

was aircraft safety, but also to meteorologists, who wished to understand and forecast such turbulence. Another external factor was the transfer of the Imperial College Department of Cloud Physics to the Office in October 1966.

The subject of this transfer was brought up in the House of Commons on 1 December 1966 by Frank Allaun, who asked the Secretary of State for Defence to make a statement. He particularly wished to know the objects and scope of the new Cloud Physics Branch and whether there were any prospects of its leading to developments in weather control. In reply, the Secretary of State, Merlyn Rees, said that the branch would take over the work on cloud physics that was already being done in the Office and would continue "the major research done by Imperial College". The work of the new branch would embrace the growth and behaviour of cloud droplets and ice crystals; the formation of rain, hail and snow; and large-scale movements of air in cloud. It would carry out laboratory and mathematical work at Bracknell and radar work at Malvern, and it would be assisted by the MRF at Farnborough. The research was of a fundamental nature, Mr Rees said, but it would "contribute to a better understanding of the possibilities of artificial rainmaking". Only nine days earlier, on 22 November, Mr Rees had told Mr Allaun that experiments in seeding clouds with chemicals as a means of influencing the weather had proved inclusive at home and abroad. For there to be further progress, he had said, a better understanding of the physics of cloud and rain was needed.

The Cloud Physics Branch, in collaboration with the Malvern group and the MRF, began to move research in cloud physics away from laboratory studies towards the interaction of cloud microphysics with cloud dynamics. This necessarily involved the use of research aircraft. One of the first tasks undertaken by the new branch was to develop a radiosonde which could be released from the Hastings aircraft, the intention being that the data transmitted by these so-called dropsondes would be received and stored by a computer on board the aircraft.

The first major venture of the new branch, *Project Scillonia*, was based on the Scilly Isles, where studies of the fronts of depressions approaching from the Atlantic could be carried out comparatively free of any topographic influences. Aircraft could also operate from the Scillies with minimal Air Traffic Control restrictions. Starting in 1968, measurements of approaching cloud systems were made with a mobile Doppler radar facility installed on the Scilly Isles, and dropsondes were released from the Varsity aircraft from a height of five kilometres to obtain data about temperature, humidity and winds within the cloud systems. In the last and most intense phase of the project, in 1971, four radars operated simultaneously on the Scillies. A major finding of the project was that organized bands of ascent and descent and associated bands of precipitation existed in the cloud systems of depressions. This was a discovery of great value to operational weather forecasters and provided an observational framework for the development and testing of numerical models of fronts.[12]

Another important development which involved radar was the demonstration by the Malvern group in the early 1970s that radar calibrated against a small number of rain-gauges could provide measurements of precipitation with sufficient accuracy for most meteorological and hydrological purposes, even in hilly areas. This was established through a project carried out in the Dee River area of North Wales. The project also showed that weather radar could be a cost-effective method for extending the coverage by telemetering rain-gauges for obtaining real-time precipitation information over large areas.

New Meteorological Services

A new service for mariners was introduced in 1967. This was 'ship routeing' (also known as 'weather routeing'), an advice prepared ashore and offered to seafarers crossing oceanic areas where various routes could be chosen, depending on the weather, icebergs, sea ice, load-line zones and state of loading. Its objectives included greater vessel safety, reduced fuel consumption, faster voyages, avoidance of heavy-weather damage and avoidance or limitation of condensation problems in holds and containers. It was concerned with the avoidance of inclement weather on a day-to-day basis and thus differed from the kind of routeing pioneered by Maury and FitzRoy, in that their work enabled climatological routeing to be practised (i.e., exploitation of

---

[12] For information about the work of the MRF in the second half of the 1960s and early part of the 1970s, see 'The current work of the Meteorological Research Flight', by D G James and J M Nicholls, published in1976 in the *Meteorological Magazine* (Vol. 105, pp. 86–99).

prevailing winds and ocean currents, avoidance of the stormy latitudes of the southern hemisphere, and so on) (see Chapters 1 and 2).

Ship routeing was pioneered in the 1950s by meteorologists of the U.S. Navy, who proved that avoidance of waves greater than about four metres in height resulted in faster voyages and reduced fuel consumption. The first commercial routeing service began in 1960, when the Royal Netherlands Meteorological Institute began to provide a routeing programme for the Holland-America Line. The Office introduced routeing on a trial basis in September 1967. Four ships took part and followed courses across the North Atlantic selected by staff of the Office's Marine Branch to provide the quickest crossing and least buffeting from heavy seas. The trial was judged successful and plans to institute a routine service in the near future were made. It was a fundamental principle of this service that it would be operated by a team of experienced Master Mariners who had all held command.

The routeing technique was to project the wave pattern ahead as far as possible along the normal path of the vessel from port of departure to destination. Weather forecasts for three to five days ahead were produced by the CFO, and staff of the Marine Branch turned these into routeing advice by converting predicted wind speeds and directions into forecasts of waves which, in turn, were used for estimating the vessel's speed.[13] An important aspect of the service was that it was merely informed advice that was issued. Whilst at sea, the Master was in charge of the vessel and able to reject routeing advice if there was good reason to do so. There was also no single recommended route for all ships in any given weather. The routeing was ship specific. The route in practice depended in part on the cargo carried (e.g., containers on deck, heavy objects, liquids, elderly passengers on cruise ships).

When regular ship routeing began, in early 1968, it was provided for the North Atlantic only. The Office introduced routeing across the North Pacific in 1972. The charge for the service in 1968 was £50 per crossing, plus the cost of telegrams to the ship. Ten years later, the charge for an Atlantic crossing was £100 and the charge for a Pacific crossing £150. In both cases, the costs of radio messages were additional. They were also, after the UK joined the EEC, subject to Value Added Tax at the standard rate (which was 8% in 1978).[14]

The Office provided a number of other services for seafarers in the late 1960s. Weather advice could be obtained, for example, from a number of forecast centres. There were sixteen centres around the UK, at Aberdeen Airport, RAF Valley (Anglesey), Belfast Airport, the RAF Record and Pay Office at Gloucester, Glasgow Weather Centre, HQ No. 19 Group RAF at Plymouth, HQ No. 18 Group RAF Pitreavie Castle (Fife), Newcastle upon Tyne Weather Centre, RAF Manby (Lincolnshire),

---

[13] See 'Weather routeing of ships', by S H Evans, published in 1968 in *Weather* (Vol. 23, pp. 2–8). See also: 'Weather routeing of ships on the North Atlantic', by R F Zobel, published in 1972 in the *Proceedings of the Royal Society of Edinburgh B* (Vol. 72, pp. 311–320).

[14] See 'Weather routeing of ships', by G V Mackie, published in 1981 in *Marine Observer* (Vol. 51, pp. 121–127).

Liverpool Airport, Manchester Weather Centre, RAF Kinloss (Morayshire), Kirkwall Airport (Orkney), the Air Traffic Control Centre at Preston (Lancashire), Southampton Weather Centre and the London Weather Centre.

Mariners could also obtain advice from Port Meteorological Officers (PMOs), of whom there were eight, at Liverpool, London, Southampton, Cardiff, Glasgow, Leith, Newcastle and Hull. An important aspect of the work of these PMOs was the supply of instruments to vessels of the Voluntary Observing Fleet and the checking of barometers. Weather bulletins for the sea areas off north-west Europe were broadcast by the BBC four times a day on the Light Programme, using the 200-kHz frequency, and gale warnings were broadcast by the same station on the same frequency at the earliest convenient time after receipt. Forecasts for coastal waters were broadcast on the BBC Home Service, which used a number of frequencies on the Medium Wave.

Services provided by the Office for the offshore industry began in the mid 1960s. Drilling for oil and gas in the North Sea began in late 1964, and meteorological services for the industry followed soon afterwards. By 1966, staff of the London Weather Centre were attending to the needs of most of the oil companies and consortia that were drilling in the North Sea, including the preparation of special forecasts as required for oil rigs whilst being towed from one location to another.[15] The overwhelming meteorological interest of rig operators was in the related elements of wind and waves, with particular reference to rig behaviour, conditions whilst rigs were being towed, and the operation of supply vessels.

For estimating wave conditions, the London Weather Centre converted wind predictions provided by the CFO at Bracknell into forecasts of wave heights and periods by means of graphically presented relationships which had been derived empirically by staff of the National Institute of Oceanography (NIO).[16] These graphs enabled wave heights and periods to be estimated from knowledge of wind speed, fetch and duration.

By 1972, as noted by the Head of the London Weather Centre, R J Ogden, in an article published that year in *Weather*, forecasts were given in five sections: viz. inference; forecast for the first twelve hours; forecast for the next twelve hours; forecast for the following twenty-four hours; forecast for 'Day 3'.[17] As Ogden said, precision over predicted wave conditions could not be guaranteed even twelve to

---

[15] The London Weather Centre moved from Kingsway to High Holborn in January 1965. When the Air Ministry became part of the Ministry of Defence in April 1964, 'Roof of the London Weather Centre' replaced 'Air Ministry Roof' as the place where meteorological observations were made in central London. When the move to High Holborn occurred, observations were not made on the roof of the new premises, Penderel House, because the exposure there was poor. Instead, observations were made on the roof of a fifteen-storey building across the road, State House. At High Holborn, as at Kingsway, the Weather Centre had a large window in which a range of current weather information for the public was displayed.

[16] KNMI collaborated with Klaus Hasselman in Germany in the 1960s and early 1970s to produce a world standard wave model which gradually replaced the NIO methods.

[17] See 'Forecasting for North Sea oil and gas rigs', by R J Ogden, published in *Weather* in 1972 (Vol. 27, pp. 336–342).

twenty-four hours ahead, let alone three days. Accordingly, he said, the inference in the standard forecast assumed a special importance, because it was from this that the current synoptic situation could be explained and the expected meteorological evolution on which the forecast was based. The practice at the London Weather Centre was to indicate degree of confidence in the chosen solution and, in addition, to discuss with clients other possible evolutions and their likely consequences at the rig's position. Thus the rig operator was given as much information as possible to make vital operational decisions.

During the 1950s and 1960s, services for aviation developed progressively, as civil aviation expanded, military flights became more numerous and aircraft flew ever faster and higher. An A-5 Vigilante bomber reached a height of 27,874 metres on 13 December 1960, and the American U2 reconnaissance aircraft operated routinely in the lower stratosphere, at an altitude of twenty to twenty-one kilometres. Even in the lower stratosphere, though, weather could not be escaped, especially in low latitudes, where, at certain times of year, the summits of cumulonimbus clouds could reach eighteen kilometres.

Civil airliners on long-haul flights typically cruised at an altitude of ten or eleven kilometres, and their pilots and passengers soon realized that the publicity which had attended the introduction of the Comet jet airliner in the early 1950s had been unduly optimistic. This aeroplane did not fly so high that it was "above the weather". Jet streams and vigorous turbulence could, and often did, occur at cruising altitudes. Efforts by meteorologists to understand weather systems and provide accurate forecasts of upper-level conditions were therefore essential. At Heathrow and other airports, staff of the Office briefed air crews on the atmospheric conditions to be expected during their flights and provided them with copies of weather charts. Advice on jet streams was particularly important, for these could assist or delay flights. Understanding of clear-air turbulence was as yet, though, rudimentary.

A new reason for studying the atmosphere at levels above the cruising altitudes of long-haul civil airliners emerged towards the end of the 1950s. The UK, France, the United States and the Soviet Union began to consider the development of supersonic air transport. Progress by the British and the French followed quickly, and on 28 November 1962 a draft treaty was signed by the governments of these two countries to construct jointly a supersonic airliner, subsequently called 'Concorde'. Its cruising altitude would be about seventeen kilometres.

When the Concorde project began, many had thought this aircraft would indeed fly above all weather, but meteorologists soon realized this would not necessarily be so. Matters of concern included the frequency with which cumulonimbus clouds penetrated the stratosphere, and there was a need to obtain information about clear-air turbulence and horizontal gradients of wind and temperature. But there was a problem: how could the required data on cloud and hail and other conditions at the cruising altitude be obtained without an aircraft to fly at those heights? The Office's annual report for 1965 shed some light on the way this problem was tackled.

A group of scientists from the Royal Aircraft Establishment had, from 21 April to 21 June 1965, visited the National Severe Storms Laboratory of the U.S. Weather Bureau at Norman, Oklahoma. A member of the Office's staff had accompanied them and taken part in an investigation of conditions in and around the upper levels of thunderclouds, using instrumented aircraft, airborne radar and ground radar. Copies of films taken by U2 aircraft from heights above eighteen kilometres in the vicinity of severe storms had also been obtained. These films, which had not previously been analysed, had revealed the presence of cloud tops reaching 17.7 kilometres. The report for 1965 further stated that formal collaboration with French and American meteorologists had continued after the visit to Norman, focusing on research requirements and the specification of meteorological parameters in connection with the certification and operation of supersonic transport aircraft. However, little factual information had been gained additional to that already obtained, as no aircraft capable of gathering information at heights of seventeen to twenty kilometres had been available.

The UK's Air Registration Board had established a committee in 1959 to investigate the airworthiness of supersonic transport. In turn, this committee had established a number of working groups, one of them its 'Structure of the Atmosphere Group'. The Office became much involved in the work of this group and in the work of a panel of experts that was established in 1968, the 'Supersonic Transport Panel of the International Civil Aviation Organization'. Attention was focused on various possible influences: turbulence, hail, heavy rain, ice crystals, temperature, wind, ozone, solar radiation, and cosmic radiation. By 1969, the year the first Concorde made its first test flight and first went supersonic (respectively, 2 March and 1 October), it had been concluded that the single most adverse atmospheric condition was the thunderstorm, which was typically accompanied by turbulence, hail, heavy rain, lightning and strong vertical air currents.

From 1966 onwards, paper planning of Concorde's operations across the North Atlantic was carried out at Heathrow Airport through the Office supplying weather data to the British Overseas Airways Corporation. A basic assumption for the provision of an operational service was that the aeroplane behaved as a subsonic aircraft when flying subsonically. Thus no special requirements were envisaged for take-off and landing that could not be met by airport weather forecasts and observations currently available. During the subsonic climbing phase, it would be necessary to avoid cumulonimbus clouds, but that was a routine precaution for subsonic jets, and during the deceleration and descent phases of Concorde flights no special requirements were envisaged. Accurate forecasts of temperature at Concorde's cruising altitude were, however, important, because ambient temperature affected both the efficiency of the engines and the material of the airframe.[18]

---

[18] A Discussion Meeting on 'Meteorological problems of supersonic aircraft' was held in the rooms of the Royal Meteorological Society on 18 June 1969. Contributions to this meeting were published in the October 1969 issue of the *Quarterly Journal of the Royal Meteorological Society* (Vol. 95). For an explanation of the atmospheric conditions that were likely to affect Concorde flights, see 'Weather and Concorde', by W T Roach, published

Radiosonde observations showed that temperatures in the lower stratosphere could vary considerably over periods of a few days. The reliability of soundings made by some countries left something to be desired, though, and the density of the network of upper-air stations, which was satisfactory over North America, Europe and much of the Soviet Union, was less so elsewhere. However, the WWW being planned by WMO was expected to provide an opportunity to address this problem.

Two other special investigations in the 1960s which involved staff of the Office were brought about by weather events characterized by strong winds.

On 16 February 1962, the city of Sheffield in Yorkshire experienced sustained winds of at least sixty-five knots, with gusts of eighty knots or more. Almost 70% of the city's housing stock was damaged, and three people died. Gales were experienced widely over the British Isles that day but had been exceptionally strong over Sheffield and much stronger there than anywhere else. An investigation by staff of the Office provided an explanation. There had been an enhancement of the strong westerly air flow over the lee slopes of the Pennines. This was a rare occurrence over the UK but common over mountainous areas on the Continent and in North and South America.[19]

The other event that was characterized by strong winds also occurred in Yorkshire. Three of a group of eight cooling towers at the Ferrybridge Power Station near Ponte-fract collapsed during a gale on 1 November 1965, and the other five were all severely damaged. The Central Electricity Generating Board (CEGB) set up a committee of inquiry to investigate the incident and make recommendations. For this inquiry, H C Shellard of the Office's Climatological Services Branch was asked to specify the wind conditions at the time of the collapse and estimate the probability of occurrence of such winds. His investigation showed that the gale had not been exceptional. Winds of the strength which had occurred (40–45 knots) could be expected on average every five years or so. Why, therefore, had the towers collapsed?

The part of the inquiry which considered the situation, specification, design and construction of the towers showed that the designers had only considered average wind speeds over one minute. They had neglected short-lived gusts, which can be very damaging to thin shell structures that can vibrate, but that was only part of the problem. Tests in wind tunnels showed that impinging turbulence from the wind added to the fluctuating air motions within the group of towers, causing stresses that triggered structural resonances which destroyed some of the towers. It was found that the grouped nature of the towers had contributed significantly to the incident

in *Weather* in 1970 (Vol. 25, pp. 254–265). See also 'Meteorological problems in the design and operation of supersonic aircraft', by R F Jones *et al*, published in 1967 as *WMO Technical Note* No. 89, 71 pp.

[19] For an analysis of the Sheffield gale, see 'The Gale of February 16, 1962, in the West Riding of Yorkshire', by C J M Aanensen and J S Sawyer, published in *Nature* in 1963 (Vol. 197, pp. 654–656). See also 'Gales in Yorkshire in February 1962', by C J M Aanensen, published in 1965 by the Meteorological Office in *Geophysical Memoir* No. 108 (Her Majesty's Stationery Office, Met.O.711c, 44 pp.).

and concluded that the disaster had been more about bad planning than the strong wind.[20]

A positive outcome of the incident was that an intensive programme of research by several British institutes was stimulated to study turbulence in the atmospheric boundary layer. These institutes included the Office, the CEGB, the NPL and the Building Research Establishment, and the research showed how important wind forces were for the lightweight buildings being erected. The need for the Office to issue warnings about the dangers of high winds to trains, lorries and other vehicles was also recognized.

During the 1960s, regular services for the public continued to evolve, including weather presentations on television. As one of the BBC TV presenters, Bert Foord, noted in an article published in *Weather* in 1994 (Vol. 49, pp. 390–393), grey and black maps covered with thick perspex were being used by 1963, the year he became a presenter. On these, isobars and fronts were drawn in black marking ink, which, he said, was "very sticky, oily, dirty stuff indeed". There were three cameras, so the routine was to have three charts. One showed the isobars and fronts over Europe and the North Atlantic at midday (drawn from a copy which the presenter brought from the London Weather Centre). Another showed a chart of the isobars and fronts expected at midday the next day over the British Isles, and the third was a summary chart.

By about 1966, Foord recalled, presentations included a chart for the British Isles which showed the international weather symbols, with dots for rain, triangles for showers, numbers for temperature, and so on. About this time, he said, he was "contacted by a firm in Sheffield who had developed a rubber material impregnated with iron filings which was magnetic when used on a steel surface". Foord and his colleagues cut the material into the shapes of the weather symbols and started to use them in TV presentations. The new steel charts were very heavy, though, so they had to be put on wheels. Moreover, the Atlantic chart measured some eight feet by twelve, so Foord needed to stand on a box to reach the Arctic regions. And this was not the only problem. The symbols tended to break after a few weeks of use and sometimes fell off during a live presentation. All in all, though, as he said, the magnetic system saved the BBC hundreds of pounds per annum, because the cleansing fluid that had been used with the marking ink was expensive. An important development came in the autumn of 1967, when the BBC1 channel began to use colour. The hues used on the weather maps now needed to be suitable for viewing in both colour and black and white.

All through the 1960s, Foord recalled, "there was pressure from the BBC to reduce the duration of weather forecasts, especially those following the early evening news".

---

[20] For an overview of the meteorological investigation, see 'Collapse of cooling towers in a gale, Ferrybridge, 1 November 1965', by H C Shellard, published in *Weather* in 1967 (Vol. 22, pp. 232–240). See also *Report of the Committee of Inquiry into collapse of cooling towers at Ferrybridge, Monday 1 November 1965*, published by the CEGB in August 1966.

Figure 12.5. BBC TV weather presentation by Norman Ellis, early 1960s. © Crown Copyright 2010, the Met Office.

The problem was, he said, that the BBC was run by two kinds of managers. There were, he said, the journalists, who wanted everything instant and definite. If the news headlines took fifteen seconds, why did the weather forecast need two minutes? There were also the people from the arts or light entertainment, who sought individuality and unusual or amusing presentations, for example, a Mexican hat if it was hot and a toboggan if snow was likely. The duration of the 6 p.m. forecast fell to sixty seconds and then for a brief period thirty seconds, for the whole of the UK! Worse was to come, Foord recalled, when a costing exercise by McKinsey & Co. caused the BBC's Head of Presentation to decide that weather broadcasts could no longer be afforded, so they ceased. However, viewers complained strongly, and in any case, taking the weathermen off the air did not, in fact, reduce costs. The absence of live TV weather presentations lasted a week.

As Foord mentioned, the TV weathermen became household names and started to appear in other programmes. The idea of civil servants becoming personalities had been anathema to the Office and the Air Ministry in 1954, when live presentations had begun (see Chapter 11), but the inevitable had happened; Foord and his colleagues

Figure 12.6. Plotting weather charts in the Central Forecasting Office at Bracknell, 1965. © Crown Copyright 2010, the Met Office.

had become celebrities. There were limits, though, as Foord said. The Office refused to allow him to appear in a Christmas show as they thought his appearance in morning dress dancing like the popular group known as the 'TV Toppers' would trivialize the weather forecasts. And perceived flippancy was frowned upon. A BBC TV weather presenter, Peter Walker, was withdrawn in 1966 after he likened a depression with two centres of low pressure to a fried egg with two yolks, calling it a "double-yolked job", but he was reinstated several months later.[21]

---

[21] The author of this book (Malcolm Walker) was given this information in 1967, by Peter Walker himself, when they met as a result of both being shortlisted for a lectureship in Cardiff. It was amusing that Peter went on, in 1968, to become a Lecturer in Communication at Heriot-Watt University, Edinburgh.

Soon after Dr Mason became Director-General, another attempt to revive Airmet was made. The context was the broadcasting of pop music, with particular reference to the activities of pirate radio stations. As reported in *The Times* on 7 December 1965, the government had been considering the possibility of local radio. There was a suggestion, *The Times* said, that licences might be granted for commercial sound broadcasting, enabling the government "to regulate the frequency and power of transmissions and to replace the pirates with respectable privateers". The assumption behind these plans was that "any available space in the jungle of wavelengths must be filled with pop". This was being "challenged furiously by people not possessed by the craving for these sounds". Some people wanted any vacant space to go to Airmet, which had, these people said, a prior claim. "The devotees of Airmet", said *The Times*, were "extremely bitter" that the spare wavelengths were "apparently to be handed over to disc jockeys and not to meteorologists, who have been queueing for them for 15 years". The hopes of Airmet devotees were dashed. Airmet was not revived.[22]

International Developments

Throughout his time as Director-General, Dr Mason played an influential and some-times controversial part in the activities of WMO. He championed, in particular, a number of ambitious research enterprises which were coordinated by WMO and involved cooperation and collaboration between the meteorological services of many nations.

In early 1966, the Office was inspected by the Parliamentary Estimates Commit-tee (now called 'Public Accounts Committee'), a group of Members of Parliament whose job it was (and still is) to scrutinize spending by government departments and thus ensure economy, efficiency, effectiveness and value for money. The inspection provided an opportunity for Mason to explain to the government the importance of a global observation scheme which had been approved in principle by WMO in 1963. This was the World Weather Watch (WWW), mentioned earlier in this chapter. It was to be important not only for the UK but also all other nations, the idea being that the scheme would become an operational system to coordinate the rapid collection and exchange of weather data among all WMO members, as well as to disseminate weather forecasts and other meteorological information from centralized processing centres. Mason was an enthusiastic supporter of the plan and keen that the Office should cooperate fully.

Credit for the WWW concept was due in no small measure to Ernest Gold. Through-out his time as President of the IMO's Commission for Synoptic Weather Information

---

[22] A letter published in the September 2010 issue of *Weather* (Vol. 65, pp. 253–254) shows that Airmet has still not been forgotten. In the opinion of the letter's author, Gordon Currie, there is a place for a broadcasting service like Airmet today.

(1919–1947), he had pressed for the creation of a worldwide meteorological system that was composed of coordinated national facilities and services provided by the Organization's member states, supplemented by other international bodies.[23] An important development had come on 25 September 1961, when the President of the United States had proposed to the United Nations (UN) the establishment of an international effort in weather prediction. The response of the UN had come in Resolution 1721(XVI)C of its General Assembly adopted on 20 December 1961, which had called on WMO and the International Council of Scientific Unions (ICSU):

(a) To advance the state of atmospheric science and technology so as to provide greater knowledge of basic physical forces affecting climate and the possibility of large-scale weather modification

(b) To develop existing weather forecasting capabilities and to help Member States make effective use of such capabilities through regional meteorological centres

The Fourth WMO Congress, held at Geneva in April 1963, had acted on this resolution by accepting the principle of a WWW and, in association with it, a Global Atmospheric Research Programme (GARP). The latter would, to some extent, continue and develop the meteorological work which had been carried out during the IGY of 1957–1958 and would also embody research techniques that had proved effective during the International Indian Ocean Expedition (IIOE) of 1959 to 1964, a scientific enterprise which had involved British oceanographers but only to a small extent British meteorologists. During the IIOE's meteorological phase, in 1963 and 1964, Indian and American meteorologists had studied the monsoon of southern Asia by means of oceanographic research vessels, instrumented aircraft and dropsondes. The Office had contributed to the expedition by setting up a radiosonde station on the Seychelles and supplying surface and upper-air data from it to the expedition's base at Bombay (now Mumbai).

It is no exaggeration to say that WMO's Fifth Congress, held at Geneva in April 1967, marked a turning point in world meteorology. It approved the WWW plan, decided to proceed with GARP on a joint basis with ICSU, and approved a Voluntary Assistance Programme which provided a means of helping countries implement the plan. The British government and the Office supported the WWW and GARP fully. Mason was a member of WMO's Executive Committee and therefore closely involved in the planning of both programmes.

As an essential part of the WWW implementation process, a global telecommunications network was created, at the heart of which there were three World Meteorological

---

[23] See 'British architects of the international meteorological system', by D G Harley, published in 1973 in the *Meteorological Magazine* (Vol. 102, pp. 249–257). Gold was honoured by WMO in 1958, when he became the first of three British recipients of the IMO Prize, the others being Dr R C Sutcliffe (in 1963) and Sir Graham Sutton (in 1968).

Centres, at Melbourne, Moscow and Washington, their functions being to collect, process, exchange and archive meteorological data from all around the world. Regional Meteorological Centres were also created to carry out similar tasks on a regional basis. In addition, Regional Telecommunications Hubs were linked to the Centres and to each other by a dedicated high-speed communications system with low-speed links that branched off to national centres.

The contribution of the Office to the WWW was the subject of a question asked in the House of Commons on 5 February 1969 by Tam Dalyell. In reply, Gerald Reynolds, the Minister of Defence for Administration, advised the House that the Office was serving as a Regional Meteorological Centre in Europe and that planning was in hand for the establishment of a Regional Telecommunications Hub at Bracknell. The first annual contribution of £30,000 for the WWW had been paid to WMO, and equipment worth £100,000 had been offered to developing countries. A British merchant ship was making routine upper-air soundings at sea after successful trials, and the number of merchant ships which made surface observations at sea had been increased.[24] Moreover, an upper-air station was in operation at the RAF's staging post on Masirah Island (off Oman). The Office's research programme had been expanded, and the Office was helping in the planning of GARP. Furthermore, three overseas meteorologists had been awarded fellowships in the Department of Meteorology of the University of Reading.[25]

GARP was indeed a global *research* programme, and it was ambitious. It involved intensive studies of physical processes in the troposphere and stratosphere, and its principal objectives were twofold: (1) greater understanding of the physical basis of climate and (2) an increase in the accuracy of forecasting over periods from one day to several weeks.[26] It was composed of sub-programmes and experiments, the former being projects of both theoretical and experimental character as were consistent with GARP objectives, while experiments comprised observational programmes for

---

[24] A bulk carrier, the *Sugar Exporter*, was the first British vessel to provide regular upper-air measurements of pressure, temperature and humidity whilst on the high seas, doing so in early 1968 on a round-trip voyage between London and Trinidad. Balloons filled with helium were released by two meteorological assistants every twelve hours, and these balloons carried radiosonde equipment to heights up to 60,000 feet. In addition, one of the Office's experienced weather forecasters was in late 1968 serving (voluntarily) aboard the Hull trawler *Orsino* and measuring wind, barometric pressure, humidity, air temperature and sea temperature. He was able to make regular contact with the CFO at Bracknell and to receive facsimile weather charts, shipping bulletins and special briefings and forecasts.

[25] The Reading department opened in 1965 with R C Sutcliffe its first Professor of Meteorology. The plan to form the department had begun with informal discussions between the Office and the University of Reading in 1960. Both Sir Graham Sutton and Dr Sutcliffe had supported the idea of forming the department, and agreement had been reached that the Office would initially provide some staff to teach some courses, pending the appointment of permanent staff. See *Meteorology at Reading: The first forty years*, edited and compiled by Jackie Hoskins, published by the University of Reading in 2006 (136 pp.).

[26] See 'First session of the Joint GARP Organizing Committee', by R V Garcia, published in 1968 in the *WMO Bulletin* (Vol. 17, pp. 126–130). For an overview of the aims and objectives of GARP, see 'Progress on the planning and implementation of the Global Atmospheric Research Programme', by B Bolin, published in 1969 in *The global circulation of the atmosphere* (the main papers presented at a conference held at the Royal Society of London from 25 to 29 August 1969, published by the Royal Meteorological Society, pp. 235–255).

ascertaining the behaviour of the whole atmosphere or some part of it, depending on the requirements of a particular sub-programme.

The first major experiment was the GARP Atlantic Tropical Experiment, known by the double acronym of GATE. It took place from June to September 1974, and its objectives were, in the words of Dr Mason:[27]

To extend our knowledge of those aspects of the meteorology of the equatorial belt that are essential for a proper understanding of the circulation of the atmosphere as a whole and, at the same time, to improve the understanding and prediction of weather in the tropics.

The reason for focusing on low latitudes was that they contain the source regions for much of the energy required to drive the atmospheric and oceanic motions of the whole globe. The area covered by GATE extended from the parallel of 10°S to the parallel of 20°N, from the western Indian Ocean to the eastern Pacific Ocean across Africa, the Atlantic Ocean and South America, with an extension to 24°N over the Caribbean Sea. Meteorologists were keen to concentrate on this area from June to September because tropical disturbances which developed into Atlantic hurricanes during the northern hemisphere's summer and autumn months were believed to originate over North Africa and intensify over the tropical Atlantic.

GATE was the largest and most complex international scientific experiment ever undertaken up to that time, with ten nations participating (Brazil, Canada, France, the Federal Republic of Germany, the German Democratic Republic, Mexico, The Netherlands, the UK, the United States and the USSR). From the UK, scientists of the Institute of Oceanographic Sciences and the Royal Navy took part, and staff of the Office contributed greatly to the experiment, working closely with fellow scientists from the other collaborating nations. Two whole issues of the *Meteorological Magazine* were devoted to the British contributions, those published in August and September 1976 (Vol. 105, pp. 221–245 and 249–288).[28]

The UK provided four of the thirty-nine ships which took part in GATE: the hydrographic survey vessel HMS *Hecla*, the oceanographic research vessel *Discovery*, and two small ships chartered by the Office. Upper-air soundings of wind were made from the small ships, and an intensive programme of meteorological and oceanographic observations was carried out from *Hecla*, including measurements of temperature and humidity in the lowest 1000 metres of the atmosphere by means of instrumental packages suspended from a tethered balloon. The C-130 aircraft of the MRF was also involved in GATE, making full use of its radar, radiometers, particle samplers and Doppler wind-finding equipment. It flew forty missions, totalling 386 hours.

---

[27] See 'The GARP Atlantic Tropical Experiment', by B J Mason, published in 1973 in the *WMO Bulletin* (Vol. 22, pp. 79–85).

[28] Sadly, the August issue also carried an obituary of Dr G M B Dobson, who had died on 10 March 1976, aged 87. He was a world authority on atmospheric ozone and when a young man had been employed by the Meteorological Office (see Chapters 7 and 8).

And throughout the field phase of GATE, observations which had been made in the GATE area reached Bracknell through a product of the WWW, the Global Telecommunication System. Using these observations, the Office made real-time forecasts every twelve hours with an eleven-level numerical model that had been developed in the Dynamical Climatology Branch for the purpose.

During GATE, atmospheric disturbances were investigated with the aid of a network of upper-air stations over Africa and South America, supplemented by radar, research aircraft, radiosonde ascents from dedicated ships, instruments suspended from tethered balloons, and observations from several meteorological satellites, one of them a new type of satellite, launched on 17 May 1974, just in time for GATE. It was Synchronous Meteorological Satellite 1 (SMS-1), and it was the first operational spacecraft to sense meteorological conditions from a fixed location above the earth. In contrast to the other satellites that were used during GATE, SMS-1 was in a geostationary orbit. It orbited the earth above the equator at the same rate as the earth below was rotating. The other satellites were in polar orbit. SMS-1 was positioned approximately 36,000 kilometres above the equator at longitude 45°W and provided visible and infra-red coverage of Central America, eastern North America, South America, Africa west of about 20°E and the whole of the Atlantic Ocean to about 65°N and S. It was intended to be in orbit at 30°W, so there was, unfortunately, less coverage of Africa than originally planned.

GATE was undoubtedly successful and produced much valuable new knowledge. Some observational difficulties were, however, encountered. For example, a few stations did not become operational, and some equipment failed to function satisfactorily. The experiences gained during GATE, and the lessons learned, soon proved helpful in the planning and execution of an even larger and more ambitious international scientific venture, the First GARP Global Experiment (FGGE), nicknamed 'Global Weather Experiment', which took place during 1979 and comprised an intense observational programme that used the synoptic and upper-air network established for the WWW, supplemented by specially equipped ships and aircraft, satellites, ocean buoys and constant-level balloons.[29] All of the 140 member states of WMO took part, and the data collected still constitute the most comprehensive compilation of meteorological variables ever assembled. The data led to new methods of analysis in operational weather forecasting and stimulated major improvements in not only the forecasting

---

[29] For an overview of the work carried out during the FGGE, see 'The Global Weather Experiment I: The observational phase through the first Special Observing Period', by R J Fleming *et al*, published in 1979 in the *Bulletin of the American Meteorological Society* (Vol. 60, pp. 649–659). See also 'The Global Weather Experiment II: The second Special Observing Period', by R J Fleming *et al*, published in 1979 in the *Bulletin of the American Meteorological Society* (Vol. 60, pp. 1316–1322). For an overview of the FGGE from a British perspective, see 'The First GARP Global Experiment', by A Gilchrist, published in 1979 in the *Meteorological Magazine* (Vol. 108, pp. 129–134). For a review of the highlights and accomplishments of the FGGE, see 'The Global Weather Experiment – 10 years on', by A Hollingsworth, published in *Weather* in 1989 (Vol. 44, pp. 278–285).

models used for weather forecasting but also the numerical models that were being developed for climate prediction.

As well as mathematical modelling, the observing systems and telecommunications facilities that were so necessary for the great international projects advanced greatly during the 1970s. Polar-orbiting weather satellites became ever more numerous and provided imagery of ever greater quality. Moreover, a second Synchronous Meteorological Satellite (SMS-2) was launched on 6 February 1975 and placed in a geostationary orbit over the equator at longitude 135°W, over the Pacific Ocean. Rather more useful to the weather forecasters of Britain and the rest of Europe, though, was the European Space Agency's first geostationary satellite, Meteosat-1, which was launched on 23 November 1977 and supplied its first images on 9 December from 36,000 kilometres above the equator close to the prime meridian. It provided visible and infra-red coverage of the whole of Europe and most of the North Atlantic and constituted a welcome additional source of information for the forecasters.

Besides GATE, there were many other GARP experiments and sub-programmes, some that involved the Office, others that did not. One that did was the Royal Society Joint Air-Sea Interaction Project (JASIN), an intensive study of the atmospheric boundary layer and upper ocean in an area of the North Atlantic 450 kilometres west of the Outer Hebrides.[30] In this project, the Office collaborated with the Institute of Oceanographic Sciences and with scientists of eight other countries. Field trials to develop and test necessary instrumentation took place in June 1970 and September 1972 near Ocean Weather Station J (52°N 20°W), and the project itself took place farther north several years later, from July to September 1978, near Rockall.

The MRF's Hercules took part, along with aircraft provided by the United States and West Germany. These three flying laboratories used sophisticated instrumentation to measure air motions, humidity, air temperature and sea-surface temperature. Fourteen ships also took part, including three from the UK: HMS *Hecla*, the RRV *John Murray* and the MV *Gardline Endurer*. From these, using instruments and automatic recording equipment from the Office's turbulence unit at Porton Down, measurements were made of temperature, humidity, wind and barometric pressure. In addition, meteorological data were transmitted from a moored buoy to *Hecla* through a VHF link, and 163 radiosonde ascents were made from *Hecla*. A tethered balloon was also flown from this ship, carrying instruments which measured fluxes of heat, momentum and water vapour between the ocean surface and a height of about 500 metres. Staff of the Office spent a total of six weeks aboard *Hecla*, stationed at 60°15'N 10°30'W. The

---

[30] The atmospheric boundary layer is the layer of the atmosphere in contact with land or sea. Its depth varies with static stability, from a few hundred metres in stable conditions to a kilometre or more in convective conditions. It is a layer which is characterized by turbulent motions, mostly generated by the wind flowing over the land or sea surface. The upper ocean also contains turbulent motions.

JASIN Project showed that the thermal structure of the upper ocean was more complex than hitherto realized, a matter of some relevance in respect of weather forecasting then and later.[31]

GARP became the World Climate Programme in 1980, but that is a story we take up in Chapter 13. We now consider other facets of the Office's work in the 1970s.

An Exciting New Era

When Dr Mason took up his appointment in 1965, he was "dismayed to discover", as he put it in an interview published in the *WMO Bulletin*, that "virtually no young scientists had been recruited for years".[32] High-quality graduates in mathematics and physics tended to be more attracted to subjects considered 'modern', such as nuclear physics and radio astronomy, than to meteorology, which was popularly thought to be a branch of geography with a veneer of classical physics. He himself led endeavours tó raise the status of meteorology as a subject worthy of the attention of physics and mathematics graduates.[33] In particular, he lectured in universities, promoting meteorology as a scientific subject with an exciting future.

By 1970, his efforts had begun to pay off. Graduate recruitment had improved greatly. He had also attracted a proven scientific leader, a full professor at the Massachusetts Institute of Technology, Raymond Hide, who had joined the Office in 1965 to establish a Geophysical Fluid Dynamics Laboratory.[34] Moreover, he had brought back to the UK from the United States a promising young physicist, Keith Browning, who was only in his late twenties but had already become an authority on the use of radar for studying the structures of depressions and thunderstorms. He had joined the Office in 1966, as a Principal Research Fellow, to head the Meteorological Research Unit in the Royal Radar Establishment at Malvern.[35]

The Presidential Address which Dr Mason delivered before the Royal Meteorological Society on 15 April 1970 constituted, he said, his "first attempt at long-range forecasting"![36] In it, he looked forward to the major scientific and technological developments he thought likely over the coming thirty years, as well as "changes in the pattern of demand for weather services by industry and the general public, and to their economic and social consequences". All of these would together, he

---

[31] For an overview of JASIN, see 'Summary of the JASIN 1978 field experiment', by R T Pollard *et al*, published in 1983 in the *Philosophical Transactions of the Royal Society* (Vol. A308, pp. 221–230).

[32] See 'The *Bulletin* interviews: Sir John Mason', published in 1995 in the *WMO Bulletin* (Vol. 44, pp. 315–325).

[33] These endeavours were part of the ten-point plan which Mason formulated shortly after he became Director-General. This plan included, besides staff recruitment, NWP and computer developments, international projects, better observations, better telecommunications, better research facilities and expertise, and a new radiosonde.

[34] This Laboratory proved a remarkable nursery for many of the later leaders of meteorology in the Office and the universities.

[35] Both men were subsequently elected Fellows of the Royal Society, Hide in 1971, Browning in 1978.

[36] See 'Future developments in meteorology: An outlook to the year 2000', by B J Mason, published in the *Quarterly Journal of the Royal Meteorological Society* in 1970 (Vol. 96, pp. 349–368).

predicted, "determine the evolution of our science and the organization of meteorological institutions, both national and international". He said he was optimistic. He believed that meteorology was entering a new and exciting era in which there would be great advances and changes in the subject both as a science and as a service, and he drew attention to numerical modelling and the research associated with the WWW and GARP as indicators and manifestations of a new spirit of hope and confidence among meteorologists. Advances in dynamical meteorology, numerical modelling and weather prediction had begun to render the "old divisions of synoptic, dynamic and physical meteorology, aerology and weather forecasting largely outmoded and irrelevant".

Mason said that the numerical model used by the Office for weather forecasting, though highly simplified, had nevertheless "reduced the frequency of really bad forecasts and provided much more reliable indications of trends over the next few days". Besides this overall improvement in the accuracy of forecasts, specific improvements were desirable, too. The gas and electricity supply industries, for example, needed quantitative forecasts of temperature and forecasts of the timing, distribution and intensity of precipitation. But at present, Mason said, it was not possible to "improve significantly on climatological averages in predicting even mean daily temperatures three days ahead". It was also not possible, he went on, to provide the accurate quantitative forecasts of rainfall required for irrigation, flood forecasting and the regulation of stream flow, dams and reservoirs. Nor was it possible to forecast heavy snowfalls accurately.

He pointed out that many weather-sensitive industries required forecasts for much more than two or three days ahead. In agriculture, for example, there would be considerable economic benefits if decisions could be made with confidence at least a week in advance for activities such as ploughing, sowing, harvesting, haymaking, crop spraying and irrigation. Other industries that would benefit from the availability of forecasts for several days ahead were building and construction, fuel and power, and a wide range of manufacturing industries. And reliable forecasts for a week ahead would be of great value for ship routeing.

Mason reported that seasonal forecasts were being made in the Office on an experimental basis by extending the techniques that were used for producing monthly forecasts. These techniques were based on analogue recognition of correlations between anomalies of sea-surface temperature in the North Atlantic and patterns of barometric pressure over Europe.[37] Eventually, though, he expected monthly and seasonal forecasting to be carried out by numerical methods. Both in the Office and the

---

[37] A discussion meeting entitled 'Long-range weather forecasting' was held in the rooms of the Royal Meteorological Society on 19 November 1969. Four contributions were published in the October 1970 issue of the Society's *Quarterly Journal* (Vol. 96). See particularly: 'Recent developments in long-range forecasting in the Meteorological Office', by R Murray (pp. 329–336); and 'Sea temperature anomalies and long-range forecasting', by R A S Ratcliffe (pp. 337–338).

United States, models of the general circulation of the atmosphere were developing to the stage where this could soon be considered a realistic possibility. Models which included ocean-atmosphere interactions were confirming what many meteorologists had long thought, that the oceans played an important role in the atmospheric general circulation.[38]

The advances in numerical modelling of weather and climate that Mason mentioned in his lecture could not be achieved without an increase in computing speed and capacity. A new computer was needed, and an IBM 360/195 machine had indeed been ordered already. It was delivered to the Office in October 1971, and site trials of it were completed successfully on 13 December. After that, it was brought into operation and used not just for testing complex mathematical models of the atmosphere which had been developed for operational forecasting but also for data extraction, chart plotting, graphical displays and various research tasks. Meanwhile, the KDF9 computer remained an essential part of the CFO at Bracknell and had to be kept fully functional until some time after the IBM machine had been brought to an acceptable operational state. From 18 December 1971 to 21 March 1972, the new computer was operated 14 hours a day, seven days a week. Thereafter, it ran continuously, and on 1 August 1972 it replaced the KDF9 as the operational machine for NWP. At the same time, a ten-level model of the atmosphere replaced the old three-level model.[39]

The statement of Sir Graham Sutton in the annual report for 1961 that the new premises at Bracknell should be adequate to meet foreseeable demands proved over-optimistic. The headquarters building was overcrowded right from the start, and the accommodation problem became more and more acute until, in 1971, it was alleviated by the opening of an extension to the headquarters building and completion of a new centre for instrument research and development at Beaufort Park, Easthampstead, four miles from the headquarters. Moreover, the Training School moved that same year to a new home in Reading. For the Office, 1971 was truly an *annus mirabilis*.[40]

---

[38] The term 'ocean-atmosphere system' was coined to express succinctly the functioning of the oceans and the atmosphere as a complex system of two coupled fluids. See *The Ocean-Atmosphere System*, by A H Perry and J M Walker, published in 1977 by Longman (160 pp.).

[39] For information about the IBM 360/195 computer and the functions it fulfilled, see 'The Meteorological Office 360/195 computing system', by G A Howkins, published in 1973 in the *Meteorological Magazine* (Vol. 102, pp. 5–14). This article contains an appendix which provides a comparison of the Meteor, Comet and IBM 360/195 machines.

[40] That same year, 1971, the Royal Meteorological Society moved from London to premises in Bracknell near the Office. As there was insufficient room in the new premises for the Society's extensive library, many items from it were deposited for safe keeping in the National Meteorological Archive, which was then at Bracknell but is now at Exeter. Rare meteorological books were among the items deposited, including more than 100 published before 1600. Other material that was deposited included the original of the diary of Admiral Beaufort, which contains his famous wind and weather scales. This had been presented to the Society in 1927 by a grandson of Admiral Beaufort, Augustus Francis Beaufort. Most of the Society's committee minutes, together with much correspondence and many other documents, were discarded in 1971 and found their way to the University of Texas at Austin, United States. Happily, though, they were recovered in 1994 and are now in the National Meteorological Archive.

Figure 12.7. The headquarters of the Meteorological Office at Bracknell after completion of the Richardson Wing. © Crown Copyright 2010, the Met Office.

The extension at Bracknell formed the fourth side of a square around a grassed court, the other three sides consisting of the original building. It had five storeys and housed the CFO, the Telecommunications Centre and the Computing Laboratory. The new computer was installed in it in the autumn of 1971, and occupation of the extension was completed in the spring of 1972. The extension was called the Richardson Wing, to honour L F Richardson, the pioneer of NWP (see Chapter 8), and it was opened officially on 6 October 1972 by the Prime Minister, Edward Heath, in the presence of six members of the Richardson family. Mr Heath had a keen interest in meteorology. He was a world-class yachtsman who had won the Sydney to Hobart yacht race in 1969 (the year before he became Prime Minister) and had captained Britain's winning Admiral's Cup team in 1971 (whilst he was Prime Minister). Over a number of years, he often benefited from Meteorological Office weather forecasts and other advice for his yachting activities.[41]

Automatic Systems Develop

The new laboratories and workshops at Beaufort Park were occupied by the Instruments and Observations Branch towards the end of 1971. Soon, to quote from the Office's annual report for 1972, they "paid dividends in terms of a higher level of activity in the research and development field". The report singled out three topics for

---

[41] Another keen yachtsman, D M Houghton, a senior forecaster in the Office, served as a member of the British yachting team at the Olympic Games held in Mexico in 1968. He advised the British team on small-scale meteorological factors which might help their performance. See 'Acapulco '68', by David Houghton, published in *Weather* in 1969 (Vol. 24, pp. 2–18). He subsequently advised the British yachting team at several more Olympic Games.

Figure 12.8. Prime Minister Edward Heath opens the Richardson Wing of the Meteorological Office at Bracknell, 6 October 1972, watched by Dr Mason. © Crown Copyright 2010, the Met Office.

mention: (1) the introduction into service of the first of a new generation of automatic climatological systems, (2) sea trials of the Office's first telemetering meteorological buoy and (3) evaluation in flight of the new Mk 3 radiosonde against reference radiosondes from two other countries.

The climatological system was specifically for recording rainfall totals and the timing of rainfall events and proved so successful that authority was given in 1972 for the equipment to be introduced into operational service. Magnetic tape was used for recording rainfall events, and data from the tapes could easily be converted to a form suitable for computer processing.

The telemetering meteorological buoy was moored in Cardigan Bay some ten kilometres off Aberporth from June to August 1972. From there, it transmitted observations of air and sea temperature, wind speed, humidity and barometric pressure, which were received at Aberporth with a data recovery rate of about 90%. Development of the buoy, in collaboration with the Royal Navy's Directorate of Meteorology and Oceanographic Services, had begun in 1968, when the need had been recognized for a small instrumented buoy to be used in air–sea interaction studies.

The buoy was first put to the test during the Atlantic Trade-wind Experiment, which took place in mid-Atlantic near 10°N in February 1969, and was also tested on the Atlantic during the field phase of the JASIN project in June 1970. The buoy was further tested in the Thames Estuary from February to April 1973 and again performed well, once problems over ingress of water into the compartment which contained the electronics had been overcome. These trials raised confidence that networks of automatic weather stations on buoys, automatic light-towers, oil rigs and other marine platforms could soon be established.[42] In the late 1970s, meteorological sensor suites were installed on Mobil's *Beryl A* and Occidental's *Piper A* oil platforms in the North Sea. Furthermore, a buoy which measured wave motions and also carried oceanographic and meteorological sensors, together with an on-board data-recording system, was deployed in Lyme Bay (off Dorset). From this buoy, data could be obtained on demand by the London Weather Centre through a modem link. Another type of data buoy was tested off Norfolk in 1975 and 1976 and deployed 160 miles south-west of the Scilly Isles in June 1978. Its meteorological data could also be retrieved through a telephone dial-up facility.[43]

In 1979, during the First GARP Global Experiment, there were more than 300 operational buoys around the world, supplied by eight nations (Australia, Canada, France, New Zealand, Norway, South Africa, the UK and the United States). The majority were in the southern hemisphere, where many parts of the ocean were (and still are) rarely visited by ships. In late 1977, a project concerned with a European network of meteorological and oceanographic buoys was established. This was Project 43 under the EEC's scheme for Cooperation in Science and Technology (COST), which involved collaboration with the Royal Navy's Department of Naval Oceanography and Meteorology and the Norwegian Christian Michelsen Institute.

During the 1970s, much of the effort at Beaufort Park was focused on automation for meteorological instruments and on the design and development of automatic weather stations. As well as buoys, automatic weather stations for use on land were developed, in particular the Meteorological Office Weather Observing System (MOWOS).[44] This system had progressed sufficiently by the end of 1969 for approval to be given for a pilot network of eight MOWOS stations to be established to assess the overall effectiveness of the system before proceeding to a larger network. Such was the success of these trials that MOWOS was in use operationally at a number of places in the British Isles by the end of the 1970s.

---

[42] For an overview of the work carried out at Beaufort Park, see 'Special topic – instrumentation for operational meteorology', published in the Office's annual report for 1974, pp. 1–9.

[43] For details of this buoy, see 'The United Kingdom National Data Buoy (DB1) – a future source of meteorological data', by M Bader *et al*, published in *Weather* in 1978 (Vol. 33, pp. 353–361).

[44] For a description of MOWOS, see 'The Meteorological Office Weather Observing System (MOWOS) Mk 2', by G J Day *et al*, published in the *Meteorological Magazine* in 1974 (Vol. 103, pp. 329–337).

Figure 12.9. Servicing a moored meteorological buoy. © Crown Copyright 2010, the Met Office.

Development of a Meteorological Office Observing System for Ships (MOSS) began in 1976, and some progress was made by the end of the 1970s, but intractable communication problems over the transmission of data from ships to the UK were encountered. There was a growing need for MOSS because difficulties over obtaining weather observations from ships on a regular basis were increasing, mainly as a consequence of the central control systems of ships being automated. This was bringing about a reduction in deck and radio officer complements such that, at certain times, no radio officer was on duty to transmit meteorological reports on schedule. Automation was needed.

In a review article entitled 'The use of automatic weather stations in the observational network of the UK', published in the *Meteorological Magazine* in 1979 (Vol. 108, pp. 1–10), D N Axford outlined the current policy for the use of such stations by the Office and described what he called "the four main guiding factors": (1) economics, (2) staff, (3) the international factor, and (4) the basic network requirements.

On the subject of economics, he pointed out that new observing systems involved considerable investment, and this investment had to be considered in the light of resources available for meteorological purposes (i.e., the Office's budget of £30 million per annum). A pragmatic or even evolutionary approach needed to be taken.

New equipment should be introduced on an 'as and when' basis, when financial circumstances were favourable.

In regard to the staff factor, Axford pointed out three aspects. First, it was no longer the case that amateurs would "necessarily make reliable routine measurements for the Office out of interest in the science". Auxiliary and amateur observing stations were closing down and not always being replaced, so it had been found necessary to consider installing automatic weather stations to fill the gaps. Second, he said, there was "an understandable reluctance on the part of users to depart from the customary". Forecasters of the future would need to know which observations had been made by observers and which had come from an automatic weather station, as the forecasters would be the people who would assess the credibility of the readings and take any necessary quality control measures. Third, Axford noted, the growing number of staff who would be required to maintain the new equipment in operational condition needed to be trained.

In respect of the international factor, Axford pointed out that the introduction of new observing practices and automatic weather stations within national boundaries implied coordination with the WMO. Observations were exchanged with many other countries for various purposes, especially climatology and weather forecasting, so compatibility of observations with those made in other countries was essential.

As regards the basic network requirements, Axford reported that there had been an agreement within the Office that there was "a need for a basic network of 'key' synoptic observing stations, each fully manned by professional observing staff". This network, he said, was required to maintain the standards, quality control and meteorological integrity of all observations in the UK. The Office had decided that a minimum of thirty such key stations would be needed, with observations made at them hourly throughout the twenty-four hours. There would be, he went on, a separate requirement for a climatological network, for which approximately ninety climatologically different zones had been distinguished in the UK. For this, there needed to be at least one fully manned 'key' climatological observing station in each zone.

In the development, testing and manufacture of the new Mk 3 radiosonde, there were many delays. Eventually, though, technical problems were overcome, and the first operational flight of the new device was made at Aughton (near Liverpool) in November 1977. Thereafter, conversion of the other seven upper-air stations in the UK was carried out, with Lerwick the last station to begin operational flights with the new device, in October 1978. As with the Mk 2 device, the new radiosonde consisted of transducers for measuring temperature, pressure and relative humidity. It also contained a package of electronics which served to convert measurements of thermometer, hygrometer and barometer into radio signals which were transmitted to a receiving station on the ground, where the data were processed automatically by a mini-computer. The Mk 3 radiosonde system was the first fully automatic upper-air measurement equipment used operationally anywhere in the world.[45]

Automation became important in the National Meteorological Library, too, during the late 1970s, with the Library staff able to call on the expertise of the Office's Systems Development Branch to help in the development of a Meteorological Office Library Accessions and Retrieval System (known as MOLARS). This system relied on a Digital Equipment Corporation PDP 11/34 mini-computer that was dedicated to Library use, and the work it carried out involved catalogue card production, the updating of catalogues and production of the Monthly Accessions List. The Office's main computer was used for major sorting of files, twice-weekly printing of catalogue cards and preparation of the Monthly Accessions List of some 800 entries. By the late 1970s, about 10,000 items were being added to the catalogue each year, including individual journal articles and conference papers. The author catalogue, which dated back to the end of the nineteenth century, now consisted of 400,000 cards. The time for computerization had certainly arrived.[46]

Automation was also increasingly important in radar meteorology, with a major objective in the Office in the latter part of the 1970s being the establishment of an automated weather radar network. Considerable emphasis was placed on 'nowcasting' (i.e., forecasting for the period up to to six hours ahead), with particular reference to amounts and intensities of precipitation.[47] At the same time, and partly in connection with this, radar-based research led by Keith Browning continued to yield new knowledge of depressions and severe storms that was valuable to forecasters and hydrologists. The eventual aim of moves to create automated networks was the formation of a National Weather Radar Network, composed of interlinked unmanned radars and funded through a cooperative venture between the Treasury, the Office and water authorities.[48] The first radar in the network was set up at Camborne in Cornwall and began to provide data in the summer of 1978, with the data sent by telephone line to the Office's Radar Research Laboratory at Malvern, as well as to a local forecast office and to a Water Authority in the south-west of England. By 1980, there were four radars in the network, the others at Upavon in Wiltshire, Clee Hill in Shropshire and Hameldon Hill in Lancashire.

[45] See 'A brief introduction to the United Kingdom Mark 3 radiosonde system', by R E W Pettifer, published in *Weather* in 1979 (Vol. 34, pp. 95–101).

[46] See 'MOLARS – automating the National Meteorological Library', by E W C Harris and T McSean, published in the *Meteorological Magazine* in 1980 (Vol. 109, pp. 18–20).

[47] The term 'nowcasting' appears to have been invented by Roderick Scofield, who used it in the title of a report he presented in 1977, viz. 'A report on the Chesapeake Bay region nowcasting experiment' (NOAA Technical Memorandum NESS, No. 94, 52 pp.). Keith Browning has used the term in many publications, including a set of symposium papers he edited, entitled *Nowcasting*, published by Academic Press in 1982 (256 pp.).

[48] For a detailed description of the concept of a Weather Radar Network and the kind of data it was expected to provide, see 'Towards an automated weather radar network', by B C Taylor and K A Browning, published in *Weather* in 1974 (Vol. 29, pp. 202–216).

By the end of the 1970s, an ambitious plan for using radar and satellite imagery to analyse and forecast precipitation patterns for periods of up to to six hours ahead had been conceived. It was called FRONTIERS, an acronym which embodied the key elements of the project (i.e., Forecasting Rain Optimized using New Techniques of Interactively Enhanced Radar and Satellite). A basic feature of the project was that images obtained from a geostationary satellite and a network of radars would be superposed automatically on a regular basis and also be combined with the outputs of a new generation of numerical models. The specific purpose of these models was to provide forecasts of mesoscale weather systems (i.e., systems with a horizontal scale of a few tens to a few hundreds of kilometres, such as thunderstorms, squall lines, the fronts of depressions, sea breezes and tropical cloud clusters).[49]

The exciting technological advances of the 1960s and 1970s included, too, a British-built instrument flown on the American *TIROS-N* series of satellites to measure stratospheric temperatures worldwide. Called a 'Stratospheric Sounding Unit', this instrument consisted essentially of an infra-red radiometer developed from 1974 onwards by the Office in collaboration with the University of Oxford and Heriot-Watt University. The first unit was launched on 13 October 1978 and provided profiles of temperature in the altitude range twenty-five to forty kilometres. The second was launched on 27 June 1979 and provided profiles in the altitude range sixteen to fifty kilometres. Launches thereafter took place every two years or so until 1992, and the profiles produced by the satellite-borne units were checked from time to time by means of sondes released from rockets launched from the Outer Hebrides.

This was not the Office's only interest in the stratosphere: a global numerical model of the stratosphere was formulated and developed in the early 1970s, and the effects of Concorde on the stratosphere were also studied. Concern that pollutants emitted by supersonic aircraft might alter the gaseous constitution of the atmosphere in the lower stratosphere led the governments of the United States, the UK and France to undertake independent research programmes to investigate the potential seriousness of the possible environmental problems that might be associated with flights in the stratosphere, particularly ozone depletion. Concern in the United States was such that the Congress decided in March 1971 to terminate development funding of commercial supersonic aircraft.

Action in the UK was instigated by the Concorde Division of the Department of Trade and Industry in 1972 and coordinated by a specially formed 'Committee on Meteorological Effects of Stratospheric Aircraft' (COMESA). The Office was commissioned to assess the impact of stratospheric flight on the composition of the stratosphere and on the climate on the ground. In the course of this investigation,

---

[49] See 'The FRONTIERS plan: A strategy for using radar and satellite imagery for very-short-range precipitation forecasting', by K A Browning, published in 1979 in the *Meteorological Magazine* (Vol. 108, pp. 161–184). See also 'Radar as part of an integrated system for measuring and forecasting rain in the UK: Progress and plans', by K A Browning, published in *Weather* in 1980 (Vol. 35, pp. 94–104).

the Office made considerable use of its numerical model of the stratosphere and also carried out laboratory research into possible photochemical reactions. The final report of COMESA was published in 1975, and the conclusions it contained were broadly similar to those reached by the American and French investigators, in particular that exhaust gases from supersonic aircraft were unlikely to have any measurable effect on amounts of ozone or other gases or on global temperature or climate. As a result, the U.S. Congress withdrew objections to Concorde. Flights from London and Paris to New York and Washington commenced on 24 May 1976.

Developments at Reading

The staff of the Office's Training School had hoped they would be able to vacate the old wartime huts at Stanmore as part of the Office's move to Bracknell. Alas, as mentioned in Chapter 11, their hopes had been dashed at the last minute, when space could not be found for the School in, or even close to, the new headquarters. Eventually, though, the unsatisfactory situation came to an end, as D H Johnson explained in an article published in the *Meteorological Magazine* in 1972 (Vol. 101, pp. 299–301):

Weight was added to the case for relocation by a recommendation of the Parliamentary Estimates Committee in the third report of the Session 1966–67. It was felt that specialists working in the various branches of the science at the Headquarters at Bracknell should be able to participate more fully in the lecturing programmes. The search for a suitable site came to an end a year or two later when the Royal Air Force vacated the Flying Training Command Headquarters at Shinfield Park, near Reading. The Meteorological Office was offered a substantial part of the site. After a period of planning, work commenced in September 1970 on the conversion and extension of existing buildings to provide accommodation for a new Meteorological Office College. Training ended at Stanmore on Friday 15 October 1971 and recommenced without break at Shinfield Park on the 18th.

In the new accommodation, there were spacious classrooms, tiered lecture theatres, laboratories, communications facilities, an instruments enclosure, a library, social and sports facilities, and common rooms for instructors and students. There was also residential accommodation for all students, which was a great improvement on the position at Stanmore, where students had needed to live in lodgings. An advantage of the new site was that it was close to the University of Reading, where there was a Department of Meteorology that was not only growing but also forging close links with the Office. In 1972, the University introduced meteorology as a single subject sandwich degree which included one year of practical training at the 'Meteorological Office College' (as it had become when the move to Shinfield Park had taken place), and opportunities grew during the 1970s for staff of the Office to read for undergraduate and postgraduate degrees of this and other universities.

A few years after the move to Shinfield Park, the College found itself adjacent to another important meteorological institution, the European Centre for Medium Range Weather Forecasts. The origins of this institution lay in a resolution adopted in October 1967 by the Council of Ministers of the EEC. The resolution had affirmed that the six Member States of the Community (Belgium, France, the Federal Republic of Germany, Italy, Luxembourg and The Netherlands) would be willing to implement a programme to promote scientific and technical research, with consideration given to opportunities for cooperation in six fields, one of them meteorology.

An expert group in meteorology was formed and submitted its first report in March 1969, suggesting *inter alia* the establishment of a meteorological centre that would be equipped with powerful computing and information-processing facilities and would carry out research into medium-range weather forecasting (i.e., forecasting for the period four to ten days ahead). Other European nations were invited to participate, with Austria, Denmark, Ireland, Norway, Portugal, Spain, Sweden, Switzerland and the UK responding positively to the invitation. Finland and Yugoslavia became involved later. Study groups investigated the potential benefits of the enterprise and the financial and human resources that would be required. Then, in October 1973, fifteen nations signed the Convention which set up the European Centre for Medium Range Weather Forecasts (ECMWF). The fifteen were Belgium, Denmark, the Federal Republic of Germany, Finland, France, Greece, Ireland, Italy, The Netherlands, Portugal, Spain, Sweden, Switzerland, the UK and Yugoslavia. Austria signed in January 1974, and Turkey acceded later.[50]

Article 1 of the Convention stated that ECMWF would be situated at Shinfield Park.[51] This particular site was chosen because it was close to the Office (where there was a Regional Telecommunications Hub) and also to several universities and institutions with strong meteorological departments, among them the University of Reading and the Meteorological Office College.[52] The British government offered to provide a suitable building on the site, and the first phase of the building work began in January 1977. Meanwhile, the first Director was appointed, a Dane, Aksel Wiin-Nielsen, on 1 January 1974, and he gathered around him a small group whose job it was to plan the Centre's programme of activities and determine the resources needed to carry out the programme. He remained Director until 31 December 1979, when he was succeeded by Jean Labrousse, a Frenchman. The new Centre was opened by His Royal Highness the Prince of Wales on 15 June 1979, and operational forecasting

---

[50] For the story of the establishment of ECMWF, see 'The European Centre for Medium Range Weather Forecasts', by E Knighting, published in 1978 in the *WMO Bulletin* (Vol. 27, pp. 14–19).

[51] See *Convention establishing the European Centre for Medium-Range Weather Forecasts with Protocol on Privileges and Immunities (Brussels, 11 October 1973 to 11 April 1974)*, presented to Parliament by the Secretary of State for Foreign and Commonwealth Affairs by Command of Her Majesty, June 1974, published by Her Majesty's Stationery Office as Cmnd.5632, Miscellaneous No. 21, 1974, 27 pp.

[52] Dr Mason pressed strongly for a location in the UK.

began on 1 August 1979, with forecasts for ten days ahead issued five days a week.[53] Forecasts were made seven days a week from 1 August 1980.

The computer system at ECMWF was linked to computers in national meteorological services by high-speed telecommunications links. The first computer possessed by ECMWF was a very powerful Cray 1-A machine, which had a performance of 100 million operations per second. This was replaced in 1983 by an even more powerful computer, a Cray X-MP/22 machine, which, in turn, was replaced at the end of 1985 by a Cray X-MP/48 machine. The throughput of the latter machine was about ten times that of the Cray 1-A.[54]

Meanwhile, the Office's computing facilities had been further upgraded. In his Foreword to the annual report for 1974, Mason stated that the IBM 360/195 computer system had become "saturated" during the year and that time for research had therefore needed to be rationed. However, the situation would soon be relieved. An IBM 370/158 computer was to be added as a front-end processor, thus effectively doubling the power of the whole system. Moreover, the new computer would serve as a back-up should the main computer break down.

Four years later, Mason reported that the combination of the 360/195 and 370/158 computers, known as COSMOS, had itself become saturated. Rationing had again proved necessary, especially for large tasks such as modelling of the global circulation. In consequence, a case had been prepared for the Office to acquire yet another new computer, this time one that would operate about ten times faster than the 360/195. Delivery of the new machine was expected early in 1980. In fact, it was installed the following year and brought into use in early 1982. It was the Office's first supercomputer, a CDC Cyber 205 machine, which had a performance of 400 million operations per second and was capable of enhancement to provide an even greater performance in future. Two IBM 3081 mainframe computers supported it, their purpose being to handle the observational database and the post-processing of forecast data and associated graphics. At the same time, the computer system was interfaced with the communications computers and thence linked to the Global Telecommunications System and other communications networks.

The Office introduced its first global operational weather forecasting model in 1982, in time for it to assist in operations associated with the Falklands War. With the Cyber 205 computer, through highly specialized software, it was possible by

---

[53] The Prince of Wales visited both ECMWF and the Office on 15 June 1979. During his visit to the Office, he was shown the CFO and the Computing Laboratory. The previous year, on 14 July 1978, Her Majesty the Queen visited the Office and the Royal Meteorological Society. She too was shown the CFO and the Computing Laboratory. The purpose of her visit to the Society was to view an exhibition of meteorological research in the UK and to open the Society's new headquarters building. See 'Royal visit to the Society', published in *Weather* in 1978 (Vol. 33, pp. 423–430). See also 'Visit of Her Majesty the Queen to the Meteorological Office, 14 July 1978', published in the *Meteorological Magazine* in 1979 (Vol. 108, p. 34).

[54] For details of the early work of ECMWF, see 'ECMWF – ten years of European meteorological cooperation', by L Bengtsson, published in the *Meteorological Magazine* in 1986 (Vol. 115, pp. 150–157). Dr Bengtsson, a Swede, directed ECMWF from 1 January 1982 to 31 December 1990.

1984 to compute a global forecast for twenty-four hours ahead at fifteen levels in the atmosphere with a resolution of 1.5° latitude and longitude in under four minutes. The fifteen-level numerical model of the atmosphere was coupled to a multi-level ocean which contained sea ice and interacted with ocean-floor topography. Such a forecasting capability would have astonished Admiral FitzRoy. Indeed, many people in the 1980s were highly impressed, both at home and abroad. In the UK, the Office had been a very early user of computers. When introduced to the Office, the KDF9 computer had been the most powerful machine in the country, and the COSMOS combination of a 360/195 with a 370/158 had been among the five most powerful.

End of an Era

The bicentenary of Kew Observatory was celebrated in 1969. Since 1842, this had been a place where instruments had been designed, developed, tested and calibrated, a place where many famous physicists and meteorologists had done their best work, a place that would surely never close. The roll-call of superintendents was impressive, and the list of others who had worked there or been associated with the Observatory's management contained many of the greatest names in the history of physics and meteorology.[55]

The story of the Observatory was told in a special issue of the *Meteorological Magazine* published in June 1969 (Vol. 98, pp. 161–196). Among the articles were some by staff of the Office who had served there. The reminiscences of Dr F J Scrase (pp. 180–186), for example, provided a glimpse of life at Kew soon after the Great War. He had gone there in 1920, as a Junior Professional Assistant, and been unimpressed with some aspects of the place, saying that the facilities for experimental work had compared unfavourably with those he had enjoyed at the Cavendish Laboratory. He mentioned that visitors had ranged from parties of local schoolchildren to the heads of geophysical observatories and meteorological services from all over the world, that there had been frequent training courses for meteorological observers, that sheep had grazed on the golf course, that the caretaker had kept hens, and that the grounds had sometimes flooded when the Thames overflowed its banks.

Others who had been involved in the work at Kew contributed articles on the development and calibration of radiosondes at the Observatory from 1938 to 1946 and on the pioneering work on the design and calibration of meteorological sensors

---

[55] The full list of those who served as Superintendent at Kew was, with year of appointment: Sir Francis Ronalds (1842), John Welsh (1852), Balfour Stewart (1859), Samuel Jeffery (1871), G M Whipple (1876), Charles Chree (1893), F J W Whipple (1925), J M Stagg (1939 and 1946), Sir George Simpson (1939), G D Robinson (1947), K H Stewart (1957), R H Collingbourne (1960), R A Hamilton (1966), S G Crawford (1968), H E Painter (1973) and W R Richardson (1974).

Figure 12.10. Kew Observatory in the 1960s. © Crown Copyright 2010, the Met Office.

for use on rockets and satellites that had been carried out from 1959–1961.[56] The author of the present book, Malcolm Walker, remembers working at the Observatory for six months of 1962, as a college-based sandwich student, testing and calibrating radiometers. He also remembers weather observations being made routinely, among them readings of instruments in a screen attached to the north wall of the Observatory, a site used continuously since 1773.

There was a hint in the Director-General's Foreword to the special issue of the *Meteorological Magazine* that the long-term future of the Observatory was not assured. Kew continued to be "an active centre of investigation", Dr Mason said, but "the encroachment of suburbia and the demands of modern research" made it unlikely that its "future would match its splendid past". By 1980, the Observatory had become surplus to requirements. The Office's centre for instrument development was now its facility at Beaufort Park. The last weather observation was made at 21:00 GMT on 31 December 1980, by Tim Donovan. All that remains today to continue the long history of meteorological observing at Kew is a small weather station two kilometres

---

[56] See 'The British radiosonde: Its debt to Kew', by D N Harrison (pp. 186–190), and 'Development of rocket and satellite experiments at Kew Observatory, 1959–61', by K H Stewart (pp. 195–196). The radiosonde work was transferred to the Instrument Branch at Harrow in 1946.

to the north-east, in the Royal Botanic Gardens, a station that is part of the Office's current climatological network.[57]

The Observatory was sold, and most of the instruments were deposited in the Science Museum.[58] Books and papers were taken to the National Meteorological Archive at Bracknell, and the papers remained unsorted until 1997, when, through a contract between the Office and Cardiff University, they were catalogued and dispersed. Some went to the Science Museum, some to the National Archives at Kew and others remained in the National Meteorological Archive.

Another era in British meteorology seemed to be coming to an end in the 1970s. Ocean weather ships were being withdrawn. These ships had once been considered essential for obtaining meteorological data and for providing services for aviation, but times were changing. The United States withdrew all of its Atlantic weather ships in 1973 and 1974, on the grounds that the expense of operating them could not be justified when most weather systems moved eastwards over the Atlantic (i.e., away from their own coast). This left five stations on the North Atlantic, all manned by European nations. Then, in 1975, the International Civil Aviation Organization discontinued its support for the network of Atlantic weather ships, the reason being that, for various practical reasons, aircraft crossing the ocean no longer needed the ships for navigational, communications or air–sea rescue purposes. However, meteorologists still required the surface and upper-air data supplied by weather ships. Data from satellites were not yet sufficient to replace the data from these ships.

The network was reduced to four stations: 'L', at 57°N 20°W, manned by the UK; 'R', at 47°N 17°W, manned by France; 'C', at 52°45'N 35°30'W, manned by the Soviet Union; and 'M', at 66°N 02°E, manned by Norway, Sweden and The Netherlands. Initially, the ships used by the Office were *Weather Reporter* and *Weather Surveyor*, but they were replaced in 1977 because they were showing signs of age. Their sister ships *Weather Adviser* and *Weather Monitor* were modified and modernized and given new names, which were, respectively, *Admiral FitzRoy* and *Admiral Beaufort*. In turn, these ships were withdrawn, in January 1982, for financial reasons. As explained by the Director-General in the Office's annual report for 1980:

The Meteorological Office could not hope to escape the cuts in spending and staff which the Government has imposed on the Civil Service as a whole. So far, we have had to take our full proportionate share of staff cuts, amounting to 10 per cent or 300 staff over the three-year period April 1979 to April 1982, with no allowance for the fact that half our staff costs are met by the provision, on payment, of services to the public and private sectors. Redundancies have been avoided largely by natural wastage and restricted recruitment. Some posts have

---

[57] For an overview of the work carried out at Kew from 1773 to 1980, see 'Kew Observatory', by J F P Galvin, published in *Weather* in 2003 (Vol. 58, pp. 478–484).

[58] In 2010, Kew Observatory was the headquarters of Belron International Ltd., a company which specialized in vehicle glass repair and replacement services. The original name of the building has lived on, though, in the company's postal address: The King's Observatory, Old Deer Park, Richmond, Surrey.

been saved by the continued introduction of automation, the contracting out of cleaning and catering at the College, and reductions in our commitment to the Ocean Weather Ship scheme. However, we have had to cease a number of long-established activities, including the issue of the *Daily Weather Report* and long-range forecasts, close Kew Observatory, and make cuts in research, including the rocket-sounding programme.

The UK's two weather ships were replaced in 1982 by one large trawler, the *Starella*, which was chartered until the end of 1985. She was fitted with the latest equipment for making regular meteorological observations at the surface and in the upper air and for transmitting them automatically by radio. She had a crew of twenty, including seven staff of the Office, and she shared the manning of Station L with a Dutch ship. Stations C, M and R continued to operate. In the early 1980s, therefore, the story of the ocean weather stations had not yet ended.

By the early 1980s, the Office had long since left behind the largely qualitative and descriptive era of meteorology. Thanks especially to Dr Mason (now Sir John Mason, knighted in 1979), the Office had exploited in no small measure the scientific and technological advances of the post-war decades and thus established an enviable reputation as a world leader in numerical modelling, satellite remote sensing, automation and applications of radar, not to mention the provision of meteorological services for a wide range of users. The death of Ernest Gold, on 30 January 1976, aged 94, served to emphasize the passing of an era. He was the doyen of British meteorology and founder of international synoptic meteorology. He had been at the forefront of British meteorology in a bygone era. He had served under Shaw, Simpson and Johnson and played important meteorological roles in two world wars. He had helped lay the scientific foundations of the Office and then, in retirement, lived to see the use of computers, radar, satellites and automation by meteorologists taken for granted. And he had retained a keen interest in meteorology well after he retired. Indeed, he had often attended meetings of the Royal Society and Royal Meteorological Society and not infrequently asked probing questions, even when into his nineties.[59]

In the years to come, the importance of numerical modelling, remote sensing and automation continued to grow, and the Office continued to develop, partly in response to new challenges. But the future of the Office was not without failures, criticism and controversy, as we will see in the next chapter.

---

[59] A number of others who had contributed to the development of the Meteorological Office (see earlier chapters) died in the 1970s. They were, in chronological order: Sydney Chapman (1 June 1970), Sir Robert Watson-Watt (5 December 1973), Sverre Petterssen (31 December 1974), James Stagg (23 June 1975), Sir Geoffrey Taylor (27 June 1975), Jacob Bjerknes (7 July 1975) and Percival Albert Sheppard (22 December 1977).

13

Winds of Change

During the 1970s, the Meteorological Office's numerical models of the global atmospheric circulation began to simulate the main features of the world's climate and its variations and show that the climatic consequence of increasing the carbon dioxide content of the atmosphere was global warming.[1] The Office's previous predictions by simpler models of a general warming much accentuated in polar regions were confirmed, but further study with improved models and a more powerful computer would be required to assess these climatic effects and their likely economic and social impact with real confidence.

The causes and the economic and social consequences of climatic change were debated in the House of Lords on 30 November 1978. The opening speaker, Lord Tanlaw, asked the government whether there was cause for concern in the current increase in atmospheric carbon dioxide and apparent change in global weather patterns. He referred to an article by the Office's Director-General that had been published in *Nature* (1978, Vol. 276, pp. 327–328). In this, Dr Mason had suggested that a doubling of carbon dioxide would lead to an increase in global temperature of two or three degrees Celsius. Though this was a modest increase, Lord Tanlaw said, it "might well lead to increased food production by prolonging the growing season and, furthermore, produce considerable savings in energy consumption". There might, though, be "unforeseen disadvantages", such as adverse effects on agriculture in some parts of the world as a result of changes in rainfall patterns. The melting of glaciers might also cause sea level to rise and thus threaten coastal communities.

There had been for some years, Lord Tanlaw went on, conflicting conclusions about the meteorological consequences of increased concentrations of atmospheric carbon dioxide. To this end, he said, WMO, ICSU and other international agencies were

---

[1] It is generally accepted that combustion of fossil fuels since the beginning of the Industrial Revolution has caused atmospheric concentrations of carbon dioxide to rise. In the middle of the eighteenth century, the concentration was about 275 parts per million (ppm). By 2010, it had risen to 389 ppm.

planning a World Climate Programme that would span the two decades from 1980 to 2000. This would be preceded by a World Climate Conference in February 1979. The Programme would have three main elements: (1) climate data and applications, (2) an investigation of the impacts of human activities and (3) research into climatic change and variability.

Baroness Birk, a Parliamentary Under-Secretary in the Department of the Environment, assured their Lordships the UK was playing "a full part in promoting and contributing to this international research effort, as well as carrying on research at home". She reported that seventeen of the thirty-three people in the Office's Climatology Research Branch were "directly concerned with developing and experimenting with climate models at an annual cost of £380,000", and much of this research was related to work on carbon dioxide and its effects. There were plans, she said, to increase the computing power available, so that work relating to carbon dioxide could be pursued more effectively.

Awareness of the economic and social consequences of climatic variability had come to the fore in Parliament some years earlier. In the House of Commons on 3 November 1972, for example, in a debate about food prices, Charles Morrison had pointed out that the prices of food for animals demonstrated how farming costs could be affected by factors completely outside the control of this country. For example, the price of protein feed materials had risen dramatically. "Who would have thought", he said, "that the disappearance of the anchovy shoals off the coast of Peru, due to changing world currents, would be of any major importance?" It was, he said, because those anchovy shoals were one of the main sources of the world's fishmeal protein supply.[2]

The subject of the anchovies was raised in the House of Commons a number of times in the 1970s. On 23 October 1973, for example, Dennis Skinner asked the Prime Minister if he would appoint an expert on worldwide weather conditions to the Central Policy Review Staff. Mr Heath said in reply that he would not because such advice was available from the Office. The riposte of Mr Skinner was to ask if the reply meant that the Prime Minister would "no longer blame unusual climatic conditions, such as snow in Siberia, sun in Africa and rain in Manchester, for the astronomical rise in

---

[2] The anchovy harvest failed in 1972 because upwelling of water in the eastern South Pacific off Peru and northern Chile failed. The importance of this upwelling is that ascending currents in the upper ocean raise nutrients to the surface and thus make the surface waters rich in floating plants on which small drifting animals depend. In turn, anchovies and other fish feed on the small animals. Every three to seven years, the upwelling is replaced by a huge eastward movement of warm water from the central Pacific that can persist for several years. This has serious ecological consequences. The anomalous behaviour of the current has long been known as the 'El Niño phenomenon' and was thought to be a purely local occurrence. In the late 1960s, however, Jacob Bjerknes showed that it was associated not only with climatic variability in the area in question but also with anomalous behaviour of the ocean-atmosphere system much farther afield. It is, in fact, a fundamental part of climatic variability around the globe. There was a new aspect to the 1972 El Niño which exacerbated the problem. Though the El Niño occurrence was entirely natural, trawlers with modern equipment for locating and exploiting shoals depleted an already reduced stock of anchovies. Overfishing occurred when fish stocks were at their most vulnerable. Consequently, the Peruvian anchovy industry collapsed (and has never fully recovered).

food prices". Mr Heath agreed that a lack of snow on the Steppes had brought about poor harvests in the Soviet Union. However, he wondered if Mr Skinner realized that problems in Britain with animal feeding stuffs were mostly "due to the drop in fish meal supplies following a sudden change in the direction of the warm current off Peru".

A Host of Environmental Concerns

By the early 1970s, global temperature series were showing that cooling had occurred since the 1940s, and this had led some to believe that another ice age was imminent.[3] The matter was raised in the House of Commons on 3 December 1974 by David Stoddart, who asked the Secretary of State for Defence what the views of the Office were on "the imminence of a new ice age". Brynmor John, the Under-Secretary of State for the Air Force, replied that he had been advised by the Office that there had been no changes in the climate, or in factors affecting it, in recent years which suggested that an ice age might be imminent. Because, however, ice ages had occurred several times already, it was to be expected that natural causes would at some stage bring about another, "probably some time in the next few thousand years".

Mr John went on to say that the BBC TV programme *The Weather Machine*, broadcast on 20 November 1974, might have given a false impression of the imminence of a new ice age because it had "given prominence to two new and, as yet, incompletely substantiated ideas", these being that "interglacial periods were shorter than previously thought" and their onset could be relatively rapid, "taking place over a few centuries". He stressed that there was no evidence an ice age was imminent and then said the following:

The Meteorological Office gives advice to a wide range of commercial organizations, national-ized industries and official bodies in regard to future climate. The recent television programme and associated newspaper articles were basically a subjective compilation from a wide range of sources and the Office was not invited to provide professional guidance on the possibility of climatic change.

This was disappointing for the Office. It was one of the world's foremost meteorolo-gical institutions but had been ignored by the makers of a programme broadcast by the BBC.

By 1974, a specialist unit for the study of climate had been established in the University of East Anglia. Called the 'Climatic Research Unit', it had been founded in 1972, with Hubert Lamb its founding director. He had spent most of his career in the Office, rising through a special merit promotion to a senior post in the Synoptic

---

[3] See 'The myth of the 1970s global cooling scientific consensus', by T C Peterson *et al*, published in 2008 in the *Bulletin of the American Meteorological Society* (Vol. 89, pp. 1325–1337).

Climatology Branch.[4] In the early 1970s, he had believed that global cooling would lead to an ice age within 10,000 years. However, he had changed his mind, a factor that had brought this about being the UK's exceptional drought and heat wave of 1975–1976.[5] He had come to the view that global warming could have serious effects within a century and warned of serious consequences for agriculture, the melting of ice caps and the flooding of major cities.

So serious did the water shortage become in England and Wales in 1976 that a Drought Act was passed by Parliament (on 5 August) and a Minister for Drought was appointed (on 24 August), the job going to Denis Howell, the Minister for Sport and Recreation. Only a few days later, the heavens opened. Thunderstorms brought rain to some places for the first time in several weeks!

The drought and heat wave brought criticism of the Office, for its day-to-day predictions of the weather and its monthly forecasts had been far from successful. Questions were asked in Parliament. On 2 July 1976, for example, Philip Goodhart stated that the drought had "clearly taken the Meteorological Office by surprise" and asked the Minister for Planning and Local Government, John Silkin, if there were "any plans to increase the resources and capacity of the Office". In reply, Mr Silkin said that the Office had been working well with the government and the water authorities, and he was "very satisfied with its performance". There was another poor long-range forecast at the end of August 1976, though, when the Office announced that there were no clear indications of a wet autumn anywhere in the British Isles. The autumn proved to be one of the wettest on record, and the wet weather continued through the winter to such an extent that Lord Paget of Northampton asked in the House of Lords on 21 March 1977 if "Mr Howell, who was so successful in turning on the rain in the drought, could be asked whether he could turn it off".

In the House of Commons on 11 October 1976, Nigel Spearing asked the Secretary of State for the Environment what his findings were "concerning the likelihood of a series of wet seasons in 1977 and 1978". In reply, Denis Howell said that the Office's forecasts were for thirty days ahead and he understood there were, "as yet, no reliable methods for predicting rainfall for periods of up to one or two years". The Office and the Climatic Research Unit of the University of East Anglia had been commissioned, he reported, "to carry out a preliminary study on the feasibility of predicting long-term changes in rainfall and air temperatures over the United Kingdom". The results of this preliminary study would not be available before March 1977.

The Earl of Selkirk asked in the House of Lords on 30 March 1977 whether the Office had "reached any general conclusions on the causes which led to the summer of 1976 being exceptionally dry and the winter of 1976–77 exceptionally wet".

---

[4] See 'Meteorologist's profile – Hubert H Lamb', by R A S Ratcliffe, published in 1992 in *Weather* (Vol. 47, pp. 263–266). Lamb is generally considered the greatest climatologist of his generation.
[5] The summer of 1976 was the hottest in southern and central England for at least 300 years and the driest since 1772.

In reply, Lord Winterbottom stated that limited studies of the summer and autumn of 1976 had been carried out "but more research involving extensive and costly use of large computers would be required in order to reach any positive conclusions".

Attention was drawn increasingly in the 1970s to a further possible human impact on climate. In the House of Commons on 3 July 1974, Bruce Douglas-Mann asked the Secretary of State for the Environment whether he had considered a report published in a recent issue of *Nature* concerning changes in ozone levels of the atmosphere since 1965 and the implications in relation to climatic changes and exposure to atmospheric pollution.[6] In reply, Denis Howell said that he had indeed considered it and went on to report that the Office was "actively pursuing a research programme on the effects of pollutants on stratospheric ozone levels and climate".[7] He added that the "preliminary results were reassuring". This did not satisfy Mr Douglas-Mann. He asked if Mr Howell was "aware that if the decline of ozone levels continued at the present rate, human life on this planet would become totally impossible". Mr Howell said that he was not concerned. The data on falling ozone levels had come from a single point in the southern hemisphere, and he had been advised that further observations by scientists of the Office and other institutions had indicated that total ozone levels in the northern hemisphere were rising. There was little doubt, he said, that these fluctuations were caused by natural causes.

Over the next few years, complacency over ozone depletion gave way to concern, and questions were asked in the House of Commons about the potentially harmful effects of the propellants used in aerosol sprays. Mr Howell informed the House on 11 February 1976 that an appraisal of the available information about the effects of chlorofluorocarbons on the ozone layer in the stratosphere was being carried out, and he said in the House on 14 April 1976 that a report on the matter had been published that day, by the Department of the Environment's Central Unit on Environmental Pollution. Again, though, he did not seem unduly concerned. But by the early 1980s, the detrimental effects of chlorofluorocarbons on the ozone layer had been widely recognized and the first steps taken towards the production of a global agreement which would restrict their use. On 1 January 1989, the 'Montreal Protocol on Substances that Deplete the Ozone Layer' came into force. Chlorofluorocarbons were to be phased out over a number of years.

And other environmental issues were addressed by the Office in the 1970s. An article published in *Nature* in 1974 raised the possibility of a large depletion of stratospheric ozone being a consequence of a nuclear war.[8] The following year, the

---

[6] See 'Trends in the vertical distribution of ozone over Australia', by A B Pittock, published in *Nature* in 1974 (Vol. 249, pp. 641–643).

[7] From the 1970s onwards, the Office contributed much to knowledge and understanding of stratospheric ozone, partly through numerical modelling and partly through collaboration with scientists of other interested bodies (notably the British Antarctic Survey).

[8] See 'Photochemical war on the atmosphere', by J Hampson, published in *Nature* in 1974 (Vol. 250, pp. 189–191).

National Research Council of the United States published a report entitled *Long-Term Worldwide Effects of Multiple Nuclear Weapons Detonations*. This confirmed the likelihood of ozone depletion and, along with it, increased ultra-violet radiation. It also stressed that oxides of nitrogen and large amounts of dust would reach the stratosphere, with significant, possibly dramatic, climatological effects.

Alarm over the long-term effects of a nuclear war increased in the 1980s. Ozone depletion was but one facet of the problem. In addition, it was argued, massive amounts of soot would be lifted into the stratosphere and thus reduce the intensity of the sunlight that reached the surface. The soot would persist in the stratosphere for many months, because this layer of the atmosphere was stable and did not contain precipitation to remove the dirt. The term 'nuclear winter' was coined in the early 1980s for the large reduction in global temperatures that was predicted. The expectation was that an extended period of extremely cold weather and greatly reduced sunlight would follow the detonation of large numbers of nuclear weapons over flammable targets such as cities and forests, and some believed that surface temperatures in the interior of continents would fall more than 20°C in summer.[9]

In the early 1970s, mainly in the context of flights by Concorde and other supersonic aircraft, the Office calculated the magnitude of the climatic effect of polluting the stratosphere and concluded that, although amounts of nitrogen oxides equivalent to the output from many supersonic aircraft had been released into the atmosphere when nuclear testing had been at its peak, the amount of ozone had not been affected.[10] In the mid 1970s, mainly through numerical modelling, the effect of reducing stratospheric ozone by 50% was investigated by the Office, along with the effect of increasing the dust content of the stratosphere sufficiently to reduce solar radiation by 4%. Cooling of the stratosphere by about 10°C was indicated, but with no significant change in surface temperature. During the 1980s, however, public anxiety over the climatic effects of a nuclear war grew, fuelled to no small extent by the media.

The attitude of the British government was noncommittal. In the House of Commons on 20 December 1984, for example, William McKelvey asked the Secretary of State for the Home Department if he intended to publish the findings of the evaluation of the nuclear winter hypothesis that had been carried out by the government. In reply, Giles Shaw, the Home Office Minister of State, said that the term 'nuclear winter' referred to an "unvalidated hypothesis" which was "still the subject of research in the United States and elsewhere". A few months later, on 4 April 1985, when Tony Lloyd accused Mr Shaw of suppressing the findings of the evaluation, Shaw's reply was similar. The hypothesis of the nuclear winter had yet to be validated properly, and

---

[9] See 'Nuclear winter: global consequences of multiple nuclear explosions', by R P Turco *et al*, published in *Science* in 1983 (Vol. 222, No. 4630, pp. 1283–1292). See also 'Climatic change induced by a large-scale nuclear exchange', by D M Elsom, published in *Weather* in 1984 (Vol. 39, pp. 268–271).

[10] See 'Nitrogen oxides, nuclear weapon testing, Concorde and stratospheric ozone', by P Goldsmith *et al*, published in *Nature* in 1973 (Vol. 244, pp. 545–551).

until it had been the government could not take a view on it. The government's view was that a nuclear winter was but one of a number of catastrophic consequences of a nuclear conflict and probably not the one of most immediate concern to those who survived.[11]

The Royal Meteorological Society held a discussion meeting on 18 December 1985 entitled 'Climatic impact of nuclear war', a choice of subject that did not please everyone, as correspondence in *Weather* showed. One reader was "appalled".[12] In his words, "anyone with time to spare to prepare papers on such a subject would be better advised to get into the political arena and put all their efforts into ensuring that a large-scale (or even small-scale) nuclear war never happens". He did not think the Society should become involved in politics. The Meetings Secretary of the Society disagreed, saying that meteorologists could not leave the nuclear winter hypothesis "to be discussed only by politicians and journalists". "Suddenly", he said, "our science had something important to input into one of the most significant debates of our time". The Office was sufficiently interested in the meeting to review it at some length in the *Meteorological Magazine* (1986, Vol. 115, pp. 118–120), and staff of the Office were by then carrying out detailed studies of various aspects of the hypothesis.[13] The Office's involvement in nuclear winter research was comparatively short-lived, though, and declined in the late 1980s. Nevertheless, meteorologists around the world are still interested in the subject, as papers published in the past few years show.[14]

All of a sudden, the attention of meteorologists across Europe and farther afield was focused on a nuclear event of a different kind which happened far from the UK: an explosion in one of the reactors at the Chernobyl nuclear power station near Pripyat in Ukraine on 26 April 1986. At first, there was no admission from the Soviet Union that the disaster had occurred. Then, on 28 April, after the discovery of radioactive material in Sweden and the application of diplomatic pressure on officials in Moscow, details of the occurrence at Chernobyl were released by Soviet authorities. The following day, the CFO at Bracknell began to calculate trajectories from Chernobyl, using predicted winds at a height of about 1500 metres. These trajectories showed that the plume might cross the British Isles within a few days. There were, however, questions. Meteorologists needed to know whether the release of radioactive material was continuing. If so, how much was being released? At what height would the plume

[11] The findings of a United Nations inquiry were essentially the same. The inquiry concluded that there would be severe climatic consequences of a nuclear war, along with many other disastrous consequences. See: *Study on the climatic and other global effects of nuclear war*, published by the United Nations in 1989 (Document A/43/351, 61 pp.).

[12] See *Weather*, 1985, pp. 261–262.

[13] See, for example, 'Importance of local mesoscale factors in any assessment of nuclear winter', by B W Golding *et al*, published in *Nature* in 1986 (Vol. 319, pp. 301–303); see also 'Climatic effects of nuclear war: The role of atmospheric stability and ground heat fluxes', by J F B Mitchell and A Slingo, published in 1988 in the *Journal of Geophysical Research (Atmospheres)*, (Vol. 93 No. D6, pp. 7037–7045).

[14] See, in particular, 'Nuclear winter revisited with a modern climate model and current nuclear arsenals: Still catastrophic consequences', by A Robock *et al*, published in 2007 in the *Journal of Geophysical Research – Atmospheres* (Vol. 112, No. D13107, 14 pages).

travel? How much would it disperse? How much deposition of radioactivity would occur over the British Isles? Where else in Europe might the plume spread and deposit radioactivity?

The response of the Office to these questions involved not only forecasters in the CFO but also specialists in atmospheric dispersion and deposition and staff whose job it was to keep the media informed. When it became clear that radioactive material would spread across parts of the British Isles, ecologists, hydrologists, the farming community and others needed to know the extent to which deposition of material would occur through precipitation and how much dry deposition there would be. In its response to the Chernobyl event, the Office acquitted itself well. Forecasts of plume trajectories and dispersion were commendably accurate and showed that considerable reliance could be placed on the numerical models which were used to predict trajectories and deposition.[15] Indeed, a consequence of the Chernobyl incident was that the Office was designated a lead agency by the Department of the Environment in the development of a model to forecast atmospheric dispersion and deposition following any future nuclear accident.

Yet another environmental concern which came to the fore in the 1970s was that commonly called 'acid rain', this being the term used to describe acid deposition by all forms of precipitation and by clouds, fog and dew. In its broadest sense, however, the acidification problem included, too, dry deposition of gases and small particles. The consequences took the form of damage to trees and other vegetation, soils, buildings, fisheries and so on. Public concern grew as the media reported serious tree damage in Germany and Scandinavia, as well as ecological damage in rivers and lakes in Scandinavia and Scotland.

Prevailing winds over the British Isles blow from a westerly point. Accordingly, the finger of blame for the acid rain problem in Scandinavia was pointed at the UK, as a result of which studies were carried out by the Office and the Central Electricity Research Laboratories, many of them jointly by these two bodies. The methods employed to investigate the transport and dispersion of pollutants included numerical modelling and the use of aircraft.[16] It was found that the UK was not the only significant source. Most of the pollution that was deposited in Scandinavia when rain fell from warm and occluded fronts originated in the industrial areas of the German Democratic Republic. Trajectories of air in these fronts are broadly from the south or south-east, not the west. However, much of the pollution that was deposited in Scandinavia in dry form and in showery precipitation originated in the British Isles, carried across the North Sea by westerly winds.

---

[15] See 'Chernobyl – the radioactive plume and its consequences', published in 1987 in the *Annual Report of the Meteorological Office for 1986* (Met.O.976, HMSO, pp. 29–32).
[16] For information about studies of acid rain in the mid to late 1970s, see 'Acid rain and the long-range transport of air pollutants', by B E A Fisher, published in *Weather* in 1981 (Vol. 36, pp. 367–369). See also 'Acid rain – cause and consequence', by B J Mason, published in *Weather* in 1990 (Vol. 45, pp. 70–79).

A New Director-General

Sir John Mason stepped down as Director-General on 30 September 1983 and returned to the Imperial College of Science and Technology the following day as part-time Director of a project called the 'Anglo-Scandinavian Surface Waters Acidification Programme', based in the College's Centre for Environmental Technology.[17] He had reached the age of 60, the normal retirement age for a civil servant.[18]

In his eighteen years as Director-General, he had transformed the Office.[19] It was now a world leader in climate modelling, NWP, instrument development and uses of weather radar. It was also a leading member of the international meteorological community, providing training for students from many countries and playing a full part in the administration and scientific activities of WMO. At home, the Office provided services for the general public and a wide range of specialist users, including local-area weather forecasts available in many parts of the UK through the Post Office's Automatic Telephone Weather Service. In addition, a British Telecom view-data system called *Prestel* had been introduced, through which the public could gain access to a computer bank of weather forecasts, meteorological reports and climatological statistics by means of an ordinary telephone line and display the information at home on a suitably modified television set. Weather presentations continued on radio and TV, broadcast by the BBC and commercial stations, and both the BBC and the Independent Broadcasting Authority had introduced *Teletext*, a television information retrieval service which employed the spare lines in the television signal to display a variety of information, including weather.

Sir John's vision and achievements were widely recognized and applauded, but he could be forthright and opinionated. Indeed, when interviewing him for WMO in 1994, Dr H Taba raised this matter.[20] "You are an outspoken person", said Dr Taba. "Has this caused you any difficulties during your career?" Sir John replied that he was frank and preferred to "tackle people directly rather than criticize them behind their backs". In the *Meteorological Magazine* in September 1983 (Vol. 112, pp. 269–273), G A Corby commented that, "in the rather different and staid corridors of the Scientific Civil Service back in 1965, the newly-appointed Director-General of the Office had seemed to some a very youthful, even perhaps brash, successor to his sedate predecessors". However, Mason had "quickly established himself as a very shrewd

[17] The project was managed by the Royal Society, the Norwegian Academy of Science and Letters and the Royal Swedish Academy of Sciences and funded by the National Coal Board and the Central Electricity Generating Board. It ran for six years. See *The Surface Waters Acidification Programme*, edited by B J Mason, published in 1990 by Cambridge University Press (522 pp.).

[18] Whilst Director-General, Sir John became Treasurer of the Royal Society (1976), President of the Institute of Physics (1976), Pro-Chancellor of the University of Surrey (1979) and President of the British Association (1983). He also received the Symons Gold Medal of the Royal Meteorological Society (1975).

[19] For his own account of his years as Director-General, see *The Meteorological Office (1965–83)* by Sir John Mason, published in 2010 by the Royal Meteorological Society (30 pp.).

[20] See 'The *Bulletin* interviews – Sir John Mason', published in 1995 in the *WMO Bulletin* (Vol. 44, pp. 315–325).

scientific administrator and forceful organizer with a great capacity for getting things done and shaping circumstances for the benefit of the Office".[21]

Those who knew Mason only superficially, Corby said, may not have been aware that he had a genuine concern for his staff and their welfare. He was distressed when a staff member or relative was "in trouble, perhaps from illness, bereavement or otherwise", and had always offered "what help or comfort lay within his power". Behind the brisk, extrovert exterior, he added, there was a "kindly, sensitive inner man".

The new Director-General was John Theodore Houghton, Professor of Atmospheric Physics in the University of Oxford since 1976 and Deputy Director of the Rutherford-Appleton Laboratory at Chilton in Oxfordshire since 1979. Born at Dyserth near Rhyl in North Wales on 30 December 1931, he had been educated at Rhyl Grammar School and proceeded from there at the early age of sixteen to Jesus College, Oxford, where he had gained a double first in mathematics and physics in 1951. After that, he had remained at Oxford and gained the degree of DPhil in 1955 for his study of radiative fluxes in the stratosphere. Later that year, he had become a Research Fellow at the Royal Aircraft Establishment, and in 1958 he had returned to Oxford as lecturer in atmospheric physics and continued his studies of radiation through both theoretical and experimental approaches. Four years later, he had been promoted to the position of Reader and become Head of the Sub-Department of Atmospheric Physics. During his years at Oxford, he had come to be recognized as an authority on the design and use of sophisticated satellite-based instrumentation to measure the temperature structure and composition of the upper atmosphere, and he had been appointed, in 1976, a member of the Joint Organizing Committee of the Global Atmospheric Research Programme. He had been elected a Fellow of the Royal Society in 1972 and served as President of the Royal Meteorological Society from 1976 to 1978. In addition to many scientific papers, he had written a definitive book, *The Physics of Atmospheres*, published by Cambridge University Press in 1977.

Dr Houghton became Director-General two months after a substantial report had been published. The report had been prepared by Lord Rayner, a director of Marks & Spencer, whose task it had been to advise the Prime Minister (Margaret Thatcher) on ways and means of improving efficiency and eliminating waste in government. To carry out this work, he had set up an Efficiency Unit which had made intensive studies of specific areas of government work. His review of the Office had taken sixteen months. The report was broadly complimentary, though critical of some aspects of the Office's work, notably bureaucratic procedures and some techniques of financial management. It was gratifying to note, said Dr Houghton in his Foreword to the Office's annual report for 1983, that Rayner had been able to conclude that the UK

---

[21] Mason was very effective in negotiations with government when seeking funding for satellite-related work, radar and computers.

Figure 13.1. John Theodore Houghton. © Crown Copyright 2010, the Met Office.

undoubtedly had in the Office a centre of excellence of which she could be proud. This had been achieved, he had found, at one of the lowest costs per head of population for meteorological services in countries of the developed world.

Rayner had concluded that the Office should continue to provide the public with a free service of weather information and forecasts financed through general taxation, and also that the Office should remain within the MoD. However, he had also concluded, to quote from the annual report for 1983, that "the Office should continue to explore ways of marketing and selling specialized meteorological and climatological services and look into new organizational ways of furthering commercial activities, possibly through the establishment of private venture companies".[22] To many in the Office and not a few outside, the latter conclusion was not well received. Was not the Office fundamentally a scientific and technological institution of world class which excelled in meteorological services and provided most of them gratis? Certainly some services were not free of charge, for example ship routeing and the Automatic

[22] See *The report of the Resource Control Review of the Meteorological Office*, published by the Ministry of Defence in 1983 (HMSO, two volumes). See also the review of this report by B J Hoskins, published in the *Quarterly Journal of the Royal Meteorological Society* in 1984 (Vol. 110, pp. 1200–1202).

Telephone Weather Service, but the idea of the Office becoming overtly a commercial body was anathema to many.

However, a wind of commercial change was blowing in the 1980s, and the Office's senior management responded to Rayner's report by giving much thought to the balance between free public services on the one hand and repayment services on the other. They also considered a further question raised by Rayner's inquiry, which was, in the words of the Office's annual report for 1983, "the extent to which restrictions needed to be placed on the general accessibility of data and forecast products generated by meteorological services in order that the Office could itself recoup some of the costs of producing the information". This was another matter which caused dismay among some groups of users. In schools and universities, for example, meteorological and climatic data were needed for projects. At the best of times, little money was available for data to be purchased, but this was especially so in a time of financial retrenchment such as the 1980s.

Two years or so after he became Director-General, Dr Houghton decided the time had come to review the Office's activities and set out primary goals. In the April 1987 issue of the *Meteorological Magazine*, he explained his priorities in an article called 'The Meteorological Office – a ten-year perspective' (Vol. 116, pp. 97–116). He stressed that the Office's main responsibility was to provide meteorological services to defence, civil aviation, shipping, other government departments, public bodies, the media, industry, commerce and the general public. The Office was widely considered, he said, one of the leading national meteorological services in the quality of its forecasting products, an excellence the Rayner review had shown to be achieved at modest cost. The main reasons for the Office's pre-eminence and commendable cost-effectiveness were, he said:

- Operational meteorology in the UK was concentrated in the Office, which was not the case in many countries, where operational meteorology for defence had been separated from that for the civil sector, or services for commerce and industry had been separated from services to the public.
- The Office had, over the years, concentrated its effort into key areas and ensured that research and development had been followed through into operational products.
- The Office had attracted some of the best graduates in mathematics and physics.

Dr Houghton went on to consider the international context, stressing that close cooperation between the Office and other national meteorological services must be maintained, as must, too, the principle of free exchange of data and products with them. The Office would continue to support WMO fully, assist small and developing countries, enhance its role as a World Area Forecast Centre for civil aviation, and cooperate with ECMWF, the European Meteorological Satellite Organization (EUMETSAT) and other international bodies. In addition, the Office attached great importance to

European COST programmes, such as COST-43 (which was concerned with the setting up of an experimental network of oceanographic and meteorological measuring stations in European waters) and COST-72 and COST-73 (concerned with weather-radar networks).[23]

Dr Houghton said that future developments in observations would be "governed by the need to improve data quality and coverage and the need to improve the cost-effectiveness of data acquisition through a programme of automation". Of particular importance would be the enhancement of space observing techniques "both in improved hardware in space and also in better methods of data retrieval and interpretation". He also reported that a major equipment replacement programme was planned, including replacement of the central computer, the Cyber 205, by the end of the 1980s.

Near the end of his article, Dr Houghton turned to the allocation and control of resources, saying that he considered it "a high priority to achieve a more rational arrangement for the allocation of resources from the MoD to the Office". He referred to the report of Sir Kenneth Sharp and Mr John Hansford into the financial management of the Office which they had submitted to the Minister of State for Defence Procurement on 12 August 1985.[24]

In the House of Commons on 26 February 1986, Jerry Wiggin drew attention to this report. He advised fellow Members of Parliament to read it, saying that its main recommendation was that "the Meteorological Office should adopt a commercial approach to all its dealings and that its charges should be based where practical on market conditions, and where not practical by free negotiation between the parties concerned". According to him, the Office charged itself out of the wrong markets and managed to put off profitable private enterprise operations in which it had a public duty to get involved.

Wiggin pointed out that Sharp and Hansford had recommended the establishment of a Meteorological Office Management Board. He asked the government to establish one and place the Office's accounting system on a proper basis. Then, he said, "many of the complaints of private enterprise and of those who were involved in working with the Office would go away". He went on to say that difficulties had been experienced by private weather services in acquiring information not only from the Office but also from ECMWF. It was cheaper (and idiotic), he said, for "a British firm to get information from America and to bring it back, at a cost in dollars, than it was to pay

[23] See 'The achievements of COST-43', by D N Axford, published in the *Meteorological Magazine* in 1988 (Vol. 117, pp. 186–193). See also 'COST-72 – European weather radar project', by G A Clift, published in 1981 in *Nowcasting: Mesoscale observations and short-range predictions*, Proceedings of the IAMAP Symposium held in Hamburg from 25 to 28 August 1981 (pp. 313–316); also 'International weather-radar networking in western Europe', by C G Collier *et al*, published in 1988 in the *Bulletin of the American Meteorological Society* (Vol. 69, pp. 16–21).

[24] See *The Meteorological Office. Financial management*, by K J Sharp and J Hansford, published in 1985 by the Office (60 pp.).

a fee in sterling to our own Meteorological Office for exactly the same information which was transmitted on an international telecommunications network". The strong views expressed by Wiggins did not fall on deaf ears, as we see later in this chapter.

Dr Houghton also pointed out in his article that military activity was, and was likely to remain, highly weather sensitive. The Falklands War in 1982 had served to underline this. Before the invasion, the Office's then experimental NWP model for the southern hemisphere did not produce forecasts south of 30°S, and the Falkland Islands lay close to 51°30'S. As a result of a major programming effort over the weekend of 3–4 April 1982, however, the two days following the invasion of the Falkland Islands and South Georgia, forecasts for the area of the conflict became available immediately (i.e., four months ahead of the planned introduction of the Office's new global model). In addition, a separate forecasting unit was set up in the CFO to provide weather forecasts and meteorological data for all who required them, including forecasters at the Fleet Weather and Oceanographic Centre at Northwood (north-west London) and meteorologists in the field on the Ascension Island staging post and in the Falkland Islands.

A further involvement of the Office in the Falklands conflict came through deployment of the Mobile Meteorological Unit (MMU) to the South Atlantic. The existence of this Unit was not then well known, even within the Office. However, it had existed since the early 1960s, as part of the RAF's Tactical Communications Wing.[25] In 1982, it had a maximum complement of twenty officer volunteers drawn from the headquarters and outstations of the Office, all holding Class CC (Civil Component) Commissions in the RAF Reserve of Officers. It was equipped with its own portable accommodation, meteorological instrumentation and communications equipment, and its staff were expected to deploy at short notice anywhere within the NATO area.

On 6 April 1982, the MMU received instructions to deploy to Ascension Island. Three days later, two forecasters, two observers and an officer in charge reached the island, their purpose being to provide meteorological support to the various military groups who were operating during the conflict in, around, and to and from the island. On 13 July 1982, a month after the Argentine surrender, a second small detachment of the MMU began to operate on the Falkland Islands, on Stanley airfield. And there the Unit remained until 29 April 1986, when a civilian meteorological office opened at the new airfield at Mount Pleasant. The MMU was withdrawn simultaneously.[26] Today, the MMU has a staff of around fifty forecasters, engineers and support staff and is deployed in fields of war wherever support for military operations is needed.[27]

[25] For the early history of the Unit, see 'The Mobile Meteorological Unit', by W R McQueen, published in *Proceedings of the Conference of Chief and Principal Meteorological Officers* held at the Meteorological Office College on 10 and 11 January 1990 (pp. F1-F2).

[26] See 'The Mobile Meteorological Unit in the South Atlantic 1982–86', by H E Brenchley, published in the *Meteorological Magazine* in 1986 (Vol. 115, pp. 257–263).

[27] For information about the work of the MMU, see 'The Mobile Meteorological Unit', by A Dickson, published in *Weather* in 1999 (Vol. 54, pp. 374–376). For an account of the MMU's work during the Gulf War of 1990–1991,

Progress and Storms

For many years before the 1980s, the name 'Meteorological Office' had often been shortened to 'Met Office'. All of a sudden, the abbreviated form was used on the cover of the annual report. The name on the cover of the 1987 report was 'Meteorological Office'. The name on the cover of the 1988 report was 'The Met.Office', with a dot after 'Met'. Some years later, the use of the dot was discontinued and the institution came to be known as the 'Met Office', which is now officially the Office's name. It will be used henceforth in this book, unless the full name needs to be given. No reason for shortening the name on the cover of the 1988 report was given, and the full name was otherwise used punctiliously throughout that report.

Dr Houghton was keen to strengthen links between the Office and universities. Thanks mainly to him, an institute was established in the Clarendon Laboratory of the University of Oxford in 1984 to promote research in atmospheric science, particularly satellite meteorology and geophysical fluid dynamics. Called the 'Robert Hooke Institute for Cooperative Atmospheric Research', it was a joint venture between Oxford University, NERC and the Met Office.[28]

Another who was keen to strengthen links between the Office and academia was Keith Browning, who became the Office's Director of Research in 1989. He played a leading role in the establishment of another collaborative research venture between the Office and a university, the Joint Centre for Mesoscale Meteorology (JCMM), which was set up in the University of Reading's Department of Meteorology in 1988 with the objective of increasing basic understanding of mesoscale weather systems and improving skill in forecasting them.[29] Though he did not know it at the time, he was soon to become a member of the Centre himself. He stepped down as Director of Research in 1991, initially to take a year's sabbatical leave to carry out research in the JCMM, but he stayed at Reading until he retired twelve years later.[30]

NWP became progressively more reliable during the 1970s and 1980s. However, spectacular failures still occurred occasionally, as in August 1979, when competitors in the biennial yacht race from Cowes (Isle of Wight) to the Fastnet Rock (off southern Ireland) and back to Plymouth (south-west England) encountered a storm which was much more severe than forecast and indeed proved the most disastrous in yachting history. Mean winds reached fifty to sixty knots, with gusts of more than seventy

see 'The Mobile Meteorological Unit and Operation Granby', by P J Wyatt, published in *Weather* in 1992 (Vol. 47, pp. 196–200).

[28] Following a review by NERC and the continuing need for economies by the Met Office, the staff of the Institute moved in 1992 to Oxford and Southampton Universities and to the Met Office.

[29] See 'Joint Centre for Mesoscale Meteorology', by A J Thorpe, published in the *Meteorological Magazine* in 1988 (Vol. 117, pp. 285–287). Browning was President of the Royal Meteorological Society for 1988–1990 and on 21 June 1989 delivered a comprehensive Presidential Address entitled 'The mesoscale data base and its use in mesoscale forecasting', published in the *Quarterly Journal of the Royal Meteorological Society* in 1989 (Vol. 115, pp. 717–762).

[30] Today, the JCMM still exists and is still based in the University of Reading's Department of Meteorology.

knots, and there were steep and breaking waves fifteen metres high. Only eighty-five of the 303 yachts which set out finished the race. Sadly, fifteen crew members died.

The origin and development of the storm were reviewed by the Office, with particular reference to the data that were available in the CFO and the guidance provided by numerical modelling. A critical contributory factor to the poor forecast was found to be a dearth of observations. For the weather analysis made twenty-four hours before the storm reached its peak, no observations were available at all from the crucial area of the Atlantic where the storm was developing. Six hours earlier, readings of barometric pressure had in fact been received from a number of ships, including three near the centre of the developing depression, and these readings had been accepted by forecasters. But all three had been in error. To compound the problems, the IBM 360/195 and 370/158 computers both broke down during the afternoon of the day when the storm was nearing its maximum intensity.

The numerical predictions for forty-eight and seventy-two hours ahead indicated that the depression was likely to deepen as it approached Ireland. However, forecasters believed these predictions were greatly underestimating the storm's intensity, so they intervened to deepen it. In his article entitled 'The Fastnet Storm – A Forecaster's Viewpoint' published in the *Meteorological Magazine* in 1981 (Vol. 110, pp. 271–287), A Woodroffe commented that this intervention was an outstanding example of the contribution that a forecaster could still make to the so-called man-machine-mix. He also stressed the vital importance of "reliable and correctly-coded ship observations, even in this era of computers and satellite data".

A feature of the storm was a narrow belt of exceptionally strong winds on its southern side. As Woodroffe pointed out, the capabilities of NWP at the time were such that a feature so limited in extent was unlikely to be forecast reliably, if at all. Detailed studies of the storm revealed, too, that the storm contained particular mesoscale and smaller-scale distributions of heat, humidity and airflow which were found to be consistent with atmospheric processes of fundamental importance in the development and maintenance of severe mid-latitude storms.[31] A close examination of the wind and barometric pressure records of yachtsmen showed that the belt of strong winds was a 'low-level jet stream', an atmospheric feature about which little was known in 1979.[32] The occurrence of such jets in extratropical depressions has since been much studied by staff of the Office and other institutions, and the jets have been found to occur commonly in depressions which develop very rapidly.[33]

[31] A rhyme attributed to L F Richardson encapsulates the nature of atmospheric systems: "Big whirls have little whirls that feed on their velocity, and little whirls have lesser whirls and so on to viscosity – in the molecular sense". Depressions are systems which contain interdependent sub-systems of various time and space scales and are themselves sub-systems of the global ocean-atmosphere system.

[32] See 'The Fastnet storm of 1979: A mesoscale surface jet', by D E Pedgley, published in *Weather* in 1997 (Vol. 52, pp. 230–242).

[33] For a detailed study of rapidly deepening depressions and their mesoscale features, see 'The sting at the end of the tail: damaging winds associated with extratropical cyclones', by K A Browning, published in the *Quarterly Journal of the Royal Meteorological Society* in 2004 (Vol. 130, pp. 375–399).

Another storm which developed rapidly and contained a low-level jet stream wreaked havoc across northern France, south-east England and East Anglia during the night of 15/16 October 1987. The winds in the early hours of the morning were terrifyingly strong. Millions of trees were lost, roads and railways were blocked, electricity and telephone lines were brought down, many buildings were damaged and the storm killed eighteen people in England and four in France. Winds reached Force 11 in many coastal regions.

A BBC TV weather presenter, Michael Fish, will long be remembered for telling viewers at lunchtime on 15 October there would be no hurricane. He was, in fact, correct. He was referring to a tropical cyclone which was over the western part of the North Atlantic, a weather system that was certainly not heading for the British Isles. Earlier in the day, when referring to the weather to be expected in southern Britain, he advised viewers to "batten down the hatches" because there was some "extremely stormy weather on the way". That was a forecast which could hardly be faulted. The storm which was to cross south-east England and northern France that night was developing over the Bay of Biscay. Many people now believe that Fish was the presenter who said in an evening forecast on the 15th that it would be "breezy up through the [English] Channel and on the eastern side of the country". He was not on duty that evening. The presenter was Bill Giles.[34]

The Great Storm of October 1987 was not, by any definition, a hurricane. It was, however, exceptional. South-east of a line extending from Southampton through north London to Great Yarmouth, gust speeds and mean wind speeds were as great as those which can be expected to recur, on average, no more frequently than once in 200 years.[35] Comparison with a storm mentioned in Chapter 2 was entirely justified, viz. the storm of 1703, which affected much the same area of the British Isles.

Four days before the Great Storm struck, Met Office forecasters predicted severe weather on the 15th or 16th. A couple of days later, however, the guidance provided by numerical modelling was somewhat equivocal. Instead of stormy weather over a considerable part of the British Isles, the modelling now suggested that severe weather would occur no farther north than the English Channel and coastal parts of southern England. In the event, forecasts of wind strengths for the sea areas of the English Channel were both timely and adequate, but the forecasts for land areas left much to be desired.

During the evening of 15 October, radio and TV forecasts mentioned strong winds but indicated that heavy rain would be the main feature, rather than strong wind. By the time most people retired to bed that evening, exceptionally strong winds had

[34] See 'Media reaction to the storm of 15/16 October 1987', by D M Houghton *et al*, published in 1988 in the *Meteorological Magazine* (Vol. 117, pp. 136–140).

[35] The word 'average' must be stressed. A 'once-in-200-years storm' is not one that recurs exactly every 200 years. It is one that can be expected to occur *on average* once in 200 years. Another storm as severe as that of October 1987 could have occurred in the following months or, indeed, in any year since 1987.

not been mentioned in national radio and TV weather broadcasts. The Office had issued warnings of severe weather, however, to various agencies and to the authorities responsible for dealing with emergencies. The London Fire Brigade had been notified at 21:50 GMT, and the first warnings for civil and military airfields had been issued as early as midday. Perhaps the most important warning of all was that issued to the MoD at 01:35 GMT on the 16th. This warned that the anticipated consequences of the storm were such that civil authorities might need to call on assistance from the military.

Journalists looking for scapegoats and a sensational story were quick to accuse the Office of failure to forecast the storm correctly. Time and again they returned to the statement by Michael Fish that there would be no hurricane. The *Sunday Telegraph* reported that Nicholas Ridley, the Secretary of State for the Environment, had condemned the "unbelievable failure" of the Office to "get it right". It mattered not that the Office's forecasters had for several days warned of severe weather, and it was hard to shake the views of some that ECMWF and the French and Dutch meteorological services had forecast the storm a great deal more accurately than their British counterparts when they had not.

Dr Houghton set up an internal inquiry and the Secretary of State for Defence asked two independent assessors, Sir Peter Swinnerton-Dyer (Chairman of the Meteorological Committee) and Professor Robert Pearce (Head of the Department of Meteorology in the University of Reading), to review the findings of the inquiry and advise as appropriate.[36] The main conclusions and recommendations of the Office and assessors were that:[37]

- Observational coverage over the ocean west of France and the Iberian peninsula, where the storm had developed, needed to be improved by increasing the quality and quantity of observations from ships, aircraft, buoys and satellites.[38]
- Certain specific refinements to the computer models used for weather forecasting needed to be made.
- Procedures for issuing to authorities warnings of severe weather should be reviewed.
- Procedures for presenting such warnings to the public should also be reviewed.
- The training given to weather forecasters needed to be improved and lengthened.
- Forecasters should have greater seniority in the Office.
- The economies imposed by the government on the Met Office in recent years had compromised the effectiveness of the CFO, especially on occasions of unusual weather situations.

[36] Sir Peter was a distinguished mathematician and Fellow of the Royal Society.

[37] See *The storm of 15/16 October 1987*, by P Swinnerton-Dyer and R P Pearce, published in 1988 by the Secretary of State for Defence (17 pp.). See also 'The Meteorological Office report on the storm of 15/16 October 1987', published in the *Meteorological Magazine* in April 1988 (Vol. 117, pp. 97–98). The whole of this issue of the *Meteorological Magazine* (pp. 97–140) was devoted to the Office's studies of the storm. The whole of the March 1988 issue of *Weather* (Vol. 43, pp. 66–142) was devoted to the storm, too, including comparisons with the tempest of 1703.

[38] In the 1970s, there had been eight ocean weather stations on the North Atlantic, including one to the west of the Bay of Biscay (at 47°N 17°W). In October 1987, there were only two, one at 57°N 20°W, the other at 66°N 02°E. In 1977, the UK had owned three weather ships. In 1987, it owned only one.

Swinnerton-Dyer and Pearce addressed the question of whether the computer-based forecasts available from Météo-France (the French national meteorological service) were superior to those from the Met Office. They found that the French forecasts were "somewhat more consistent" than those from the Office, mainly because they used a more powerful computer. However, they said, the forecasts "equally underestimated the storm's intensification". The Cyber 205 computer used by the British was, they pointed out, significantly less powerful than the Cray machines available to ECMWF and Météo-France and possessed a smaller memory. A consequence was that the model used by the French had a resolution twice that of the model used by the Office. They stressed the importance of ensuring that the Office always had at its disposal the most powerful computer available. In their words:[39]

Underprovision of computing power would indeed be a false economy, because it would undermine the campaign to increase the Met Office's commercial income – and this campaign is essential to the Met Office's future funding strategy. We are relieved to hear that the Met Office will be provided with an ETA 10 supercomputer in the spring, even though the cost has had to be found by internal economies. We are not in a position to comment on the damage done by these economies beyond saying that it will have to be endured because the new computer is essential.

A storm of intensity comparable to that of the 1987 storm occurred on 25 January 1990 and has been dubbed the 'Burns' Day Storm' because it occurred on the anniversary of the birth of Scotland's national poet. It caused widespread damage over England and Wales and the loss of forty-seven lives. The centre of the depression moved across southern Scotland, and the severity of the gales on its southern and western flanks was well forecast. On this occasion, no blame was placed on the Office's weather forecasters.[40]

More Independence for the Met Office

Sir Robin Ibbs of ICI succeeded Lord Rayner as the Prime Minister's adviser on ways and means of improving efficiency and eliminating waste in government. He was asked in 1987 to assess progress that had been achieved towards improving management in the Civil Service. The response came in February 1988, in a report prepared by the Efficiency Unit entitled *Improving Management in Government: The Next Steps*. Mrs Thatcher announced in the House of Commons on 18 February that she had accepted the report's recommendation that, to the greatest extent practicable, the executive functions of government, as distinct from policy advice, should be carried out by units clearly designated within Civil Service departments and known

---

[39] See 'Summary and conclusions from the Secretary of State's enquiry into the storm of 16 October 1987', by Sir Peter Swinnerton-Dyer and R P Pearce, published in the *Meteorological Magazine* in 1988 (Vol. 117, pp. 141–144).

[40] See 'The Burns' Day Storm, 25 January 1990', by E McCallum, published in *Weather* in 1990 (Vol. 45, pp. 166–173).

as 'agencies'. Each agency would be staffed by civil servants and directed by a chief executive, who would be responsible for day-to-day control. Agencies would be given specific briefs so that overlaps between them would be avoided. They would be bound by 'Framework Documents', and chief executives would be accountable directly for performance, in some cases to Parliament. Agencies would be expected, so far as possible, to pursue a commercial approach.

The Under-Secretary of State for Defence Procurement, Tim Sainsbury, announced in the House of Commons on 18 March 1988 that the MoD had in mind for agency status the Met Office and the non-nuclear research establishments, and this was reiterated by the Secretary of State for Defence, George Younger, on 19 October 1988. A committee had been formed by the MoD to examine options for the management of the Office. Four were considered: abolition, privatization, contractorization and agency. A member of the committee, the Office's Director of Services, Dr Peter Ryder, put the committee's conclusion thus when reporting to colleagues in January 1990, "Agency status was clearly the thing that offered the best advantages for an organization such as the Met Office".[41]

The Government agreed in October 1989 that the Office would become an Executive Agency on Monday 2 April 1990, and the Under-Secretary of State for the Armed Forces, Michael Neubert, made the following statement in the House of Commons on 18 October:

The Met Office has worldwide renown for its contribution towards meteorological and environmental matters. It is well known to millions, but not everyone is aware that it is in fact part of the Ministry of Defence. It is part of this Ministry because much of its work involves direct support of all three services, a role which would become particularly crucial in wartime when weather conditions can be of vital operational importance. Therefore, we have decided that the new agency will remain within the ambit of the Ministry of Defence. Nevertheless, I am anxious that the Office should have every opportunity, within that necessary constraint, to manage its own affairs. It will have a new charter, and we will give it performance targets to encourage a new management approach and to stimulate greater flexibility and openness toward commercial opportunities. I intend that the Met Office should, as necessary, recruit business and commercial talent from outside. That will complement the unique scientific expertise that the Office already has. Our aim in establishing the Office as an agency is to improve its management and commercial skills, while retaining the high technical and scientific standards that have placed it as a front runner in its field. This will provide a better service for all its customers, civil and military, and give the Ministry of Defence better value for the operating costs of the Office, which amount to some £80 million a year.

In a letter to the Office's staff, Dr Houghton gave reassurance there would in many ways be no change to the Office's functions, and he stressed that the Office would remain

---

[41] See 'Agency – the Met Office into the 90s', by P Ryder, in *Proceedings of the conference of Chief and Principal Meteorological Officers*, held at the Meteorological Office College on 10 and 11 January 1990, pp. B1–B14 (copy held in the National Meteorological Library, Exeter). The steering committee was chaired by Mr T Knapp, Assistant Under-Secretary of State for Supply and Organization (Air).

the state meteorological service.[42] He went on to say that Executive Agency status would give the Office greater autonomy over its manpower and financial resources and thus enable it to work more efficiently, react to circumstances more quickly and, in particular, benefit from commercial opportunities. Instead of a Director-General, the Office would be headed by a Chief Executive, who would lead a management team which consisted of Directors of Operations, Research, Finance and Administration, and Commercial Services. He, Dr Houghton, would become the Office's first Chief Executive.

The Secretary of State for Defence, Tom King, announced in the House of Commons on 2 April 1990 that the launch of the first of the MoD's executive agencies had taken place in Bracknell that day. He praised the Office. As the national weather service, it had played a leading role in the international structure on which modern weather forecasting depended. Its staff included "some of the world's foremost experts in climate prediction and research".[43] Utilizing these resources, one of the Office's key aims would be "to provide timely and authoritative advice on climate change to Ministers nationally and internationally through the auspices of the Intergovernmental Panel on Climate Change".[44] The initial targets set for the Office during the period 1990–1995 were, Mr King went on:[45]

- To devise and implement a system of performance measurement in each of the major activity areas based on output quantity, quality and timeliness and on input costs, to enable publication by 31 March 1991 of targets which required improvements to efficiency that would make a significant contribution to the efficiency targets of the MoD as a whole
- To improve the efficiency of commercial services, measured as the ratio between revenue earned and the cost of provision of such services
- To reduce to 15% the standard error of forty-eight–hour forecasts of mid-atmosphere pressure fields over the North Atlantic and north-west Europe
- To achieve measurable improvement in the quality of public warnings and services to the MoD and the Civil Aviation Authority (CAA), within available resources
- To achieve measurable improvement in the productivity of operational support services through new technology, automation and the integration of functions
- To reduce net operating costs in real terms by 1% in 1990–1991 and to continue to achieve this performance with the aim of doubling it by the end of the period

---

42 See 'The Meteorological Office to become an Executive Agency', an anonymous report on p. 267 of the December 1989 issue of the *Meteorological Magazine* (Vol. 118).

43 During the 1980s, the expression 'climate change' replaced 'climatic change' and will be used henceforth in this book. According to legend, the use of the noun 'climate' as an adjective came about because so many people did not appear to know the difference between 'climatic' and 'climactic'!

44 The decision to form this Panel was made at the Congress of the WMO held at Geneva in May 1987. The Secretary-General of WMO was authorized to consult with the Executive Director of the United Nations Environment Programme (UNEP) to establish what became, in 1988, the Joint WMO-UNEP Intergovernmental Panel on Climate Change.

45 The objectives, organization, financial arrangements, etc. of the Met Office as an Executive Agency were set out in *The Meteorological Office Executive Agency: Framework Document*, published in April 1990 by the Ministry of Defence (11 pp. plus four annexes).

- To achieve a 6% per annum cumulative increase in the uptake of meteorological services as measured by revenue generated from commercial services offered to the public, commerce and industry, exclusive of services to the CAA
- To achieve a 10% per annum cumulative increase in the contribution of commercial service revenue to offsetting core costs

The increased emphasis on targets had been anticipated by the Office. This was clear from the Office's report for the period 1 January 1989 to 31 March 1990, where it was stated that commercial activities had been guided by the 'Meteorological Office Marketing Plan 1989–90'.[46] This plan had set revenue targets for each of sixteen market sectors. Overall, the report said, "the revenues earned from commercial services were expected to be on, or above, target, with the Offshore and Media and TV sectors likely to be significantly above target". In contrast, less growth than expected had been experienced in the Retail sector (owing to delays in launching new services) and in the Professional and Legal sector (owing to lower levels of demand than expected).

During the year, the report continued, there had been a field trial of thirty-day forecasts for a small number of customers which had "demonstrated successfully its high potential value to companies in the private sector". A product development group also existed, whose activities included "the assessment of the sensitivity of various industries to weather and the use of weather radar to determine conditions of flood and drought". For the Media sector, the main success of the year had been the start of a new daily weather presentation which incorporated state-of-the-art television graphics. This presentation had been sponsored by Powergen, an electricity supply company, through Independent Television News, and was the first TV sponsorship deal of its kind in the UK.

The report for 1989–1990 went on to say that there had been a considerable increase in the range of weather information available by telephone, including forecasts for holiday resorts in the Mediterranean and ski resorts in Europe and North America. So far as land transport was concerned, the 'Open Road' service had been introduced to county authorities, this being a weather information service which helped local authorities cope with snow and ice problems in winter. Through its links with the Finnish company Vaisala TMI Ltd, the Office was able to supply the 'Open Road' service by means of direct computer links between Weather Centres and ice prediction systems.[47]

---

[46] For three decades, the Office's annual report had covered calendar years. From 1990, it reverted to being a report from 1 April to 31 March.

[47] 'TMI' stands for Thermal Mapping International. Thermal mapping is a process whereby high-resolution infra-red thermometers are used to measure the temperatures of road surfaces. For a description and assessment of the Open Road service, see 'A preliminary performance and benefit analysis of the UK national road ice prediction system', by J E Thornes, published in 1989 in the *Meteorological Magazine* (Vol. 118, pp. 93–99).

After a period of decline in previous years, revenues from the Offshore sector had increased in 1989–1990, largely because of increased exploration activity in the North Sea, and this sector had been identified, the annual report stated, "as a major development area for the electronic distribution of services by computer-to-computer links, automated document facsimile distribution, and access to databases, etc". For the Utilities sector, plans had been made to provide quantitative short-period rainfall forecasts to certain water authorities from the Office's FRONTIERS system (see Chapter 12). For the Press sector, improved graphics capabilities had been developed to provide national and regional newspapers with an improved service and a better source of weather text and maps for publication. During the year, Bracknell had been confirmed as one of two World Area Forecast Centres and given increased international responsibilities in respect of services for civil aviation.

Dr Houghton was able to say in the annual report for 1 April 1990 to 31 March 1991 that the move to Agency status had been one of the most important events in the recent history of the Office. It had provided a stimulus for reorganizing the Office's activities for the benefit of its customers and for streamlining its operations and placing emphasis on the quality of its services. Commercial revenue targets had been exceeded, with the income from commercial services £13.1 million, which was 17% higher than for the previous year. A Director of Commercial Services had been recruited from the private sector, and this had led to the introduction of a number of new services and products, among them 'The Weather Initiative', a consultancy service that was "aimed at helping managers in the retail industry identify and exploit the sensitivity of their businesses to the weather".

Further Important Developments for the Met Office

Three events in 1988 assisted greatly in bringing the issue of man-made climate change to the notice of politicians.[48] The first was a World Ministerial Conference on Climate Change that took place in Canada in June. The second was a speech in September by the British Prime Minister, Margaret Thatcher, at the annual dinner of the Royal Society, when she spoke about the science of anthropogenic climate change and stressed the importance of taking action to combat this change.[49] The third was the first meeting of the Intergovernmental Panel of Climate Change (IPCC), held in November 1988.

A speech of fundamental importance was delivered by Mrs Thatcher in New York on 8 November 1989, to the General Assembly of the UN. She shocked the political

---

[48] An article by C K Folland, D J Griggs and J T Houghton about the history and antecedents of the Met Office Hadley Centre was published in a special issue of *Weather* in 2004 to mark the sesquicentenary of the Office (Vol. 59, pp. 317–323).

[49] Mrs Thatcher was a trained scientist, having read chemistry at the University of Oxford and been employed as a research chemist before qualifying as a barrister and, in 1959, becoming a Member of Parliament.

world by devoting the whole of it to the subject of the global environment, and she was typically forthright. Of all the challenges faced by the world community in the four years since she had last addressed the Assembly, she said, the threat to the global environment had "grown clearer than any other in both urgency and importance". She focused on the "vast increase in the amount of carbon dioxide reaching the atmosphere" and the damage to the ozone layer that was resulting from the widespread use of chlorofluorocarbons.[50]

Mrs Thatcher said that Britain had some of the world's leading climate experts. She was pleased to be able to announce that the UK would be establishing a new centre for the prediction of climate change. It would provide the advanced computing facilities that scientists needed and "would be open to experts from all over the world, especially from the developing countries". The Minister for the Environment and Countryside, David Trippier, announced in the House of Commons on 10 November 1989, during a debate on 'World Climate Change', that the centre would be a focus for basic research on climate models. His department would provide more than £5 million a year to fund it.

Later in the debate, in response to sceptical comments over the findings and claims of climate scientists, Mr Trippier said:

The Government are spending over £15 million on research on climate change and the ozone layer through the Natural Environment Research Council, the Science and Engineering Research Council, the Met Office and Government Departments. We have also more than doubled Britain's contribution to the United Nations Environment Programme, and I echo the Prime Minister's call for other countries to do the same. The quality of British scientific work on climate change was given international recognition last year, when we were chosen to chair the working group on scientific assessment of the Intergovernmental Panel on Climate Change. This body, which was set up jointly by the United Nations Environment Programme and the World Meteorological Organization, is the principal international focus for research and action on climate change. I stress once again that we are fully committed to the work of the IPCC and we are providing £750,000 to support its work.

The Under-Secretary of State for the Environment, David Heathcoat-Amory, confirmed that the new centre would be annexed to the Met Office. This was controversial, as some had expected it to be attached to a research institute. However, the consensus inside and outside Parliament was that the Office within the MoD was the appropriate place to locate the centre because it possessed not only the necessary computing facilities but also more experience and expertise in climate science and, indeed, all aspects of atmospheric science, than any other British institution.[51]

The first report of the IPCC's Scientific Assessment Working Group, entitled *Climate Change: The IPCC Scientific Assessment*, was published in September 1990

---

[50] Britain's Ambassador to the United Nations, Sir Crispin Tickell, helped Mrs Thatcher write her speech. He was (and remains) a committed environmentalist, with a particular interest in climate change.

[51] One of the Met Office's computers had been funded by the MoD, the other by the Department of the Environment.

Figure 13.2. The Prime Minister, Mrs Margaret Thatcher, inspecting a poster display at the Hadley Centre after she had opened the Centre, 25 May 1990. © Crown Copyright 2010, the Met Office.

(by Cambridge University Press), and Mrs Thatcher invited Dr Houghton, the group's first chairman, to present to her Cabinet on 21 May 1990 the scientific findings which were to be set out in the report. Four days later, she opened the building at Bracknell close to the headquarters of the Office which had been secured to house the new centre.[52]

The centre has flourished and is today the foremost climate change research institute in the UK and one of the world's leading institutes in climate science, climate change research and climate modelling. Originally called the 'Hadley Centre for Climate Prediction and Research', in honour of George Hadley (mentioned in Chapter 1), it is now known as the 'Met Office Hadley Centre'. Its aims are:

- To understand physical, chemical and biological processes within the climate system and develop computer models of the climate which represent them
- To use computer models to simulate differences between global and regional climates and changes seen over the last 100 years

[52] Two other buildings close to the main headquarters building at Bracknell were acquired at the same time. Called the Sutton and Johnson Buildings, after past heads of the Met Office, the former housed the Office's Commercial Section, the latter the staff responsible for data provision and consultancy. Other named buildings in Bracknell included the FitzRoy, Dines, Richardson and Napier Shaw Wings of the main building and the Simpson Building at Western Road (which housed the stores and test facilities). Also in Bracknell, at Eastern Road, new customized air-conditioned and smoke-protected premises for the National Meteorological Archive were occupied in the autumn of 1991. Outside Bracknell, at Easthampstead, the Experimental Site was called Beaufort Park, after Admiral Beaufort.

- To use computer models to predict changes over the next 100 years
- To monitor global and national climate variability and change
- To attribute recent changes in climate to specific factors.

A major event of the year which was covered by the Office's annual report for 1990–1991 was the Gulf War, which began in August 1990. During this conflict, the Office provided much support, not least through the MMU and through advanced numerical modelling techniques, which produced forecasts of low-level winds of crucial operational value. During the conflict, the Office was asked by the Department of the Environment to assess the environmental consequences of the oil wells in Kuwait being ignited. The Minister of State for the Armed Services, Archie Hamilton, announced in the House of Commons on 17 January 1991 that a note prepared by the Office in response to this request was being placed in the Library of the House that day.[53]

Several references to the note were made in a debate in the House of Commons on 15 March 1991 concerned with a range of ecological and other environmental consequences of the Gulf War. Critical comments about the note's contents were made by Tam Dalyell, and derogatory remarks about the Office's weather forecasting ability were made by a few other Members. However, some speakers in the debate spoke up for the Office, among them Jacques Arnold, who said that Britain could take pride in the work of the Office, which had adapted its computer models to give the best possible predictions of the likely behaviour of the oil slick caused by the deliberate release of 300,000 tonnes of oil into the Arabian Gulf by Iraqi troops on 26 January 1991. It was also analysing, he said, the impact of the smoke clouds of the oil fires. He wanted to know if the Secretary of State for the Environment and his Department would "find ways of assisting the Meteorological Office with the practicalities to enable it to observe those filthy clouds". He believed the Office needed to "take measurements of this unprecedented phenomenon".[54]

About 700 wells were set alight in early 1991 and many burned for several months. Thus the conclusions of the note's authors (Keith Browning and colleagues) could be checked. The arguments of those who predicted that a 'nuclear winter' might ensue were shown to be fallacious. There were certainly unpleasant downwind consequences within a few hundred kilometres of the source, but the global effects of the oil fires proved to be negligible, as forecast. Observations made whilst the fires were burning verified almost all of the predictions made by the Office.[55]

---

[53] An updated version of the note was written by Keith Browning and a dozen or so Met Office colleagues and published in *Nature* on 30 May 1991. See 'Environmental effects from burning oil wells in Kuwait' (Vol. 351, pp. 363–367).

[54] The C-130 aircraft of the Meteorological Research Flight was duly sent to the Gulf region. See 'Aircraft measurements of the Gulf smoke plume', by G J Jenkins *et al*, published in *Weather* in 1992 (Vol. 47, pp. 212–220).

[55] See 'Environmental effects from burning oil wells in the Gulf', by K A Browning *et al*, published in *Weather* in 1992 (Vol. 47, pp. 201–212).

A further development in the Office in 1990 was the achievement of a 'Unified Model', the name given to a numerical modelling system through which different configurations of the same model could be used to produce all weather forecasts and climate predictions. This was an extremely important development which was not only a major technical advance in its own right but also brought financial economies through a reduction of the overheads of running two forecast models and a climate model separately.

Operational implementation of the Unified Model began in June 1991 on a super-computer purchased by the Office with MoD funding the previous year, this being a Cray Systems C90 Y-MP-8 machine, which replaced the Cyber 205 machine. It was capable of one thousand million calculations per second. A year later, the Office acquired another Y-MP-8 machine, this one specifically for the Hadley Centre's climate research programme and funded by the Department of the Environment. It was twice as fast as the machine purchased the previous year. In addition to the two Cray machines, the Office also possessed a general purpose mainframe computer, an EX100 processor from Hitachi Data Systems, which performed a wide range of functions, among them data preparation for the Cray machines and processing of data generated by these machines.

A New Chief Executive

Sir John Houghton stepped down as Chief Executive on 31 December 1991, the day after his sixtieth birthday.[56] He had achieved much scientific and commercial success during his eight years at the helm of the Office and proved an outstanding leader, especially when guiding the Office through the difficult transition to Agency status. Thanks to his perseverance and persuasion, the Hadley Centre had been formed, and his encouragement of greater European cooperation in meteorology in general and the supply of meteorological services in particular was to bear fruit a few years later in the form of ECOMET, the Economic Interest Grouping of the National Meteorological Services of the European Economic Area.[57]

In Sir John's honour, the new lecture hall in the headquarters building at Bracknell was called the 'Houghton Lecture Theatre'; and when he opened it (on 7 July 1992) he presented a new award, the 'John Houghton Trophy', for contributions to weather

---

[56] Sir John's knighthood was announced in June 1991.
[57] ECOMET was formed in 1995, with the Met Office one of the first seven signatories, and it still exists today, with its primary objectives being, as stated in a document discussed at a WMO meeting held at Brussels in 2009 (Paper XV-RA VI/Doc.6), to preserve the free and unrestricted exchange of meteorological information between national meteorological services for their operational functions within the framework of WMO regulations and to ensure the widest availability of basic meteorological data and products for commercial applications. Further objectives of ECOMET are to guarantee access to meteorological data and products for public or private sectors and to recover part of the infrastructural expenses of the European national meteorological services by a contribution from all commercial users in order to maintain and improve the meteorological infrastructure.

Figure 13.3. Julian Charles Roland Hunt. © Crown Copyright 2010, the Met Office.

forecasting.[58] Like most former heads of the Office, he continued to be active after he stepped down. He chaired the Royal Commission on Environmental Pollution from 1992 to 1998, served as a member of the British Government Panel on Sustainable Development from 1994 to 2000, and chaired or co-chaired the Scientific Assessment Working Group of the IPCC from 1988 to 2002. He received the Nobel Peace Prize in December 2007, as part of the IPCC delegation, together with the former vice-president of the United States, Al Gore.

The new Chief Executive, Julian Charles Roland Hunt, took up his appointment on 1 January 1992. He had been born on 5 September 1941 and educated at Westminster School, after which he had proceeded to Trinity College, Cambridge, to study mechanical sciences. He had graduated in 1963 with First Class Honours and gone on to gain a Cambridge PhD in 1967 for research in magnetohydrodynamics. He had then been a visiting lecturer at the University of Cape Town in South Africa for three months and, after that, with a Fulbright Scholarship, a research associate at Cornell University in the United States. From 1968 to 1970 he had worked with the CEGB on the atmospheric factors that had contributed to the collapse of cooling towers at Ferrybridge in 1965 (see Chapter 12). Since 1970, he had been an academic in the Department of Applied Mathematics and Theoretical Physics of the University of Cambridge, first as a Lecturer in Applied Mathematics, then, from 1978 to 1990, as Reader in Fluid Mechanics and latterly, since 1990, as Professor in Fluid Mechanics. Since 1989, he had been involved in a collaborative project between the University of Cambridge, the Met Office and National Power to advance methods of calculating atmospheric dispersion. He also possessed commercial experience as director of

---

[58] The first winners of the trophy were the Met Office's Offshore Forecasting Team at Aberdeen.

Cambridge Environmental Research Consultants, a company he had helped form in 1986. He had been elected a Fellow of the Royal Society in 1989.

In the first instance, Professor Hunt was appointed Chief Executive for five years, on a fixed-term contract, and this was extended to 30 June 1997. Thereafter, he returned to academic life, spending time at L'Institut Mécanique des Fluides de Toulouse in France and universities in the United States (Arizona State and Stanford), before becoming, in 1999, Professor in Climate Modelling at University College, London. He was created a 'Working Peer' in the House of Lords in May 2000 with the title Baron Hunt of Chesterton in Cambridgeshire.

An important development during Professor Hunt's time as Chief Executive came in 1996, when the Office became a 'Trading Fund' within the MoD. Now, instead of an annual vote from the MoD and other government departments, the Office's revenue came from the customers of its meteorological services through individual contracts. The Office was required to operate on a commercial basis and meet agreed performance targets as set by the MoD. A draft 'Meteorological Office Trading Fund Order 1996' was laid before the House of Commons on 21 February 1996 and approved by the House on 6 March. Then, in accordance with the Government Trading Funds Act 1973 as amended, the Order was made on 8 March and came into force on 1 April, with terms, conditions and financial arrangements as set out in Statutory Instrument 1996 No. 774.

As Professor Hunt commented in the April/May 1996 issue of *Outlook* (a controlled-circulation magazine which was distributed by the Office free of charge to decision makers in industry and commerce for whom weather and climate inform-ation affected their business), the change to Trading Fund status would make little difference to the Office's operations with existing customers but would affect the way its services were supplied to other government departments. The new status meant, he said, that the Office would operate "just like a commercial company – by preparing budgets, monitoring cash flow, invoicing customers, and producing balance sheets and profit and loss accounts". The change presented "great opportunities to be competitive in the growing markets for meteorology and environmental services required at the most demanding international level".

In reply to a question asked by Iain Duncan Smith in the House of Commons on 8 May 1996, the Minister of State for Defence Procurement, James Arbuthnot, stated that the Office had been set a range of quality, financial and efficiency targets to ensure it delivered progressive improvements in the provision of weather-related services in 1996–1997. The key quality target for the financial year 1996–1997 was to achieve 80% of the designated business-plan external targets for customer satisfaction, forecast accuracy and timeliness. In addition, the Office was to devise a value-for-money index to measure the quality of its services, and the accuracy of the Office's numerical weather prediction model was to be improved by at least 2% in 1996–1997, as measured by an index of internationally exchanged model performance data.

The 1996–1997 efficiency target was to meet the key quality target within gross expenditure, before interest charges, of £140.3 million. For future efficiency targets, the agency was to develop an auditable efficiency index.

These changes concerned the staff greatly, as surveys showed. Team-briefing was introduced to ensure rapid and direct communication between management and staff at every level. Professor Hunt was asked to write a paper about the changes in the Office for a House of Lords committee in 1996. He also appeared before the Public Accounts Committee in 1995 and was asked, as always, whether there was sufficient progress in weather forecasting for the Office's budget to be justified. This was accepted, but he was criticized for rounding up the percentage of correct forecasts (defined statistically) from 79.5% to 80%, which was the target. This was, in fact, standard procedure in most government statistics but deemed not to be appropriate for deciding whether a target had been achieved. The newspapers rejoiced in seeing the bonus of the Office's Chief Executive removed!

In the event, most of the Office's key performance targets for 1996–1997 were met but some were missed, among them the improvement of 2% in the accuracy of weather forecasting. A number of factors contributed to this failure, the most significant being the improved use of satellite data, which had led, the Office's annual report for 1996–1997 said, to "greater detail in the description of the atmosphere in the tropics and southern hemisphere and thereby set a much harder target for forecast verification". When judged against other forecasting centres, however, the Office was among the best in the world, if not actually the best.

Another important development in the 1990s was the introduction of the World Wide Web as a publicly available service on the Internet. From late 1991 onwards, with the development of user-friendly browsers and efficient search engines, use of the Web by the public grew rapidly, and weather services around the world soon began to make satellite images available, along with other meteorological information. Here was a means by which anyone with a computer and a connection to the Internet could receive such information from anywhere in the world. The Web soon became an indispensable means of finding and disseminating meteorological information. However, issues of data availability to the public arose. In particular, in an increasingly commercial world, what could or should be made available free of charge?

This matter was addressed at WMO's Twelfth Congress, held in Geneva from 30 May to 21 June 1995, the outcome being a set of guidelines and international agreements that were codified in Resolution 40: *World Meteorological Organization policy and practice for the exchange of meteorological and related data and products, including guidelines on relationships in commercial meteorological activities to facilitate worldwide co-operation in the establishment of observing networks and to promote the exchange of meteorological and related information in the interest of all nations*. This codification did not, however, resolve the matter fully, as the policies of the various meteorological services over the commercial value of data were not consistent. Tensions developed. Why, for example, were the Americans more ready to make current weather observations available via the Web than the British, and why were facsimile versions of Met Office weather analyses and forecasts made available by the Americans but not by the Office? To what extent were the Americans contravening Resolution 40 by scanning the facsimile charts they had received legitimately via the Global Telecommunications System and then re-transmitting them via the Web? In the fullness of time, these inconsistencies were resolved, and most members of the public who obtain meteorological information from the Web today are satisfied with arrangements over what is freely available from the Office and what is not.

During Professor Hunt's time as Chief Executive, weather services provided by the Office expanded greatly. This can be seen from, for example, the April/May 1997 issue of *Outlook*, in which there were features about:

- A weather service for providers of buildings and contents insurance in the UK
- Weather-training courses for the offshore industry
- The daily weather forecasts issued by the Office's International Forecasting Unit for engineers who were building a bridge in Bangladesh
- Thunderstorm forecasts for electricity boards
- The Meteorological Information Self-briefing Terminal (MIST), through which business and industry could obtain up-to-the-minute weather information direct to a personal computer
- Weather forecasts from the Birmingham Weather Centre for golf's Ryder Cup, which was contested at Valderrama in Spain in September 1997

For the public, a Charter Standard was published which listed the various services provided by the Office and summarized arrangements for monitoring the Office's performance against key corporate targets.

A publicity leaflet called 'Introducing the Met Office' provided details of the Office's activities in 1997. The annual revenue was more than £150 million, with the largest customer groups the MoD and the Public Meteorological Service. Other key customers included the Civil Aviation Authority, the Department of the Environment, Transport and the Regions (DETR) and commercial customers. The Office employed 2200 people in more than eighty different locations, with just over half at the Bracknell headquarters. There were 430 forecasters; 200 observers; 260 staff in the research branches; 750 in computing and technical areas; 155 in sales, marketing and other commercial areas; and just over 220 in finance and support services. The latest mainframe supercomputer, a Cray T3E-900, was jointly funded by the Office and the DETR and had a peak processing power of 80,000 million calculations per second, making it five times faster than the Cray C90 it replaced.[59] The Office's computer models produced forty operational forecasts a day. Met Office staff made more than 1500 surface observations per day and more than forty upper-air observations at

---

[59] The Cray C90 replaced the two Cray Y-MP-8 machines in May 1994.

fourteen locations. With the help of other weather observers and automatic weather stations, more than 8500 weather observations were made in the UK each day from more than 3000 sites. The Office met 13% of the cost of ECMWF, partially funded fifteen rainfall radars around the UK, and contributed 13% to the cost of geostationary satellite images produced by EUMETSAT. Moreover, sounding instruments were provided by the Office on polar-orbiting satellites.

During Professor Hunt's time as Chief Executive, publication of the *Meteorological Magazine* ceased, with the last issue being that for December 1993. Commonly known as *Met Mag*, this was primarily an internal publication much liked by many in the Office and others for its mixture of research reports, review articles, Office announcements and personnel news. For the more technical kinds of papers that used to be published in *Met Mag*, a new journal was launched in 1994 by the Royal Meteorological Society. Called *Meteorological Applications*, it was, and continues to be, a fully peer-reviewed journal with an international readership.[60] Information about Office developments, staff and their activities is now provided by corporate magazines that have replaced *Outlook* (e.g., today, *Barometer*).

Professor Hunt's time as Chief Executive saw, too, the withdrawal of the UK's last Ocean Weather Ship, *Cumulus*, which had replaced *Starella* in late 1985 as the ship which manned the station at 57°N 20°W. She was decommissioned on 7 June 1996 and then returned to her original Dutch owners. She had been handed over to the UK by the Royal Netherlands Meteorological Institute at Hull on 18 December 1985 and the symbolic sum of £1 paid for her, the amount agreed by the consortium of European nations represented by WMO's North Atlantic Ocean Stations Board. The understanding had been that *Cumulus* would be returned to The Netherlands in due course. Unlike *Starella*, a converted trawler, *Cumulus* was a purpose-built weather ship, constructed in The Netherlands and launched in 1963. She contained accommodation and facilities not only for the regular meteorological crew but also for guest scientists who were carrying out special research projects. As a weather ship, she was now surplus to requirements. The information needed by weather forecasters and others was provided by satellites, aircraft, commercial ships and fixed and drifting buoys.

A further development in the early 1990s was the introduction of the Royal Meteorological Society's Chartered Meteorologist (CMet) accreditation scheme, which aimed to provide a professional qualification in meteorology that would satisfy clients and employers its holders had reached and continued to maintain a high level of knowledge and experience and were conversant with current best practice. The CMet qualification continues today, its role being to assure the public and customers of

---

[60] This journal reported on the new series of European meetings which Professor Hunt opened at Oxford in 1993 and which led to the formation of the European Meteorological Society, which is, in fact, a federation of meteorological societies across Europe. The Royal Meteorological Society is a member.

Figure 13.4. Professor Hunt presenting a barograph to Captain M Bechley, September 1995, continuing a tradition which began in the days of FitzRoy of rewarding mariners whose work at sea for the Meteorological Office was exemplary. © Crown Copyright 2010, the Met Office.

weather services that the advice they seek is provided by persons who have attained the highest level of professional performance available in the meteorological industry and are maintaining their expertise through an active programme of Continuing Professional Development. The availability of professional accreditation was extended to weather forecasters and meteorological observers in the late 1990s through National and Scottish Vocational Qualifications which are, like the CMet qualification, recognized by the European Union. The Office has encouraged members of its staff to take advantage of these accreditation schemes.

Professor Hunt built on the progress towards total quality management made under Sir John Houghton and was particularly keen to improve the quality of weather forecasts. By the end of his time as Chief Executive, the accuracy of the Office's forecasts was second to none. He was a very 'visible' Chief Executive, who visited the CFO most days and took a great interest in the work of the forecasters. A particular concern of his was natural disasters, especially those associated with tropical cyclones. By 1997, authorities in the United States and other sensitive areas were receiving reliable hurricane forecasts made by the Office.

Professor Hunt was also a keen supporter of the WMO, pressing continually for the regular publication of climatic data for decision makers and the general public. He pushed urban meteorology firmly onto the agenda of WMO at the UN Habitat II Conference held at Istanbul in 1996, arguing that more than 50% of the world's population would soon be living in cities. This has now become an important area of work in the Met Office for several government departments, including the Department of Health.

So far as the change to Trading Fund status was concerned, Professor Hunt brought about necessary administrative reorganizations within the Office, including changes to the Office's pay and grading structure which saved the Office some millions of pounds. However, some staff were uncomfortable with one change he brought about. He introduced daily critiques of the previous day's weather forecasts, and these stopped within a month of his departure.[61]

Further Changes of Direction for the Met Office

The new Chief Executive, Peter David Ewins, took up his appointment on 1 August 1997. He was 54 years of age, born on 20 March 1943. He was neither an academic nor a meteorologist. He was an engineer, with a BSc(Eng) degree from the Imperial College of Science and Technology and a Master of Science degree gained at the Cranfield Institute of Technology. From 1966 to 1978, he had been a research scientist at the Royal Aircraft Establishment and had since then held a range of posts in the MoD, apart from 1981 to 1984, when he had been Head of the Helicopters Research Division of the Royal Aircraft Establishment, and 1984 to 1987, when he had been seconded to the Cabinet Office. From 1988 to 1991, he had been Director of the Admiralty Research Establishment and since 1994 Chief Scientist of the MoD. He was a Fellow of the Royal Academy of Engineering and a Fellow of the Royal Aeronautical Society.

When appointed, Mr Ewins said that his aims for the coming few years were twofold: first, to ensure that customers were provided with the range of quality services they needed, in the most efficient manner and at a price they could afford, and second, to make the Office a source of pride to its staff, owner and the public.[62] The Office had made a good start, he said, with its successful transition to an Agency and Trading Fund and now had to build on that success and ensure its future. He did not mention the Office's pre-eminent position in the world in weather and climate prediction. The impression given by the statement is that the Office was now foremost

---

[61] In conjunction with his retirement as Chief Executive, Professor Hunt commissioned a play called 'The ostrich and the dolphin', which was written by Juliet Lacey and featured the relationship between Charles Darwin and Robert FitzRoy during their voyage on the *Beagle* and the impact of *The Origin of the Species* on FitzRoy later in his life. For information about this relationship and these effects, see Chapter 2.

[62] See 'Appointment of the new Chief Executive of the Met Office', *Marine Observer* (1998, Vol. 68, p. 32).

Figure 13.5. Peter David Ewins. © Crown Copyright 2010, the Met Office.

a commercial body, and two developments during his time as Chief Executive lend support to that view. The Office's logo and Internet domain name were both changed. Out went the weather cock motif that had been the Office's logo since the time of Sir Napier Shaw, and in came a new corporate identity which included a rather dreamy-looking undulatory design with apparent allusions to two types of waves, viz. those of all kinds in the ocean-atmosphere system and also the electromagnetic waves by means of which meteorological information was communicated. The change of domain name, from metoffice.gov.uk to metoffice.com, occurred in early 2001 and sent a signal to many people that the Office was now primarily a commercial concern.[63]

Soon after Mr Ewins became Chief Executive, changes in the Office's organization and management structure were made, with effect from 2 February 1998.[64] So far as organization was concerned, there were three changes: (1) a clearer separation of business management and service delivery, (2) the creation of a new Technical Division and (3) a greater emphasis on human resource issues. The changes to the management structure were designed to make a clearer distinction between strategic management and operational management. The Management Board was replaced by a new 'Met Office Board', responsible for policy setting and planning, and a new 'Executive Committee' was made responsible for the overall management of the Office and operational issues.

---

[63] Prior to 2001, there had been a number of domain names. For email, meto.govt.uk was changed to meto.gov.uk in the spring of 1996, met-office.gov.uk in early 2000 and metoffice.gov.uk in late 2000. For the Web, met-office.gov.uk was introduced in early 1997 and the hyphen dropped in late 2000. The change in 1996 was part of a government-wide change from govt.uk to gov.uk. Reversion to metoffice.gov.uk occurred on 16 July 2004, just two weeks after Mr Ewins stepped down as Chief Executive.

[64] Staff were informed of the various changes in a special issue of an in-house newsletter, *Mercury*, that was published in January 1998.

Figure 13.6. Met Office logos.

The position of the Office as a world leader in weather and climate prediction took pride of place in the review of the year written by Mr Ewins for the 1997–1998 annual report. In the first paragraph, he reported that the global NWP target had been missed "by only the narrowest of margins", and in the second he stated that the Office had gained through the Hadley Centre an enviable reputation for its work in climate change and climate prediction. However, the balance of his review and, indeed, the whole report reflected the Office's strong emphasis on commercial activities now that it was a Trading Fund. The Office had, Mr Ewins said, "substantially exceeded the financial targets" and maintained a "strong and diverse product range". It had continued to "rank among the very best for the provision of weather services" and, through a simplification of its "still rather bureaucratic processes and procedures", was aiming for ISO 9000 accreditation for quality management. He also mentioned that efforts were being made to ensure Year 2000 (Y2K) compliance for the Office's computing activities. Resources had been made available to run a major project, called 'Project 2000', to ensure that the so-called Millennium Bug would not affect any of the Office's operations adversely. In the event, the progression from 1999 to 2000 occurred without any obvious problems.

A problem of a different kind in 1999 focused attention on the state of the headquarters building. Thundery weather brought heavy rain to the Bracknell area on 29 May. As the next day was a Sunday and the day after that a public holiday, no one visited the store of the National Meteorological Library before 1 June, but when a member of the library's staff did so that morning he found water dripping from the ceiling and more than forty books damaged by water. They were in the part of the store which held the oldest books and documents of the national collection. They were rare and valuable books. The immediate thought of the Library's staff was to blame the storm that had occurred on the Saturday afternoon. However, when water continued to drip into the store after no rain had fallen for some days, the problem was investigated and the real cause found. The inlet pipe to the Chief Executive's toilet had sprung a leak! His office suite was immediately above the library store.[65]

The state of the headquarters building was but one of the factors that led to the question of new premises being addressed. A relocation team was formed in 1999 and commissioned property advisers Rogers Chapman to conduct a review of the Office's Bracknell accommodation and development options. Meanwhile, Lorien Consulting Solutions carried out a survey of staff to ascertain attitudes to relocation. The overall first choice of the 80% who responded was for relocation in or near Bracknell, which was not surprising, as a preference for the status quo was to be expected. The second most popular region was the south-west of England and the third the south-east, and the survey showed that about 30% of the staff might decide not to move with the Office if it left the Bracknell area, which would have major implications for recruitment, training and quality of service.

At the beginning of February 2000, staff were informed that the Executive Committee had approved the Outline Business Case for relocation of the Office's Bracknell area functions, the key findings being that:

- The main headquarters building had failing infrastructure, which inhibited business development and increasingly prejudiced operational effectiveness.
- The workforce was split over several sites, including expensive leasehold properties.
- New accommodation was required to provide a modern, efficient working environment.
- Relocation offered the most cost-effective solution to the long-term accommodation needs of the Office.

The 'do nothing' option had been rejected. So, too, had the costly option of refurbishing the headquarters building. The option of a new building near Bracknell or at Shinfield Park in Reading remained open, and so also did the option of a relocation some considerable distance from Bracknell. The Executive Committee agreed there were potential benefits of being close to ECMWF but advised staff a long-distance move was still an option. The relocation date was expected to be March 2003.

---

[65] This incident was reported on page 6 of the July/August 1999 issue of *Mercury*.

Staff were informed that the Office's Relocation Team was "working on developing the specification for the new accommodation, with a view to entering into negotiations with potential contractors during the summer". They were also assured that further consultation with them would take place, and opportunities to learn more about the Office's plans would be provided in the coming months through seminars at Bracknell, visits to outstations and colloquia at the Met Office College at Shinfield Park. Regular updates would also be provided in *Mercury*.

Questions were asked in the House of Commons, from which it became clear that rumours were circulating. On 17 April 2000, for example, Andrew MacKay, the Member for Bracknell, rebuked Jane Griffiths, the Member for Reading East, for stating on local radio and in the House that all Met Office jobs would be lost and that the Office should relocate to Liverpool. Was that, MacKay asked, "official Labour party policy"? In his reply, the Under-Secretary of State for Defence, Dr Lewis Moonie, said that the current intention was to consider Shinfield Park the Office's preferred option.

Four months later, another possible destination for the Office was revealed. A Devon newspaper, the *Western Morning News*, carried the following headline on 17 August 2000: "Met Office may move headquarters to Exeter". This city, the paper reported, had made the final shortlist of four, along with two sites in the Reading area and one at Norwich. The possibility of a move to the city had been made public the previous day. However, it did not come as a surprise to Met Office staff. A relocation news item on the Office's intranet on 26 July had already informed them that four potential sites had been identified: Exeter Business Park West, Norwich Research Park, Beaufort Park at Bracknell and Shinfield Park at Reading.

The Bracknell Forest District Council Planning Committee approved the Office's planning application for Beaufort Park on 14 September 2000, but Wokingham District Council objected. The proposed development would encroach on the 'green-field' separation space between Bracknell and Wokingham, and Beaufort Park was not in a planned development area. Eventually, a decision was announced. Dr Moonie made a statement in the House of Commons on 9 November 2000. The Exeter Business Park had been chosen. The site at Norwich clearly had points strongly in its favour, including its close proximity to the University of East Anglia and its Climatic Research Unit, but the site was considered by the Office somewhat isolated on the fringe of Norwich, and there were doubts regarding whether the site was actually available.

The decision to move to Exeter was queried by Lord Hunt of Chesterton, the former Chief Executive. In the House of Lords on 27 November 2000, he suggested that a location for the Office next to ECMWF might be the best option. In reply, Baroness Symons of Vernham Dean agreed that it was important that "the Met Office should continue to develop its numerical weather prediction capability in concert with ECMWF". However, with modern communications "and the undoubted desire of both parties to strengthen the already close relationship", this could be achieved

Figure 13.7. The headquarters of the Met Office at Exeter. Photograph by Steve Jebson, by kind permission of the Met Office.

very well from Exeter "and without raising concerns amongst our European partners which would have almost certainly been the case had the Met Office relocated to Shinfield Park". Lord Hunt then asked if consultations had taken place with the public and private sector customers of the Office. Baroness Symons replied that no direct consultation had taken place, but the choice of Exeter had been made in the best interests of its customers in the public and private sectors.

There were, perhaps inevitably, some rumblings of discontent within the Office over the move to Exeter, but it is clear from the collection of material concerned with the move which is held in the National Meteorological Archive that the Met Office Board and Relocation Team consulted and assisted staff and went to great lengths to keep them fully informed about the relocation. Besides presentations, seminars, leaflets, announcements on the intranet and information in newsletters, the Office also arranged reconnaissance visits to Exeter in the summer of 2001.

The new high-tech environmentally controlled building at Exeter was handed over to the MoD officially on 18 December 2003 by the Stratus Consortium, which had built it. The first staff had arrived in June 2003, and almost all staff had moved to Exeter by December of that year. The Bracknell sites were completely vacated by February 2004, apart from the National Meteorological Archive, which moved to Exeter at the end of 2004 and opened to the public in March 2005. The Archive did not, however, move to the new headquarters building. It moved to Great Moor

Figure 13.8. In the National Meteorological Archive, Exeter. Photograph by Steve Jebson, by kind permission of the Met Office.

House, a newly completed purpose-built archive building which was shared with the Devon Record Office about 400 metres from the Office's new headquarters. The move to Exeter proceeded smoothly, including the transportation of two supercomputers. Since 1999, the Office had possessed not one but two Cray T3E machines. Their

Figure 13.9. The National Meteorological Library, Exeter. Photograph by Malcolm Walker, by kind permission of the Met Office.

Figure 13.10. The instrument development enclosure at the Met Office, Exeter, 2010.
© Crown Copyright 2010, the Met Office.

transfer was entirely successful, and the forecasting operation continued seamlessly. Another part of the Office that did not move to Exeter in 2003 was the College. It vacated its premises at Shinfield Park in the summer of 2002 and moved temporarily to the South Devon College at Torquay before moving to the headquarters at Exeter in the spring of 2004.

Today, there is nothing left of the headquarters building at Bracknell. Apartments have been built on the site. The only reminder of the Office is the large traffic island at the very busy junction where London Road meets Millennium Way, Warfield Road, Park Road, Church Road and Weather Way. The feature is known locally as the 'Met Office Roundabout' and is so described on the signs on approach roads.

Uncertain Times

For some time, the relocation to Exeter dominated life in the Met Office. However, the Office's work of course continued, and there were some notable developments. In particular, as Mr Ewins noted in the annual report for 2000–2001, recognition of the importance of the Internet to the development of the Office's business led to a major update of the Office's website and the introduction of many new features. On the new site, launched on 14 November 2000, there was a much expanded education area, along with a much wider range of current weather and climate information than hitherto, as well as free observations, forecasts and charts for registered aviation users. Moreover, there were new web pages for energy traders (providing temperature and rainfall data from the UK and continental Europe), and there was a new service for the offshore sector (providing site-specific forecasts and weather charts). And an innovation for the Office came in December 2000, when banner advertising on some web pages began.

An era ended on 3 April 2001, at 10:42 GMT, when the last radio-facsimile chart was transmitted from the Office, this being one showing a forecast of sea-level barometric pressure over the North Atlantic and Europe. Thus ended a service which had been provided for almost half a century, during which analyses, forecasts and other products from the National Meteorological Centre (as the CFO was now called) had been transmitted by radio-facsimile, including charts of particular value to mariners. Many ships now had access to the Internet via satellite communications, and various digital meteorological products were available by that means. Seafarers who did not have access to the Internet were not left without weather information, though. They continued to be able to obtain products by means of radio-facsimile broadcasts which continued from the German Meteorological Service, the Royal Navy's Fleet Weather and Oceanographic Centre, stations in North America and various places around the world.[66]

Another era for seafarers ended in July 2003, when publication of *Marine Observer* ceased. "Sadly", said the Office's Head of Observations Supply, Keith Groves, in the preface to this last issue, the Office was "no longer able to support the editorial, printing, publishing and distribution costs involved in producing the journal". And yet another era came to an end on 14 August 2003, when production of large-size paper charts by the National Meteorological Centre ceased. At one time, said a report in the September 2003 issue of *Mercury*, the National Meteorological Archive received more than sixty charts a day. With space in the Archive becoming limited, the decision had been made to store charts electronically. Paper copies could be produced for the public and Met Office staff when required.

Some of the changes proposed or made by the Office around the turn of the century formed the subjects of questions in the House of Commons. On 30 June 1999, for example, Eric Martlew asked the Secretary of State for Defence what plans he had for the weather centre in Newcastle. The ensuing letter from Mr Ewins informed Mr Martlew that a comprehensive review of the Office's forecasting processes had taken place, as a result of which plans to rationalize the national network of weather centres had been announced. The opportunities offered by improved technology would, Mr Ewins said, allow the Office to make greater use of automation in the production of weather forecasts and concentrate activities at fewer locations, thereby maintaining the quality

---

[66] There is still a Met Office fleet of Voluntary Observing Ships, numbering about 350, and the Office acts as one of two Global Collecting Centres for data from these ships, the other being the German Meteorological Service.

of the services the Office offered at a much reduced cost. Five weather centres would be closed over the next two and a half years, including Newcastle by the end of 2001. In fact, the Newcastle centre ceased operations in June 2001, having served the public, local authorities, utilities, the media, commercial companies and aviation for thirty-four years.[67]

In the House of Commons on 2 April 2001, Robert Key asked the Secretary of State for Defence to make a statement on the future of the Office's research aircraft. In reply, the Under-Secretary of State for Defence, Dr Moonie, said that the Hercules was owned by the Defence Evaluation and Research Agency (DERA) but had been used exclusively by the Office using RAF crews. The current contract with the Office had expired on 31 March 2001, and DERA had begun to seek alternative customers for the aircraft. In future, the Office would use a British Aerospace 146–300 jet aircraft to carry out meteorological research, in partnership with NERC and the University of Manchester Institute of Science and Technology. In his written reply to a question asked by Mr Key in the House of Commons on 27 June 2001, Mr Ewins said that the BAe 146–300 would be operated with these partners for at least ten years, under a service agreement with them. The consequential savings to the Office of moving to shared use of the BAe 146–300 would be £8 million over the coming ten years. A report in *Mercury* in September 2001 provided the additional information that £16.4 million in funding had been secured from the Wellcome Trust and the government's Joint Infrastructure Funding Scheme to cover the costs of converting the aircraft and installing new instrumentation.

Another change of Chief Executive occurred in 2004. Mr Ewins retired on 30 June and was succeeded by Dr David Rogers. The new Chief Executive was 47 years of age, born on 20 March 1957. He possessed a BSc degree with Honours from the University of East Anglia and a PhD gained in 1983 at the University of Southampton. He had been a Navigating Officer in the British Merchant Navy for a time but had otherwise pursued an academic career in the United States. He had held research appointments in the Desert Research Institute of the University of Nevada and the Scripps Institution of Oceanography (SIO) at San Diego, and he had been Associate Director of the California Space Institute. Then, from 1989–2000, he had been Director of the Joint Institute for Marine Observations, an institute sponsored by the SIO. From 2000–2003, he had been Director of the Office of Weather and Air Quality of the U.S. National Oceanographic and Atmospheric Administration and since 2003 Vice President Meteorological and Oceanographic Services at the Science Applications International Corporation, Virginia.

---

[67] There was anger in 2005 when closure of the meteorological office at Aberdeen was announced. Indeed, an Adjournment Debate on the subject was held in the House of Commons on 12 July that year. Malcolm Bruce complained that the office in question was "the only civilian weather forecasting operation in Scotland" and was critical of the way the Office had announced the closure. His challenge was unsuccessful. A 90-day consultation took place, and the Aberdeen office subsequently closed.

Figure 13.11. David Rogers. © Crown Copyright 2010, the Met Office.

Dr Rogers did not remain with the Office for long. He announced on 21 April 2005 that he would step down from his post in the next few months, for personal reasons. His recent marriage had, he said, brought a change in his domestic circumstances which had led him to reconsider his personal and professional future. In the event, he stepped down on 15 July 2005, after only thirteen and a half months in the post.

When he stepped down, an internal promotion was made. The Chief Finance Officer, Mark Hutchinson, became Acting Chief Executive. A number of candidates for the post of Chief Executive were interviewed in January 2006, but the MoD's selection panel concluded that none had the all-round calibre expected of a holder of this post. The appointment of Mr Hutchinson as Chief Executive *pro tempore* was announced by the Under-Secretary of State for Defence on 27 January 2006, and he retained that post until 14 September 2007. By then, a new Chief Executive had been found. He was John Raymond Hirst, aged 55, born on 9 August 1952, who took up the

Figure 13.12. Mark Hutchinson. © Crown Copyright 2010, the Met Office.

Figure 13.13. John Raymond Hirst. © Crown Copyright 2010, the Met Office.

post on 17 September 2007. He was not a meteorologist. He had studied Economics at the University of Leeds and graduated with a BA degree in 1973. Since then, he had pursued a career in business, including nineteen years working for Imperial Chemical Industries (a British chemical company) and seven as Chief Executive of Premier Farnell (a global distributor of electronics components).[68]

An inquiry into the work of the Office was announced on 20 December 2005, to be conducted by the Government's Defence Committee, as part of that Committee's scrutiny of the MoD's executive agencies. It was focused on the MoD's role as owner of the Met Office, not on the quality of the Office's work or staff. The Committee visited the headquarters in Exeter and there took evidence from the Office's senior management, saying in their report that they had been struck by the pride in the Office of those who worked there or were associated with its work. Evidence sessions were also held at Westminster in May 2006 with Peter Ewins, the former Chief Executive, and with the Parliamentary Under-Secretary of State at the MoD, as well as with senior staff of the Office and officials of the MoD. Written submissions from the MoD and others were also received.[69]

The Committee found that the MoD had provided support to the Office and enabled it to develop its research capabilities. The Committee were also "impressed at the importance placed on the work of the Office by the MoD and Armed Forces" and stressed the importance of the MMU to these Forces. However, they also noted the

[68] John Hirst was the first head of the Met Office to admit to being a drummer in a rock band whilst at university. Moreover, he still had a passion for percussion and collected percussion instruments.

[69] See *The work of the Met Office*, House of Commons Defence Committee, Tenth Report of Session 2005–2006, published on 26 July 2006 (London: The Stationery Office, HC 823, 33 pp. plus 63 pp. of evidence). The remit of the Committee is to examine the expenditure, administration and policy of the MoD and its associated public bodies.

concern of Mr Ewins that there appeared to be confusion between the MoD's role as owner of the Met Office and its role as a principal customer, also his concern that he was not sure the MoD fully understood the role of the Met Office internationally. The weather forecasting ability of the Office was now so good, Mr Ewins said, that the Office was the forecaster of choice for all countries, including the USA. Furthermore, he believed the MoD's ownership of the Office probably added to the credibility of the Office's work among the UK's military allies. The Committee looked to the MoD to address concerns over the Office's important international role.

After considering the Office's performance against its Key Performance Targets, the Committee expressed themselves "not convinced that these were the most effective way of providing direction to the Met Office, and agreed with witnesses that the customer-supplier relationship was an important part of improving performance". A major aspect of that performance, the Committee agreed, was the ability of the Office to generate commercial income. Two notable criticisms of the Office were made. The Committee urged it to develop its commercial abilities "and to resist the temptation to become 'risk-averse', following a recent unsuccessful foray into the commercial market". The Committee also "found no reasonable grounds for the MoD's decision to down-grade the post of Chief Executive of the Met Office" and recommended that the MoD should reconsider that decision. The idea of privatization was rejected. The Committee found "no compelling reason to remove the Met Office from public ownership at present" and also recommended that the MoD should "consider further whether to co-locate the UK Hydrographic Office and Met Office in Exeter". Both were executive agencies and trading funds owned by the MoD.

The Government's response to the Defence Committee's report came in a document published on 16 October 2006.[70] The MoD had considered carefully the recommendation that the grade of the Chief Executive post be increased but remained firmly of the view that the current grade was right. As far as co-location of the UK Hydrographic Office with the Met Office in Exeter was concerned, work was in hand to explore this as one of a number of possible options for the future location of the Hydrographic Office. This work would take some months to complete, after which the MoD would take a final decision "based on the best overall interests of the taxpayer". A decision came on 8 October 2007, when the Under-Secretary of State for Defence, Derek Twigg, announced that the Hydrographic Office would remain at Taunton.

The reference in the Defence Committee's report to an "unsuccessful foray into the commercial market" concerned a venture by the Office with an unhappy financial outcome. The Office formed a joint venture company called 'weatherXchange Limited' in May 2001 to provide tools and expertise for the weather risk management

---

[70] See *The work of the Met Office: Government response to the Committee's Tenth Report of Session 2005–06*, House of Commons Eleventh Special Report of Session 2005–2006, published on 16 October 2006 (London: The Stationery Office, HC 1602, 5 pp.).

market. At the time, it seemed an attractive proposition, and some major investors clearly agreed; Zions Bank and the world's largest mining company, BHP Billiton, became involved. However, the collapse of the U.S. energy giant Enron in late 2001 damaged confidence in the market. Enron was also in the weather derivatives business. Mark Hutchinson told the Defence Committee that the Office had failed to manage the investment properly and had not applied the appropriate governance checks. The company never made any money and was placed into administration on 3 October 2005. The Office wrote off the value of its investment amounting to £1,533,000 and, in addition, incurred support costs of approximately £3 million. Thus the Office lost about £4.5 million in a venture the Defence Committee called "incredibly naive, if not amateurish". Mr Hutchinson was inclined to agree but said that the experience had not made the Office more reluctant to go into joint ventures as a principle. The view of the Defence Committee was that the MoD and the Office should not overreact to the weatherXchange experience but instead learn lessons from it. The Government concurred.[71]

Controversy Continues

Under Mr Hirst, the Office has gone from strength to strength and avoided any further commercial embarrassment. In the year 2009–2010, all targets were met or exceeded, among the latter the targets against four different forecast accuracy measures: viz. the Global Numerical Weather Prediction Index, the UK Numerical Weather Prediction Index, the Combined Maximum and Minimum Temperature Index, and the Probability of Precipitation Index. Indeed, the capabilities of numerical modelling were now astonishing. Precipitation features only a few kilometres across could be forecast with some confidence. The computers used by the Office had become ever more powerful. The IBM machine in use in 2010 was capable of more than 120 times ten to the power twelve calculations per second, and the Office was planning to install a computer capable of ten to the power fifteen calculations per second in 2011.

Given the capabilities of numerical modelling today, one might think that storms like those of 1979 and 1987 will never again take forecasters by surprise. In reality, however, caprices of the atmosphere may occur at any time. Models can never represent completely the complexity of the real atmosphere. Therefore, failures of weather forecasters may still occur occasionally. Extreme weather will always be particularly difficult to forecast. However, the use of 'ensemble forecasting' has reduced the likelihood of spectacular failures. This is a numerical prediction technique which has been developed since the 1980s and is now used routinely at most of the operational weather prediction facilities around the world, including the Met Office and ECMWF.

---

[71] The weatherXchange company was acquired by Speedwell Weather Derivatives Ltd in April 2006 to complement their existing weather derivatives activities.

In it, the model generates several forecasts, each based on a slightly different set of initial conditions. If the various forecasts are consistent, then they are considered reliable. If they are divergent, their reliability is considered lower. In the development of ensemble forecasting, and also in ways and means of assimilating data into models, staff of the Office and ECMWF have long been in the vanguard.

Since 1990, the Unified Model (UM) has developed greatly, with advantage taken of increased computer power, along with improved understanding of atmospheric processes and an increasing range of observational data sources. It is now highly versatile and capable of modelling a wide range of time and space scales, including kilometre-scale mesoscale nowcasts, limited-area weather forecasts, global weather forecasts (including the stratosphere), seasonal forecasts, and global and regional climate predictions. In addition, the UM can be coupled to other models which represent the different aspects of the Earth's environment that influence the weather and climate, such as the ocean and ocean waves, sea-ice, land surface, atmospheric chemistry and the carbon cycle. This allows the UM to be used for Earth System Modelling applications.

The Internet has revolutionized the delivery of services to the public and customers. In education, for example, the Office's information service in the 1990s consisted mostly of leaflets, booklets and posters, many of them uninspiring. Now, there is a wide range of attractive and well-designed resources, lesson plans and other educational material for schools on the Office's website, including a feature on the extraordinarily hot summer of 2003 and case studies of notable storms, such as that which brought devastating floods to the Cornish village of Boscastle in 2004. In addition, the education pages provide current and historic weather data for places in Britain and around the world. Moreover, thousands of members of the public around the world have taken part in Internet experiments such as *climateprediction.net*, which is a distributed computing project designed to produce predictions of the Earth's climate up to the year 2100 and test the accuracy of climate models. The project has been supported by the Office, with climate models developed by the Office at the heart of the project. By 2010, services were being provided by the Office through the most modern of communication means, including a free iPhone application and weather widgets.

In the past few years, however, the Office has not avoided controversy. After a series of destructive floods occurred in many parts of England and Wales in the summer of 2007, for example, questions were asked by the media, the general public and others about the accuracy and timeliness of the Office's weather forecasts.

The Government asked Sir Michael Pitt to carry out an independent review of the flood events as a whole. This was wide-ranging and considered all available evidence on the flooding that occurred in England during June and July 2007. A key recommendation of his review was that the Environment Agency and Office should work together to a much greater extent, through a joint centre, to improve their technical capability to forecast, model and warn.[72] Enlarging on this recommendation, Sir Michael said that the relationship of the Environment Agency with the Office was particularly important. Weather prediction formed a crucial part of flood risk management and the Office was a world leader. There was, he said, room for improvement, particularly in relation to increased lead times for predicting events, probabilistic forecasting and more accurate local-scale forecasts at a city or town level, but closer working should deliver real changes in technical capability. This would improve the usefulness and reliability of extreme rainfall forecasts and warnings, which were essential for providing effective warnings for rapid response catchments and surface water flooding. He and his team believed this closer working would best be achieved through a joint centre.

In its response to Sir Michael's review, the Government stated that the Department for Environment, Food and Rural Affairs (Defra) would establish a new joint forecasting and warning centre (run by the Environment Agency and the Met Office) to improve the modelling and warning of flood risk.[73] The response also noted that the Office and the Environment Agency were trialling an Extreme Rainfall Alert Service, and the Office had upgraded its National Severe Weather Warning Service, which now triggered at a lower threshold of probability. In addition, the Office had improved its support to emergency responders through a web-based Emergency Support Service, which provided free access to weather observations and forecasts (including rainfall), as well as information specifically generated for emergency events. It was further stated in the response that new climate projections had been commissioned by Defra and the UK Climate Impacts Programme from the Met Office Hadley Centre and others. These would provide information at a more local level than previously possible on how the UK's climate might change over the twenty-first century and would help significantly in assessing climate change risk at key points during the course of the century.

The power of the modern media and the Internet to create mass ridicule was displayed in 2009 after the Office issued a seasonal forecast at the end of April which indicated that temperatures across the UK were likely to be higher than average and rainfall near or below average for June, July and August. The Office used the expression 'odds on for a barbecue summer'. When the coming season proved to be anything but warm and dry, scorn and derision followed. This time, the Office was partly to blame for its own predicament. Use of the word 'barbecue' was unwise.

The affair focused attention on the need for clear communication, especially for journalists and others who did not appreciate that seasonal forecasts could not convey certainty. A statement issued by the Office on 31 July 2009 reminded readers that in

---

[72] See *Learning lessons from the 2007 floods*, The Pitt Review, final report published in June 2008 (Cabinet Office, 462 pp.).

[73] See *The Government's response to Sir Michael Pitt's review of the summer 2007 floods*, published by the Department for Environment, Food and Rural Affairs in December 2008 (127 pp).

April the seasonal forecast for the summer had stated that there was a 65% probability of it being warmer than average and near or drier than average. There had been, therefore, a 35% probability that the summer would *not* be warmer and drier than average, and this had been made clear by the Office. The probability element of the forecast had been stressed at the time and indeed reported clearly in some newspapers. Either through ignorance or in search of headlines, though, some sections of the media had been selective and distorted the press release issued in April. Use of the term 'barbecue' had not helped in this respect.

The Office became involved in another controversy in 2009, one that became known as 'Climategate'. The problem began on 17 November with the illegal release of thousands of emails and other documents from the University of East Anglia's Climatic Research Unit which appeared to show that climate scientists in many institutions had colluded to withhold scientific information, interfere with the peer-review process to prevent dissenting scientific papers from being published, delete raw data or manipulate data to make the case for global warming appear stronger than it was. The matter was quickly publicized by the media and seized upon by deniers of climate change and, indeed, all who were sceptical of the evidence for climate change. Concerned that the credibility and honesty of climate scientists were being impugned, Mr Hirst and the Office's Chief Scientist urged their colleagues in the Office and other institutions to sign a statement circulated by the Office "to defend our profession against this unprecedented attack to discredit us and the science of climate change". More than 1700 British scientists signed. Investigations by the University of East Anglia, the House of Commons Science and Technology Committee, the Royal Society and the IPCC absolved climate scientists of any unprofessional conduct but did find that careless language had been used in emails and advised that greater care should be taken in future.

Climategate illustrated the power of the modern media and the Internet to arouse mob mentality. The level of ignorance about climate science revealed by blogs, newspaper articles and comments on radio and TV was breathtaking. Experts and fair-minded commentators struggled to make their voices heard. And many of the comments made about climate scientists were scornful, some hostile and unpleasant.

The eruption of the Eyjafjallajökull volcano in Iceland in the spring of 2010 brought further criticism of the Office, this time all undeserved. Once again, much of the blame for the spread of the criticism lay at the doors of ill-informed journalists and bloggers who did not take the trouble to establish the facts. Significant eruptions of the volcano began on 20 March and increased in intensity on 14 April to such an extent that an ash cloud was created which brought about the closure of most of Europe's air space the following day. Many flights within, to and from Europe were cancelled for several days, and further disruption to flights occurred in late April and early May, though on a lesser scale than in mid-April. An unfortunate aspect of the problem was that north-westerly winds on the eastern side of a persistent anticyclone carried the ash towards the British Isles, France and central Europe. This being so, the Office, in its role as one of the world's nine Volcanic Ash Advisory Centres, issued a 'Volcanic Ash Advisory'. As a result, aviation authorities closed air space and many people heaped opprobrium on the Office. But what were the facts?

The Office's forecasts of the trajectories of plumes of ash proved to be accurate, but, in the absence of data from the source region, forecasts of vertical profiles of ash concentration could not be made in anything more than general terms. To forecast the spread of the ash, the Office used its sophisticated atmospheric dispersion model, called NAME, which stands for Numerical Atmospheric-dispersion Modelling Environment. Day after day, forecasters predicted the spread of ash correctly. However, the Office was not responsible for closing air space. It merely advised the Civil Aviation Authority and National Air Traffic Services where ash would be, and did so accurately. Closure of air space was based on safety thresholds set by aircraft engine manufacturers, and the Office's forecasts reflected these thresholds. To observe the ash cloud and verify forecasts, the Office used satellite imagery, balloons, research aircraft and ground-based radar systems, and in so doing damaged the engines of the research aircraft it now shared, thus confirming that volcanic ash did indeed pose a threat to aircraft engines.

Epilogue

The first weather forecast issued to the public by the Meteorological Department of the Board of Trade in August 1861 proved correct. That may have been a tactical mistake! Perhaps it would have been better if that first forecast had been wildly incorrect. Subsequent expectations might have been lower. In reality, the forecasts that are issued by the Met Office today are so accurate that almost all of the anticipation has gone. Rarely do newsworthy failures occur nowadays. For that, the dedication, skill and professionalism of all those scientists and others who have contributed over the years to the success of today should be recognized. The weather forecasting capability of the Office is phenomenal.

Admiral FitzRoy would surely be disappointed that the Office no longer provides a ship routeing service but probably astonished at the products available commercially from the companies that now serve seafarers. Sophisticated on-board software tools are used to gain access to weather routeing advice and to the latest wind, wave, swell, ice and weather forecasts, and masters of ships can now select optimum tracks themselves by obtaining tailor-made routes based on data entered by them into on-board systems. Moreover, on-board ship behaviour sensors are now fitted to vessels. FitzRoy's memory is perpetuated in the name given to a sea area used in shipping forecasts. Since 4 February 2002, the area west of the Bay of Biscay has been called 'FitzRoy'.

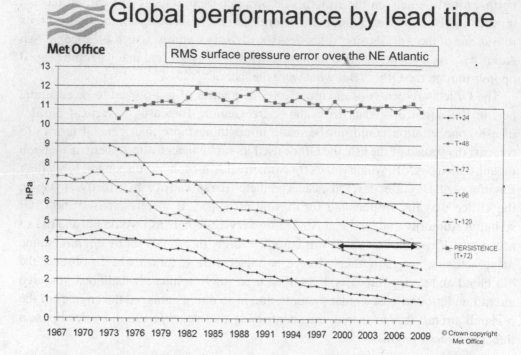

Figure 13.14. How weather forecast accuracy has improved from 1967 to 2009, shown in terms of root mean square barometric pressure error in hectopascals (millibars) over the north-east Atlantic for forecasts 24, 48, 72, 96 and 120 hours ahead, compared with the 72-hour persistence forecast (which has been consistently approximately 11 hPa). The arrow indicates that the forecast for 96 hours ahead is now as accurate as the forecast for 72 hours ahead was in 1999. © Crown Copyright 2010, the Met Office.

The Office began in an international movement after the Brussels Conference of 1853 and has been science-based from the start. The international context continues to this day. The Met Office is a highly respected member of the international meteorological community. At home, it provides a wide range of services and continually adds to them, with the innovative weather and health service that has been developed in recent years an example of the latter. The Office now plays an extremely important, and often literally vital, role in the provision of Britain's environmental services. An editorial in *The Times* in 1953 when Sir Nelson Johnson retired (see Chapter 11) stated that the Office was "everybody's servant". It was indeed, and remains so.

The coat of arms of the Met Office bears a motto, *Per Scientiam Tempestates Praedicere*, which means "To predict the weather through knowledge".[74] What more can one say about the Office? That sums up its mission throughout its history.

---

[74] The coat of arms was granted on 13 November 1991. For details, see *Marine Observer* (1992, Vol. 62, pp. 88–89).

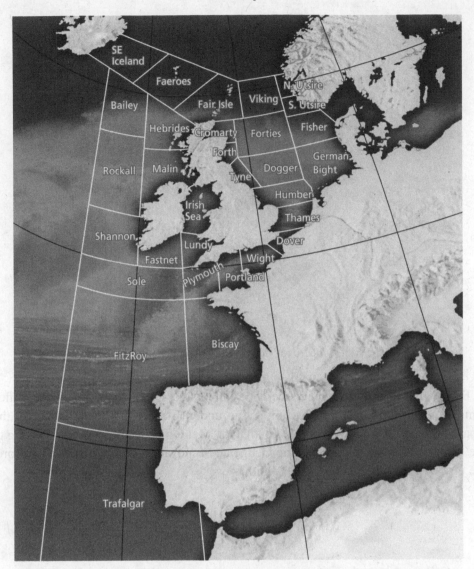

Figure 13.15. Shipping forecast sea areas, 2010. © Crown Copyright 2010, the Met Office.

Postscript

This book tells the story of the Meteorological Office to the end of 2010. However, an important subsequent development needs to be recorded. It was announced in the House of Commons on 18 July 2011, in a written statement by the Prime Minister, that responsibility for the Office had passed that day from the Ministry of Defence to the Department for Business, Innovation and Skills (BIS). Thus came to an end, after more than ninety years, the ownership of the Office by a government department which was concerned primarily with the armed services.

Figure 13.16. The Met Office coat of arms, granted in 1991. © Crown Copyright 2010, the Met Office.

By moving to a department concerned with business, the ownership of the Office almost moved full circle, for the unit set up under FitzRoy in 1854 was part of the Board of Trade. It was noteworthy, too, that the Met Office would now be reporting to the Science Minister within the Knowledge and Innovation group of BIS. Therefore, the motto of the Office remained as valid as ever.

Index

Printed in the United States
by Lonmasten

Printed in the United States
By Bookmasters